Mastercam 2022 for
SolidWorks Black Book

By
Gaurav Verma
Matt Weber
(CADCAMCAE Works)

Edited by
Kristen

ISBN # 978-1-77459-041-6

NOTICE TO THE READER

DEDICATION

To teachers, who make it possible to disseminate knowledge
to enlighten the young and curious minds
of our future generations

To students, who are the future of the world

THANKS

To my friends and colleagues

To my family for their love and support

Training and Consultant Services

At CADCAMCAE Works, we provide effective and affordable one to one online training on various software packages in Computer Aided Design(CAD), Computer Aided Manufacturing(CAM), Computer Aided Engineering (CAE), and Computer programming languages. The training is delivered through remote access to your system and voice chat via Internet at any time, any place, and at any pace to individuals, groups, students of colleges/universities, and CAD/CAM/CAE training centers. The main features of this program are:

Training as per your need

Highly experienced Engineers and Technician conduct the classes on the software applications used in the industries. The methodology adopted to teach the software is totally practical based, so that the learner can adapt to the design and development industries in almost no time. The efforts are to make the training process cost effective and time saving while you have the comfort of your time and place, thereby relieving you from the hassles of traveling to training centers or rearranging your time table.

Basic and Advanced Training Programs

CAD/CAM/CAE: CATIA, Creo Parametric, Creo Direct, SolidWorks, Autodesk Inventor, Solid Edge, UG NX, AutoCAD, AutoCAD LT, EdgeCAM, MasterCAM, SolidCAM, DelCAM, BOBCAM, UG NX Manufacturing, UG Mold Wizard, UG Progressive Die, UG Die Design, SolidWorks Mold, Creo Manufacturing, Creo Expert Machinist, NX Nastran, Hypermesh, SolidWorks Simulation, Autodesk Simulation Mechanical, Creo Simulate, Gambit, ANSYS and many others.

Computer Programming Languages: C++, VB.NET, HTML, Android, Javascript, and so on.

Game Designing: Unity.

Civil Engineering: AutoCAD MEP, Revit Structure, Revit Architecture, AutoCAD Map 3D and so on.

We also provide consultant services for Design and development on the above mentioned software packages

For more information you can mail us at:
cadcamcaeworks@gmail.com

Table of Contents

Chapter 1 : Introduction

Chapter 2 : Workpiece Setting

Chapter 7 : Milling Toolpaths-II

Chapter 8 : Milling Toolpaths-III

Chapter 9 : Multiaxis Milling Toolpaths

Preface

Mastercam for SOLIDWORKS combines the world's leading CAD software with the world's most widely used CAM software so you can create NC program for parts directly in SolidWorks, using toolpaths and machining strategies preferred by most of the manufacturing firms around the world. SOLIDWORKS users will feel at ease with the Mastercam machining tree, which delivers quick access to any point in the machining process. Mastercam users will recognize the industry-tested parameter screens and options which they are already familiar with in CAD SolidWorks. Mastercam for SolidWorks provides complete toolpath associativity which means toolpath data is stored directly in the SolidWorks file.

The **Mastercam 2022 for SolidWorks Black Book** is the 3rd edition of our series on Mastercam for SolidWorks. With lots of additions and thorough review, we present a book to help professionals as well as learners in creating some of the most complex NC toolpaths. The book follows a step by step methodology. In this book, we have tried to give real-world examples with real challenges in designing. We have tried to reduce the gap between university use of Mastercam and industrial use of Mastercam. In this edition of book, we have included many new topics of Mastercam 2022 for SolidWorks like Unified Toolpaths, Toolpath Preview, Port Expert, and so on. **There are about 25 topics newly added or thoroughly updated in this edition**. The book covers almost all the information required by a learner to master Mastercam for SolidWorks. The book starts with basics of machining and ends at advanced topics like Multiaxis Machining Toolpaths. Some of the salient features of this book are :

In-Depth explanation of concepts

Every new topic of this book starts with the explanation of the basic concepts. In this way, the user becomes capable of relating the things with real world.

Topics Covered

Every chapter starts with a list of topics being covered in that chapter. In this way, the user can easy find the topic of his/her interest easily.

Instruction through illustration

The instructions to perform any action are provided by maximum number of illustrations so that the user can perform the actions discussed in the book easily and effectively. There are about 700 small and large illustrations that make the learning process effective.

Tutorial point of view

At the end of concept's explanation, the tutorial make the understanding of users firm and long lasting. Almost each chapter of the book has tutorials that are real world projects. Moreover most of the tools in this book are discussed in the form of tutorials.

Project

Projects and exercises are provided to students for practicing.

New

If anything is added or enhanced in this edition which is not available in the previous editions, then it is displayed with symbol New in table of content.

For Faculty

If you are a faculty member, then you can ask for video tutorials on any of the topic, exercise, tutorial, or concept. Instructor resources are available for the book which contains answers to self-assessment, Models with Machining setup create, model files of examples discussed in book, and Lesson plan for the book.

Formatting Conventions Used in the Text

All the key terms like name of button, tool, drop-down etc. are kept bold.

Free Resources

Link to the resources used in this book are provided to the users via email. To get the resources, mail us at ***cadcamcaeworks@gmail.com*** with your contact information. With your contact record with us, you will be provided latest updates and informations regarding various technologies. The format to write us mail for resources is as follows:

Subject of E-mail as ***Application for resources of _____ book***.
Also, given your information like
Name:
Course pursuing/Profession:
E-mail ID:

Note: We respect your privacy and value it. If you do not want to give your personal informations then you can ask for resources without giving your information. Instructor resources cannot be provided without full information proving the eligibility.

About Authors

The author of this book, Gaurav Verma, has authored and assisted in more than 16 titles in CAD/CAM/CAE which are already available in market. He has authored **AutoCAD Electrical Black Books** which are available in both **English** and **Russian** language. He has also written **Creo Manufacturing 4.0 Black Book** which covers Expert Machinist module of Creo Parametric. He has provided consultant services to many industries in US, Greece, Canada, and UK. He has assisted in preparing many Government aided skill development programs. He has been speaker for Autodesk University, Russia 2014. He has assisted in preparing AutoCAD Electrical course for Autodesk Design Academy. He has worked on Sheetmetal, Forging, Machining, and Casting products in Design and Development department.

The author of this book, Matt Weber, has authored many books on CAD/CAM/CAE available already in market. **SolidWorks Simulation Black Books** are one of the most selling books in SolidWorks Simulation field. The author has hands on experience on almost all the CAD/CAM/CAE packages. Besides that he is a good person in his real life, helping nature for everyone. If you have any query/doubt in any CAD/CAM/CAE package, then you can contact the author by writing at cadcamcaeworks@gmail.com

For Any query or suggestion

If you have any query or suggestion, please let us know by mailing us on *cadcamcaeworks@gmail.com*. Your valuable constructive suggestions will be incorporated in our books and your name will be addressed in special thanks area of our books on your confirmation.

Chapter 1

Introduction

Topics Covered

The major topics covered in this chapter are:

- *Introduction to manufacturing.*
- *Types of Machines.*
- *Applications of CAM.*
- *Installing Mastercam for SolidWorks.*
- *General Approach in Mastercam.*
- *Walkthrough of Mastercam for SolidWorks*

INTRODUCTION TO MANUFACTURING

Manufacturing is the process of creating a useful product by using a machine, a process, or both. For manufacturing a product, there are some steps to be followed:

- Generating Layout of final product.
- Raw material/Work piece; selection of raw material depends on the application of the product.
- Forging, Casting or any other pre-machining method for creating outlines for final shape.
- Roughing Processes.
- Finishing Processes.
- Quality Control.

As the "Generating Layout of final product" is above all the steps, it is the most important step. One should be very clear about the final product because all the other steps are totally dependent on the first step. The layout of final product can be a drawing or a model created by using any modeling software like SolidWorks. (Refer to our another title **SolidWorks 2021 Black Book** for concepts of modeling)

The next step is "Selection of Raw material/Workpiece". This step is solely dependent on the first step. Our final product defines what should be the raw material and the workpiece. Here, workpiece is the piece of raw material to be used for the next step or process.

The next step is "Forging, Casting or any other process for creating outline of the final shape". The outline created for the final shape is also called Blank in industries. In this step, various machines like Press, Cutter or Moulding machines are used for creating the blank. In some cases of Casting, there is no requirement of machining processes. For example, in case of Investment casting most of the time there is no requirement of machining process. Machining processes can be divided further into two processes:

- Roughing Processes
- Finishing Processes

These processes are the main area of discussion in this book. An introduction to these processes is given next.

Roughing Process

Roughing process is the beginning of machining process. Generally, roughing process is the removal of large amount of stock material in comparison to finishing process. In a roughing process, the quantity of material removed from the workpiece is more important than the quality of the machining. There are no close tolerances for roughing processes. So, these processes are relatively cheaper than the finishing processes. In manufacturing industries, there are three principle machining processes called Turning, Milling, and Drilling. In case of roughing process, there can be turning, milling, drilling, combination of any two, or all the processes. In addition to these machining processes, there are various other processes like shaping, planing, broaching, reaming, and so on. But these processes are used in special cases.

Finishing Process

Finishing process can include all the machining processes discussed in case of roughing processes but in close tolerances. Also, the quality of machining at required accuracy level is very important for finishing. In addition to the above discussed machining processes, there are a few more machining processes like Electric Discharge Machining(EDM), Laser Beam Machining, Electrochemical Machining, and so on. These processes are called unconventional machining processes because of their cutting method. In unconventional machining processes, the tool life is much higher than the conventional machining. Different Machines used for machining processes are discussed next.

TYPES OF MACHINES

There are various types of machines for different type of machining process. For example- for turning process, there are machines like conventional lathe and CNC Turner. Similarly for milling process, there are machines called Milling machine, VMC or HMC. Some of the machines are discussed next with details of their functioning.

Turning Machines

Turning machine is a category of machines used for turning process. In this machine, the workpiece is held in a chuck (collet in case of small workpieces). This chuck revolves at a defined rotational speed. Note that the workpiece can revolve in either CW(Clockwise) or CCW(Counter-Clockwise) direction but cannot translate in any direction. The cutting tool used for removing material can translate in X (D) and Y (Z) directions. The most basic type of turning machine is a lathe. But now a days, lathes are being replaced by CNC Turning machines, which are faster and more accurate than the traditional lathes. The CNC Turning machines are controlled by numeric codes. These codes are interpreted by machine controller attached in the machine and then the controller commands various sections of the machine to do a specific job. The basic operations that can be done on turning machines are:

- Taper turning
- Spherical generation
- Facing
- Grooving
- Parting (in few cases)
- Drilling
- Boring
- Reaming
- Threading

Milling Machines

Milling machine is a category of machines used for removing material by using a perpendicular tool relative to the workpiece. In this type of machine, workpiece is held on a bed with the help of fixtures. The tool rotates at a defined speed. This tool can move in X, Y, and Z directions. In some machines, the bed can also translate and rotate like in Turret milling machines, 5-axis machines, and so on. Milling machines are of two types; horizontal milling machine and vertical milling machine. In Horizontal milling machine, the tool is aligned with the horizontal axis (X-axis). In Vertical milling machine, the tool is aligned with the vertical axis (Y-axis). The Vertical milling machine is generally used for complex cutting processed like contouring, engraving, embossing and so on. The Horizontal milling machines are used for cutting slots, grooves, gear teeth, and so on. In some Horizontal milling machines, table can moved up-down by motor mechanism or power system. By using the synchronization of table movement with the rotation of rotary fixture, we can also create spiral features. The tools used in both type milling machine have cutting edge on the sides as well as at the tip.

Drilling Machines

Drilling machine is a category of machines used for creating holes in the workpiece. In Drilling machine, the tool (drill bit) is fixed in a tool holder and the tool can move up-down. The workpiece is fixed on the bed. The tool goes down, by motor or by hand, penetrating through the workpiece. There are various types of Drilling machine available like drill presses, cordless drills, pistol grip drills, and so on.

Shaper

Shaper is a category of machines, which is used to cut material in a linear motion. Shaper has a single point cutting tool, which goes back-forth to create linear cut in the workpiece. This type of machine is used to create flat surface of the workpiece. You can create dovetail slots, splines, key slots, and so on by using this machine. In some operation, this machine can be an alternative for EDM.

Planer

Planer is a category of machines similar to Shaper. The only difference is that, in case of Planer machine, the workpiece reciprocates and the tool is fixed.

There are various other special purpose machines (SPMs), which are used for some uncommon requirements. The machines discussed above are conventional machines. The unconventional machines are discussed next.

Electric Discharge Machine

Electric Discharge Machine is a category of machines used for creating desired shapes on the workpiece with the help of electric discharges. In this type of machines, the tool and the workpiece act as electrodes and a dielectric fluid is passed between them. The workpiece is fixed in the bed and tool can move in X, Y, and Z direction. During the machining process, the tool is brought near to the workpiece. Due to this, a spark is generated between them. This spark causes the material on the workpiece to melt and get separated from the workpiece. This separated material is drained with the help of dielectric fluid. There are two types of EDMs which are listed next.

Wire-cut EDM

In this type of EDM, a brass wire is commonly used to cut the material from the workpiece. This wire is held in upper and lower diamond shaped guides. It is constantly fed from a bundle. In this machine, the material is removed by generating sparks between tool and workpiece. A Wire-cut EDM can be used for a plate having thickness up to 300 mm.

Sinker EDM

In this type of EDM, a metal electrode is used to cut the material from the workpiece. The tool and the workpiece are submerged in the dielectric fluid. Power supply is connected to both the tool and the workpiece. When tool is brought near the workpiece, sparks are generated randomly on their surfaces. Such sparks gradually create impression of tool on the workpiece.

Electro Chemical Machine

Electro Chemical Machine is a category of machines used for creating desired shape by using the chemical electrolyte. This machining works on the principles of chemical reactions.

Laser Beam Machine

Laser Beam Machine is a category of machines that uses a beam a highly coherent light. This type of light is called laser. A laser can output a power of up to 100MW in an area of 1 square mm. A laser beam machine can be used to create accurate holes or shapes on a material like silicon, graphite, diamond, and so on.

The machines discussed till now are the major machines used in industries. Some of these machines can be controlled by numeric codes and are called NC machines. NC Machines and their working are discussed next.

NC MACHINES

An NC Machine is a manufacturing tool that removes material by following a predefines command set. An NC Machine can be a milling machine or it can be a turning center. NC stands for Numerical Control so, these machines are controlled by numeric codes. These codes are dependent on the controller installed in the machines. There are various controllers available in the market like Fanuc controller, Siemens controller, Heidenhain controller, and so on. The numeric codes change according to the controller used in the machine. These numeric codes are compiled in the form of a program, which is fed in the machine controller via a storage media. The numeric codes are generally in the form of G-codes and M-codes. For understanding purpose, some of the G-codes and M-codes are discussed next with their functions for a Fanuc controller.

Code		Function
G00	-	Rapid movement of tool.
G01	-	Linear movement while creating cut.
G02	-	Clockwise circular cut.
G03	-	Counter-clockwise circular cut.
G20	-	Starts inch mode.
G21	-	Starts mm mode.
G96	-	Provides constant surface speed.
G97	-	Constant RPM.
G98	-	Feed per minute
G99	-	Feed per revolution
M00	-	Program stop
M02	-	End of program

M03	-	Spindle rotation Clockwise.
M04	-	Spindle rotation Counter Clockwise.
M05	-	Spindle stop
M08	-	Coolant on
M09	-	Coolant off
M98	-	Subprogram call
M99	-	Subprogram exit

These codes as well as the other codes will be discussed in the subsequent chapters according to their applications.

As there is a long list of codes which are required NC programs to make machine cut workpiece in desired size and shape, it becomes a tedious job to create programs manually for each operation. Moreover, it take much time to create a program for small operations on a milling machine. To solve this problem and to reduce the human error, Computer Aided Manufacturing (CAM) was introduced. Various applications of CAM are discussed next.

APPLICATIONS OF COMPUTER AIDED MANUFACTURING

Computer Aided Manufacturing (CAM) is a technology which can be used to enhance the manufacturing process. In this technology, the machines are controlled by a workstation. This workstation can serve more than one machines at a time. Using CAM, you can create and manage the programs being fed in the workstation. Some of the applications of CAM are discussed next.

1. CAM with the combination of CAD can be used to create complex shapes by machining in a small time.
2. CAM can be used to manage more than one machines at the same time with less human power.
3. CAM is used to automate the manufacturing process.
4. CAM is used to generate NC programs for various types of NC machines.
5. 5-Axis Machining

CAM is generally the next step after CAD (Computer Aided Designing). Sometimes CAE (Computer Aided Engineering) is also required before CAM. There are various software companies that provide the CAM software solutions. CNC Software is one of those companies which publishes Mastercam software. Mastercam is one of the most popular software for CAM programming. On the other side, SolidWorks is one of the most used CAD software. Hence, the combination of these two leading software is the best possible combination of CAD CAM capabilities.

DOWNLOADING STUDENT VERSION OF MASTERCAM FOR SOLIDWORKS

• Open your internet browser and reach the link :

https://signup.mastercam.com/demo-hle-mms

- On reaching the link, a web-page will be displayed as shown in Figure-1. Type your e-mail id in **Enter your e-mail to start** edit box and click on the **GET STARTED** button. The **Create mastercam account** web page will be displayed.
- Enter your information in the fields and click on the **Create Account** button. An e-mail will be sent to your email id for verification. Open your e-mail account and click on the verification link sent by Mastercam. The user validation page will be displayed in web browser.
- Click on the **PROCEED TO LOGIN** button. The user login page will be displayed.
- Enter your email id and password specified when creating account, and click on the **LOGIN** button. The demo request page will be displayed; refer to Figure-2.
- Enter your student/educator information in the fields and click on the **Submit** button. The download page for HLE of software will be displayed; refer to refer to Figure-3. Note that your license number and activation code will also display on this page. Keep this information safe and ready because you will need it after installing the software. This information is also sent to your e-mail id.
- Click on the **Mastercam for SOLIDWORKS Demo/HLE** link button. The setup file will start downloading.

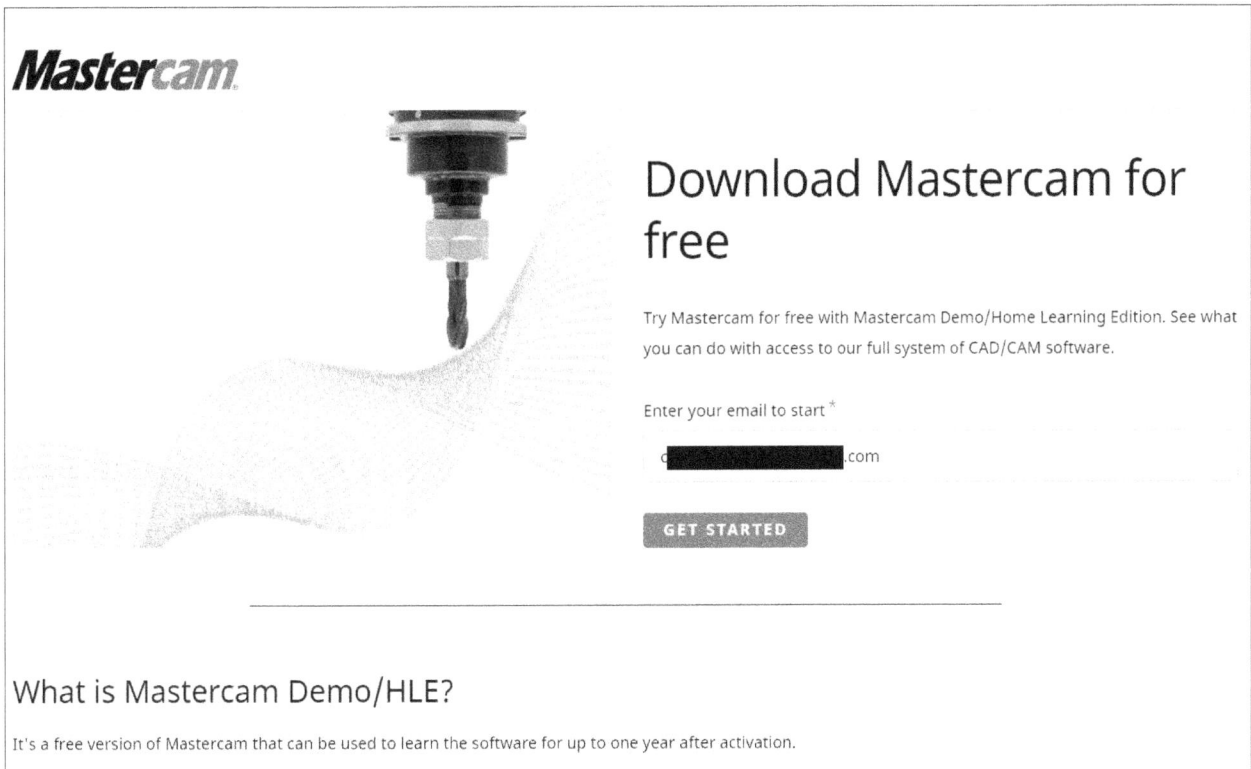

Mastercam

Download Mastercam for free

Try Mastercam for free with Mastercam Demo/Home Learning Edition. See what you can do with access to our full system of CAD/CAM software.

Enter your email to start *

▮▮▮▮▮▮▮▮.com

GET STARTED

What is Mastercam Demo/HLE?

It's a free version of Mastercam that can be used to learn the software for up to one year after activation.

Figure-1. Mastercam Home Learning Edition webpage

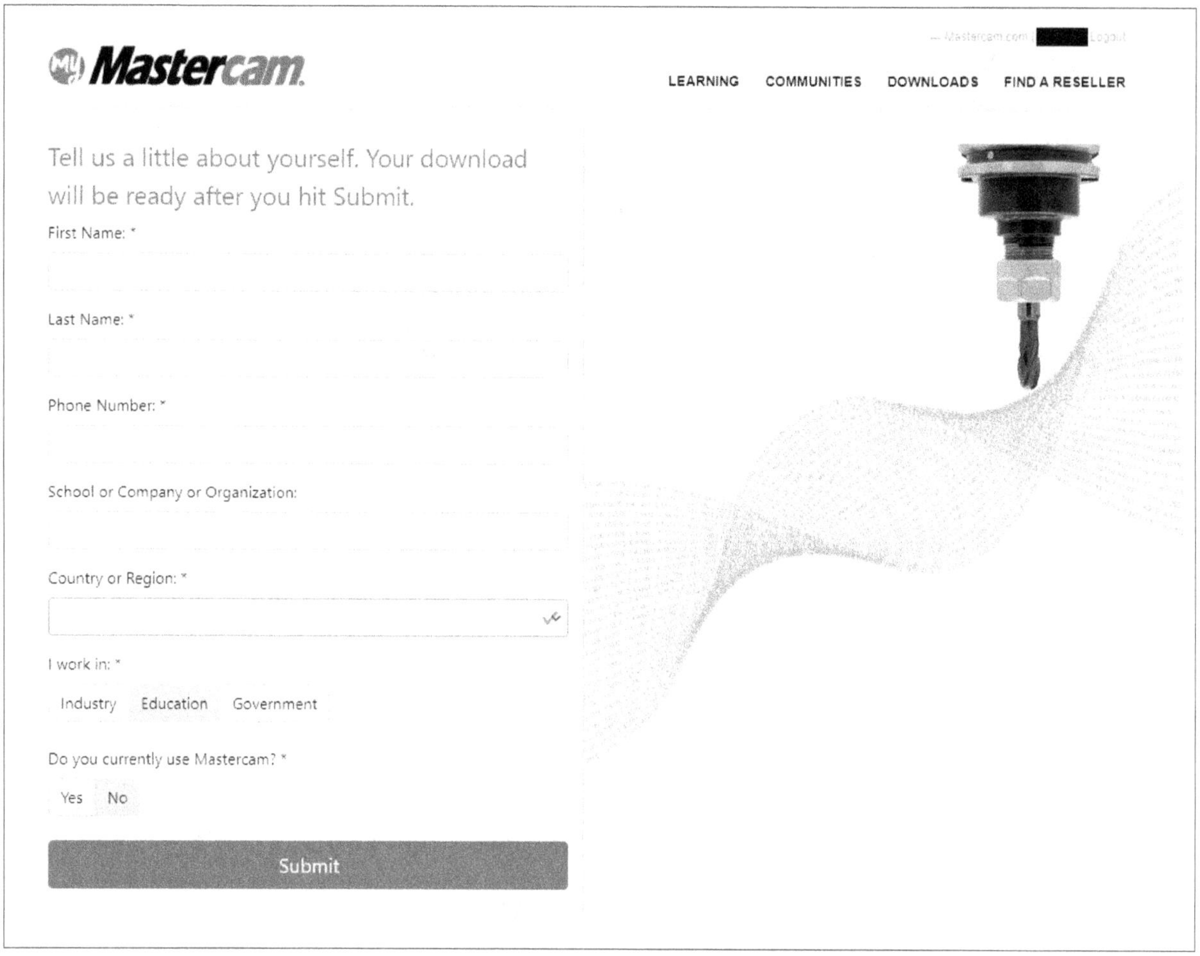

Figure-2. Demo request page

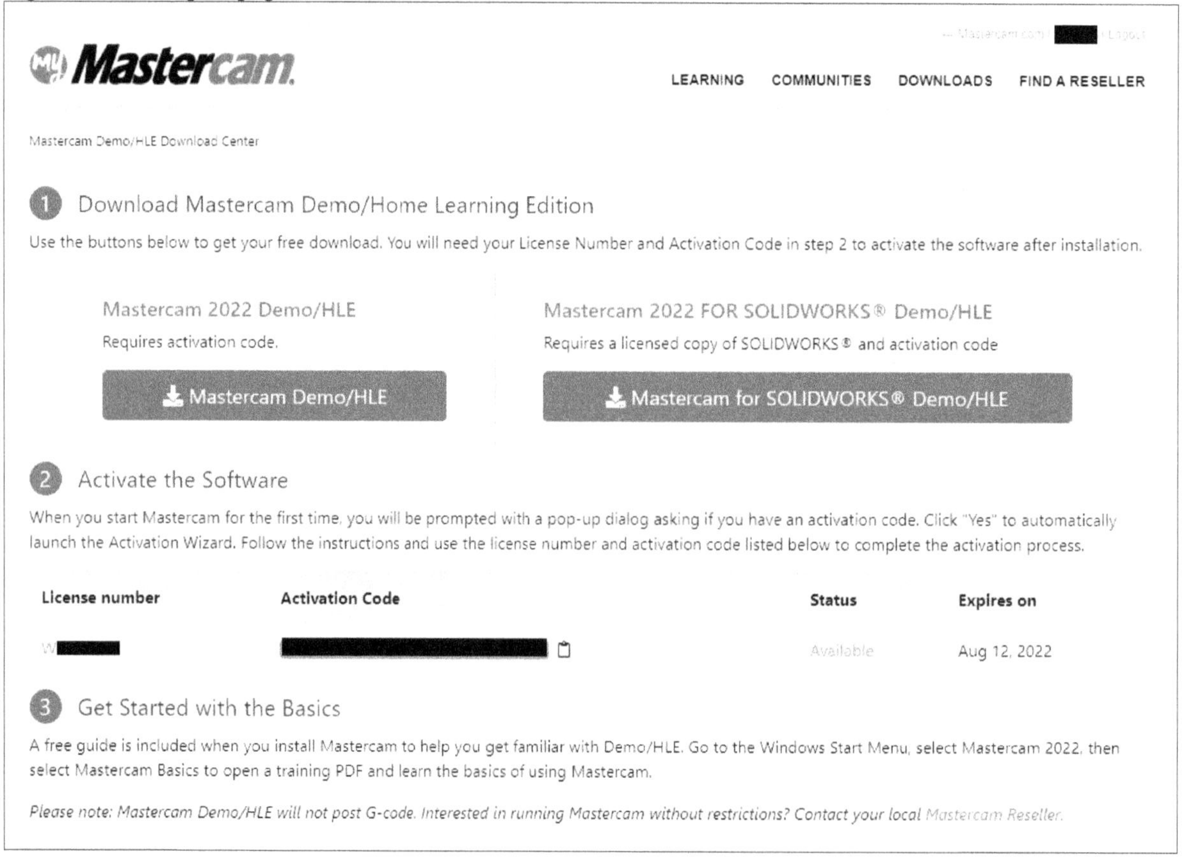

Figure-3. Download page for Mastercam HLE

INSTALLING MASTERCAM 2022 FOR SOLIDWORKS

- Download the setup file of Mastercam 2022 for SolidWorks from the Mastercam website as discussed earlier.
- Right-click on downloaded .exe file from the location where you downloaded the file.
- Select the **Run as Administrator** option from the shortcut menu and follow the instructions to install.
- After installation, click on the **Activation Wizard** option from **Mastercam** folder in **Start** menu; refer to Figure-4. The **Mastercam Product Activation Wizard** will be displayed; refer to Figure-5.

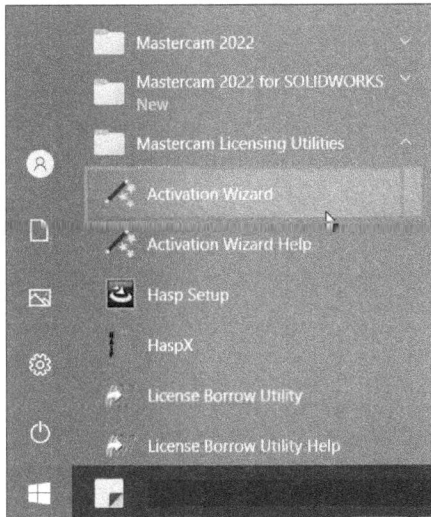

Figure-4. Activation Wizard option

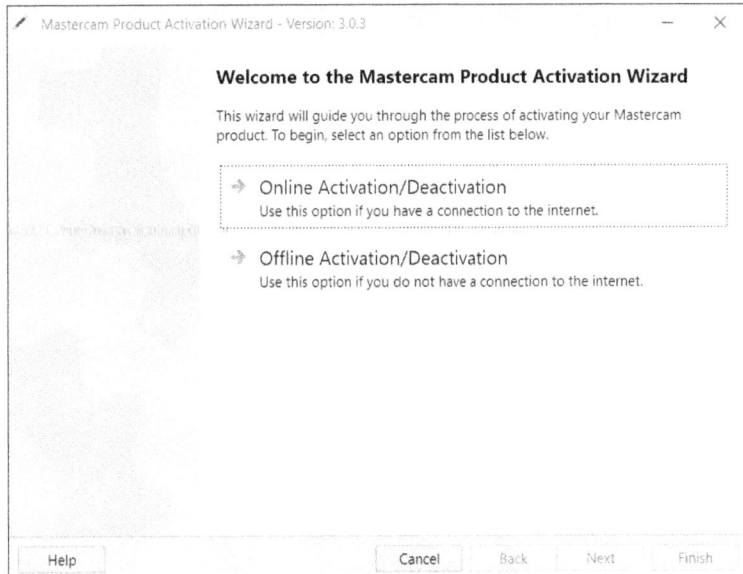

Figure-5. Mastercam product activation wizard

- Click on the Help button at the bottom left corner of the dialog box to learn about activation methods. Activate your mastercam license using License number and activation code provided to you.
- Start SolidWorks by using the Start menu or icon on Desktop.
- Click on the down button next to **Options** button in the Quick Access toolbar; refer to Figure-6 and click on the **Add-Ins** button or click on the **Tools > Add-Ins** button from the Menu bar. The **Add-Ins** box will be displayed; refer to Figure-7.

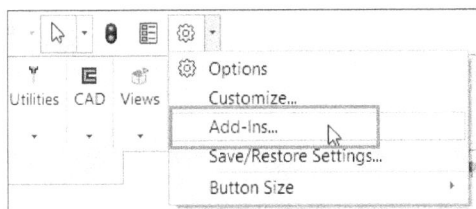

Figure-6. Quick Access toolbar

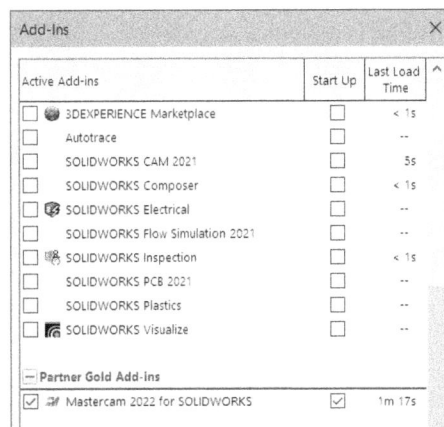

Figure-7. Add-Ins box

- Scroll-down in the box and select the check box before the **Mastercam 2022 for SOLIDWORKS** option to activate the application. If you want Mastercam to start with SolidWorks each time then select the check box after the **Mastercam 2022 for SOLIDWORKS** option in this box. Make sure the **SOLIDWORKS CAM** is not active otherwise Mastercam may not function properly.

- Click on the **OK** button from the box. The **Mastercam Customer Feedback Program** dialog box will be displayed on the first run. Select desired option and click on the **OK** button from the dialog box. After a while (may be 5-8 minutes for first run) the application will load and a new **CommandManager Mastercam 2022** will be added in the **Ribbon** and menu refer to Figure-8. Note: Make sure Code Meter and Cmcontainer processes are running in background of your system otherwise license will not activate.

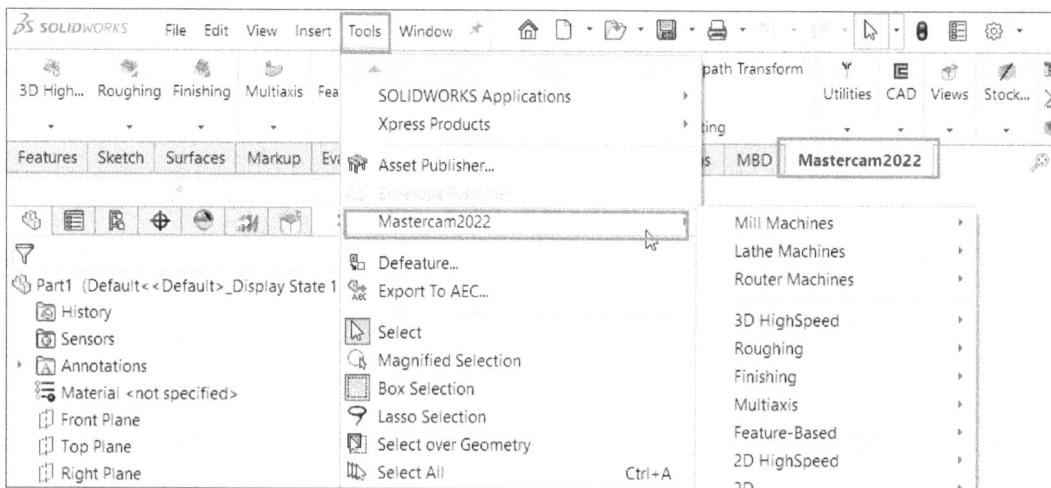

Figure-8. Mastercam 2022 CommandManager added in the Ribbon

- Note that the **Mastercam Info Center** tab is also added in the task pane at the left; refer to Figure-9. In the left area of the drawing window, the **Mastercam Toolpath Manager** is added; refer to Figure-10. You will learn about these interface elements later in this chapter.

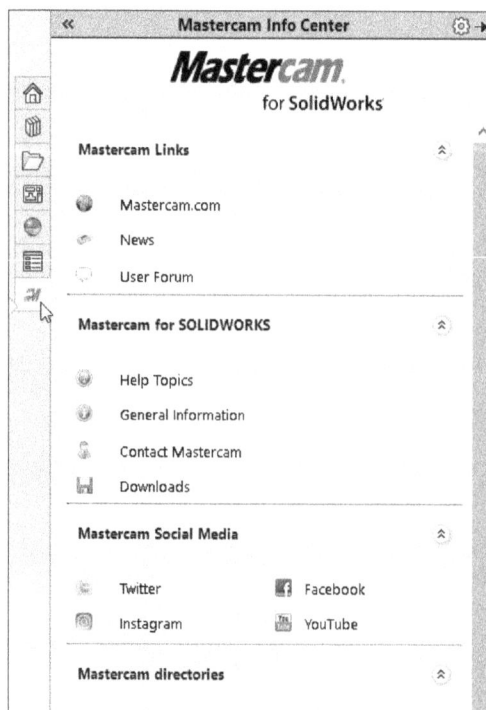

Figure-10. Mastercam Toolpath manager

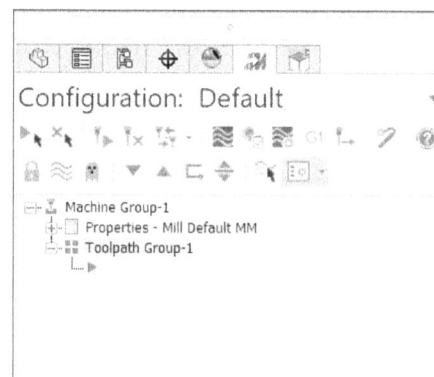

Figure-9. Mastercam Info Center

BASIC APPROACH IN MASTERCAM

Whether you use the stand-alone program of Mastercam or the integrated one with SolidWorks, the approach for creating NC programs is same. First, you need to import or create the CAD model of the product. Then, create stock of material (workpiece) from which the product will be manufactured after machining. Apply settings related to machine. Apply parameters related to tools. Create the tool paths for operations to be performed on the machine. Simulate the machining process and check whether it is as per the requirement. Generate the output of the machining which is NC codes. Refer to Figure-11.

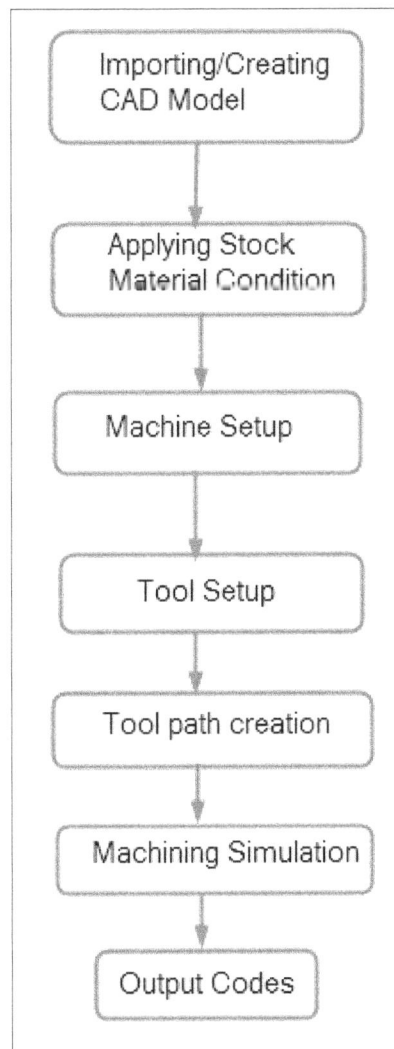

```
┌─────────────────────────────┐
│   ┌─────────────────────┐   │
│   │  Importing/Creating │   │
│   │     CAD Model       │   │
│   └─────────────────────┘   │
│              │              │
│              ▼              │
│   ┌─────────────────────┐   │
│   │  Applying Stock     │   │
│   │  Material Condition │   │
│   └─────────────────────┘   │
│              │              │
│              ▼              │
│   ┌─────────────────────┐   │
│   │   Machine Setup     │   │
│   └─────────────────────┘   │
│              │              │
│              ▼              │
│   ┌─────────────────────┐   │
│   │     Tool Setup      │   │
│   └─────────────────────┘   │
│              │              │
│              ▼              │
│   ┌─────────────────────┐   │
│   │  Tool path creation │   │
│   └─────────────────────┘   │
│              │              │
│              ▼              │
│   ┌─────────────────────┐   │
│   │ Machining Simulation│   │
│   └─────────────────────┘   │
│              │              │
│              ▼              │
│   ┌─────────────────────┐   │
│   │    Output Codes     │   │
│   └─────────────────────┘   │
└─────────────────────────────┘
```

Figure-11. Workflow in Mastercam

OVERVIEW OF THE MANUFACTURING THROUGH MASTERCAM FOR SOLIDWORKS

Now, we will go through the whole process of manufacturing step by step. Later, we will learn minute details of the tools in Mastercam add-in of SolidWorks.

Creating/Importing Model

- Create desired model by using the tools available in the SolidWorks Modeling environment.

Or,

- Click on the **Open** button from the **File** menu; refer to Figure-12. The **Open** dialog box will be displayed; refer to Figure-13.

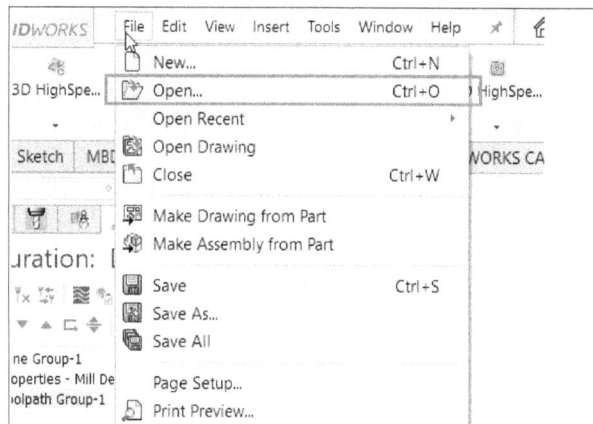

Figure-12. Open button

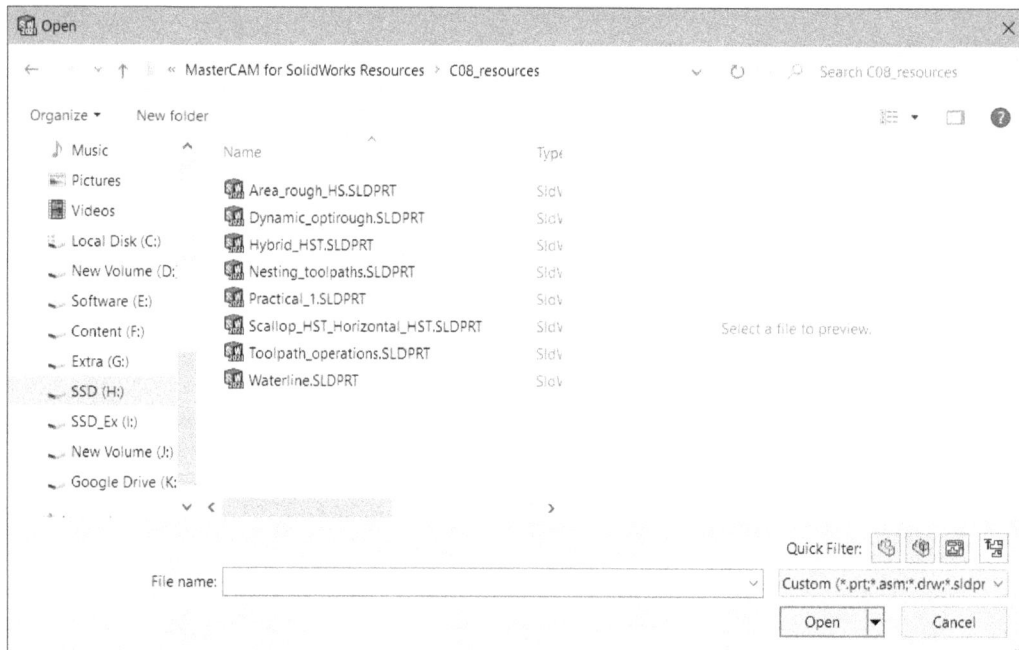

Figure-13. Open dialog box

- Browse to the location of desired file and double-click on it. The file will open in SolidWorks; refer to Figure-14.

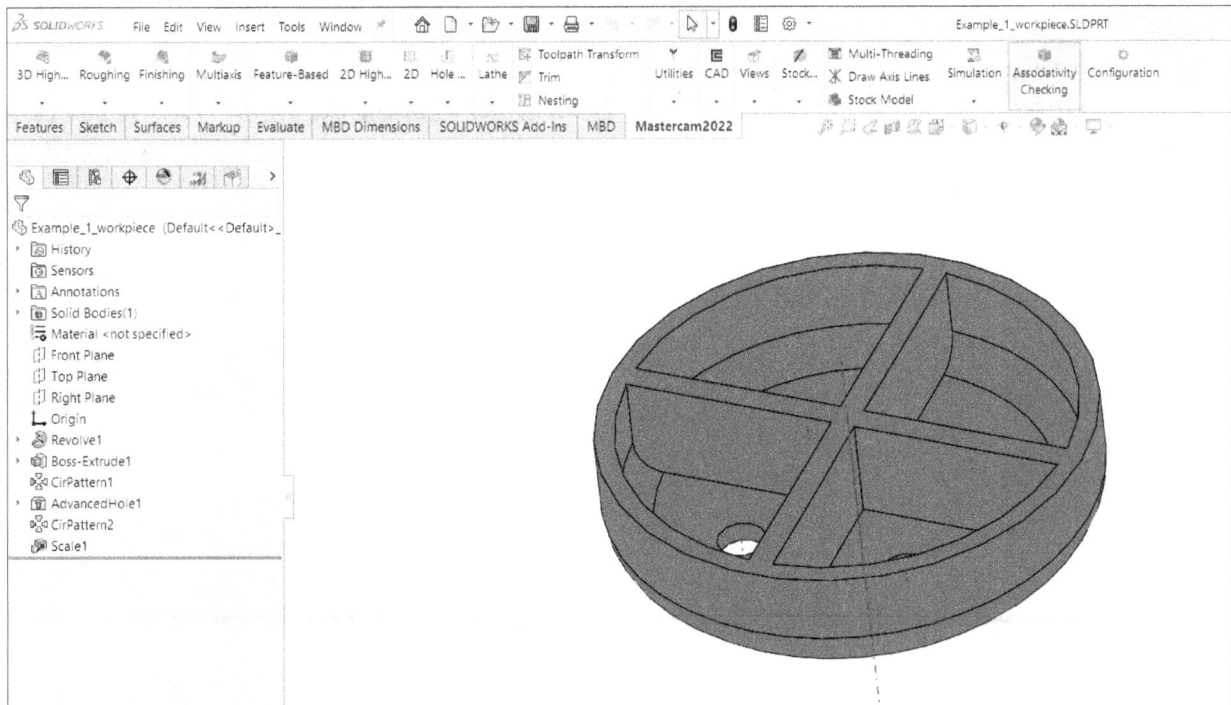

Figure-14. File opened in SolidWorks

Specifying the Top Plane

- Click on the **Planes Manager** tab from the **Model Tree** at the left in the interface. The **Planes Manager** will be displayed as shown in Figure-15.

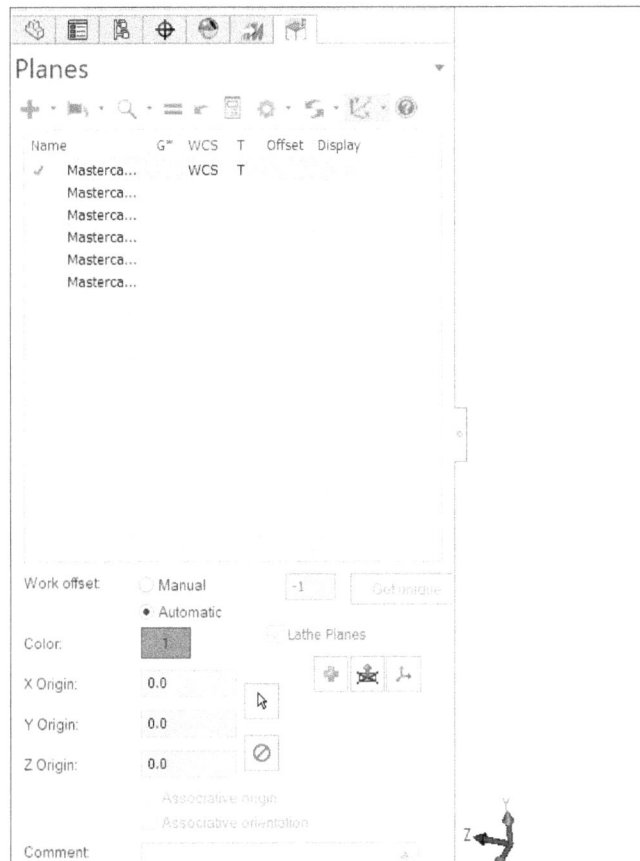

Figure-15. Planes Manager

- By default the **MASTERCAM TOP** option is selected in the list. If not selected in your case, then select it and click on the **Set current WCS and tool plane** ⊟ button from the **Planes Manager**.
- Click on cursor icon button ⓚ from the bottom in **Planes Manager**. The **Selection PropertyManager** will be displayed and you will be prompted to specify the location of the origin for current selected plane; refer to Figure-16.
- Click at desired location to specify the origin. Note that the coordinates for NC program will be measured from this origin. Generally, reference cutting tool is manually touched to the workpiece for finding this location on machine.

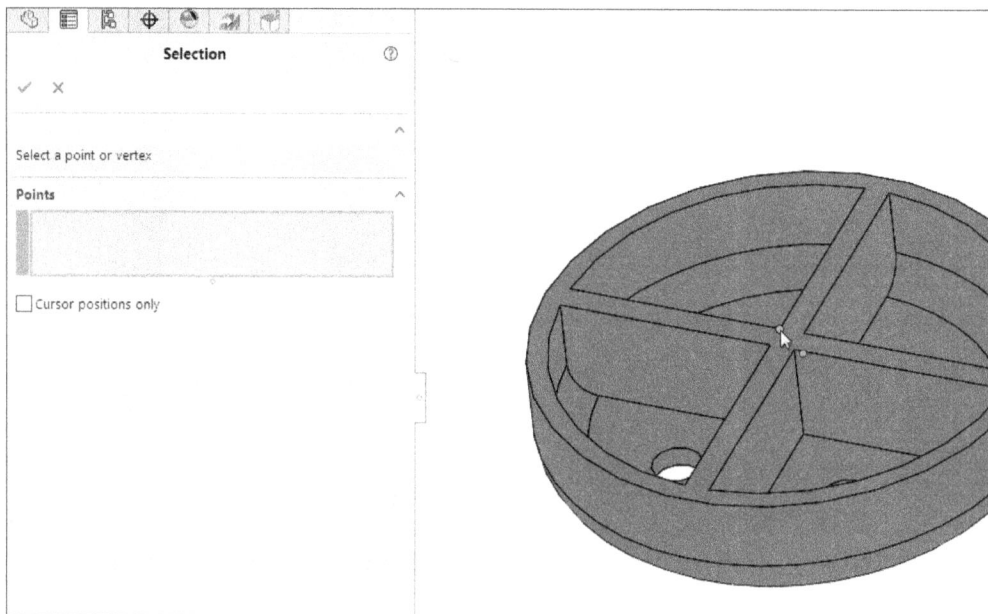

Figure-16. Selection PropertyManager

- Click on the **OK** button from the **Selection PropertyManager**. The origin will move to the selected location.
- You will learn about all the parameters of the **Planes Manager** later in the book.

Specifying Stock of Material

- Click on the **Mastercam Toolpath Manager** tab from the **PropertyManager**; refer to Figure-17. The **Mastercam Toolpath Manager** will be displayed; refer to Figure-18.

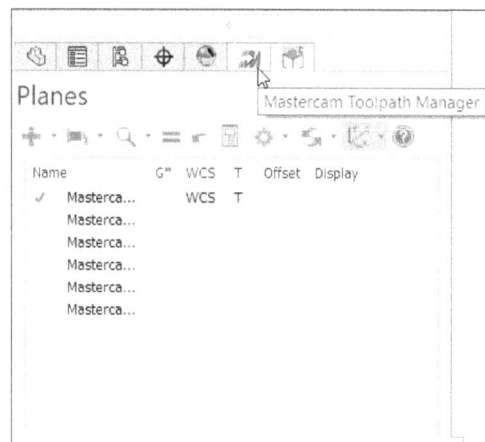

Figure-17. Mastercam Toolpath manager tab

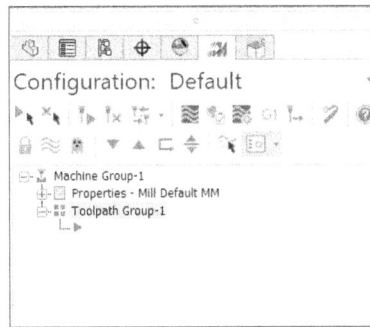

Figure-18. Mastercam Toolpath Manager

- Click on the **+** sign next to **Properties-Mill Default MM** node in the **Toolpath Manager**. The expanded node will be displayed.
- Click on the **Stock Setup** option in the expanded node. The **Machine Group Properties** dialog box will be displayed; refer to Figure-19.

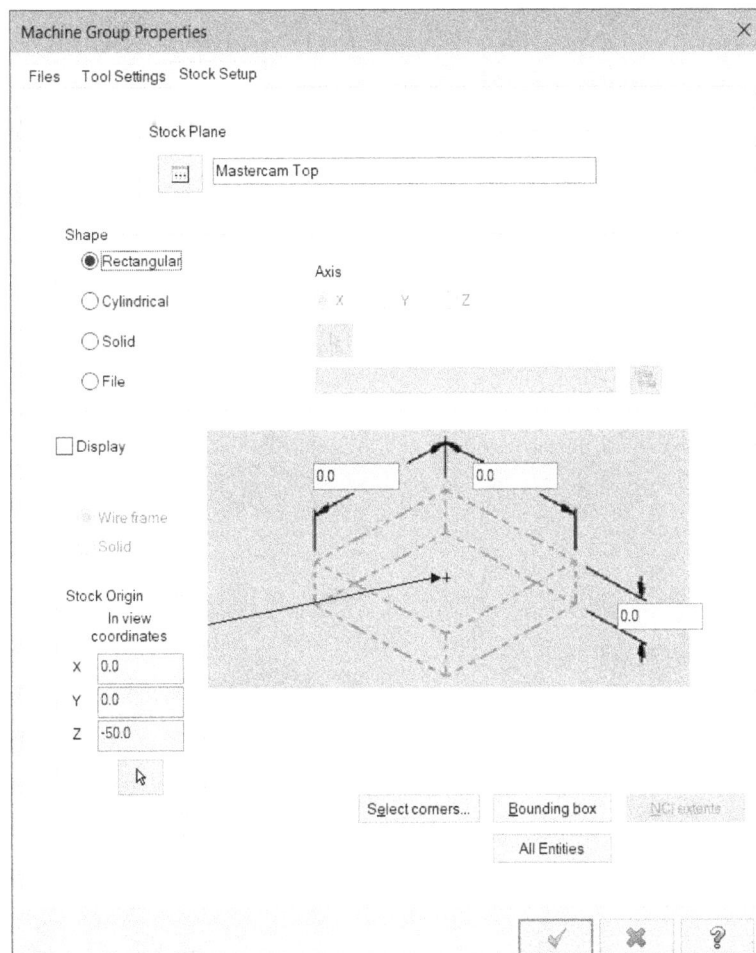

Figure-19. Machine Group Properties dialog box

- Click on the **Bounding box** button from the bottom in the dialog box. The **Selection PropertyManager** will be displayed; refer to Figure-20.

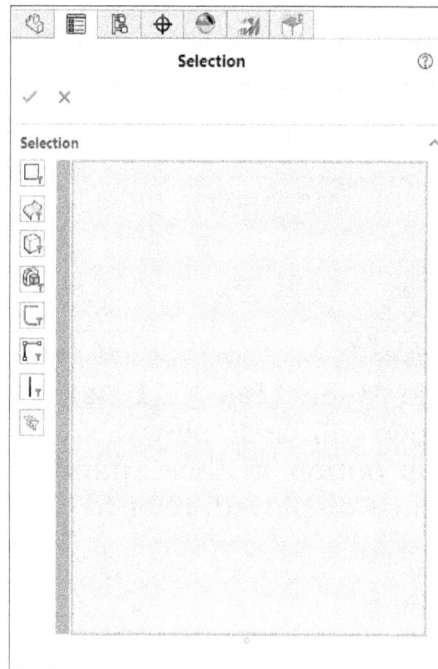

Figure-20. Selection PropertyManager

- Select desired selection filter from the left area in the **PropertyManager** if you want to use a selection filter and then click on the model; refer to Figure-21. For example, if you want to select complete body then select the **Body** selection filter button from the left in the **PropertyManager**. After selecting object, click on the **OK** button from the **PropertyManager**. The **Bounding Box Manager** will be displayed; refer to Figure-22.

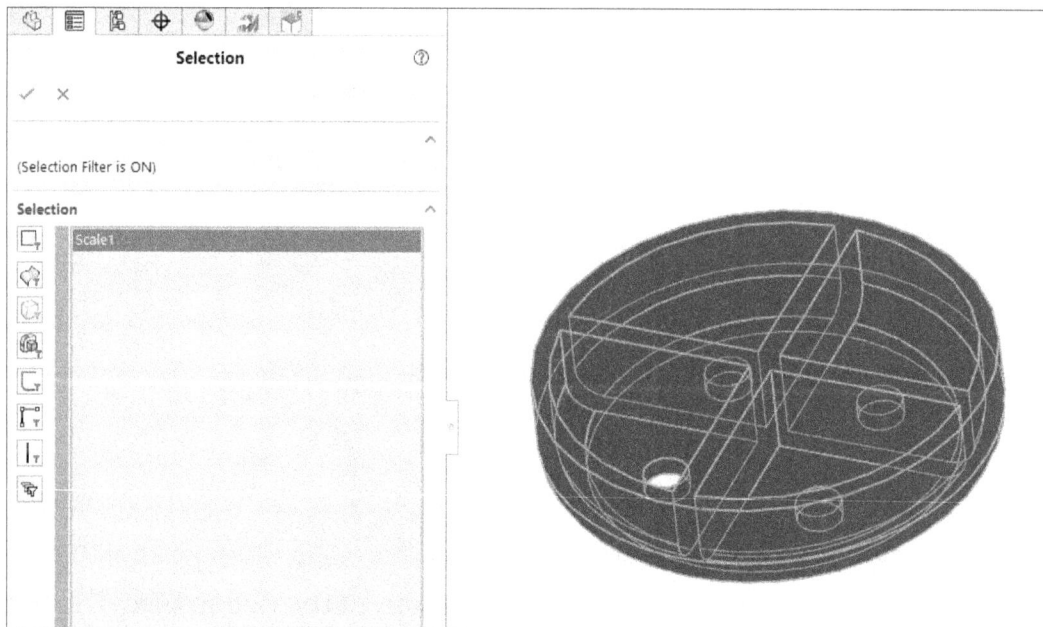

Figure-21. Selecting model

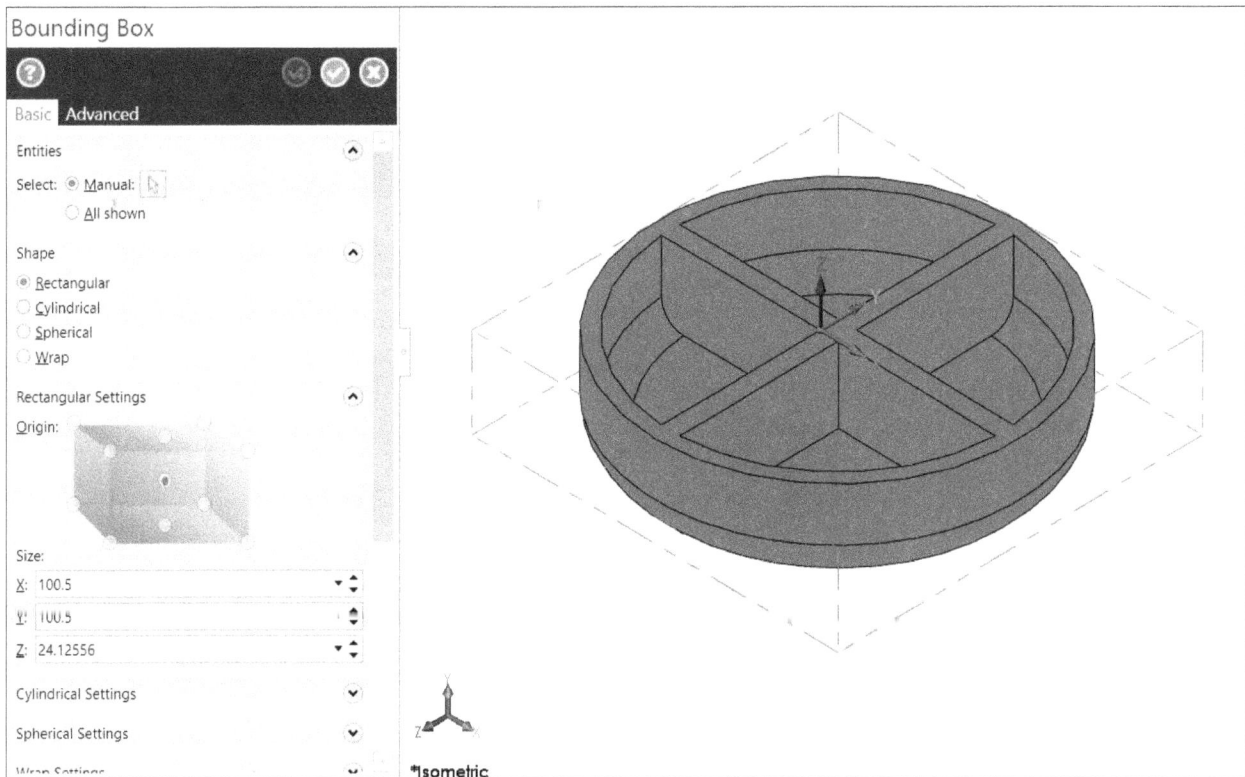

Figure-22. Bounding Box PropertyManager

- Select the **Cylindrical** radio button from **Shape** rollout and **Z** radio button from **Axis** section in the **PropertyManager**. Click on the **OK** button from the **PropertyManager**. The **Machine Group Properties** dialog box will be displayed again.
- Select the **Shaded** radio button below the **Display** check box and then click on the **OK** button. The stock of material will be displayed; refer to Figure-23.

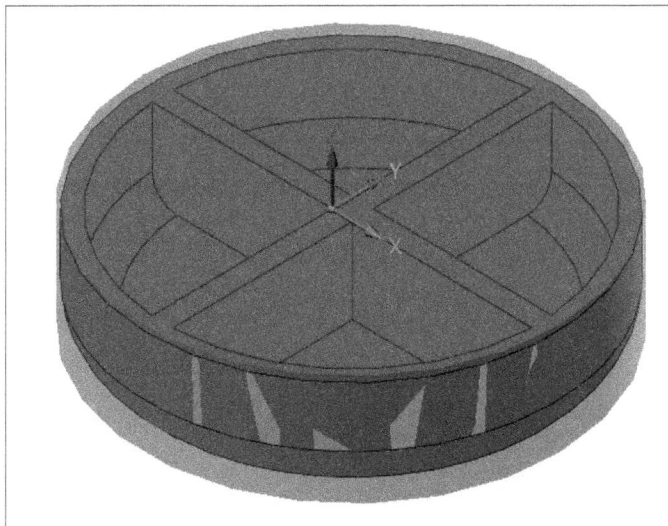

Figure-23. Stock of material

Machine Setup

- Click on the down arrow below **Utilities** tool from the **Ribbon**. A menu of tools will be displayed; refer to Figure-24.

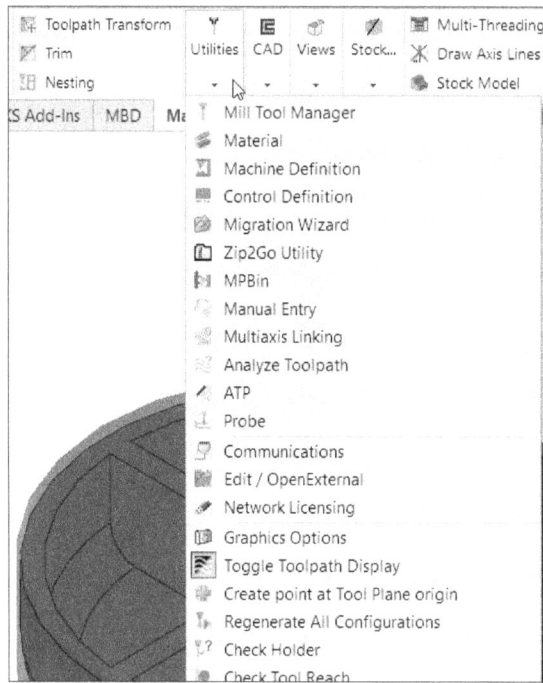

Figure-24. Utilities drop-down

- Click on the **Machine Definition** tool. The **Machine Definition File Warning** dialog box will be displayed; refer to Figure-25.

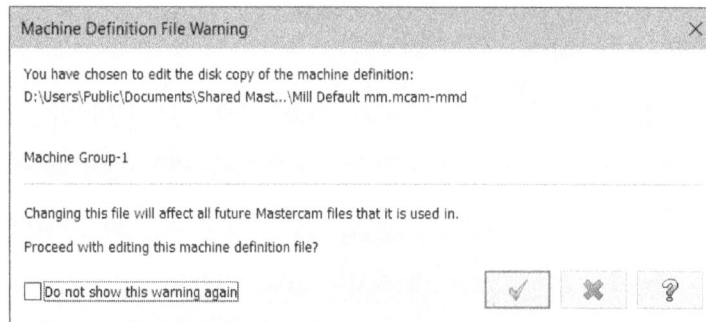

Figure-25. Machine Definition File Warning dialog box

- Click on the **OK** button from the dialog box. The **Machine Definition Manager** dialog box will be displayed; refer to Figure-26.

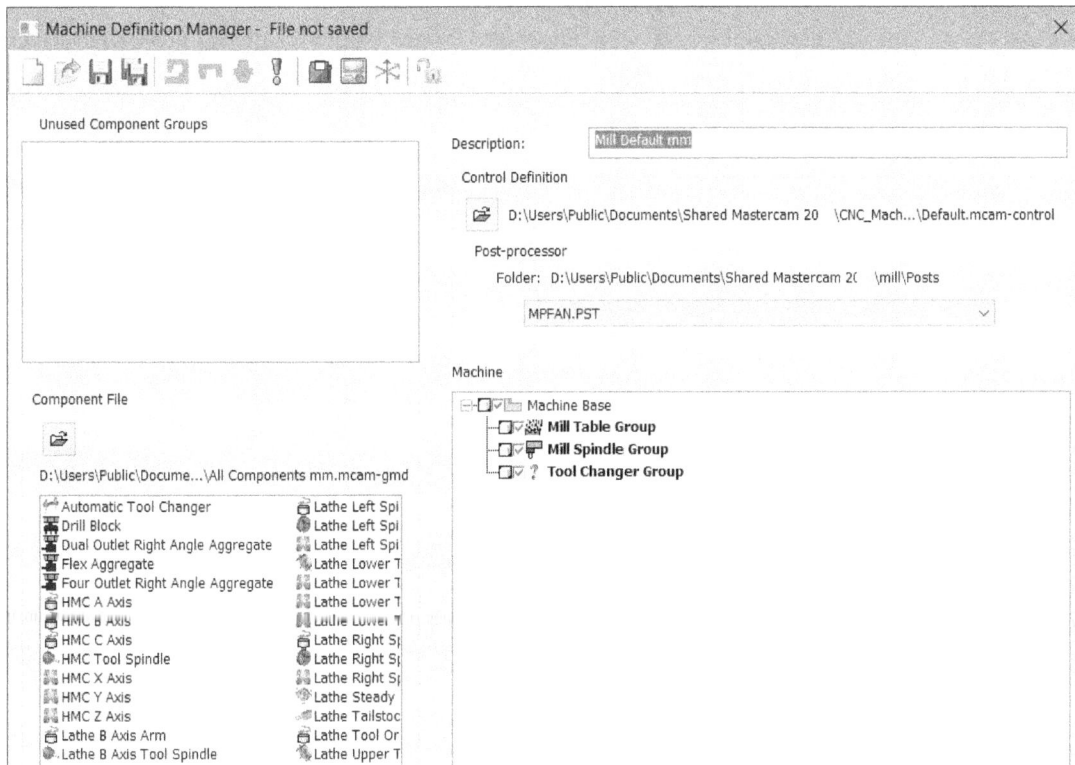
Figure-26. Machine Definition Manager dialog box

- Click on the **Open** button 📂 from the toolbar at the top in the dialog box. The **Open Machine Definition File** dialog box will be displayed; refer to Figure-27.

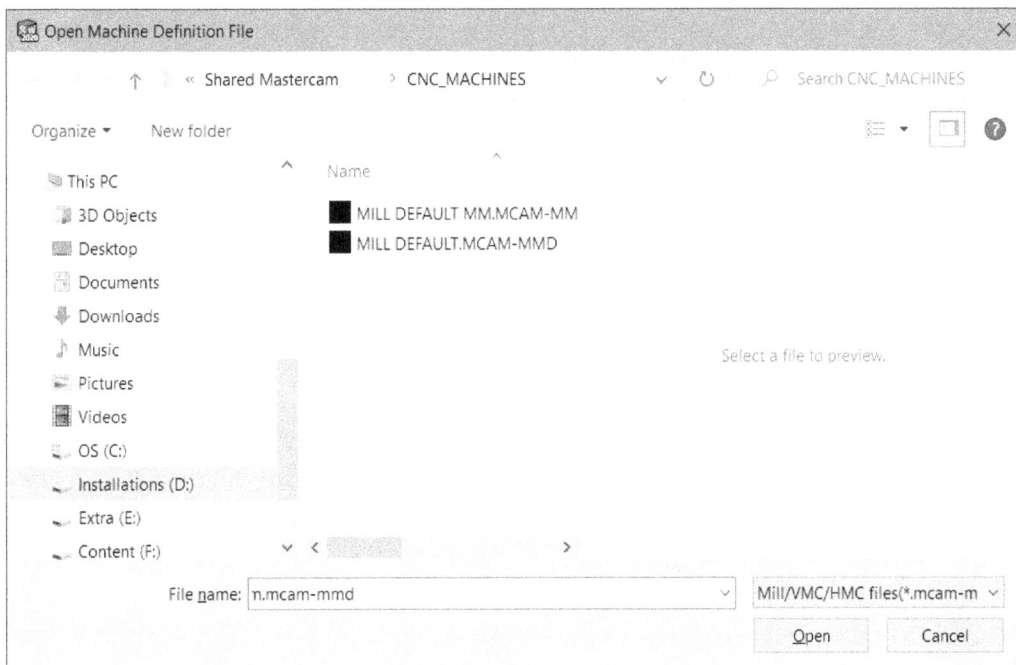
Figure-27. Open Machine Definition File dialog box

- Select the machine definition as required and click on the **Open** button from the dialog box.
- Click on the **OK** button from the **Machine Definition Manager** dialog box and choose the **Yes** button from the next dialog box (if displayed). New machine will be set as default.

Creating Toolpaths

- Click on the down arrow below **2D** button from the **Ribbon**. The menu of tools related to 2D toolpaths will be displayed; refer to Figure-28.

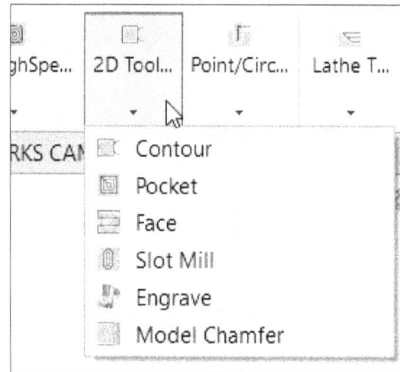

Figure-28. 2D toolpaths menu

- Click on the **Pocket** tool from the menu displayed. The **Chain Manager** will be displayed; refer to Figure-29.

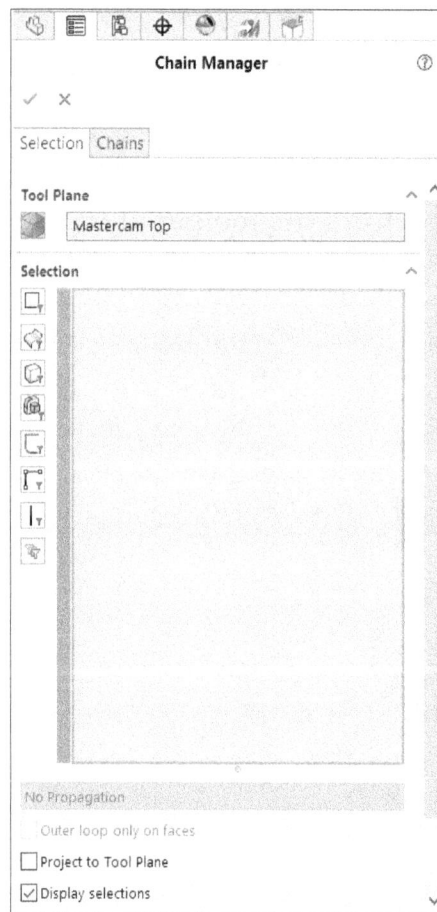

Figure-29. Chain Manager

- Select the **Face** selection filter button from the left in the **Chain Manager** and select the pocket faces of model to be machined; refer to Figure-30.
- Select the **Outer loop only on faces** check box to machine full pocket without considering holes on the face. Click on the **OK** button from the **Chain Manager**. The **2D Toolpaths - Pocket** dialog box will be displayed; refer to Figure-31.

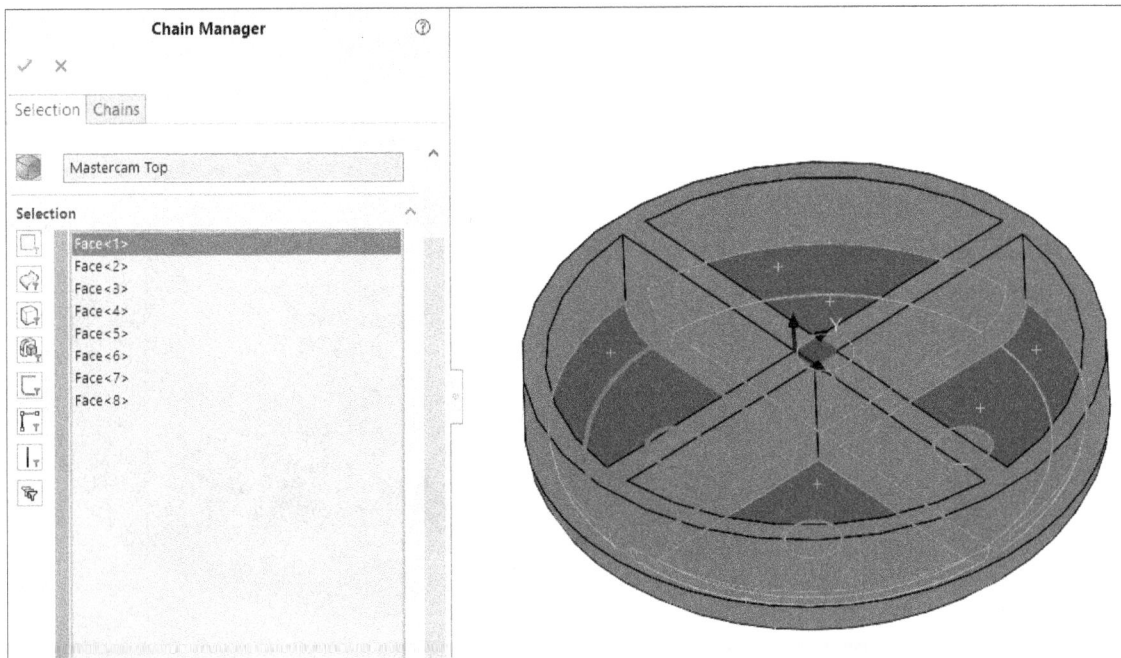

Figure-30. Face to be selected

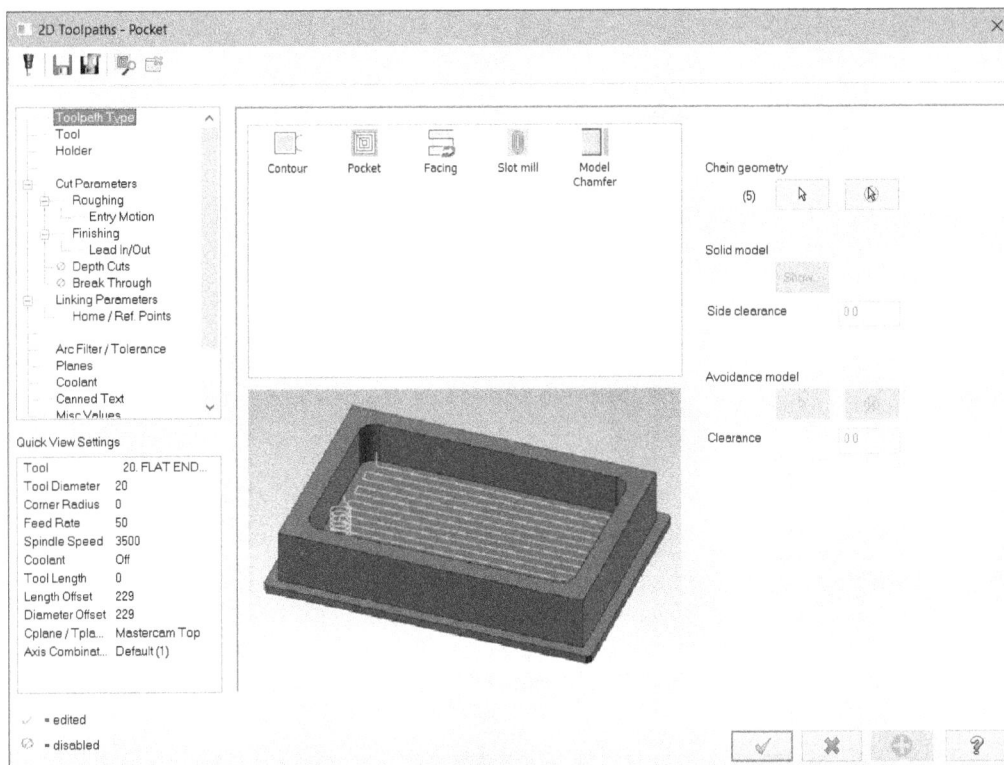

Figure-31. 2D Toolpaths-Pocket dialog box

- Click on the **OK** button from the dialog box to create the pocket toolpath. (We will learn about the settings for toolpaths later in the book.)
- On doing so, the toolpath will be generated; refer to Figure-32. If toolpath is not displayed by default after generating then select the **Toggle Toolpath Display** button from **Utilities** drop-down in the **Ribbon**; refer to Figure-33.

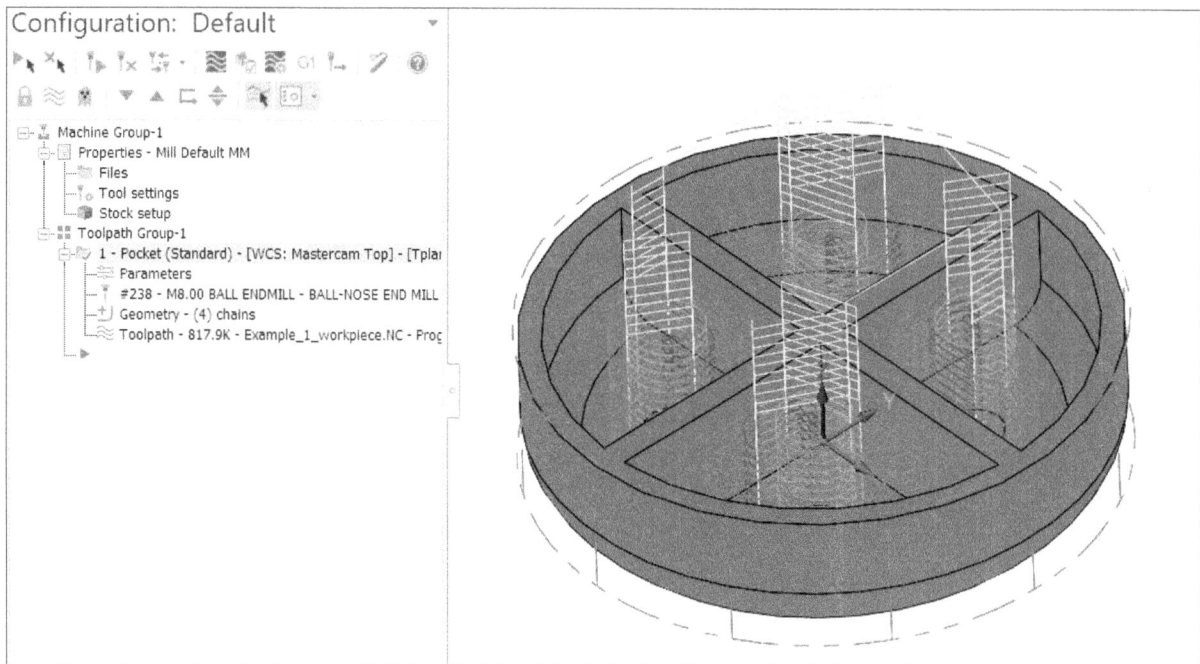

Figure-32. Toolpath generated

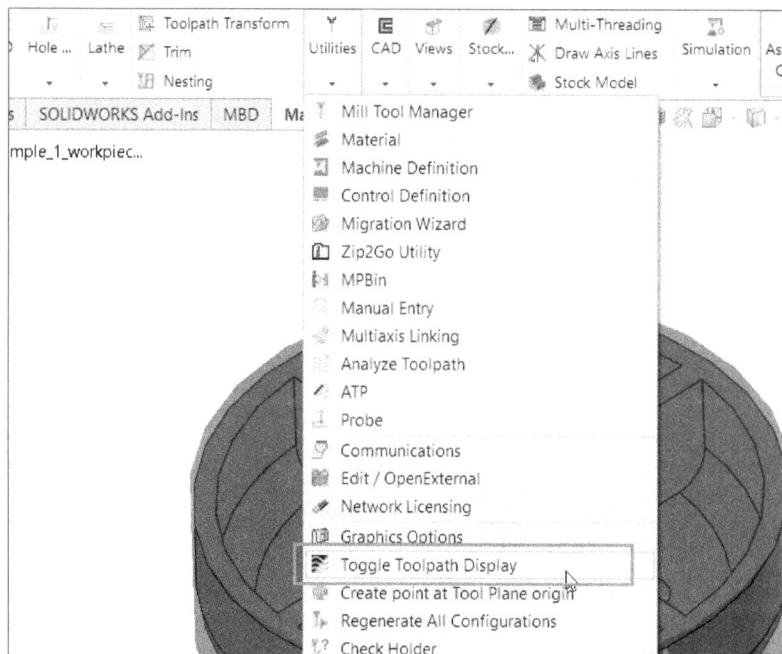

Figure-33. Toggle Toolpath Display button

Simulating the cutting operation

• Click on the down arrow below **Simulation** button in the **Ribbon**. The list of tools will be displayed; refer to Figure-34.

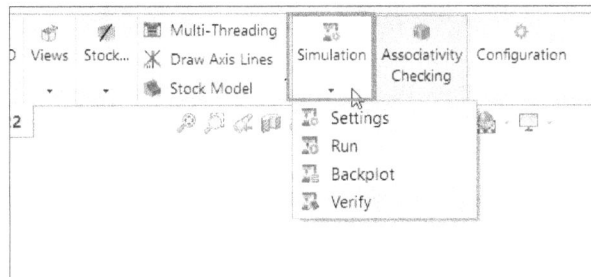

Figure-34. Tools for machine simulation

- Click on the **Settings** button to select a machine for simulation. The **Machine Simulation** dialog box will be displayed. Select desired machine from the **Machine** drop-down at top in the dialog box and click on the **Simulate** button. The **Machine Simulation** window will be displayed; refer to Figure-35.

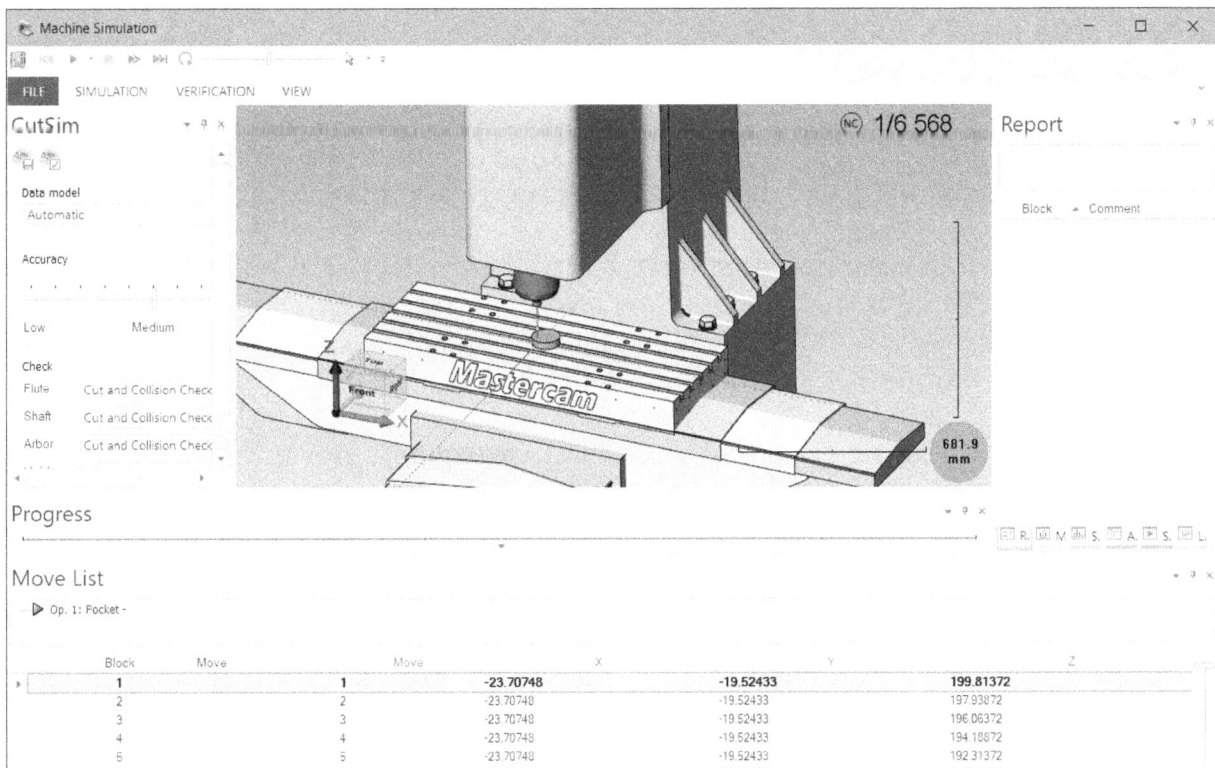

Figure-35. Machine Simulation application window

- Click on the **Run** button in the application window to play the simulation.
- Close the window by click on the **Close** button at the top-right corner of the application window.

This chapter was an overview of the Mastercam 2022 for SolidWorks. In the successive chapters, we will learn minute details of the tools in Mastercam 2022 for SolidWorks.

SELF-ASSESSMENT

Q1. Manufacturing is the process of creating a useful product by using a machine, a process, or both. (T/F)

Q2. Roughing process in machining requires close tolerances in material cutting to get good surface finish. (T/F)

Q3. In Shaper machine cutting tool goes back and forth to make linear cuts while workpiece is fixed on table. (T/F)

Q4. In Planer machine cutting tool goes back and forth to make linear cuts while workpiece is fixed on table. (T/F)

Q5. Which of the following machines is economically feasible for machining open end dovetail slots?

a. Drilling Machine   b. Shaper Machine
c. Milling Machine   d. Electric Discharge Machine

Q6. Which of the following machines is used to create accurate holes in diamonds?

a. Wire-cut EDM   b. Sinker EDM
c. Laser Beam Machine   d. Drilling Machine

Q7. Which of the following G-codes is used to define rapid movement of cutting tool from current position to a specified coordinate?

a. G00   b. G01
c. G02   d. G03

Q8. .... software is used to generate NC programs for various types of NC machines using computer.

Q9. When selecting geometries of 2D pocket toolpaths, select the .... check box from **Chain Manager** to machine full pocket without considering holes on the face.

Q10. If toolpath is not displayed by default after generating then select the .... button from **Utilities** drop-down in the **Ribbon** to display it.

Chapter 2

Workpiece Setting

Topics Covered

The major topics covered in this chapter are:

- *Inserting Part file of model.*
- *Setting the work plane.*
- *Creating stock of material.*
- *Setting Material.*
- *Setting library and tool setup.*

INTRODUCTION

The first and very important step for a CAM software is to insert the real representation of product in the CAM software. In other words, you are required to insert the model of the product with all the required details. The details of a product indicate the mechanical properties like material, mechanical strength, and so on. The method to insert a model is discussed next.

INSERTING/IMPORTING A MODEL

- Start SolidWorks with the MasterCAM add-in.
- Click on the **Open** button from the Quick Access toolbar or **File** menu. The **Open** dialog box will be displayed; refer to Figure-1.

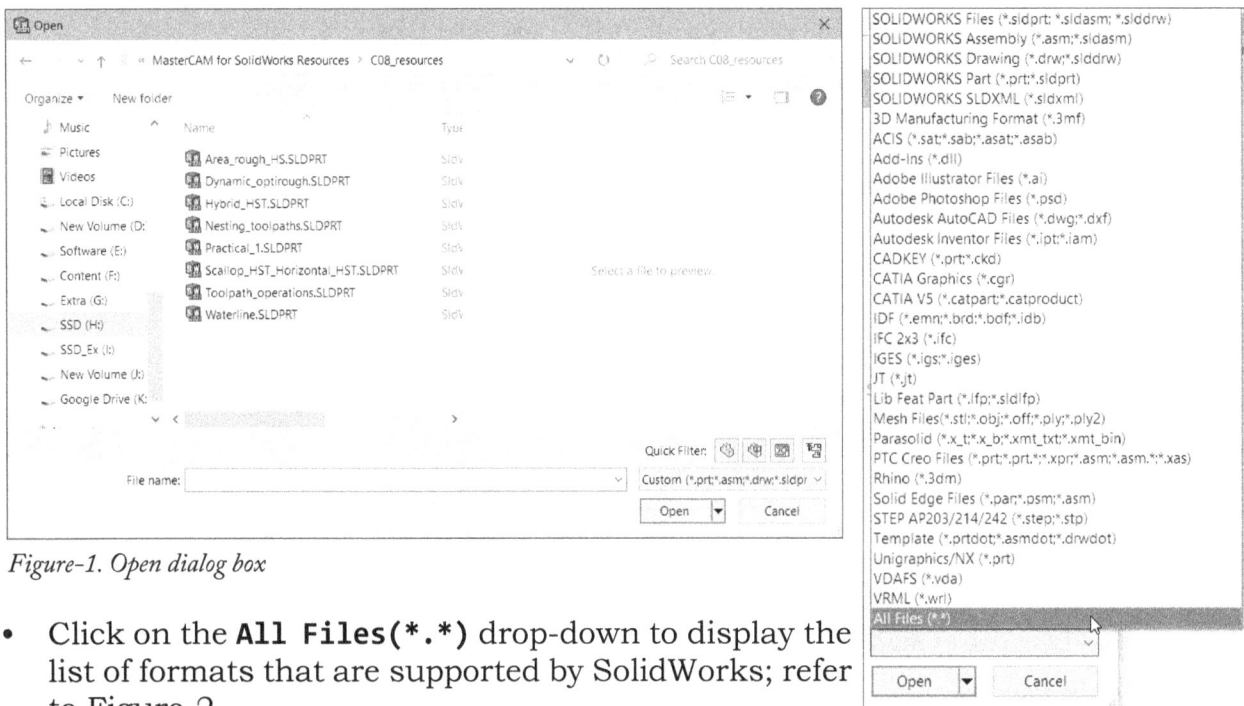

Figure-1. Open dialog box

Figure-2. All Files drop-down

- Click on the **All Files(*.*)** drop-down to display the list of formats that are supported by SolidWorks; refer to Figure-2.
- Select the format of the file that you want to open and browse to the location of file.
- Select the file that you want to open and click on the **Open** button from the dialog box. The model will be displayed in the graphics area of SolidWorks.

DEFINING RELATION BETWEEN SOLIDWORKS AND MASTERCAM PLANES

You can set Top plane of SolidWorks as default Top plane of Mastercam by using the system configuration options of Mastercam. The procedure to do so is given next.

- Click on the **Configuration** tool from the **Mastercam2022 CommandManager** in the **Ribbon**. The **System Configuration** dialog box will be displayed.
- Select the **Mastercam 2022 for SOLIDWORKS** option from the left in the dialog box. The options in the dialog box will be displayed as shown in Figure-3.

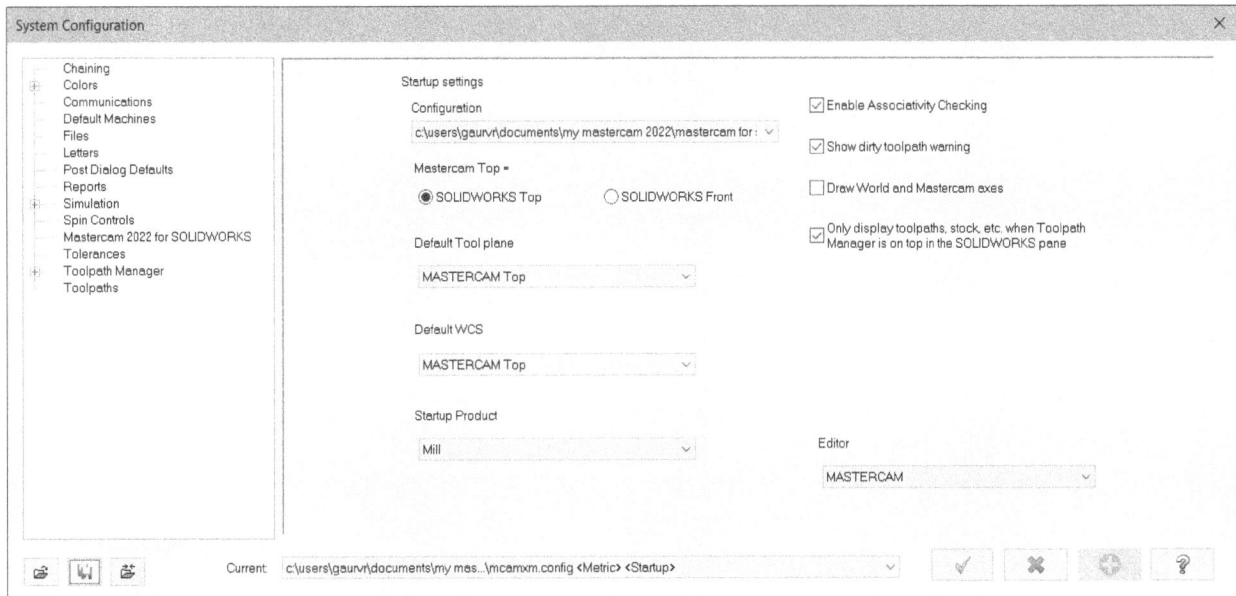

Figure-3. System Configuration dialog box

- Select the **SOLIDWORKS Top** radio button from **Mastercam Top =** area to set Top plane of SolidWorks as Mastercam Top plane. If you want to set Front plane of SOLIDWORKS as Top plane of Mastercam then select the **SOLIDWORKS Front** radio button.
- Click on the **OK** button from the dialog box to apply changes. You will learn about options of this dialog box later the book.

SETTING PLANES

In MasterCAM, we need to create various planes for machining and orienting the model as per the tool direction. The **Planes Manager** is used to perform this task. The procedure to use **Planes Manager** is given next.

- Click on the **Planes Manager** tab in the **Design Tree**. The **Planes Manager** will be displayed; refer to Figure-4. Note that there are five default column next for each plane name in the list box; **G**, **WCS**, **T**, **Offset**, and **Display**. You can add more columns by right-clicking on the header of list box and selecting desired option; refer to Figure-5. Click in the **G** column for a plane to make it parallel to screen. Click in the **WCS** column next to the plane to make it construction plane and create geometries on it. The selected plane will be set as WCS for modeling. Click in the **T** column next to a plane to make Tool plane. The tool will be in perpendicular direction to the selected plane. The **Offset** field is used to define work offset number for the current tool plane. (Note that you may have multiple parts on the machine bed lying next to each-other and want to machine them with the same tool. If you do not set an Offset number then the coordinates will be taken from the first selected plane.)

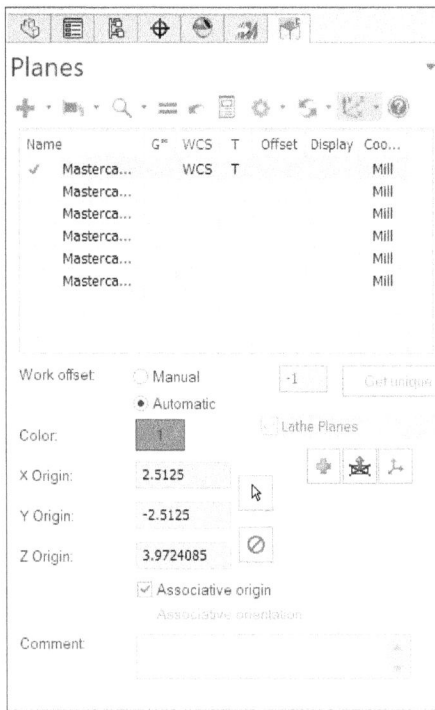

Figure-4. Planes Manager dialog box

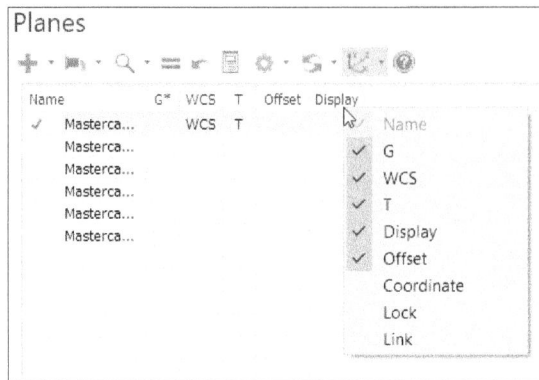

Figure-5. Shortcut menu for adding columns

- By default, six planes of mastercam are displayed for orientation of workpiece on the machine. But you can create more by using **Create a new Plane** drop-down at the top in the **Planes Manager**. You will learn about these options later.

Creating Copy of Plane

- To create more copies of the selected plane, right-click on it and select the **Duplicate** option from shortcut menu displayed. The copy of selected plane will be created; refer to Figure-6.

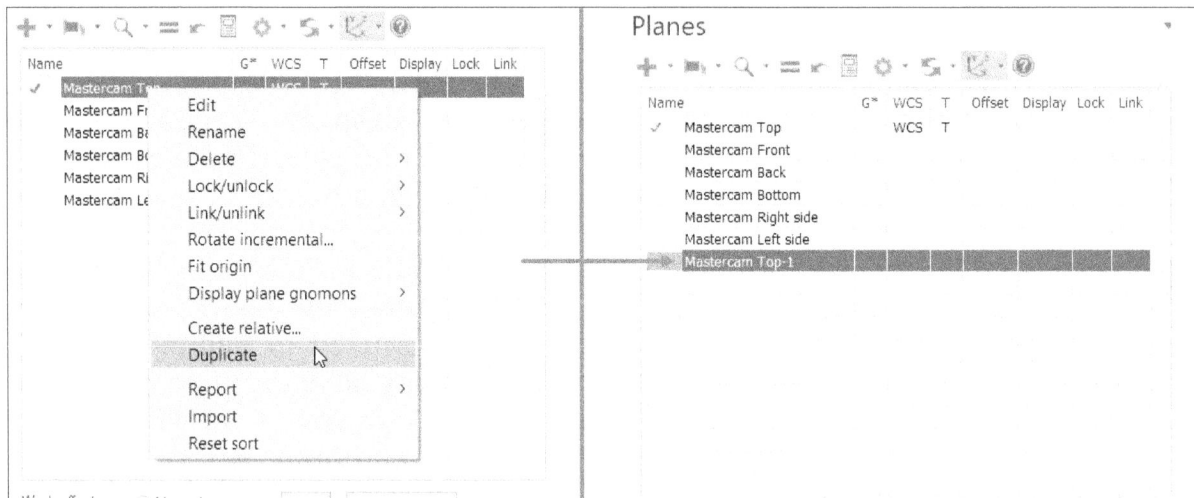

Figure-6. Copy of Top plane created

Creating Relative Planes

- To create copies of all the planes associated with part, right-click in the empty area of Planes Manager list box and select the **Create Relative** option from shortcut menu. The **Create Relative Planes** dialog box will be displayed; refer to Figure-7.

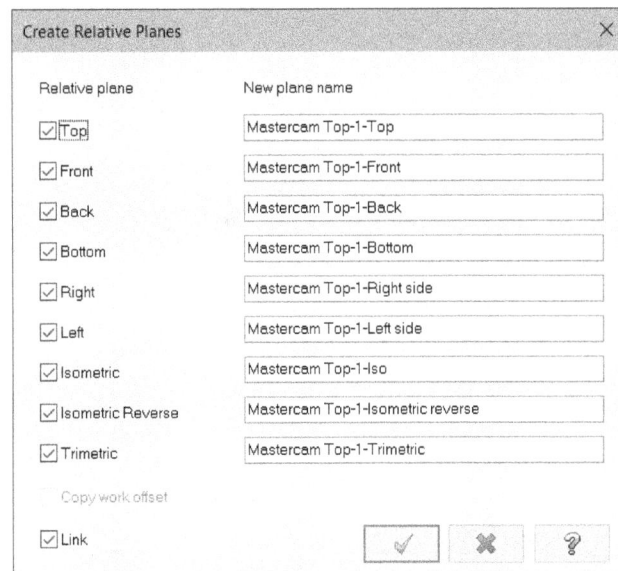

Figure-7. Create Relative Planes dialog box

- Select the check boxes next to desired planes and click on the **OK** button from the dialog box. The respective planes will be added in the list with their respective origin positions.

Setting Work offset

When you perform machining on milling machines, you will try to machine as many parts as possible on one go of the tool. If there are three parts on the machine bed and you want to perform contouring operation on all the three parts by same tool. Then by default, Mastercam creates coordinates for all the three parts with respect to one coordinate system. Like machining for part 1 will start at X0, machining for part 2 will start at X20, and machining for part 3 will start at X60. But this is not generally desired by an NC programmer. You would like to have an origin for each part at its corner. To do so, we apply codes like G54, G55, G56, and so on in NC program. To perform this operation in CAM, we assign **Work Offset Number** to each part linked by its separate plane.

- The options in **Work Offset** section at bottom in the **Planes Manager** are used to specify work-offset number. If **Automatic** radio button is selected for Work offset then a work offset number will be assigned to the tool plane automatically. If you want to manually specify work offset number then select the **Manual** radio button and specify desired value in the **Work Offset** edit box or click on the **Get unique** button.

Creating Plane from Geometry

- The **From geometry** option from the **Create a new plane** drop-down in the **Planes Manager** is used to create the Mastercam plane from existing geometries of the model; refer to Figure-8. On clicking this option, the **Define Mastercam Plane PropertyManager** will be displayed; refer to Figure-9.
- Select the edge or face to specify reference for the view. Or, you use select the **Two lines only** check box and select the two perpendicular lines to define X and Y direction for Mastercam plane.

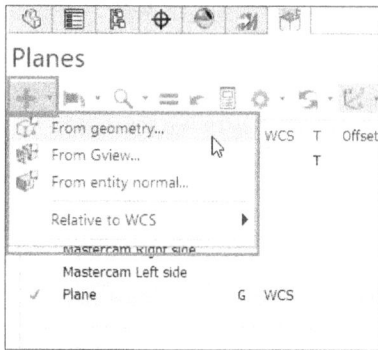

Figure-8. Create a new plane drop-down

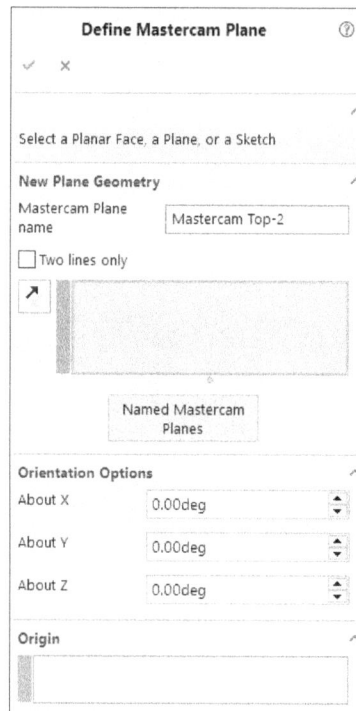

Figure-9. Define View with Geometry PropertyManager

- Click in the **Origin** selection box to specify the origin of the plane.
- Specify the other parameters like angle from X, Y, or Z axis in the respective spinners; refer to Figure-10 and then click on the **OK** button from the **PropertyManager** to create the plane. The plane will be created and will be listed in the dialog box.

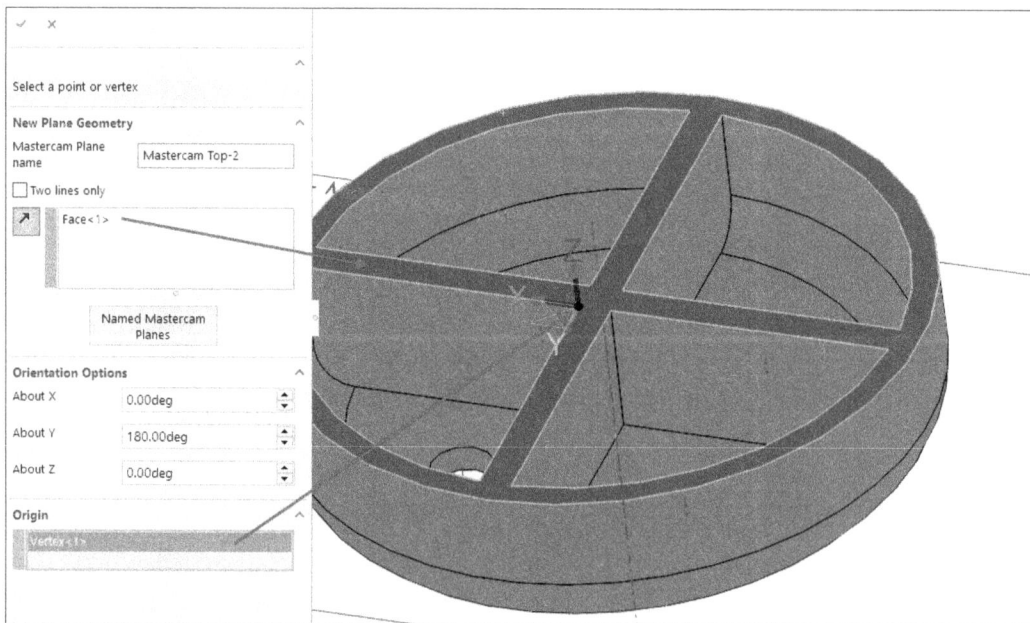

Figure-10. Creating Mastercam plane

Creating Planes based on Current View

The **From GView** option in **Create a new plane** drop-down is used to create Mastercam plane parallel to current view in the screen. The procedure to use this option is given next.

- Click on the **From Gview** option from the **Create a new plane** drop-down in the **Planes Manager**. The **Define Mastercam Plane PropertyManager** will be displayed as discussed earlier.
- Specify the name of plane in **Mastercam Plane Name** edit box and then click in the **Origin** selection box.
- Click at desired location on the model to place origin of the plane; refer to Figure-11. Do not specify **Orientation Options** and do not select any geometry for referencing the plane.

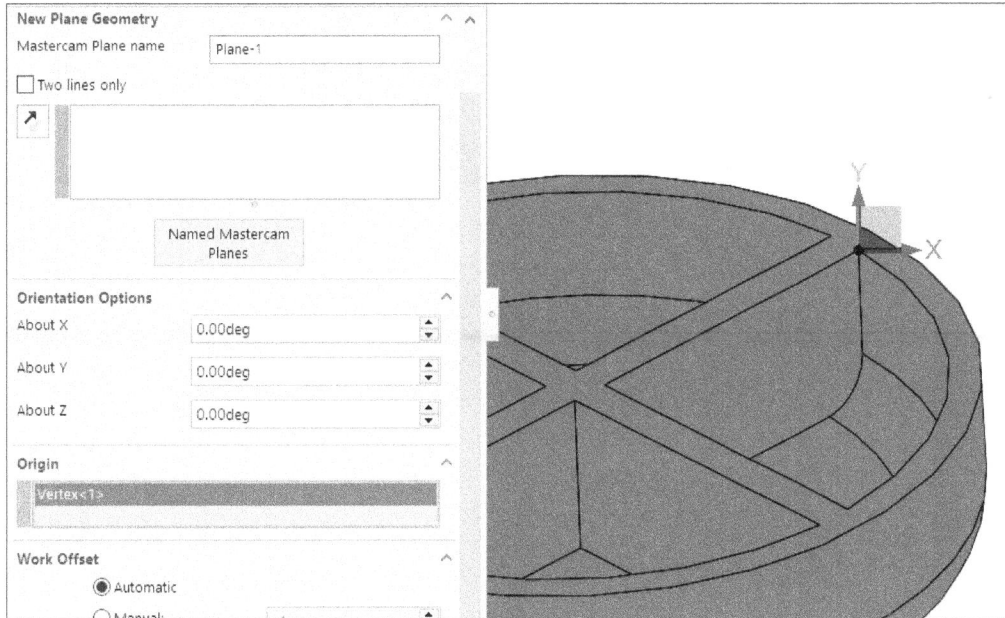

Figure-11. Plane parallel to screen

- Click on the **OK** button from the **PropertyManager**. A new plane will be created parallel to current view.

Modifying Origin Location of Plane

- To change the position of the origin for any plane, click on the **Origin** button next to **X Origin** edit box at the bottom in the **Planes Manager**; refer to Figure-12. You are asked to select a reference point to specify the position of the origin.

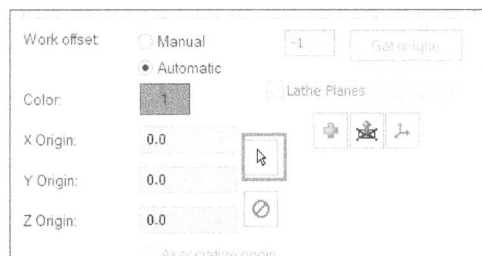

Figure-12. Origin button

- Select desired point/vertex from the model to specify the reference. Click on the **Reset Origin** button ⊘ to move origin of plane at (0,0,0) coordinates in World Coordinate System.
- Select the **Associate origin** check box to keep the origin of plane attached to selected point so that if location of selected point is changed then origin of plane will also get changes accordingly.

Other options of the **Planes Manager** will be discussed later in relevant chapters.

CREATING STOCK OF MATERIAL

In MasterCAM, you can create two type of stocks.

1. You can create a stock for the workpiece.
2. You can create stock after specific any machining operation.

The steps to create both the stocks are given next.

Creating Stock for Workpiece

- Click on the **+** sign next to **Properties** node in the **Mastercam Toolpath Manager** in the left of the modeling area; refer to Figure-13. The list of options will display.

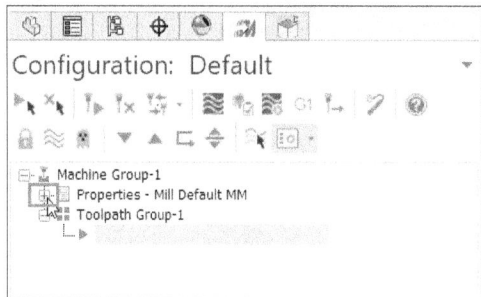

Figure-13. Node for Properties

- Click on the **Stock setup** option from the **Mastercam Toolpath Manager**. The **Machine Group Properties** dialog box will be displayed; refer to Figure-14.

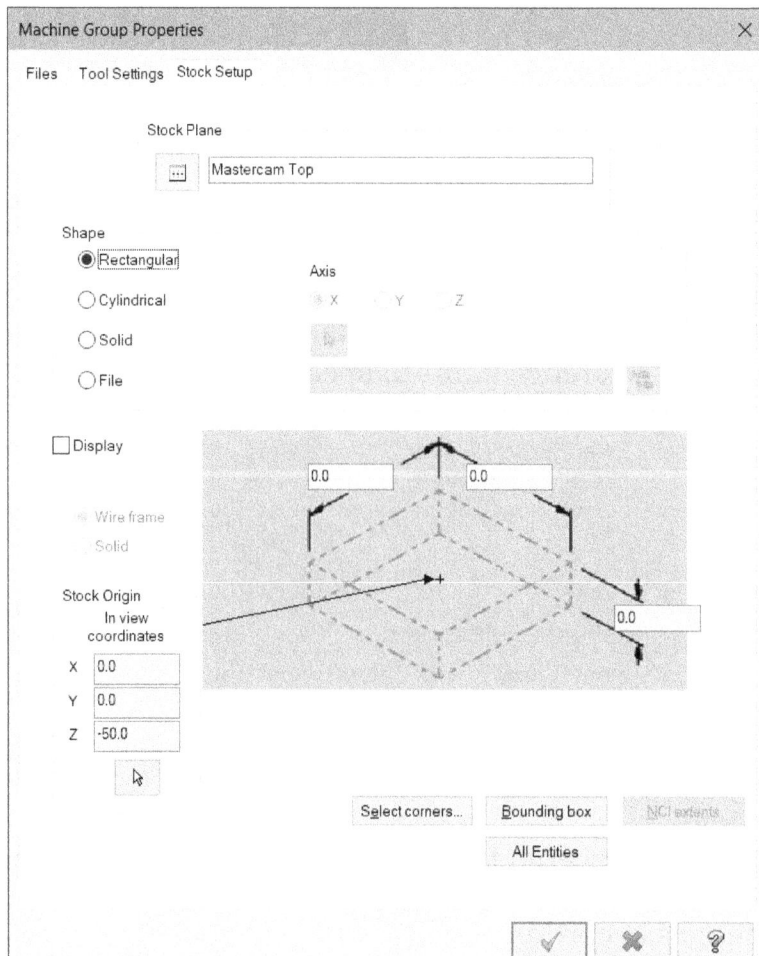

Figure-14. Machine Group Properties dialog box for Stock setup

- Click on the **Stock Plane** button ⊡ from the dialog box to set the plane for stock. The **Plane Selection** dialog box will be displayed; refer to Figure-15.

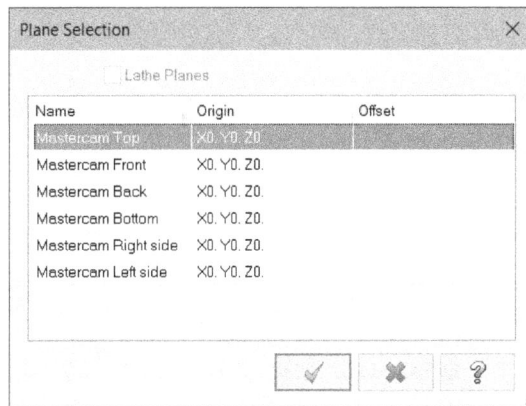

Figure-15. Plane Selection dialog box

- Select desired plane from the dialog box and click on the **OK** button. The selected plane will be used as the base plane for the creating stock.
- The radio buttons in the **Shape** area of the dialog box are used to specify the shape of the stock model. By default, the **Rectangular** radio button is selected hence the rectangular block is created.
- Select the **Cylindrical** radio button to create a cylindrical stock if your workpiece is cylindrical. After selecting this radio button, you might need to select the **X**, **Y**, or **Z** radio button to decide the axis of cylinder stock.
- If you have an existing model in the modeling area that you want to define as stock, then select the **Solid** radio button and click on ⊡ the button next to it. You are asked to select the model.
- Select desired model (body) and click on the **OK** button from the **Selection PropertyManager**. The selected model will be used as the stock model.
- If you have a model file that you want to use as stock material then select the **File** radio button from the dialog box and click on the ⊡ button. The **Open** dialog box will be displayed; refer to Figure-16.

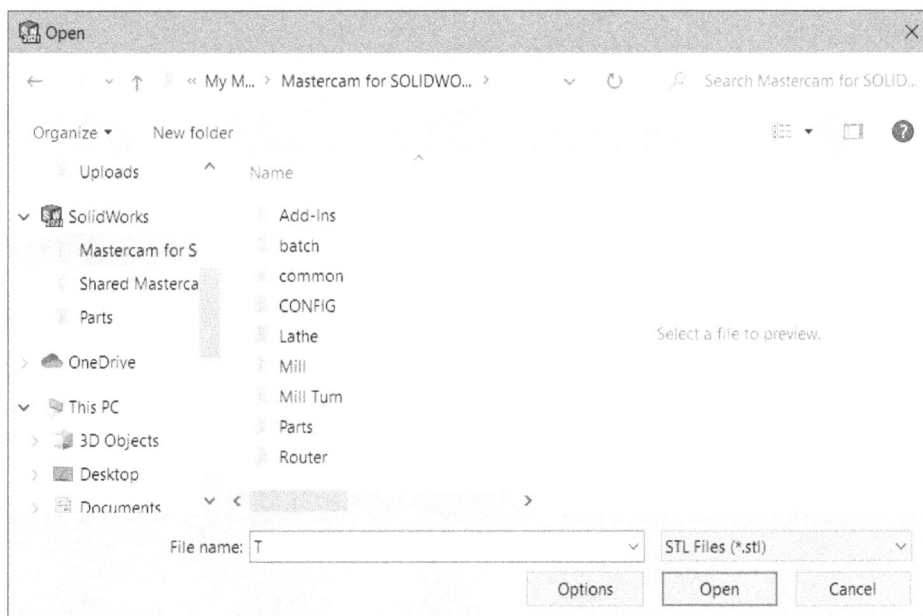

Figure-16. Open dialog box for stock material

- Browse to the location of desired file and double-click on the model file. Note that you can use only STL file format for this.
- The name of the model will displayed in the field next to the **File** radio button; refer to Figure-17.

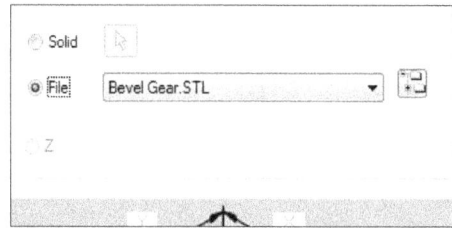

Figure-17. Name of file displayed

- Select the **Display** check box and select the display style by selecting the **Wire frame** or **Solid** radio button.
- After specifying desired settings, click on the **OK** button from the dialog box.

Bounding Box Stock

- To create a bounding box, you need to select the **Bounding box** button after selecting the **Rectangular** or **Cylindrical** radio button. On selecting the **Bounding box** button, the **Selection PropertyManager** will be displayed; refer to Figure-18. Select desired selection filter if you want to select a specified type of geometry from the model; refer to Figure-19. After selecting the geometry (mostly body of model), click on the **OK** button from the **Propertymanager**. The **Bounding Box PropertyManager** will be displayed; refer to Figure-20.

Figure-18. Selection PropertyManager

	Face	Limits your geometry selection only to faces.
	Surface	Limit your geometry selection only to surfaces.
	Bodies	Limits your geometry selection only to bodies.
	Features	Limits your geometry selection only to features.
	Sketches	Limits your geometry selection only to sketches.
	Sketch segments	Limits your geometry selection only to sketch segments.
	Edges	Limits your geometry selection only to edges.
	Clear filters	Clears all filters for selecting geometry.

Figure-19. Selection filters

- You can modify shape of model by selecting desired radio button from the **Shape** rollout. Note that in **PropertyManager**, you have the option to create a Spherical or Wrap shape of stock as well.

- Based on radio button selected in **Shape** rollout, the Settings rollout will become available for modifying size of stock. For example if you have selected **Cylindrical** radio button from **Shape** rollout then you can specify size parameters in **Cylindrical Settings** rollout.

- By default, stock is oriented as per construction plane or X, Y, and Z axis used for orientation of stock match WCS. If you want to change the orientation of stock then select the **Face** radio button from **Orientation** rollout and click on the **Select** button next to it. The **Selection PropertyManager** will be displayed. Select desired face to be use for orienting the base of stock and click on the **OK** button. Note that this option only affects Cylindrical type of stock as all other options generate symmetric stock.

- Select check boxes for geometries that you want to create along with stock from **Create Geometry** rollout at the bottom in the **PropertyManager** and then click on the **OK** button. The bounding box will be displayed along with selected geometries; refer to Figure-21.

- Click on the **OK** button from the **Machine Group Properties** dialog box to create the stock and exit the dialog box.

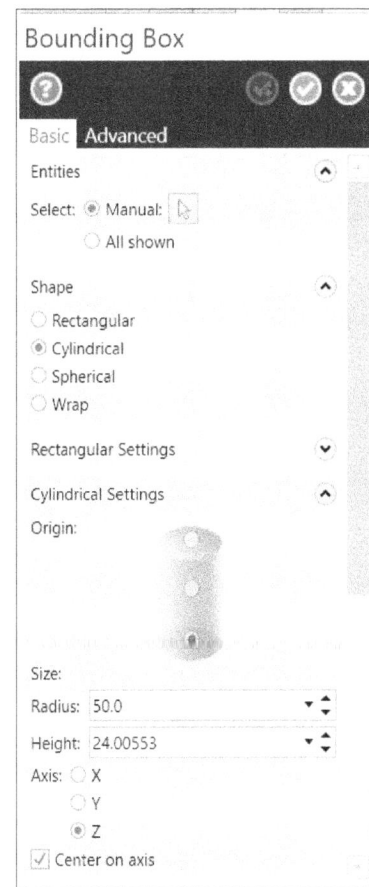

Figure-20. Bounding Box PropertyManager

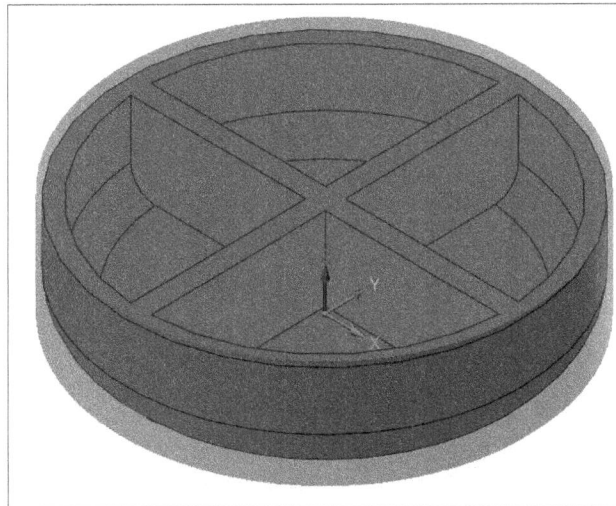

Figure-21. Bounding box stock

Creating Stock after Machining Operation

- After performing machining operations on the stock created for workpiece (refer to Figure-22), click on the **Stock Model** tool from the **Mastercam2022 CommandManager**. The **Stock model** dialog box will be displayed; refer to Figure-23.

- By default, the **Stock Definition** page of the dialog box is displayed. In this page specify the parameters of stock, you used while creating stock for workpiece as discussed earlier.

- Specify the name of stock as desired in the **Name** edit box.
- Click on the **Source Operations** option from the left box. All the operations performed on the part will be displayed in a list box; refer to Figure-24.

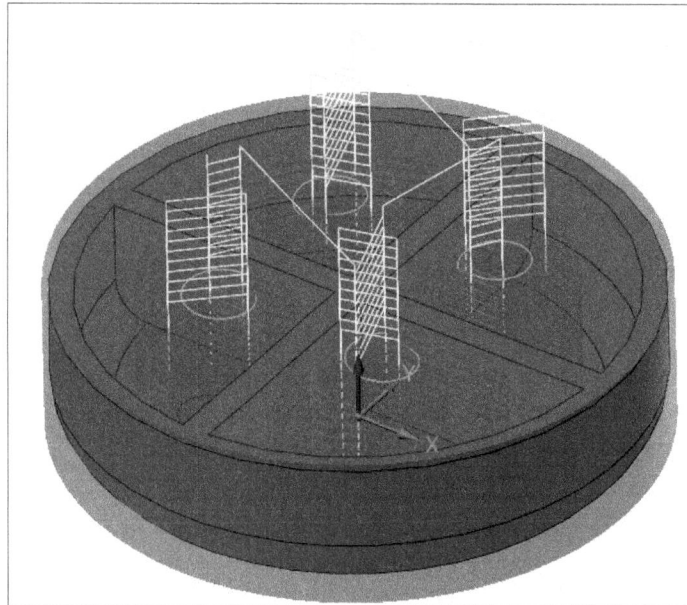

Figure-22. Model with pocket operation for generating stock

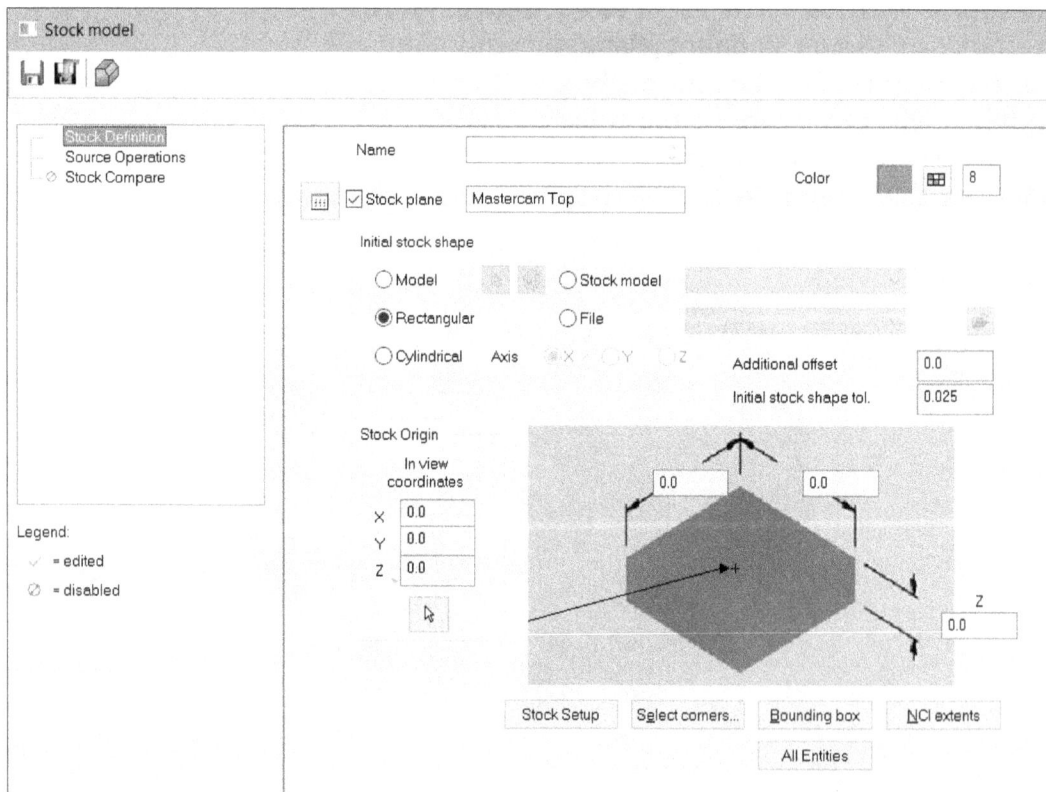

Figure-23. Stock model dialog box

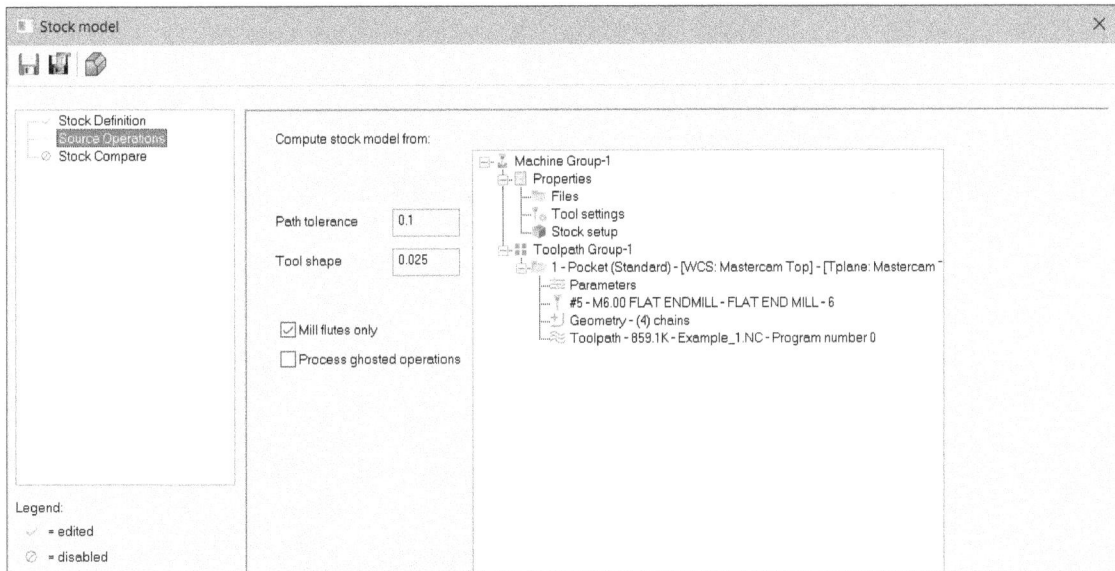

Figure-24. Source Operations page of Stock model dialog box

- Select the operations from the list box after which you want to check the stock left.
- Click on the **OK** button. The stock model will be created; refer to Figure-25. Note that to display stock model, you need to hide the initial stock by selecting **Stock Display** toggle button from the **Stock Display** drop-down in the **Ribbon**. Now, select the stock model recently created from **Mastercam Toolpath Manager** and click on the **Only display selected toolpath** button; refer to Figure-26. To display all the machining objects again, select the objects and click on the **Toggle display** button ≋ from the **Toolpath Manager**.

Figure-25. Stock model created

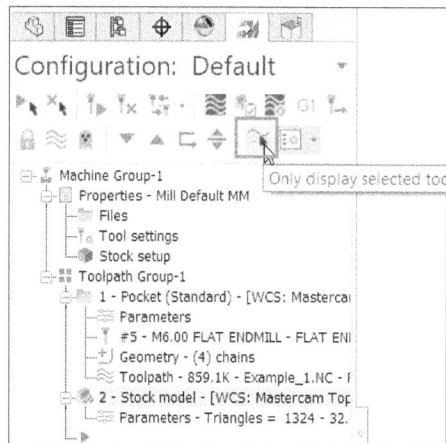

Figure-26. Only display selected toolpath button

APPLYING MATERIAL

Material is an important feature of any CAD model. This feature helps to link the model with real world objects. In machining, we need to specify the material of the stock so that the cutting parameters can be decided. The procedure to apply the material is given next.

- Click on the down arrow below **Utilities** tool in the **Ribbon**. The list of tools will be displayed; refer to Figure-27.
- Click on the **Material** button from the list. The **Material List** dialog box will be displayed; refer to Figure-28.

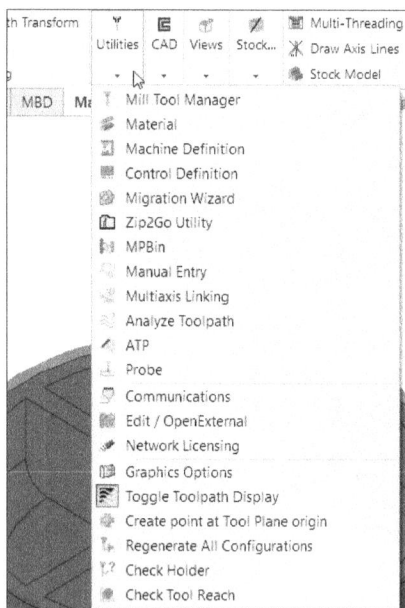

Figure-27. Utilities drop-down

Figure-28. Material List dialog box

- Select desired radio button to specify the unit system of material.
- Click in the **Source** drop-down. The list of material sources will be displayed; refer to Figure-29.

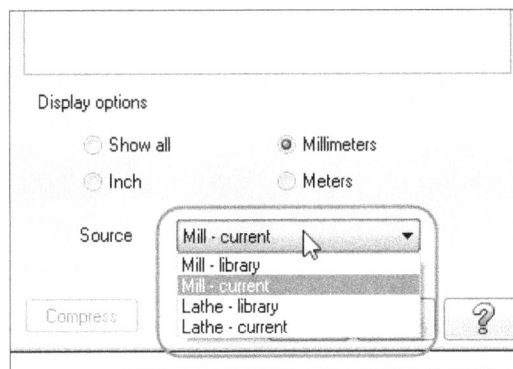

Figure-29. Material sources

- Select the **Mill - library** or **Lathe - library** option from the drop-down to display the list of materials in selected library.
- Select desired material from the list and click on the **OK** button from the dialog box to apply the material.

Editing Material or Creating New Material

You can edit the material by using the **Material List** dialog box displayed earlier. The steps to edit material are given next.

- Invoke the **Material List** dialog box as discussed earlier.
- Double-click on the name of the material in the list that you want to edit. The **Material Definition** dialog box will be displayed. If you want to create a new material then right-click in the list box of **Material List** dialog box and select the **Create new** option from the shortcut menu; refer to Figure-30. The **Material Definition** dialog box will be displayed; refer to Figure-31, where you can specify desired material name and parameters.

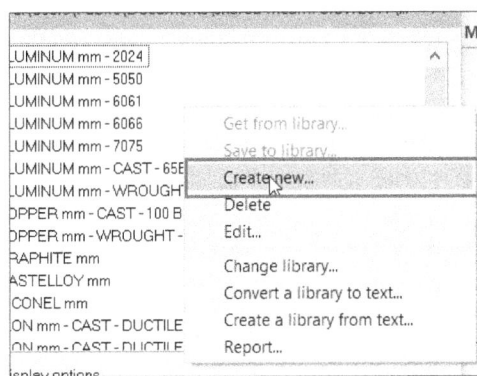

Figure-30. Create new option in right-click shortcut menu

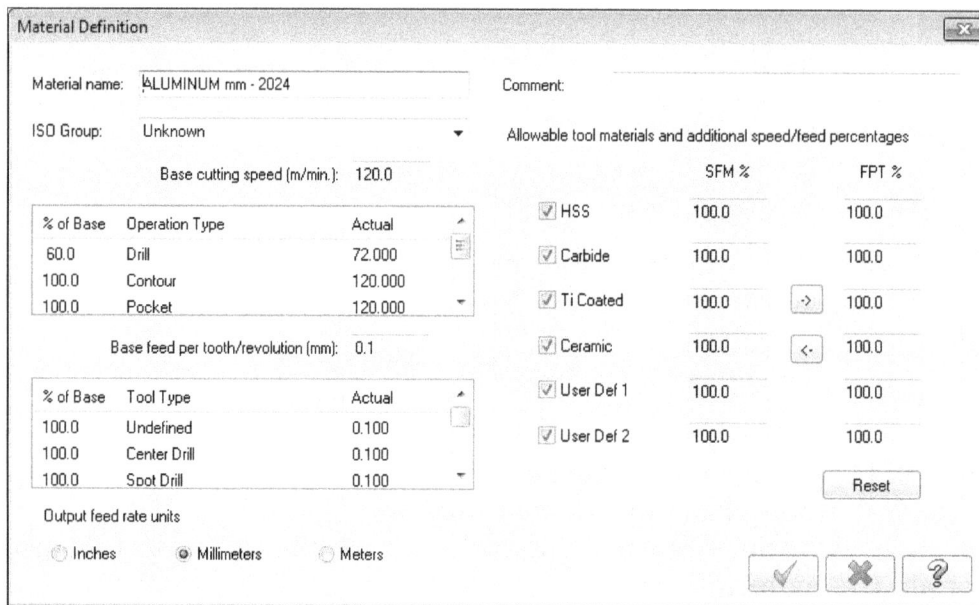

Figure-31. Material Definition dialog box

- Specify desired cutting speed and feed rate in the respective boxes. Note that the edit-able check boxes are displayed filled with yellow color.
- You can also specify the tool specific feed rates by using the options in the right of the dialog box. Mastercam combines these parameters with information from the tool definition to calculate the most accurate possible feeds and speeds. You may ask how to decide the values of these feed rates. The answer is you can not. You receive these parameters from your tool supplier. Your tool supplier performs experiment on the tool for various operations and gives you the optimum feed rates.
- After specifying desired parameters, click on the **OK** button from the dialog box. The material will be modified/created.

SPECIFYING BASIC SETTINGS
OF MACHINING GROUP

The basic settings of machining group like program number, feed rate of machining, and so on are specified in the **Machine Group Properties** dialog box as discussed earlier. The steps to specify these properties are given next.

- Click on the **Tool settings** option in the **Properties** node of **Mastercam Toolpath Manager**; refer to Figure-32. The **Machine Group Properties** dialog box with **Tool Settings** tab selected.

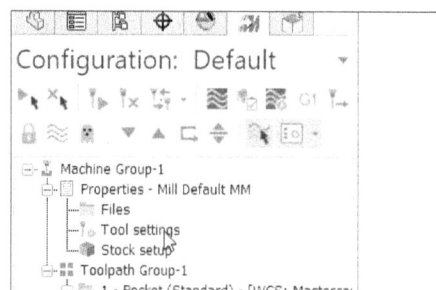

Figure-32. Tool settings

- Specify the program number for your reference in the **Program #** edit box.
- Select desired radio button from the **Feed Calculation** area. Select the **From tool** radio button if you want to use feed rates set for tools (Like we have specified feed parameters for material, we will learn to specify feed parameters for tools also). The **From material** radio button is selected if you want to use feed rates set for the material. To use the default feed rates, select the **From defaults** radio button. If you want to specify desired feed rates then click on the **User defined** radio button and specify desired values in the respective edit boxes; refer to Figure-33.

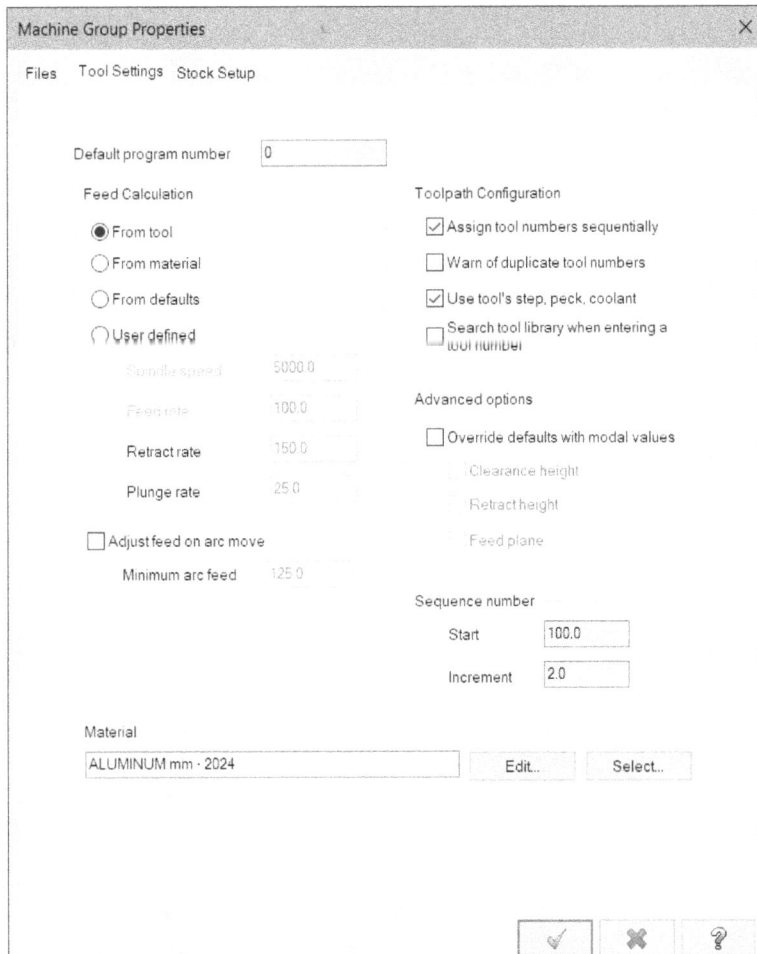

Figure-33. Machine Group Properties dialog box for tool settings

- Select the **Adjust feed on arc move** check box and specify desired value in the **Minimum arc feed** edit box to adjust the feed rate at arcs in the workpiece.
- Using the options in the **Material** area, you can set the material for machining.
- Using the options in the **Toolpath Configuration** area, you can set the options related to tool path. Select the **Assign tool numbers sequentially** check box to sequentially give tool numbers to the tools being used for different tool paths. Select the **Warn of duplicate tool numbers** check box to display an alert message if there is a duplicate tool number in programming. Select the **Use tool's step, peck, coolant** check box if you want to use the parameters of tool specified in the tool library. Similarly, select the **Search tool library when entering a tool number** check box if you want the system to search for tool parameters after entering the tool number.
- Select the **Override defaults with modal values** check box to change clearance height, retract height, and feed plane distances based on previous operation

perform in software. Mastercam stores previous values of these parameters as defaults. Select check box for parameter to which you want to apply modal default.

• You can specify the increment value for sequence of program number by using the edit boxes in the **Sequence number** area. Note that you should specify an increment of at least 10 so that you can later on insert a piece of code manually in this block if required. Like if you need to insert a code between sequence #10 and sequence #20 then you can make the new code at sequence #11, 12, 13 or any number between the two.

Files tab

• Click on the **Files** tab to display the options related to locations of various libraries and setup files; refer to Figure-34.

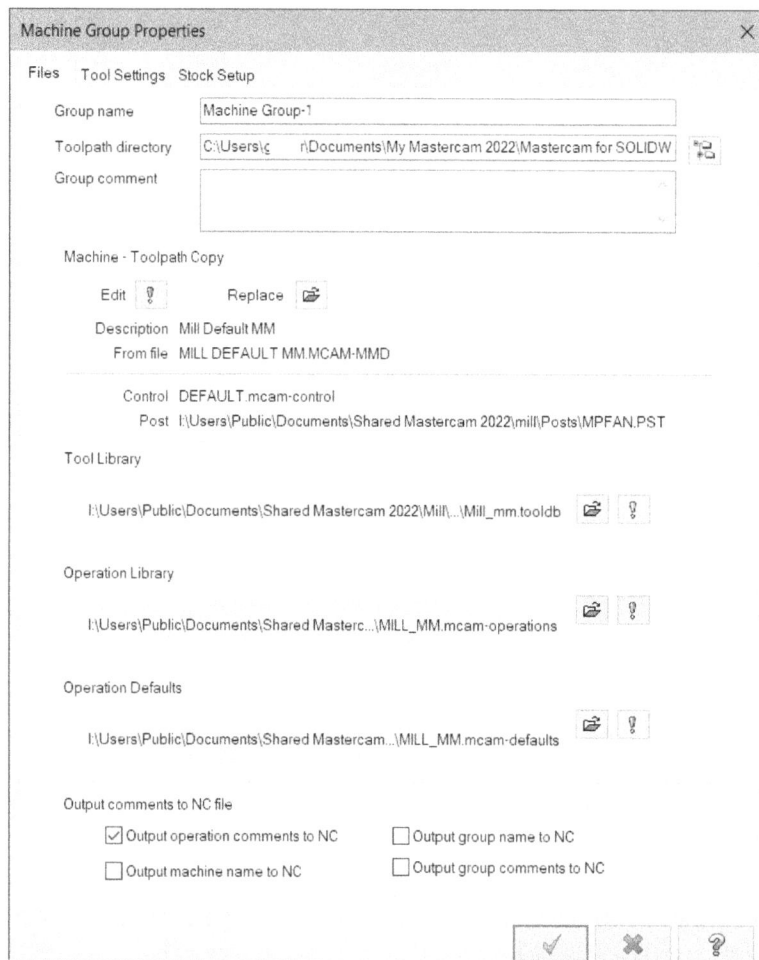

Figure-34. Machine Group Properties dialog box for files setup

• Click on the **Open** button in the **Machine-Toolpath Copy**, **Tool Library**, **Operation Library**, or **Operation Defaults** area to change the machine- toolpath copy, tool library, operation library, or default options of operation respectively.

• On clicking at the **Open** button for a library or directory, the respective dialog box will be displayed and you will be asked to select a replacement file; refer to Figure-35.

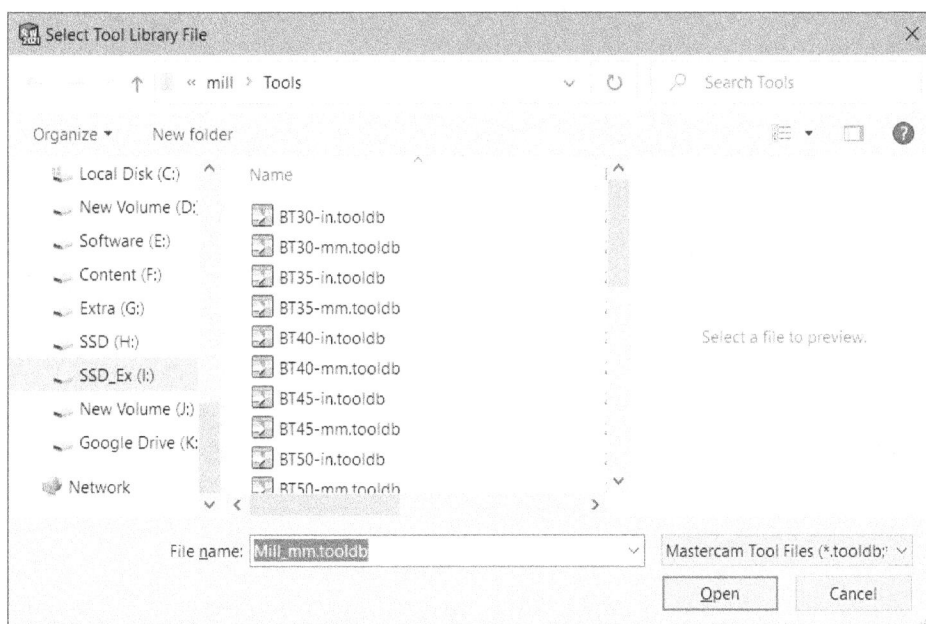

Figure-35. Select Tool Library File dialog box

- Select desired file and click on the **Open** button to use the selected file.
- From the **Output comments to NC file** area, select the check boxes for the options that you want to be included in the NC file.
- Click on the **OK** button from the dialog box to exit.

MASTERCAM TOOLPATH MANAGER OPTIONS

Mastercam Toolpath Manager organizes all the machine and toolpath information into groups; refer to Figure-36. You can rename these groups based on requirements like front machining, side machining and so on. Note that machining programs will follow the same sequence in which you have created operations in the **Toolpath Manager**. There are various tools available in **Toolbar** of **Mastercam Toolpath Manager** to organize toolpaths and perform related tasks. These tools and options are discussed next.

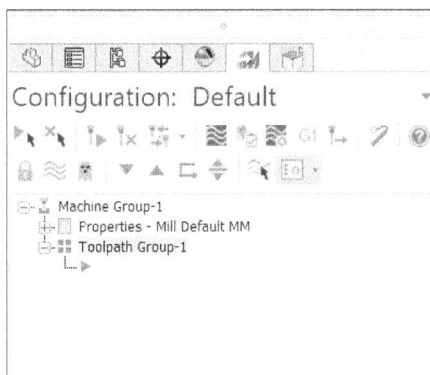

Figure-36. Mastercam toolpath manager

Icon	Name	Description
	Select All Operations	Selects all the operations in Toolpath Manager
	Select all dirty operations	Select all operations which need regeneration due to modifications in toolpaths

	Regenerate all selected operations	Recalculates toolpaths for all the selected operations
	Regenerate all dirty operations	Recalculates toolpaths for all operations which have been modified and marked as dirty.
	Dependencies	Select the **Selection** option if you want to select all dependent operations of selected operation Select the **Regeneration** option if you want to recalculate toolpaths for the dependent operations.
	Backplot selected operations	Displays simulation of toolpath in light version. No material cutting simulation displayed.
	Verify selected operations	Displays simulation of toolpaths with material removal graphics. Useful for collision detection.
	Simulator options	Used for defining parameters of Toolpath simulator displayed on select the **Verify selected operations** option.
G1	Post selected operations	Generates NC program using the post processor. Not available in HLE edition of software.
	High feed	Changes feed rates of selected operations for higher efficiency based on tool and material definitions.
	Delete all operations, groups and tools	Deletes all operations and data in Toolpath Manager and reverts to default state. Note that there is no Undo for this operations so once data goes, its gone!!
	Help	Opens Mastercam Help file in local drive with Toolpath Manager topic.
	Toggle locking on selected operations	Locks and Unlocks an operation. Locking an operation ensures no accidental changes in the toolpath.
	Toggle display on selected operations	Displays or hides selected operation and its toolpaths from graphics area.
	Toggle posting on selected operations	Toggles the ghost state of an operation. An operation marked as ghost is not sent for post processing. Generally Stock Model operation is marked as ghost.
	Move insert arrow down one item	Moves next operation insertion point down in the sequence.
	Move insert arrow up one item	Moves next operation insertion point up in the sequence.
	Position insert arrow after selected operation or after selected group	Moves next operation insertion point after selected operation or group.

	Scroll window so insert arrow is visible	Select this button to display insertion point.
	Only display selected toolpaths	Select this button to display only selected toolpath
	Advanced display	Select this button to display advanced information in toolpath like entry motion, exit motion, and so on. You can set the motions to be displayed toolpath by using Advanced Display drop-down options.

You can change the display style of **Toolpaths Manager** by using the options displayed on clicking down arrow at the top in the **Toolpaths Manager**; refer to Figure-37. Select the **Background color** option to change background color of **Toolpaths Manager** which is by default white. Select the **Line color** option to change color of dotted lines showing design tree of operation in **Toolpaths Manager**. Select the **Font** option from drop-down to change font of text displayed in the **Toolpaths Manager**.

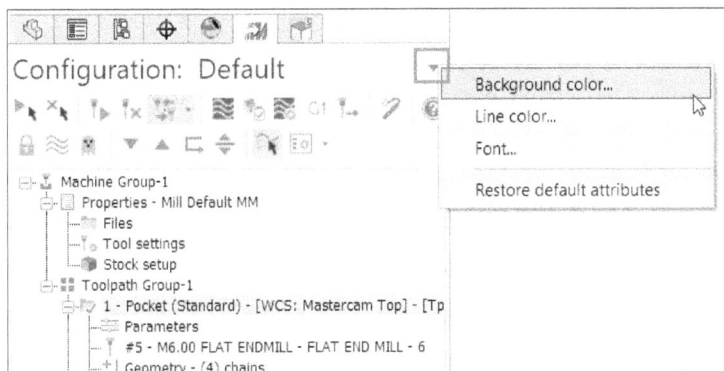

Figure-37. Options for changing Toolpaths Manager style

Note that all the parameters specified in the **Machine Group Properties** dialog box are used to define the machine group parameters. Machine group is a collection of machine definition, tool configuration, work piece and other libraries required by system to generate NC program for your applied toolpaths. If you change any of the parameters in this dialog box then it will affect all the NC program accordingly.

In this chapter, you learned about the workpiece setting as per the machine. In the next chapter, you will learn about the Machine setting and tool setting.

SELF-ASSESSMENT

Q1. Material properties and mechanical strength of model are important parameters when generating NC program using CAM software. (T/F)

Q2. Files created in Adobe Photoshop can be imported in Mastercam for machining on model surface. (T/F)

Q3. By default, SolidWorks Top plane is considered as Mastercam Top plane. (T/F)

Q4. Click in the **Offset** column next to a plane in **Planes Manager** to make it construction plane and create/project geometries on it.

Q5. To create copies of all the planes associated with part, right-click in the empty area of **Planes Manager** list box and select the .... option from shortcut menu.

Q6. .... are assigned to multiple parts on milling table to link the parts to their separate program origins.

Q7. The .... option in the **Create a new plane** drop-down of the **Planes Manager** is used to create Mastercam plane parallel to current view.

Q8. The .... check box in **Planes Manager** is used to keep the origin of plane attached to selected point on model so that if location of selected point is changed then origin of plane also gets changed accordingly.

Q9. Which of the following CAD formats can be used to import stock model in Mastercam?

a. IGES   b. SLDPRT
c. STL   d. OBJ

Q10. Changing the material of stock can cause change in cutting parameters of toolpaths. (T/F)

Q11. An operation marked as .... in Mastercam Toolpaths Manager is not sent for post processing and hence no NC program is generated for it.

Chapter 3

Machine and Tool Settings

Topics Covered

The major topics covered in this chapter are:

- *Machine Setting*
- *New Machine Creation*
- *Editing Controllers*
- *Setting Machine parameters*
- *Selection of cutting tools*
- *Creating New cutting tools*
- *Editing cutting tools*
- *Editing Post-Processors*

INTRODUCTION

In this chapter, we will learn about the settings related to machines and tools that play major role on operations and finishing of the product. There are various NC machines available in the market to perform various cutting operations. To make the CAM software understand the features of your machine, we need to edit some details of the machines so that the NC generated perform the way it is supposed to do. The options to edit the machine details are available in the **Machine Definition Manager** which will be discussed now. But, before that it is better to understand the components of CNC machine and various parameters which are used to define properties of machine.

CNC MACHINE STRUCTURE

As discussed in the starting of book, CNC machines are the cutting machines which use numeric codes to perform action. These numeric codes are understood by controller installed on the machine which gives command to various motors in the machine. But, this is not our concern now. Here, our concern is the components of machine that we should know before we create or change a machine in Mastercam. Figure-1 shows a 5 axis VMC with some of the components. Various major components milling and lathe machines are discussed next.

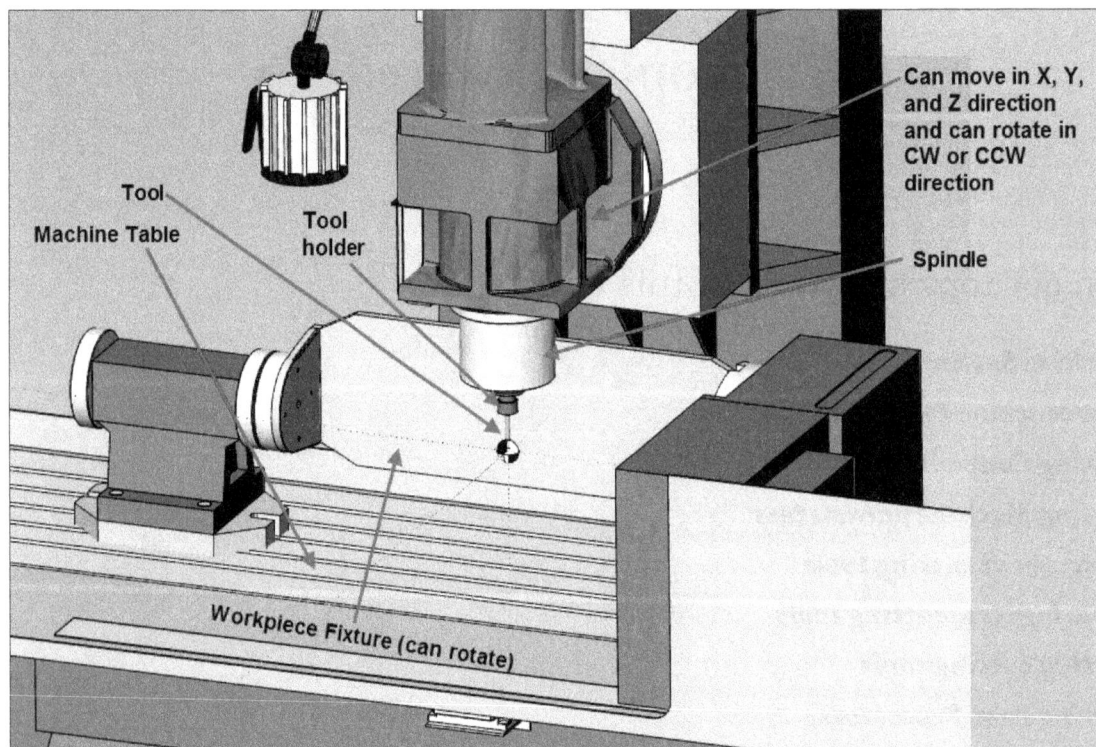

Figure-1. Five axis VMC

Tool Spindle

Machine tool spindles are rotating components that are used to hold and drive cutting tools or work pieces on lathes, milling machines and other machine tools. They use belt, gear, motorized, hydraulic or pneumatic drives and are available in a variety of configurations. Various specifications for selecting tool spindle are given next.

* Select spindle as per the required spindle Speed.
* Make sure spindle orientation correct as per your application.

- Consider the gage length of the tool. Doubling the gage length of a tool can increase the deflection at the end by a factor of 8. A way to compensate for this would be to go from a 40 taper to a 50 taper spindle and tool holder.
- Choose a spindle that can transmit the required amount of power/torque.
- When boring, select a spindle that has a nose bearing ID larger than the bore being machined
- Select the nose bearing arrangement suited for the application

Tool Changer or Turret

Tool changer, tool indexer or tool turret is used to automatically change the tool in spindle; refer to Figure-2. In some CNC milling and Lathe machines, you can load more than one tool at a time and then use the NC codes to use them in different toolpaths. While purchasing the machine, you should keep a note of time taken by machine to automatically change the tool and direction in which turret can rotate.

Figure-2. Tool Spindle with indexer

Translational and Rotational Limits

The translational and rotational limits are important aspect of cnc machines. You can not make a job which required machining length or angle more than the translational or rotational limits of machine. You can find this information in catalog of machine.

Tail Stock

Tail stock is generally found in lathe machines but can also be seen with rotary table of a milling machine. Tail stock is used to support long workpiece at its end; refer to Figure-3.

Figure-3. Tail Stock in lathe.jpg

MACHINE DEFINITION MANAGER

The **Machine Definition Manager** is used to edit the parameters related to machine. The procedure to use the **Machine Definition Manager** is discussed next.

- Click on the **Machine Definition** tool from the **Utilities** drop-down. The **Machine Definition File Warning** dialog box will be displayed; refer to Figure-4.

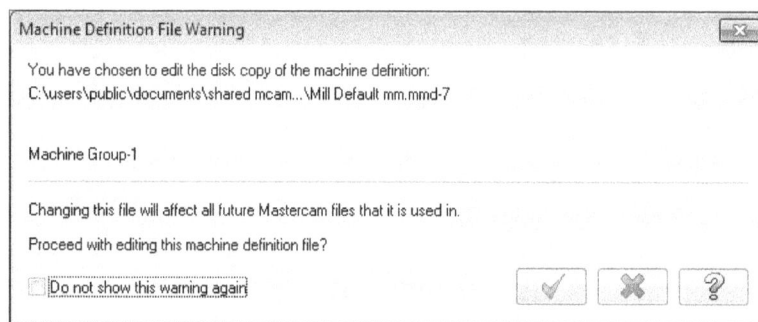

Figure-4. Machine Definition File Warning dialog box

- Click on the **OK** button to accept that we are changing the machine definition. The **Machine Definition Manager** will be displayed; refer to Figure-5.

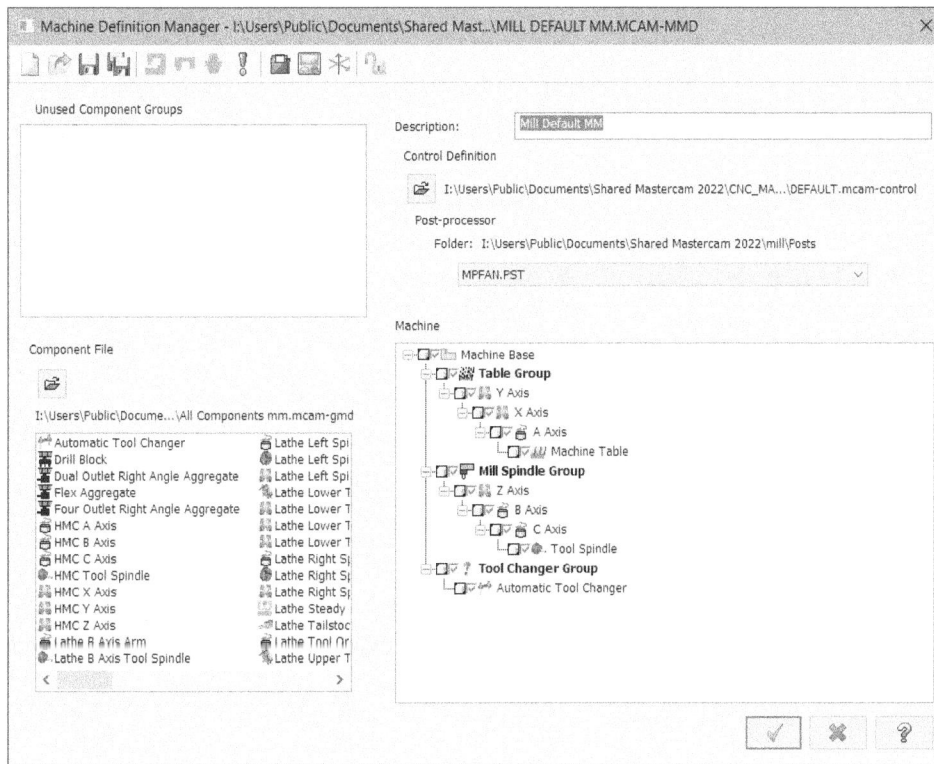

Figure-5. Machine Definition Manager

- Click on the **Open** button from the toolbar to use desired machine. The **Open Machine Definition File** dialog box will be displayed; refer to Figure-6. (Note that only those machine definitions will be displayed for which you have purchased Mastercam. To get more machine definitions, you need to contact your Mastercam Reseller.)
- Select desired file type from the **File Type** drop-down. The machine files will be displayed accordingly. Select desired machine definition file and click on the **Open** button from the dialog box. The machine definition will be displayed; refer to Figure-5.

Figure-6. Open Machine Definition File dialog box

- Select the check boxes in the **Machine Configuration** area to display the respective components of machine in the simulation.
- The icons displayed in the **Component File** area are used to display the components of the machine. Click on the **Open** button in this area to change the components of the machine.
- Click in the **Description** edit box to change the name of the machine.
- After changing the name of the machine, click on the **Save As** button. The **Save Machine Definition File** dialog box will be displayed; refer   to Figure-7.

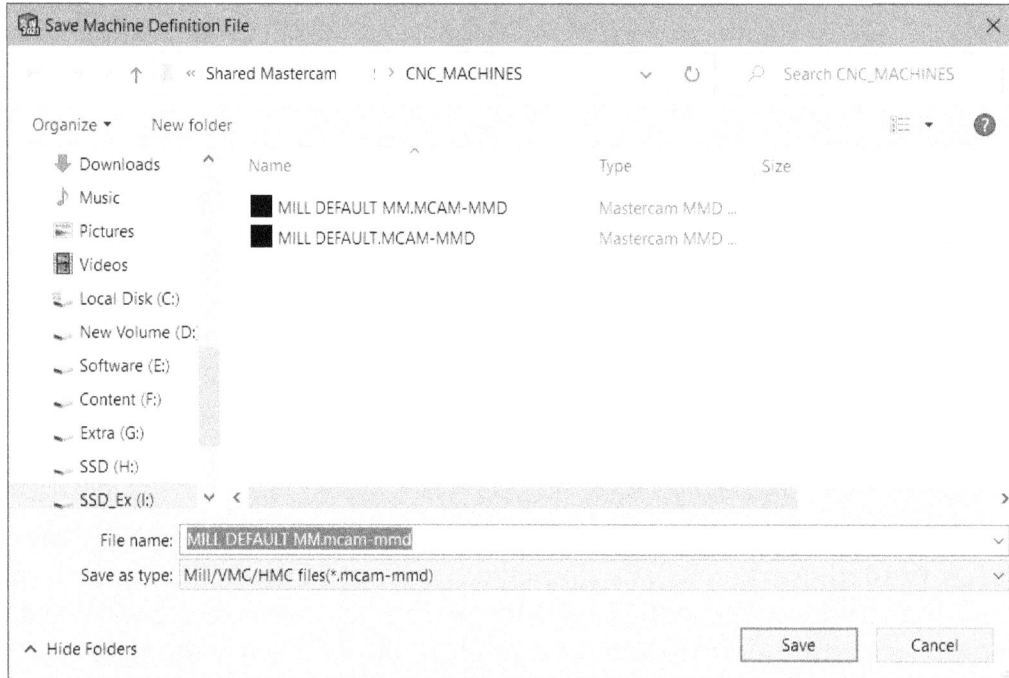

Figure-7. Save Machine Definition File dialog box

- Specify desired name and click on the **Save** button to save the configuration.

New Machine Definition

- Click on the **New** button from the **Machine Definition Manager** dialog box. The **CNC Machine Types** dialog box will be displayed; refer to Figure-8.

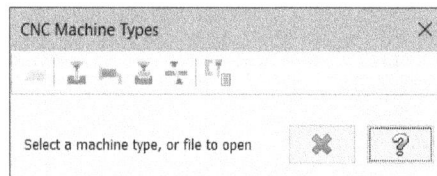

Figure-8. CNC Machine Types dialog box

- Click on desired button to create respective type of machine definition. In our case **Mill/VMC/HMC** button is selected.
- On selecting the button, the **Machine Definition Manager** dialog box will displayed as shown in Figure-9.

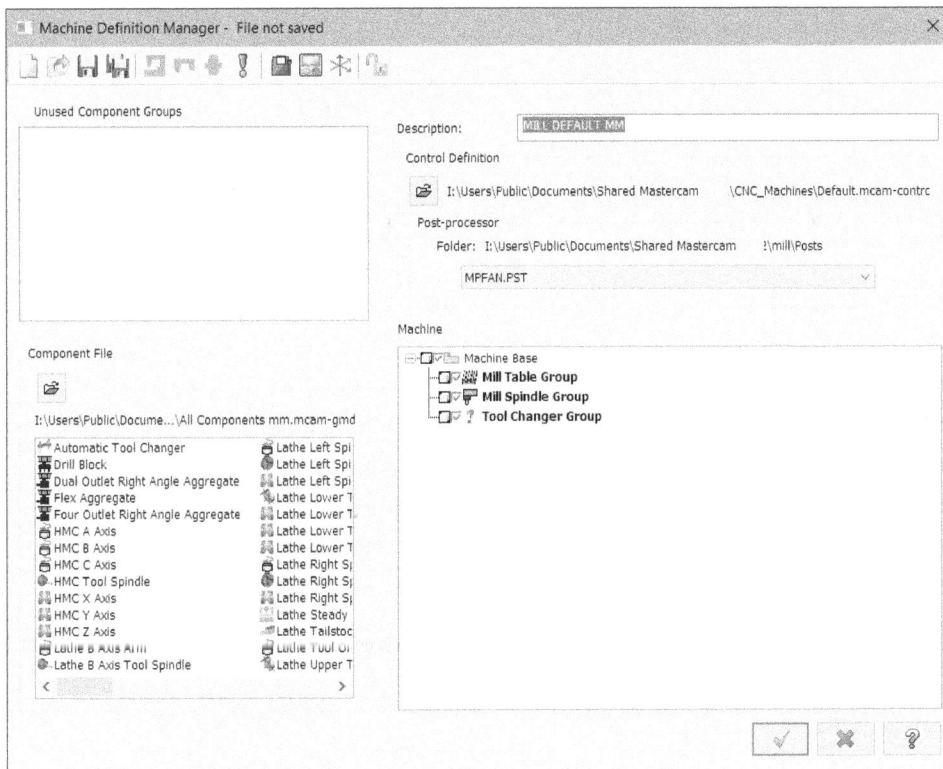

Figure-9. Machine Definition Manager dialog box with new definition

- Drag desired component from the **Component File** area and place it in desired category of **Machine Configuration** area to display it in simulation. There are various type of components like; linear axis components, rotary axis components, tool spindle, tool changer, and tail stock that can be added to machine definition. These components represent physical drives in your machine. If table in your machine can move in X and Y directions then add the **X Axis** and **Y Axis** components to the **Table Group** in **Machine Configuration**. If table can also rotate about A axis then you can drag and drop **A Axis** component to **Table Group** in configuration. Similarly, you can define configuration for **Mill Spindle Group** and **Tool Changer Group**. Save the file as discussed earlier.
- To edit a component, double-click on it. Respective dialog box will be displayed. The options for editing various components are discussed next.

Editing Linear Axis Components

- Double-click on linear axis moving components like X Axis, Y Axis or Z Axis in the **Machine Configuration** area; refer to Figure-10. The **Machine Component Manager-Linear Axis** dialog box will be displayed; refer to Figure-11.

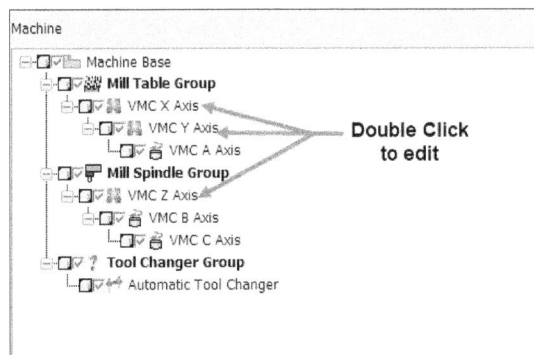

Figure-10. Linear axis components of machine

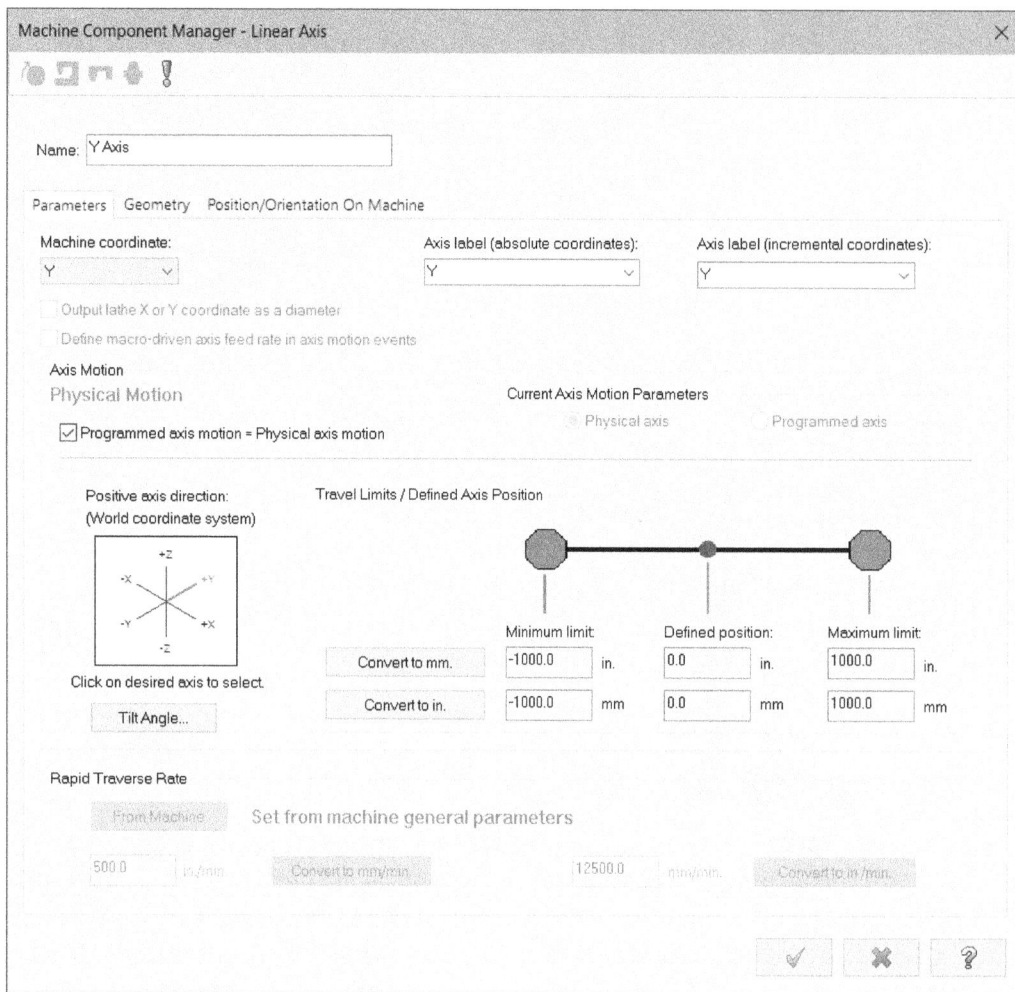

Figure-11. Machine Component Manager-Linear Axis dialog box

- Specify desired name for the axis in the **Name** edit box or you can leave it to its default name.
- Set the machine coordinate, axis label absolute and axis label relative as required in your programming.
- Enter desired value of minimum limit, defined position and maximum limit of travel in the respective edit boxes of **Travel Limits/Defined Axis Position** area of the dialog box. If you have specified the values in inch edit boxes then click on the **Convert to mm** button to automatically fill the values in mm edit boxes in the dialog box or you can do vice-versa.
- Select an axis from the **Position axis** direction box to define the axis of travel in WCS. If you want to tilt the axis then click on the **Tilt Angle** button below **Position axis** direction box in the dialog box; refer to Figure-12. The **Tilt Angle (Linear Axis)** dialog box will be displayed; refer to Figure-13.

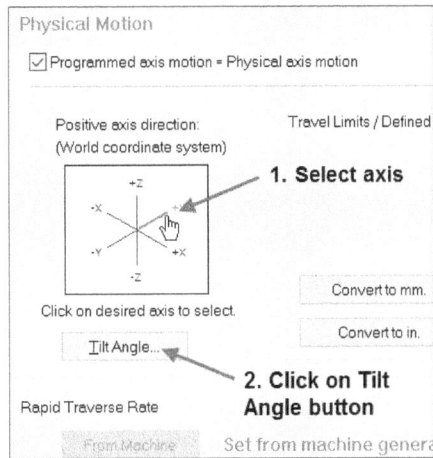

Figure-12. Tilt Angle button

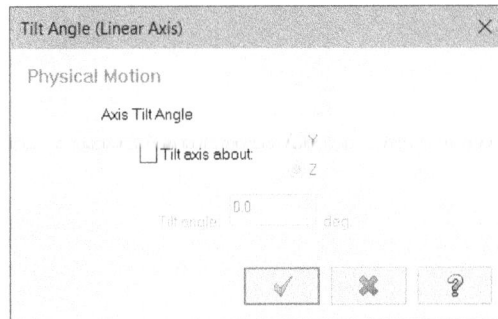

Figure-13. Tilt Angle (Linear Axis) dialog box

- Select the **Tilt axis about** check box and select the radio button for axis about which you want to tilt the current axis.
- Specify desired angle value and click on the **OK** button to tilt the axis.
- Click on the **OK** button from the **Machine Component Manager** dialog box to apply the settings.

Editing Rotary Axis Components

- Double-click on the **A Axis**, **B Axis**, or **C Axis** component in the **Machine Configuration** area of the **Machine Definition Manager** dialog box. The **Machine Component Manager - Rotary Axis** dialog box will be displayed; refer to Figure-14.
- Select desired machine coordinates and axis labels as done for linear axis components in previous topic.
- Select the axis of rotation and **0** degree position from the boxes in the **World coordinate system** area of the dialog box. Specify the tilt angle if required by using the **Tilt Angle** button.
- Set the Counter Clockwise(**CCW**) or Clockwise(**CW**) direction of rotation from the Direction area of the dialog box.
- Similarly, specify the travel limits in the **Travel Limits** area of the dialog box and position of center of rotation in the **Center of Rotation** area of the dialog box.

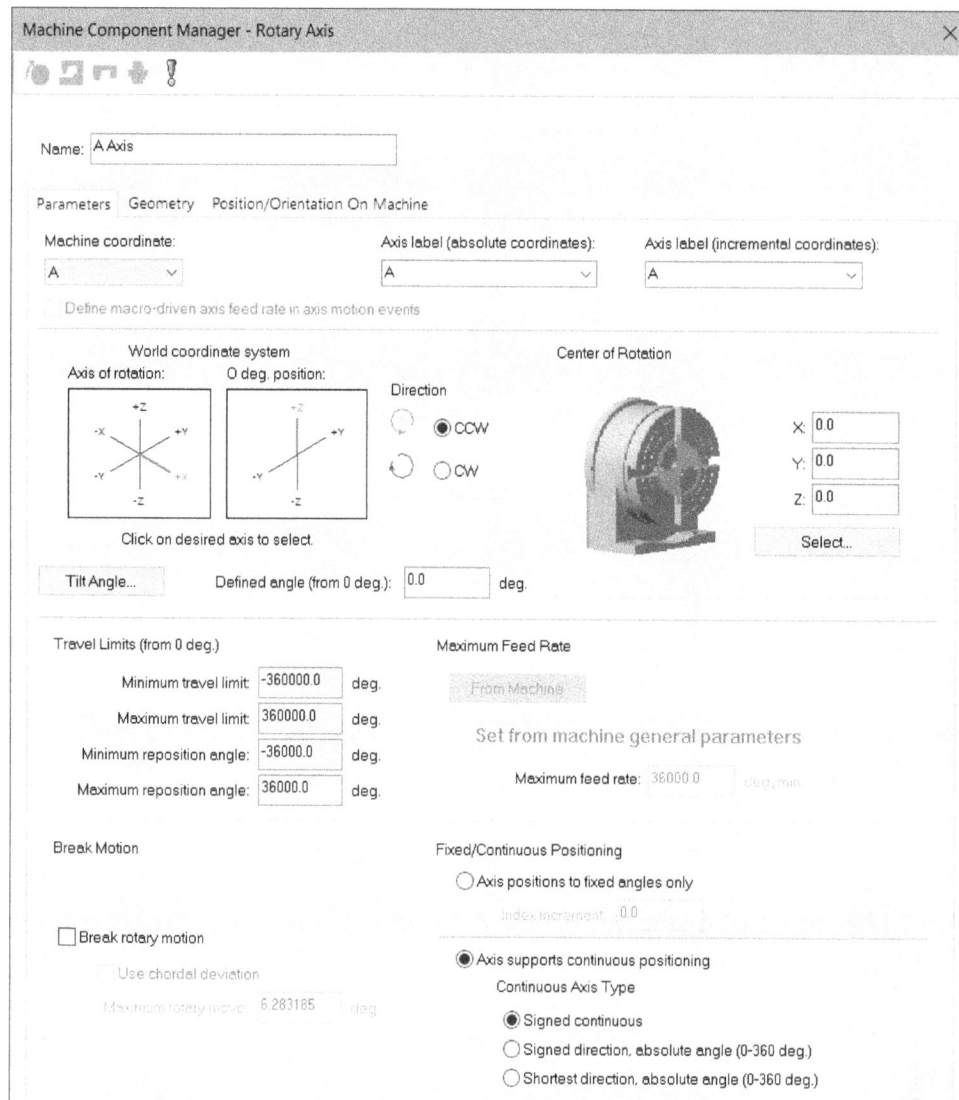

Figure-14. Machine Component Manager-Rotary Axis dialog box

- Click on the **OK** button from the dialog box to apply changes.

Editing Tool Spindle

- Double-click on `Tool Spindle` from the `Machine Configuration` area of the `Machine Definition Manager` dialog box. The `Machine Component Manager - Tool Spindle` dialog box will be displayed; refer to Figure-15.
- Set the minimum and maximum spindle rotation speed in `Minimum spindle speed` and `Maximum spindle speed` edit boxes, respectively.
- Similarly, specify the tool orientation, and other parameters of the tool spindle. Click on the **OK** button from the dialog box to apply the changes.

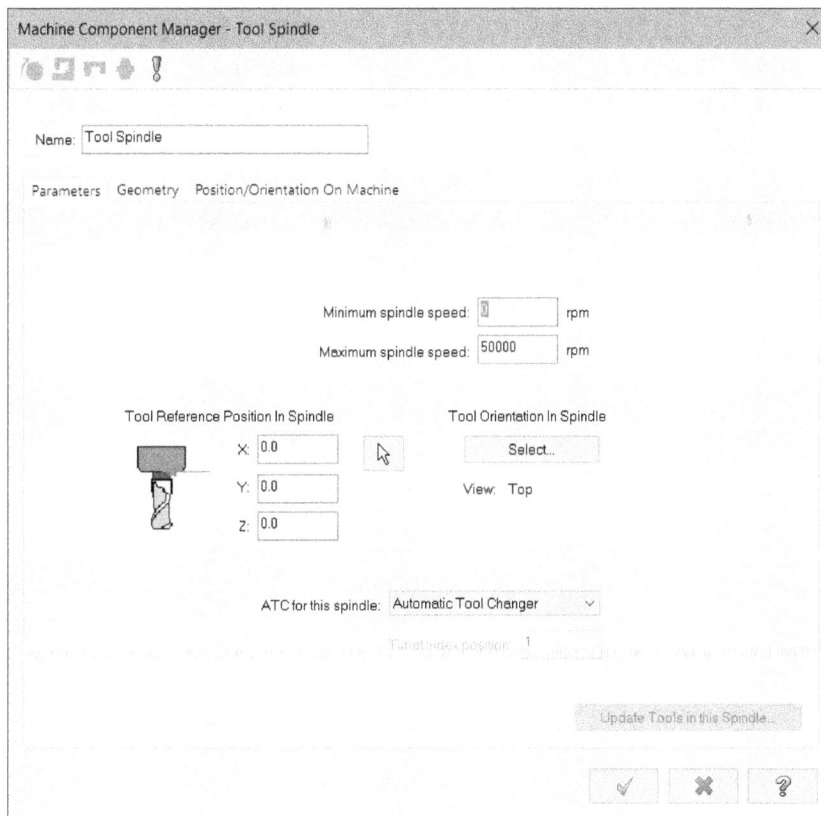

Figure-15. Machine Component Manager-Tool Spindle dialog box

Similarly, you can edit the other components of machine by double-clicking on them in the **Machine Configuration** area of the **Machine Definition Manager** dialog box.

Setting Controller

- Click on the **Open** button 🖼 in the **Control Definition** area of the dialog box. The **Control Definition Files** dialog box will be displayed; refer to Figure-16.
- Select the file of desired control system and click on the **Open** button. The control will be applied to the machine.

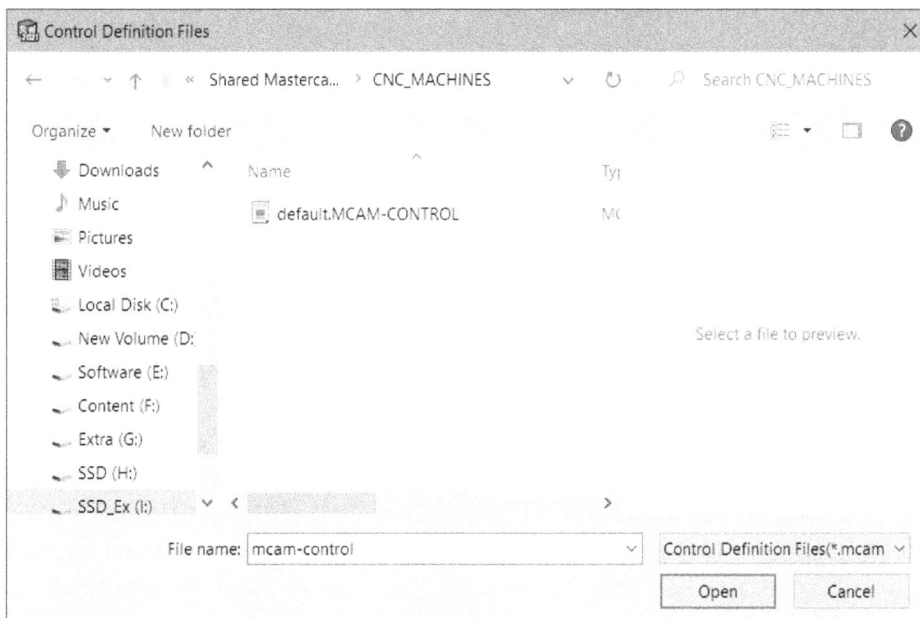

Figure-16. Control Definition Files dialog box

Editing Control Parameters

- Click on the **Edit Control Definition** button 🖥 from the toolbar. The **Control definition** dialog box will be displayed; refer to Figure-17. Using this option, you can define how codes will be output for your physical machine components.
- Click on desired topic from the **Control topics** area of the dialog box. The parameters related to the selected topic will be displayed; refer to Figure-18.

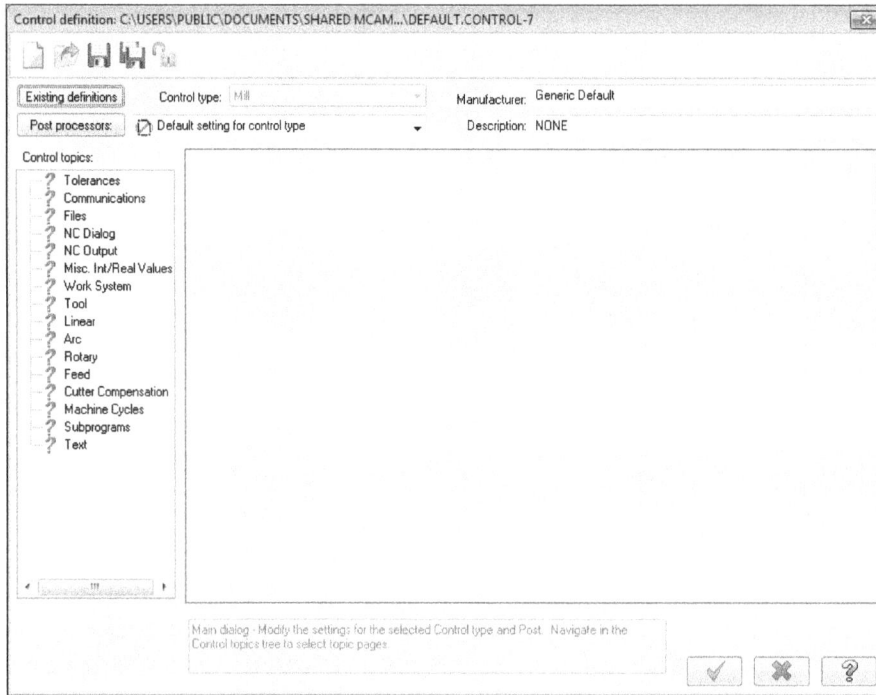

Figure-17. Control definition dialog box

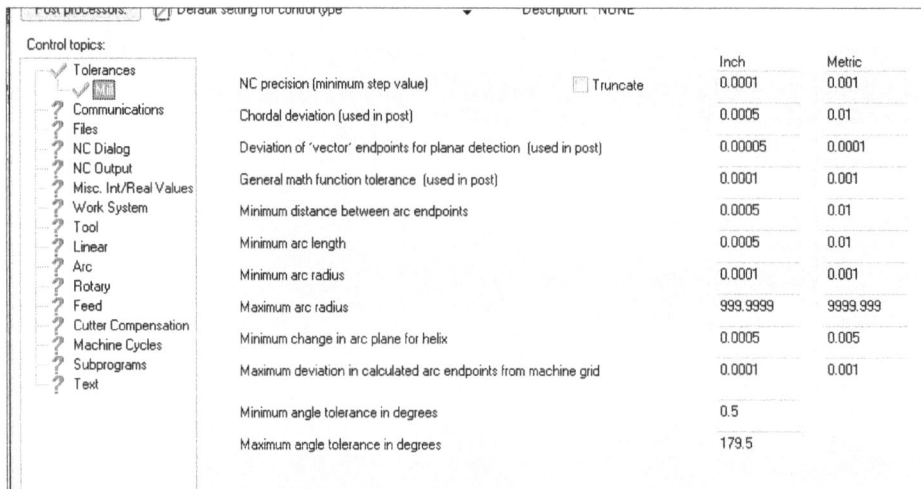

Figure-18. Parameters related to selected topic

- In **Tolerances** page, you can define the tolerance values up to which your machine is capable to perform various movements.
- In **Communications** page, you can define the connection capabilities of your machine. If your machine is capable to connect with computer using **CimcoDNC** then set it in **Communications** drop-down. Similarly, you can define other connection types.
- In the **Files** page, you can set path for various libraries used by machine and directory where output file will be saved. You can also set machine to post error files generated while performing machining operations.

- In the **NC Dialog** page, you can define whether check boxes for **Reference point** and **Tool Display** will be available in **Tool Parameters** dialog box.
- In the **NC Output** page, you can define the NC output capabilities of Machine in the software. You can define various default parameters for output like generate incremental output, add comments in NC output file, starting sequence number and increments in sequence for codes in the file, and so on.
- In the **Misc. Int/Real Values** page, you can set various integers and real numbers to represent different machine functions. For example, you can set 2 to represent G54 code for Work Coordinate system.
- In **Work System** page, you can work coordinate for an operation will be selected for performing an operation in the machine. If you want work coordinates of operations change based on Work offsets then select the option from **Work coordinate selection** drop-down on this page. Similarly, you can set Tool planes to change based on work offsets.
- In **Tool** page, you can define offsets for cutting tools used in the machine. For example if you have two end mills of 10 mm diameter and one of them is worn out the you can compensate the wear using tool offsets.
- In **Linear** page, you can define how cutting tool will move in linear directions for rapid and feed passes.
- In **Arc** page, you can define parameters related to arc moves (G02 and G03) of the machine. If there is a plane in which your machine does not support arc moves then clear respective check box from this page.
- In **Rotary** page, you can define whether rotary moves of machine will be broken into smaller sections or it will perform large continuous rotary moves.
- In **Feed** page, you can define unit to be used for various feed moves.
- In **Cutter Compensation** page, you can set whether machine supports cutter compensation or not. If machine supports cutter compensation then you can define the modes in which cutter compensation will be applied.
- In **Machine Cycles** page, you can define whether machine supports milling peck cycles and drill cycles or not. If machine supports these cycles then what are the sub types supported by it.
- In **Subprograms** page, you can define if machine supports subprograms (small repeating code blocks which are generally called by reference name in main program).
- In the **Text** page, you can define custom names (values) for various functions which can be performed on the machine.
- Specify desired manufacturer name and description for the control file in respective edit boxes at the top in the dialog box.
- Click on the **Post processors** button at the top in the dialog box to select desired post processor for the machine. The **Control Definition Post List Edit** dialog box will be displayed; refer to Figure-19. Click on the **Add files** button to add more posts in the list that can be used with this machine control definition. Similarly, you can use **Delete files** button to delete posts from list. Click on the **OK** button from the dialog box to apply changes. Note that it is important to add post processors in the list of control definition as they are linked in Mastercam for proper output.

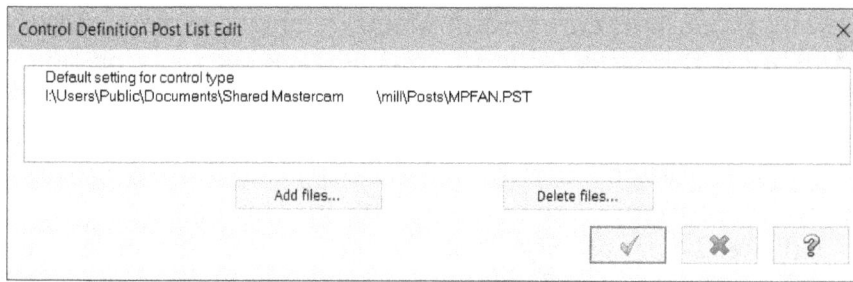

Figure-19. Control Definition Post List Edit dialog box

- Click on the **Save As** button and save your machine control definition by desired name. (You will not be able to do it in Home Learning Edition of software).
- Click on the **OK** button from the dialog box to use the created control definition. The **Machine Definition Manager** dialog box will be displayed again. Note that the parameters specified in the dialog box are used to specify the limits of your machine controller. If something is not specified in this dialog box then NC codes will not be generated for that function in the toolpath codes.

Editing General Parameters of Machine

- Click on the **General Machine Parameters** button 📠 from the **Machine Definition Manager** dialog box to define feed rates and M code function capabilities of machine. The **General Machine Parameters** dialog box will be displayed; refer to Figure-20.

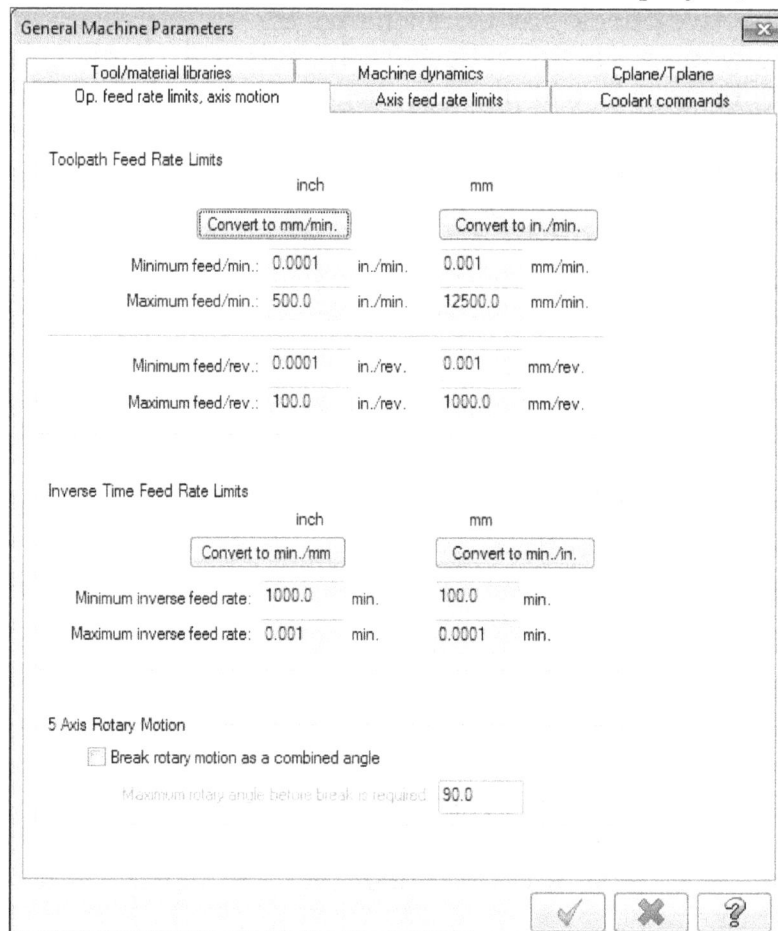

Figure-20. General Machine Parameters dialog box

- Specify desired parameters for the machine. To switch to other functions of machine, you need to select respective tabs in the dialog box.

- After specifying desired parameters, click on the **OK** button from the dialog box. The options in various tabs are discussed next.

Axis feed rate limits tab

The options in this tab are used to specify the limits of machine for axis feed rate; refer to Figure-21. Specify desired values in the edit boxes available in the tab. Note that the values entered higher than the specified values during toolpath creation will be displayed as error.

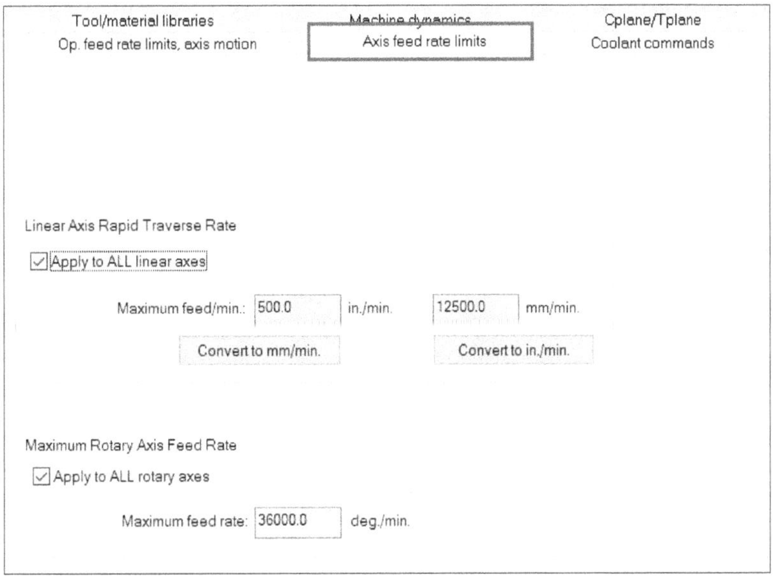

Figure-21. Axis feed rate limits tab

Op. feed rate limits, axis motion tab

The options in this tab are used to specify the limits of feed rate during cutting operations. Specify desired values in the respective edit boxes. Note that the values entered higher than the specified maximum limit or lower than the minimum limit will be displayed as error during toolpath creation.

Coolant commands tab

In older machines there were separate switches in the panel to On/Off coolants during machine but in modern machines you can control the coolant by using the NC codes. In this tab, you can set the parameters related to coolant. By default, the **Support coolant using coolant value in post processor** check box is selected which means only those parameters which are compatible with your controller will be displayed. Clear this check box to define extra options related to coolant.

Tool/material libraries tab

The options in this tab are used to set tool libraries and material libraries for the machine. During the machining simulation, system will search these directories for tools and materials.

Machine Dynamics tab

The options in this tab are used to change parameters related to machine dynamics like, acceleration while cornering, feed rate change per block, and so on. These parameters are used by high feed toolpaths. Note that high feed machining cannot be used with multi-axis or lathe toolpaths.

- After setting desired parameters, click on the **OK** button from the **General Machine Parameters** dialog box to apply all the new settings created.
- Choose the **OK** button from the **Machine Definition Manager** dialog box to apply specified machine settings. Save the machine files and replace the machine definition using the dialog boxes displayed after clicking **OK** button.

TOOL MANAGER

The **Tool Manager** is used to manage the data related to tools. The procedure to use the **Tool Manager** is given next.

- Click on the **Mill Tool Manager** tool from the **Toolpath Utilities** drop-down in the **Ribbon** while you are working with milling machines. The **Tool Manager** dialog box will be displayed; refer to Figure-22.

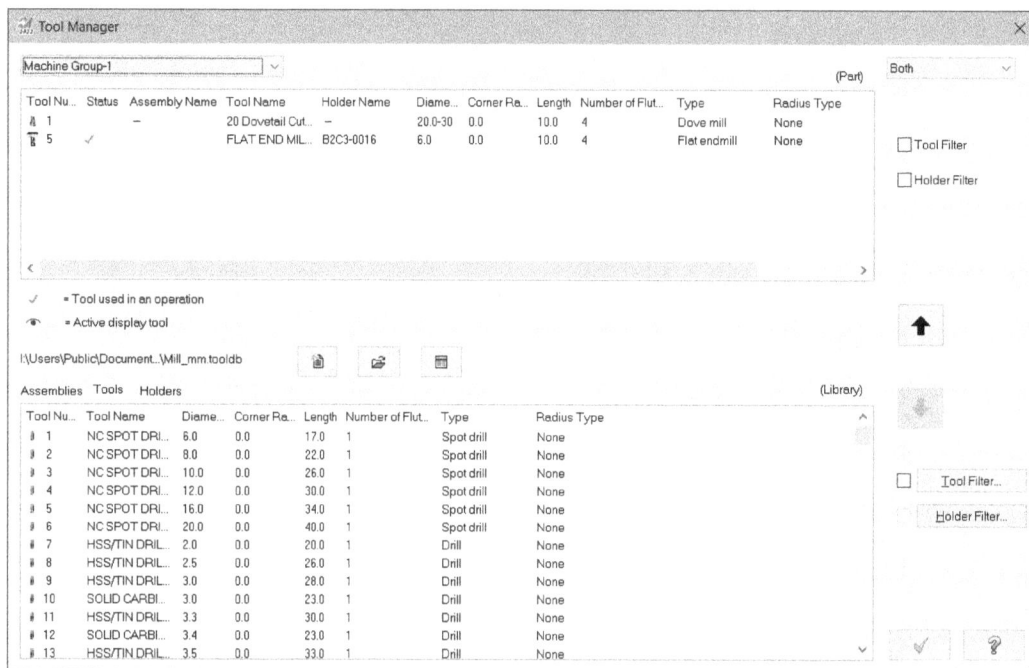

Figure-22. Tool Manager

- Select the tools that you want to include in current machine group. Note that to select multiple tools, you need to press and hold the **CTRL** key while selecting tools.
- Click on the **Up** arrow ⬆ to include the tools in the current machine group.
- Click on the **OK** button from the dialog box to accept the selected tools.

Creating New Tool

- Right-click in the top area of the dialog box to display shortcut menu; refer to Figure-23.

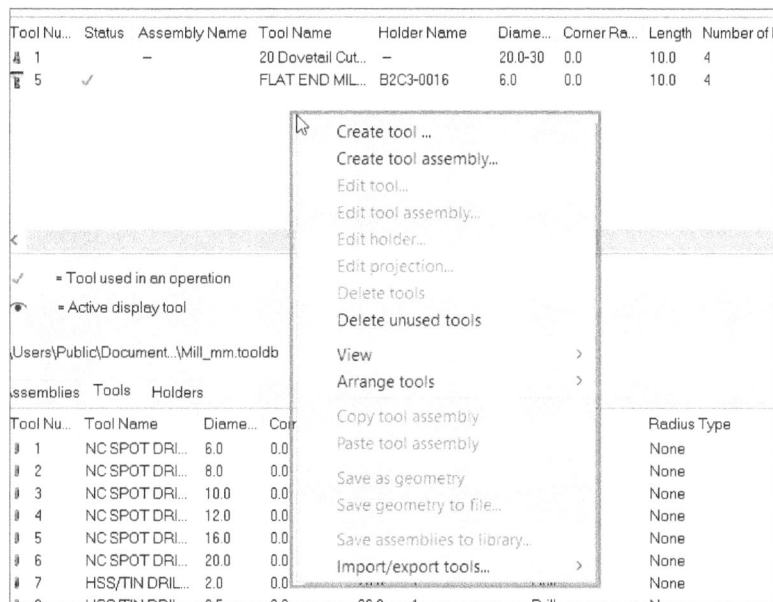

Tool Nu...	Status	Assembly Name	Tool Name	Holder Name	Diame...	Corner Ra...	Length	Number of f
1	–		20 Dovetail Cut...	–	20.0-30	0.0	10.0	4
5	✓		FLAT END MIL...	B2C3-0016	6.0	0.0	10.0	4

Create tool ...
Create tool assembly...
Edit tool...
Edit tool assembly...
Edit holder...
Edit projection...
Delete tools
Delete unused tools
View   >
Arrange tools   >
Copy tool assembly
Paste tool assembly
Save as geometry
Save geometry to file...
Save assemblies to library...
Import/export tools... >

✓ = Tool used in an operation
👁 = Active display tool

\Users\Public\Document...\Mill_mm.tooldb

Assemblies   Tools   Holders

Tool Nu...	Tool Name	Diame...	Cor		Radius Type
1	NC SPOT DRI...	6.0	0.0		None
2	NC SPOT DRI...	8.0	0.0		None
3	NC SPOT DRI...	10.0	0.0		None
4	NC SPOT DRI...	12.0	0.0		None
5	NC SPOT DRI...	16.0	0.0		None
6	NC SPOT DRI...	20.0	0.0		None
7	HSS/TIN DRIL...	2.0	0.0		None

Figure-23. Shortcut menu for tools

- Click on the **Create tool** option from the shortcut menu. The **Define Tool** dialog box will be displayed; refer to Figure-24.

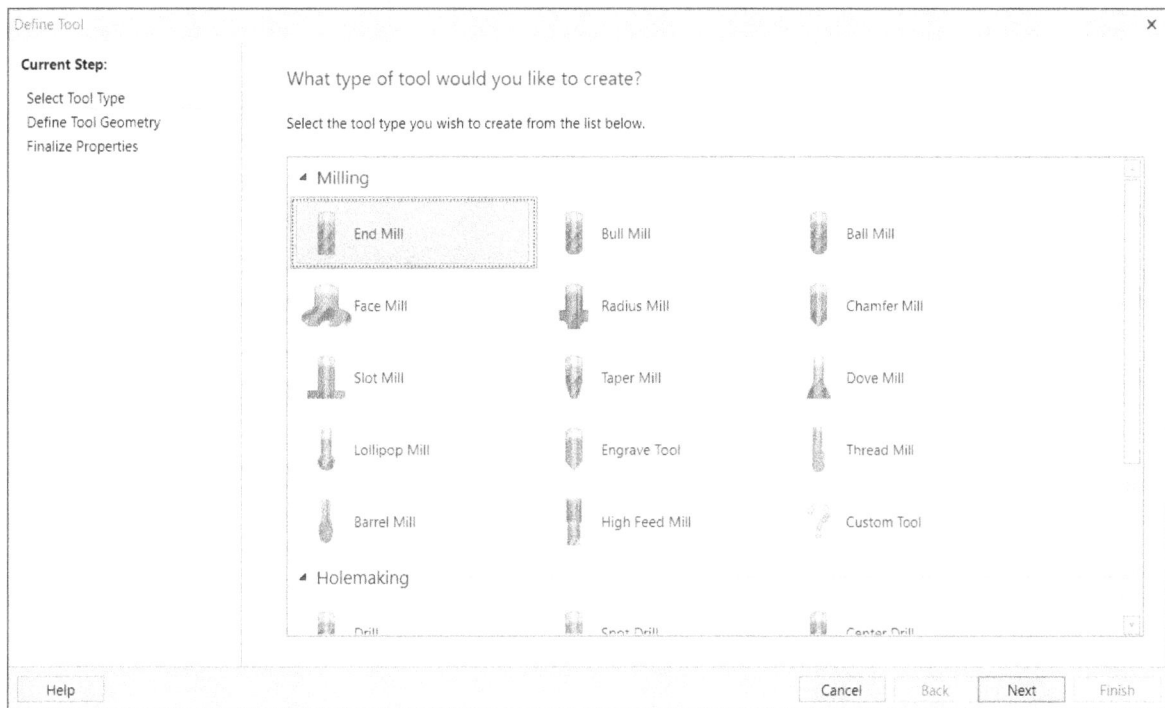

Figure-24. Define Tool dialog box

- Select the button for desired type of tool. (Dove Mill in our case).
- Click on the **Next** button from the box. The **Define Tool Geometry** page will be displayed in the dialog box; refer to Figure-25.
- Specify the parameters for the tool and click on the **Next** button from the box. The **Finalize properties** page of the **Define Tool** dialog box will be displayed; refer to Figure-26.

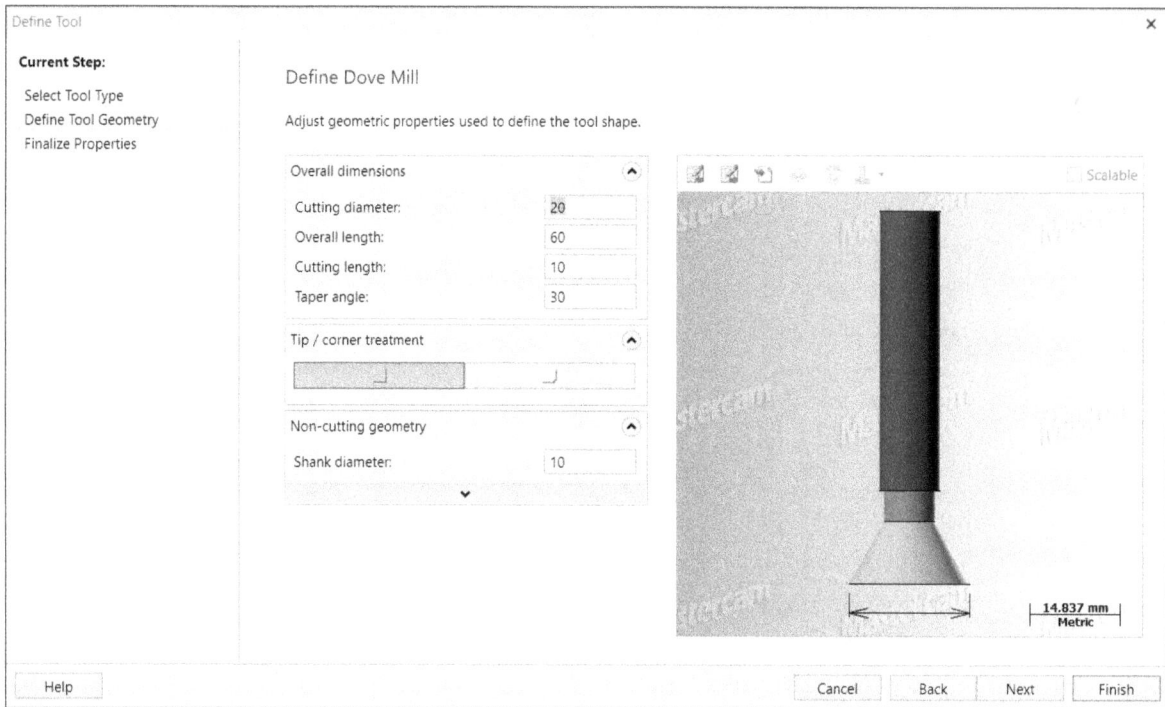

Figure-25. Create New Dove Mill dialog box

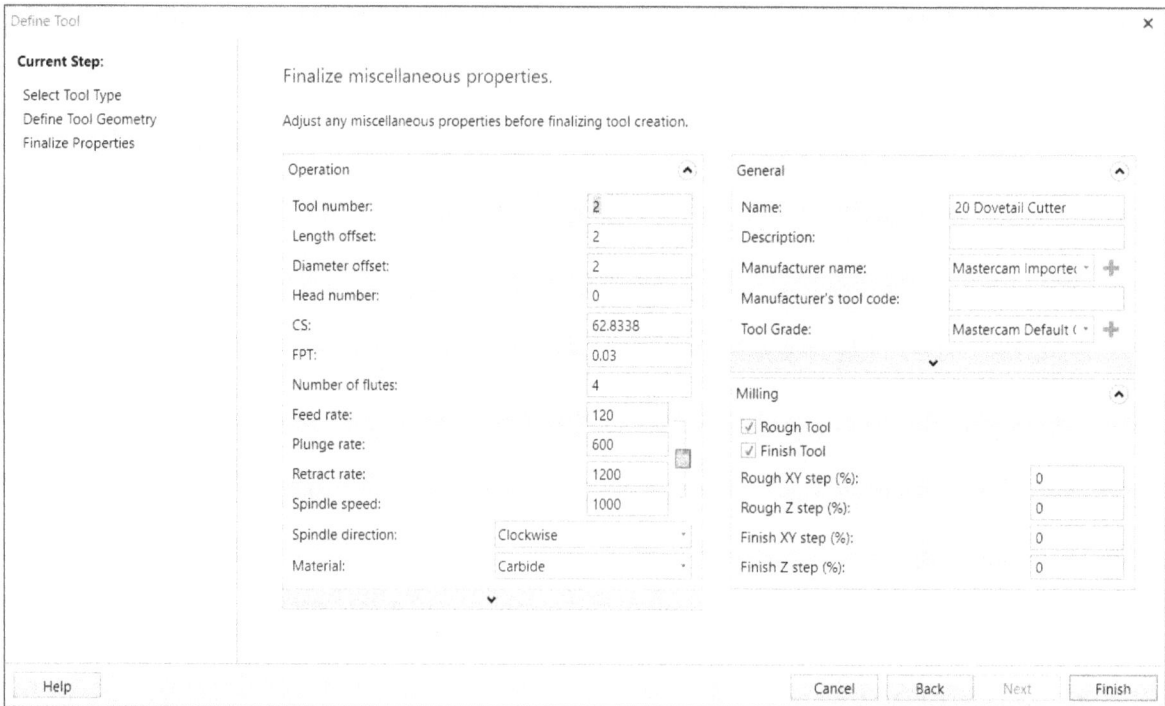

Figure-26. Finalize miscellaneous properties page

- Specify desired parameters. Click on the **Coolant** button to specify settings related to coolant. The **Coolant** dialog box will be displayed; refer to Figure-27.
- After specifying desired coolant conditions, click on the **OK** button from the dialog box to exit.
- Click on the **Finish** button from the **Create New Tool** box to create the tool.
- Select the tool from the top list and click on the down arrow ⬇ to add the tool in the library.

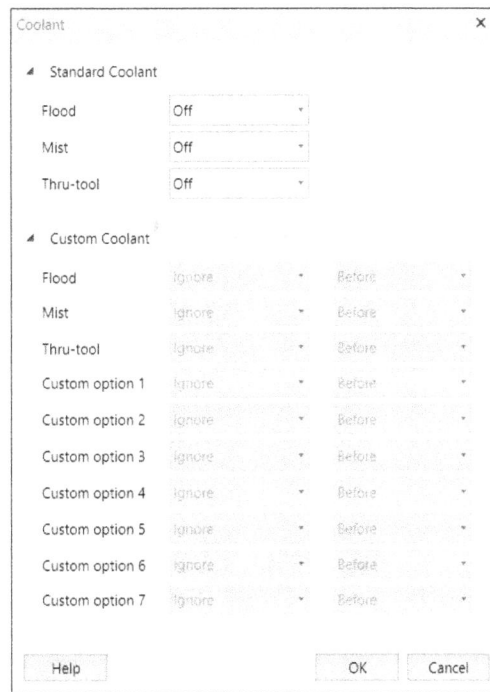

Figure-27. Coolant dialog box

Editing Tool

- Select the tool from the top list and right-click on it. A shortcut menu will be displayed as discussed earlier.
- Click on the **Edit tool** option from the shortcut menu. The **Edit Tool** dialog box will be displayed; refer to Figure-28.

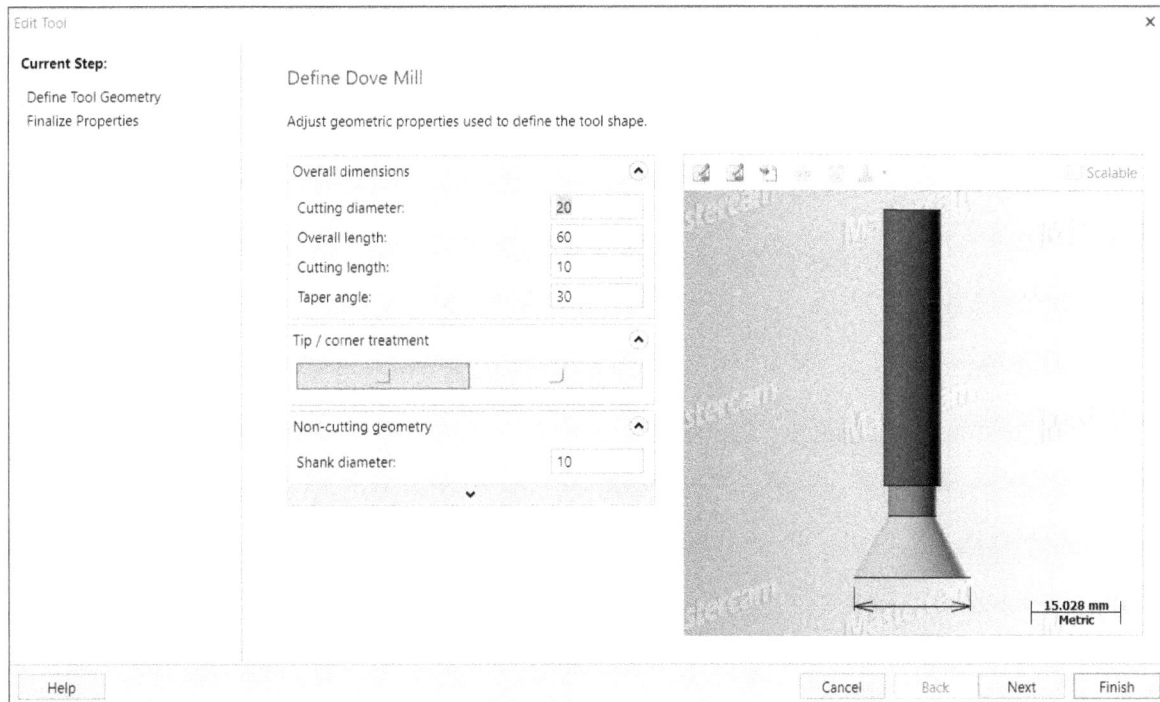

Figure-28. Edit Tool dialog box

- Specify desired parameters related to the tool and the click on the **Next** button. The parameters displayed in the dialog box are same as displayed in **Define Tool** dialog box discussed earlier.

- After changing the parameters in the **Finalize Properties** page of the dialog box, click on the **Finish** button.

- Click on the **OK** button from the **Tool Manager** to set the specified settings. Click on the **Yes** button to save changes in the library.

TOOLS USED IN CNC MILLING AND LATHE MACHINES

The tools used in CNC machines are made of cemented carbide, High Speed Steel, Tungsten Alloys, Ceramics, and many other hard materials. The shapes and sizes of tools used in Milling machines and Lathe machines are different from each other. These tools are discussed next.

Milling Tools

There are various type of milling tools for different applications. These tools are discussed next.

End Mill

End mills are used for producing precision shapes and holes on a Milling or Turning machine. The correct selection and use of end milling cutters is paramount with either machining centers or lathes. End mills are available in a variety of design styles and materials; refer to Figure-29.

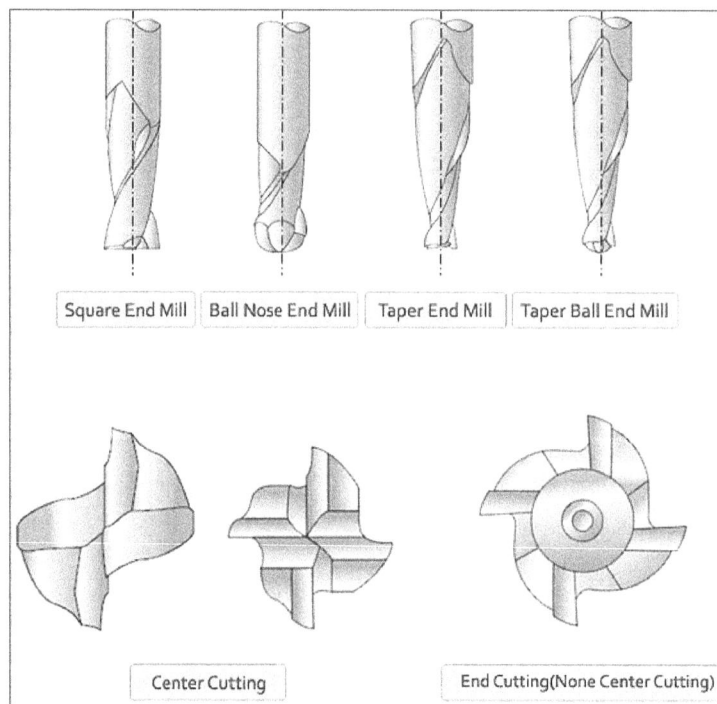

Figure-29. End Mill tool types

Titanium coated end mills are available for extended tool life requirements. The successful application of end milling depends on how well the tool is held (supported) by the tool holder. To achieve best results an end mill must be mounted concentric in a tool holder. The end mill can be selected for the following basic processes:

FACE MILLING - For small face areas, of relatively shallow depth of cut. The surface finish produced can be 'scratchy".

KEYWAY PRODUCTION - Normally two separate end mills are required to produce a quality keyway.

WOODRUFF KEYWAYS - Normally produced with a single cutter, in a straight plunge operation.

SPECIALTY CUTTING - Includes milling of tapered surfaces, "T" shaped slots & dovetail production.

FINISH PROFILING - To finish the inside/outside shape on a part with a parallel side wall.

CAVITY DIE WORK - Generally involves plunging and finish cutting of pockets in die steel. Cavity work requires the production of three dimensional shapes. A Ball type end mill is used for the finishing cutter with this application.

Roughing End Mills, also known as ripping cutters or hoggers, are designed to remove large amounts of metal quickly and more efficiently than standard end mills; refer to Figure-30. Coarse tooth roughing end mills remove large chips for heavy cuts, deep slotting and rapid stock removal on low to medium carbon steel and alloy steel prior to a finishing application. Fine tooth roughing end mills remove less material but the pressure is distributed over many more teeth, for longer tool life and a smoother finish on high temperature alloys and stainless steel.

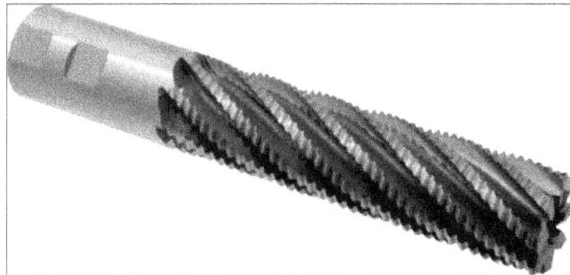

Figure-30. Roughing End Mill

Bull Nose Mill

Bull nose mill look alike end mill but they have radius at the corners. Using this tool, you can cut round corners in the die or mold steels. Shape of bull nose mill tool is given in Figure-31.

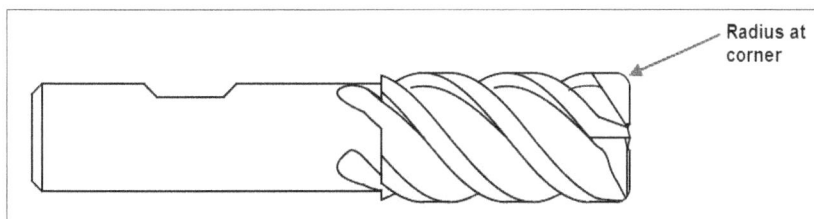

Figure-31. Bull Nose Mill cutter

Ball Nose Mill

Ball nose cutters or ball end mills has the end shape hemispherical; refer to Figure-29. They are ideal for machining 3-dimensional contoured shapes in machining centres, for example in moulds and dies. They are sometimes called ball mills in shop-floor slang. They are also used to add a radius between perpendicular faces to reduce stress concentrations.

Face Mill

The Face mill tool or face mill cutter is used to remove material from the face of workpiece and make it plane; refer to Figure-32.

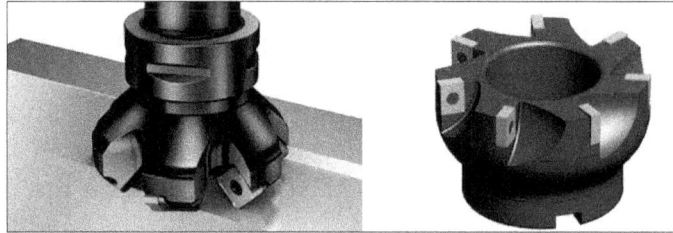

Figure-32. Face milling tool

Radius Mill and Chamfer Mill

The Radius mill tool is used to apply round (fillet) at the edges of the part. The Chamfer mill tool is used to apply chamfer at the edges of the part. Figure-33 shown the radius mill tool and chamfer mill tool.

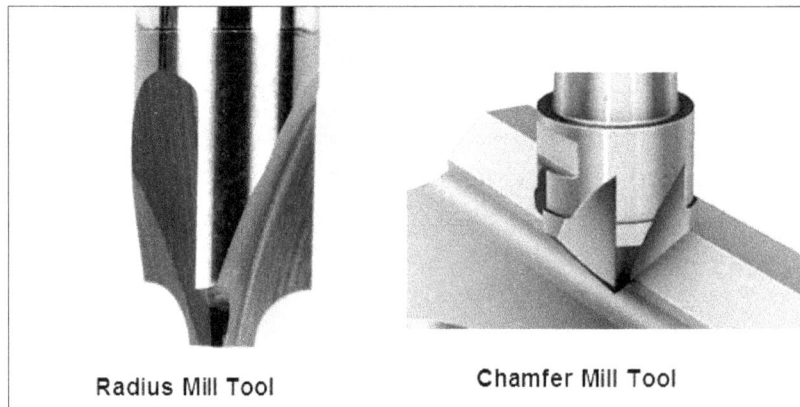

Radius Mill Tool   Chamfer Mill Tool

Figure-33. Radius mill and Chamfer mill tool

Slot Mill

The Slot mill tool is used to create slot or groove in the part metal. Figure-34 shows the shape of slot mill tool.

Figure-34. Slot mill tool

Taper Mill

In CNC machining, taper end mills are used in many industries for a large number of applications, such as walls with draft or clearance angle, tool and die work, mold work, even for reaming holes to make them conical. There are mainly two types of taper mills, Taper End Mill and Taper Ball Mill; refer to Figure-29.

Dove Mill

Dove mill or Dovetail cutters are designed for cutting dovetails in a wide variety of materials. Dovetail cutters can also be used for chamfering or milling angles on the bottom surface of a part. Dovetail cutters are available in a wide variety of diameters and in 45 degree or 60 degree angles; refer to Figure-35.

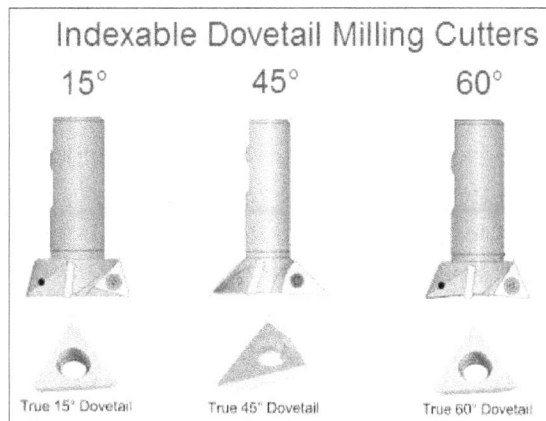

Figure-35. Dovetail milling cutters

Lollipop Mill

The Lollipop mill tool is used to cut round slot or undercuts in workpiece. Some tool suppliers use a name Undercut mill tool in place of Lollipop mill in their catalog. The shape of lollipop mill tool is given in Figure-36.

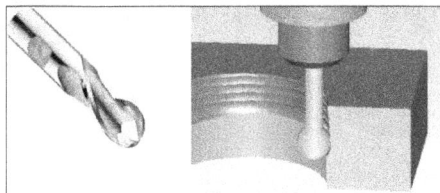

Figure-36. Lollipop mill tool

Engrave Mill

The Engrave mill tool is used to perform engraving on the surface of workpiece. Engraving has always been an art and it is also true for CNC machinist. You can find various shapes of engraving tool that are single flute or multi-flute; refer to Figure-37. You can use ball mill/end mill for engraving or you can use specialized engrave mill tool for engraving. This all depends on your requirement. If you want to perform engraving on softer materials or plastics then it is better to use ball end mill but if you want an artistic shade on the surface then use the respective engrave mill tool. Keep a note of maximum depth and spindle speed mentioned by your engrave mill tool supplier.

Figure-37. Engrave mill tools

Thread Mill

The Thread mill tool is used to generate internal or external threads in the workpiece. The most common question here is if we have Taps to create thread then why is there need of Thread mill tool. The answer is less machining time on CNC, tool cost saving, more parts per tool, and better thread finish. Now, you will ask why to use tapping. The answer is low machine cost. Figure-38 shows thread mill tools.

Figure-38. Thread Mill

Barrel Mill

Barrel Mill tool is the tool recently being highly used in machining turbine/impeller blades and other 5-axis milling operations. Barrel Mill has conical shape with radius at its end; refer to Figure-39. Note that earlier Ball mill tools were used for irregular surface contouring but Barrel Mill tools give much better surface finish so they are highly in demand for 5-axis milling now a days.

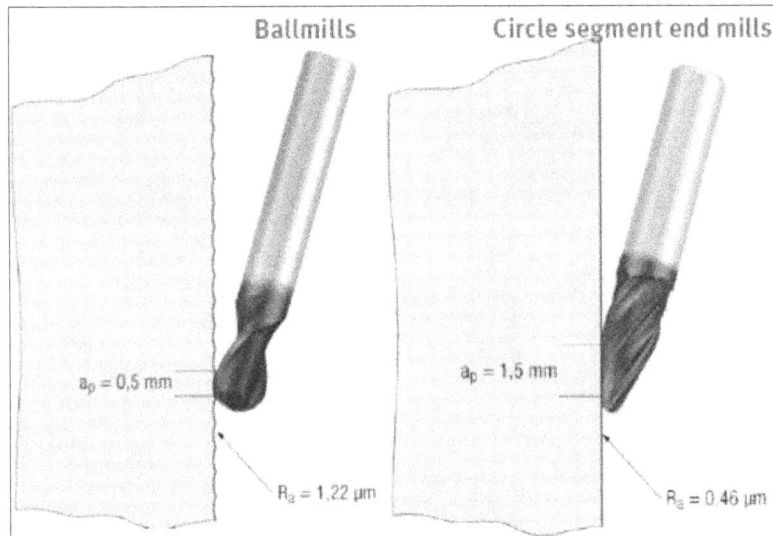

Figure-39. Barrel Mill versus Ball Mill Tool

Drill Bit

Drill bit is used to make a hole in the workpiece. The hole shape depends on the shape of drill bit. Drill bits for various purposes are shown in Figure-40. Note that drill is the machine or holder in which drill bit is installed to make cylindrical holes. There are mainly four categories of drill bit; Twist drill bit, Step drill bit, Unibit (or conical bit), and Hole Saw bit (Refer to Figure-41). Twist drill bits are used for drilling holes in wood, metal, plastic and other materials. For soft materials the point angle is 90 degree, for hard materials the point angle is 150 degree and general purpose twist drill bits have angle of 150 degree at end point. The Step drill bits are used to make counter bore or countersunk holes. The Unibits are generally used for drilling holes in sheetmetal but they can also be used for drilling plastic, plywood, aluminium and thin steel sheets. One unibit can give holes of different sizes. The Hole saw bit is used to cut a large hole from the workpiece. They remove material only from the edge of the hole, cutting out an intact disc of material, unlike many drills which remove all material in the interior of the hole. They can be used to make large holes in wood, sheet metal and other materials.

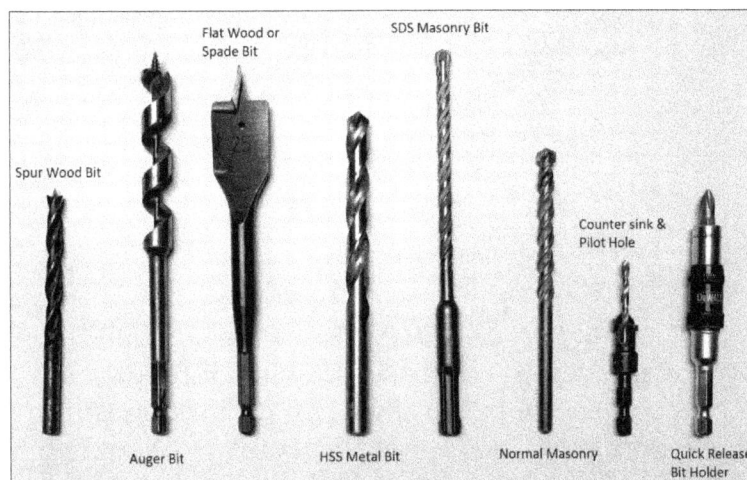

Figure-40. Drill Bits for different purposes

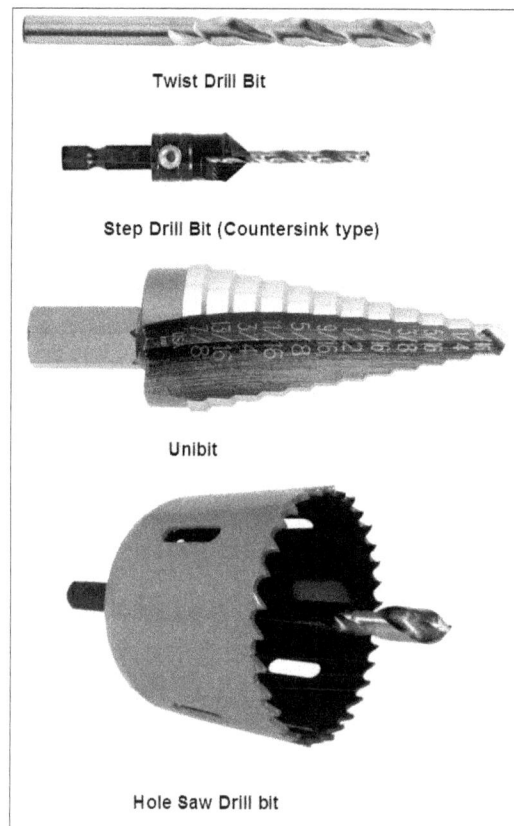

Figure-41. Types of drill bits

Reamer

Reamer is a tool similar to drill bit but its purpose is to finish the hole or increase the size of hole precisely. Figure-42 shows the shape of a reamer.

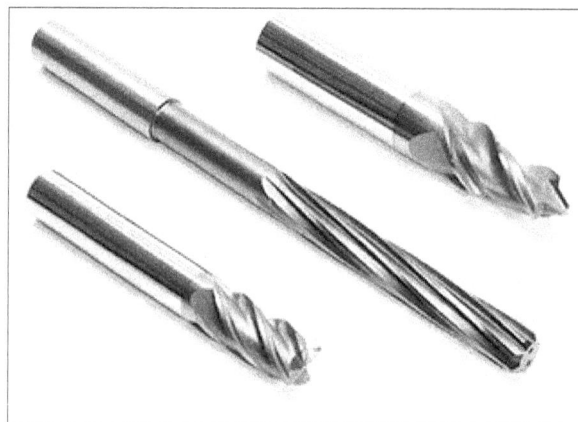

Figure-42. Reamer tool

Bore Bar

Bore Bar or Boring Bar is used to increase the size of hole; refer to Figure-43. One common question is why to use bore bar if we can perform reaming or why to perform reaming when we have bore bar. The answer is accuracy. A reamer does not give tight tolerance in location but gives good finish in hole diameter. A bore bar gives tight tolerance in location but takes more time to machine hole as compared to reamer. The decision to choose the process is on machinist. If you need a highly accurate hole then perform drilling, then boring and then reaming to get best result.

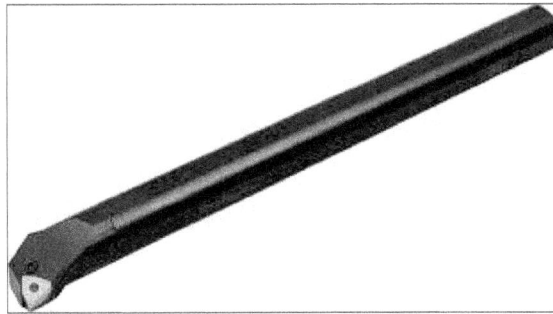

Figure-43. Boring Bar

Lathe Tools or Turning Tools

The tools used in CNC lathe machines use a different nomenclature. In CNC lathe machines, we use insert for cutting material. The Insert Holder and Inserts have a special nomenclature scheme to define their shapes. First we will discuss the nomenclature of Insert holder and then we will discuss the nomenclature of Inserts.

Insert Holders

Turning holder names follow an ISO nomenclature standard. If you are working on a CNC shop floor with lathes, knowing the ISO nomenclature is a must. The name looks complicated, but is actually very easy to interpret; refer to Figure-44.

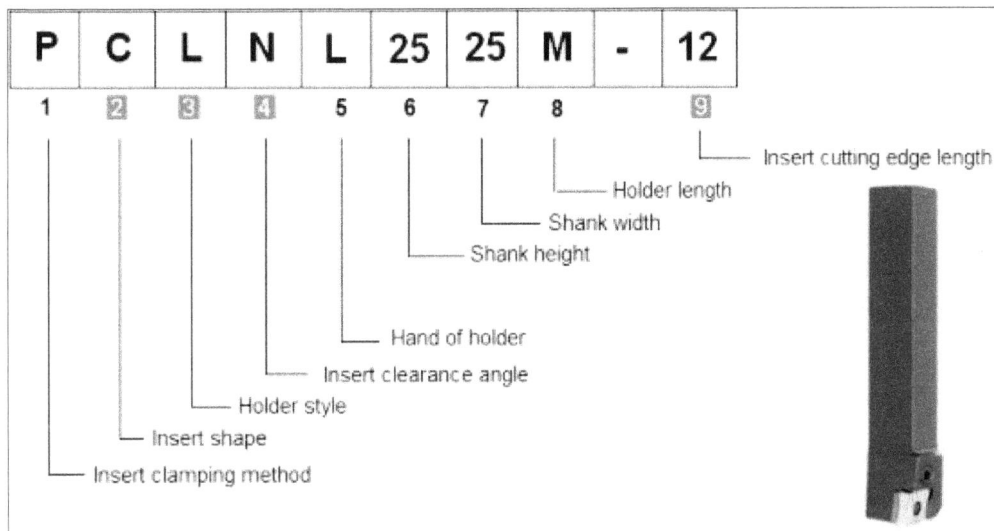

Figure-44. CNC Lathe Insert Holder nomenclature

When selecting a holder for an application, you mainly have to concentrate on the numbers marked in red in the above nomenclature. The others are decided automatically (e.g., the shank width and height are decided by the machine), or require less effort. In fFigure-45, the rows with the question mark indicate the parameters that require the decision by machinist based on job.

	Parameter		How is this decided ?
1	Insert clamping method		Select based on cutting forces. Top clamping is the most sturdy, screw clamping the least.
2	Insert shape	?	Decided by the contour that you want to turn.
3	Holder style	?	Decided by the contour that you want to turn.
4	Insert clearance angle	?	Positive / Negative, based on application.
5	Hand of holder		Decided based on whether you want to cut towards the chuck or away from the chuck, and on turret position - turret front / rear
6	Shank height		Decided by holder size.
7	Shank width		Decided by machine.
8	Holder length		Decided by machine.
9	Insert cutting edge length	?	Decide based on depth of cut you want to use.

Figure-45. CNC Lathe Insert Holder nomenclature parameters

Figure-46 and Figure-47 show the options available for each of the parameters.

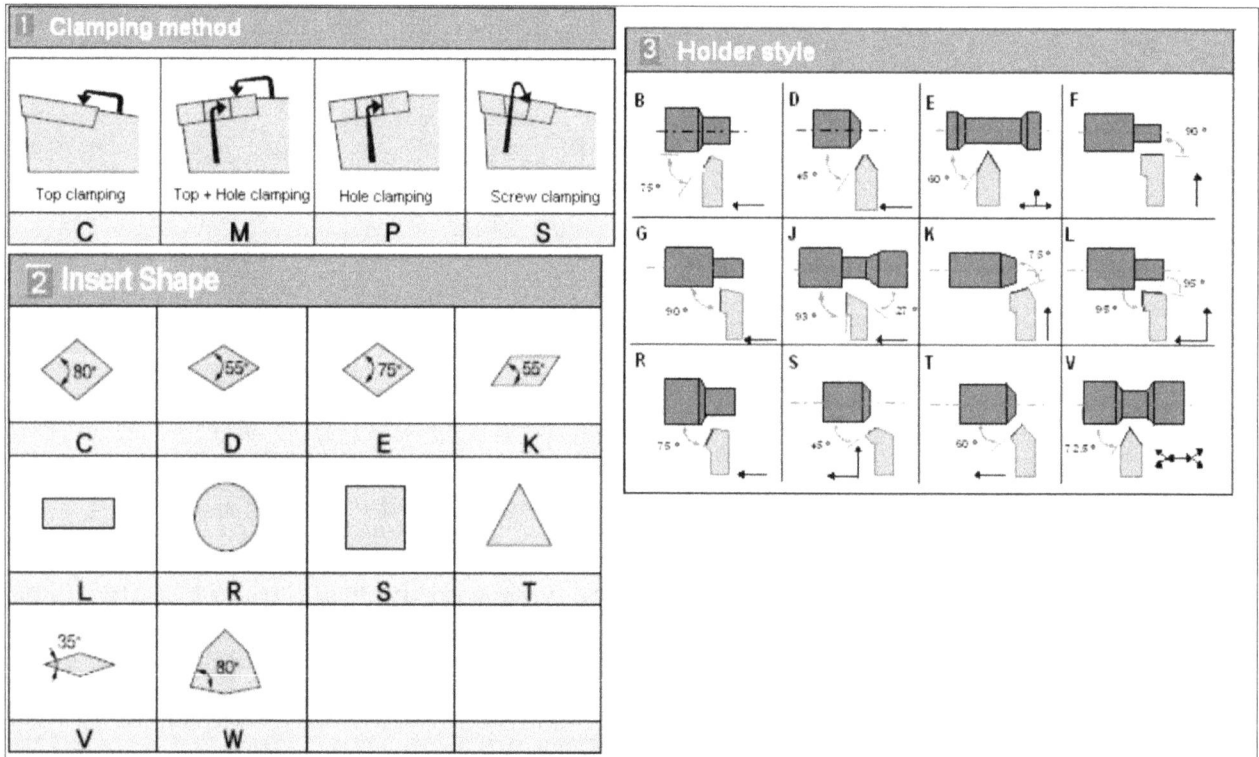

Figure–46. Clamping Method, Insert Shapes, and Holder Style

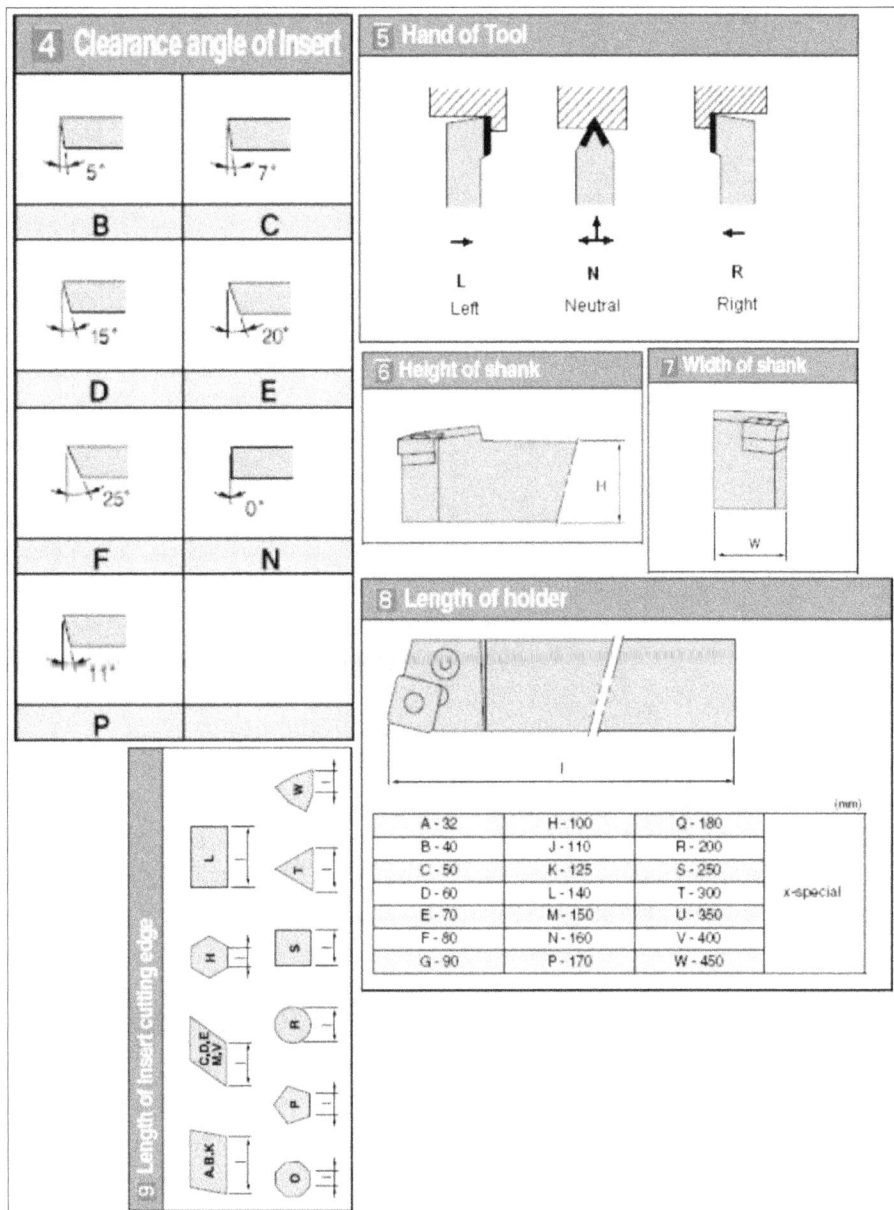

Figure-47. Insert Holder Parameters

CNC Lathe Insert Nomenclature

General CNC Insert name is given as

Meaning of each box in nomenclature is given next.

1 = Turning Insert Shape

The first letter in general turning insert nomenclature tells us about the general turning insert shape, turning inserts shape codes are like C, D, K, R, S, T, V, W. Most of these codes surely express the turning insert shape like

C = C Shape Turning Insert
D = D Shape Turning Insert
K = K Shape Turning Insert

R = Round Turning Insert
S = Square Turning Insert
T = Triangle Turning Insert
V = V Shape Turning Insert
W = W Shape Turning Insert

Figure-48 shows the turning inserts shapes.

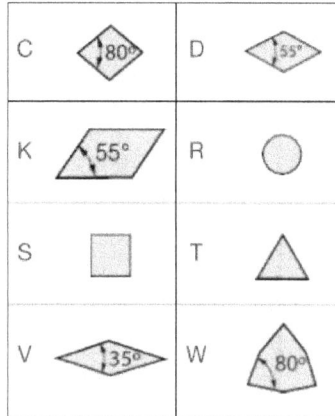

Figure-48. Turning Insert Shapes

The general turning insert shape play a very important role when we choose an insert for machining. Not every turning insert with one shape can be replaced with the other for a machining operation. As C, D, W type turning inserts are normally used for roughing or rough machining.

2 = Turning Insert Clearance Angle

The second letter in general turning insert nomenclature tells us about the turning insert clearance angle.

The clearance angle for a turning insert is shown in Figure-49.

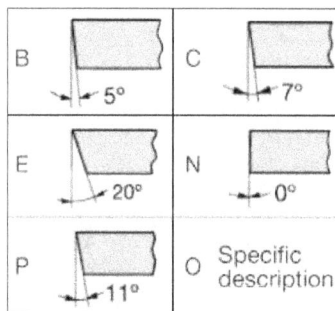

Figure-49. Turning insert clearance angle

Turning insert clearance angle plays a big role while choosing an insert for internal machining or boring small components, because if not properly chosen the insert bottom corner might rub with the component which will give poor machining. On the other hand a turning insert with 0° clearance angle is mostly used for rough machining.

3 = Turning Insert Tolerances

The third letter of general turning insert nomenclature tells us about the turning insert tolerances. Figure-50 shows the tolerance chart.

Code Letter	Cornerpoint (inches)	Thickness (inches)	Inscribed Circle (in)	Cornerpoint (mm)	Thickness (mm)	Inscribed Circle (mm)
A	.0002"	.001"	.001"	.005mm	.025mm	.025mm
C	.0005"	.001"	.001"	.013mm	.025mm	.025mm
E	.001"	.001"	.001"	.025mm	.025mm	.025mm
F	.0002"	.001"	.0005"	.005mm	.025mm	.013mm
G	.001"	.005"	.001"	.025mm	.13mm	.025mm
H	.0005"	.001"	.0005"	.013mm	.025mm	.013mm
J	.002"	.001"	.002-.005"	.005mm	.025mm	.05-.13mm
K	.0005"	.001"	.002-.005"	.013mm	.025mm	.05-.13mm
L	.001"	.001"	.002-.005"	.025mm	.025mm	.05-.13mm
M	.002-.005"	.005"	.002-.005"	.05-.13mm	.13mm	.05-.15mm
U	.005-.012"	.005"	.005-.010"	.06-.25mm	.13mm	.08-.25mm

Figure-50. Insert tolerance chart

4 = Turning Insert Type

The fourth letter of general turning insert nomenclature tells us about the turning insert hole shape and chip breaker type; refer to Figure-51.

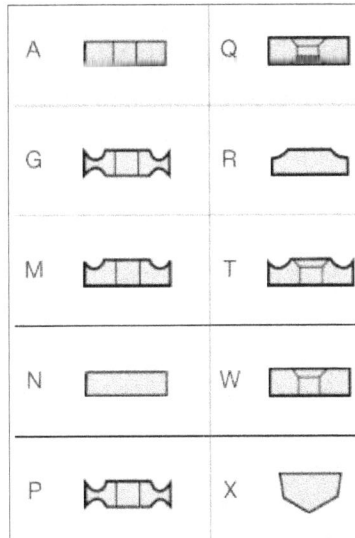

Figure-51. Turning Insert hole shape and chip breaker

5 = Turning Insert Size

This numeric value of general turning insert tells us the cutting edge length of the turning insert; refer to Figure-52.

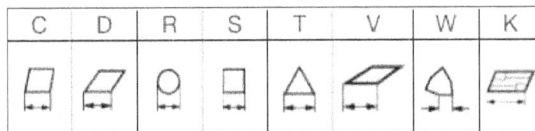

Figure-52. Turning Insert Cutting Edge Length

6 = Turning Insert Thickness

This numeric value of general turning insert tells us about the thickness of the turning insert.

7 = Turning Insert Nose Radius

This numeric value of general turning insert tells us about the nose radius of the turning insert.

Code	=	Radius Value
04	=	0.4
08	=	0.8
12	=	1.2
16	=	1.6

You can know more about tooling from your tool supplier manual.

EDITING POST-PROCESSORS

Editing a post-processor is a job which requires good attention. Adding a code for wrong functionality can cause great losses. The procedure to edit a post processor of MasterCAM is given next.

- Open the directory of post-processors. Generally, it is *C:\Users\Public\Documents\ Shared Mastercam 2022\mill\Posts*.
- Double-click on the post-processor that you want to edit and open it with any word processor application (WordPad, MS Word). You can right-click on the PST file and open it with WordPad. The post-processor will open in **WordPad** or other selected application; refer to Figure-54.

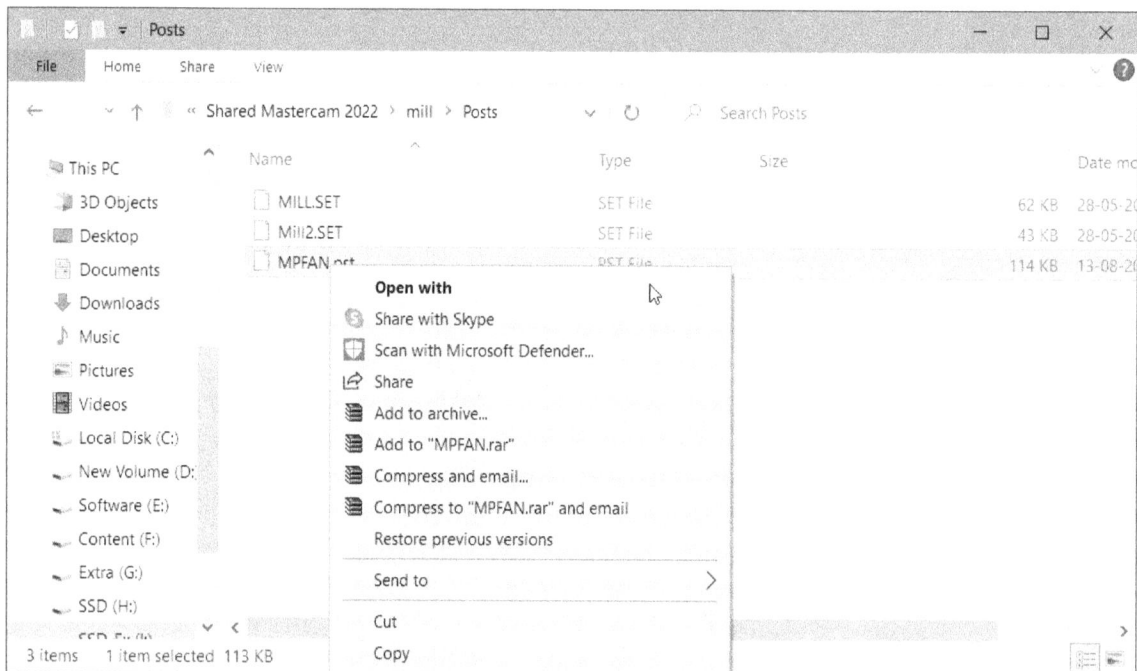

Figure-53. Opening post processor in WordPad

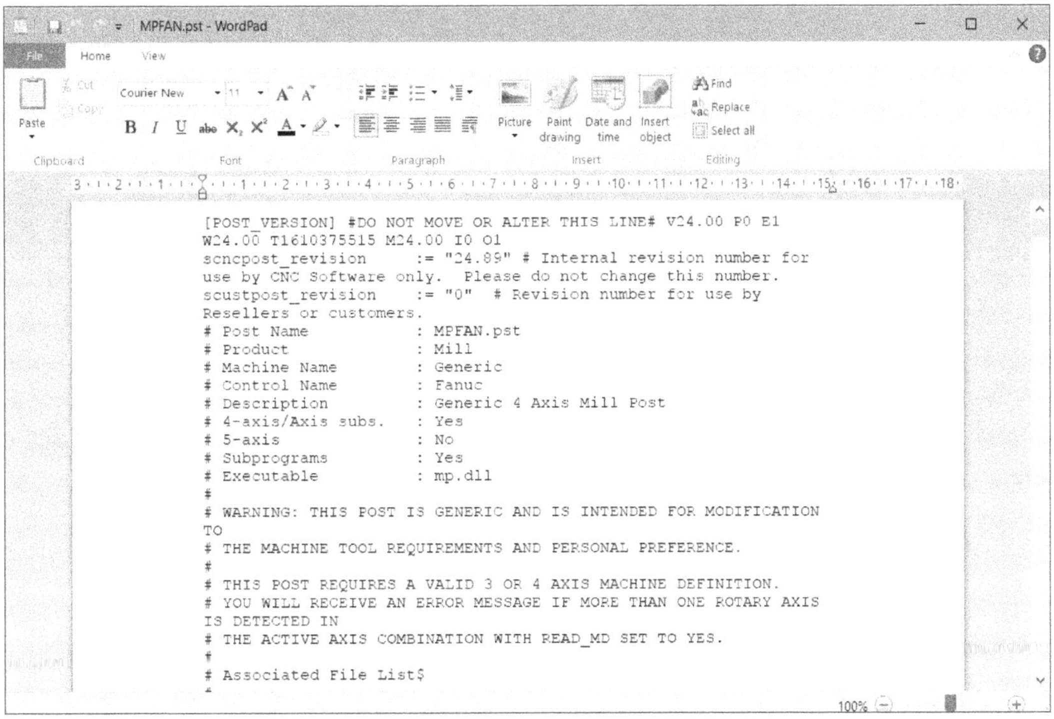

Figure-54. Post processor in WordPad

- Set desired values and click on the **Save** button to save the file as per your requirement.
- Click on the **Close** button to exit the application.

SELF-ASSESSMENT

Q1. When performing boring operation, select a spindle that has a nose, bearing ID larger than the bore diameter being machined. (T/F)

Q2. The tool indexer or tool turret is used to automatically change the cutting tool in spindle. (T/F)

Q3. .... limit of CNC machine is used to define maximum size of workpiece that can be machined on the table.

Q4. Tail stock cannot be installed in Milling machine because this is part of lathe machines only. (T/F)

Q5. Which of the following axes is by default rotational axis of milling machine table?

a. A axis   b. X axis
c. D axis   d. Z axis

Q6. Subprograms are small repeating code blocks which are generally called by reference name in main program. (T/F)

Q7. The parameters specified in **Machine Dynamics** tab of **General Machine Parameters** dialog box are used by .... toolpaths for dynamic cutting operations.

Q8. Which of the following milling tools can be used to apply round on sharp edges of the workpiece?

a. Flat end mill   b. Ball Nose end mill
c. Taper end mill   d. Face mill

Q9. Which of the following setups is optimum strategy for machining pocket walls of a mould used for casting? (Draft is applied to walls in mould for easy ejection.)

a. Using ball end mill with toolpath adjustments for machining walls.
b. Using taper end mill with straight line toolpaths for machining walls.
c. Using flat end mill with tilted tool axis to machine walls.
d. Using bull nose end mill with toolpath adjustments for machining walls.

Q10. Which of the following tools is also called Undercut mill tool?

a. Dovetail milling cutter   b. Lollipop Mill
c. Slot mill   d. Barrel Mill

Q11. Why barrel mills are better than ball end mill in 5 axis milling?

Chapter 4

Lathe Machining and Toolpaths

Topics Covered

The major topics covered in this chapter are:

- *Introduction to Toolpaths*
- *Lathe Toolpaths*
- *Loading Lathe Machines*
- *Loading Lathe Tools*
- *Performing Lathe Operation*
- *Generating NC from Lathe Operations*

INTRODUCTION TO TOOLPATH

As the name suggests, toolpath is the path followed by tool while performing a machining operation. MasterCAM categories toolpaths in following ways:

- Lathe Toolpaths
- Point/Circle Toolpaths
- 2D Toolpaths
- 2D High Speed Toolpaths
- Feature-Based Toolpaths
- Multi-Axis Toolpaths
- Finishing Toolpaths
- Roughing Toolpaths
- 3D HighSpeed Toolpaths

The toolpaths in the above categories are discussed next with their respective tools.

LATHE MACHINE SETUP

Lathe toolpaths are meant for machining a work piece on lathe or CNC turning machine. By default, the Lathe Toolpaths are deactivated. To activate these tool paths, you need to select a lathe machine. The procedure is given next.

Adding Lathe Machine in setup

- Click on the **Tools > Mastercam2022 > Lathe Machines > Manage List** tool from the Menu; refer to Figure-1. The **Machine Definition Menu Management** dialog box will be displayed; refer to Figure-2.

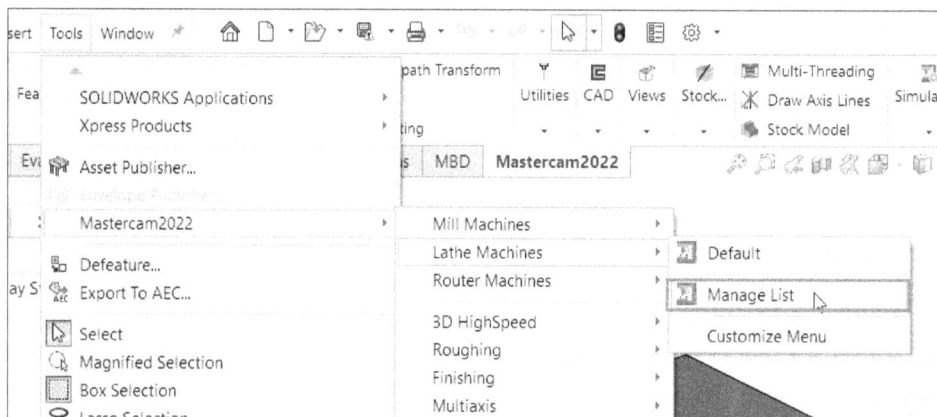

Figure-1. Manage List tool for Lathe Machine

- Select the machines from the left box and click on the **Add** button to add them in the list.
- Click on the **OK** button from the message box; refer to Figure-3.

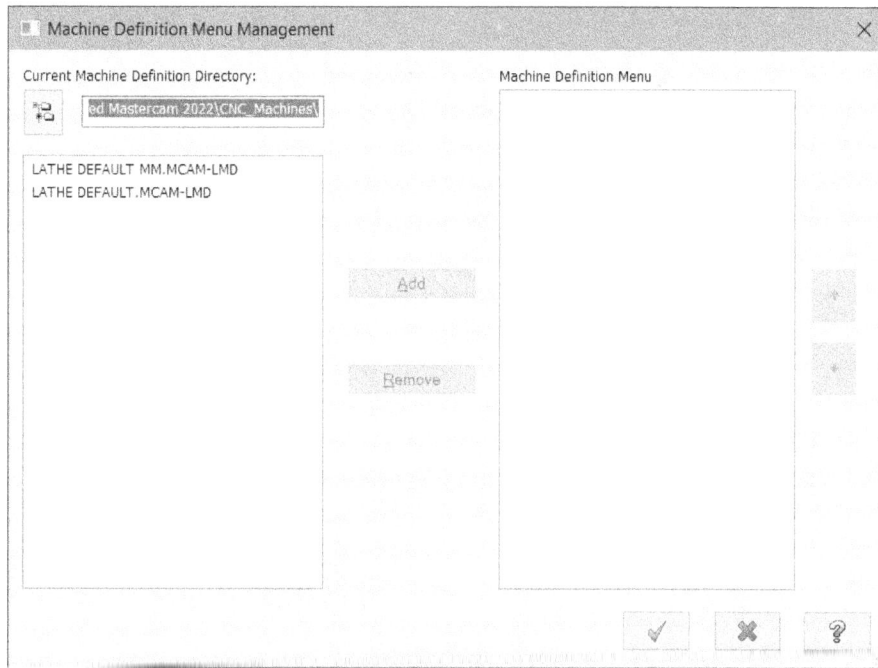

Figure-2. Machine Definition Menu Management dialog box

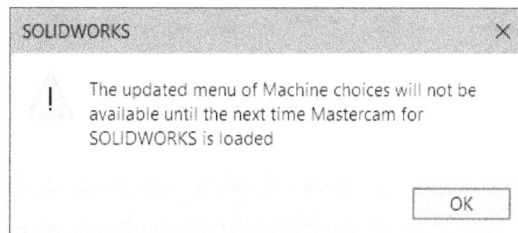

Figure-3. SolidWorks dialog box

- Restart Mastercam2022 add-in after closing it from Add-Ins box.
- Open the **Mastercam2022** menu again and hover the cursor on **Lathe Machines** option. A list of machines newly added will be displayed.
- Select desired machine from the list. The machine will added as active machine group in **Mastercam Toolpath Manager** and toolpaths in **Lathe** drop-down of **Ribbon** will become active; refer to Figure-4.

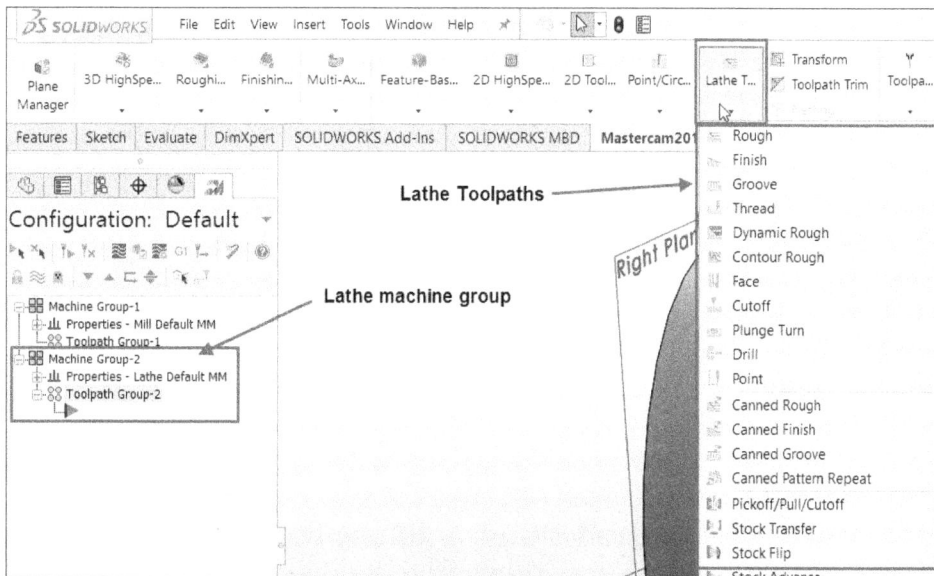

Figure-4. Machines in Mastercam Toolpath Manager

- The red triangle indicates machine active for generating toolpaths. You can drag the red arrow to make another machine active; refer to Figure-5.

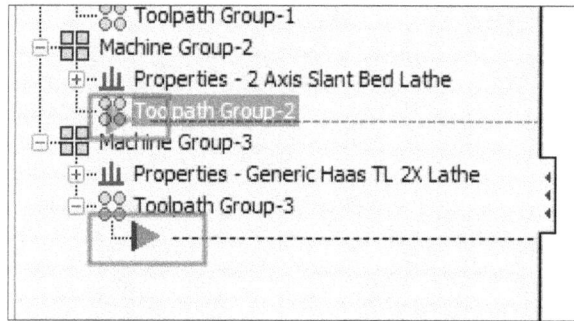

Figure-5. Active machine

WORKPIECE SETUP ON LATHE

- Expand the **Properties** node of Lathe machine group and click on the **Tool settings** option from the **Mastercam Toolpath Manager**. The **Machine Group Properties** dialog box for lathe machine will be displayed; refer to Figure-6.

Figure-6. Machine Group Properties dialog box for lathe

Tool Settings

- Enter desired program number in the **Program #** edit box. This program number is for your reference.

Feed Calculation Options

- Set desired feed rate in the **Feed Calculation** area of the dialog box. Select the **From Tool** or **From Material** radio button to use the parameters specified in tool definition or material definition, respectively. Select the **From Defaults** radio button if you want to use the parameters specified in the machine definition. You want to specify the feed parameters then select the **User Defined** radio button and specify desired parameters in the edit boxes displayed below the radio button. (Note: **Spindle speed** is the rotation speed of spindle in RPM. **Feed rate** is the movement of tool while cutting in X, Y, or Z direction and is specified in inch per minute or mm per minute unit depending on the machine definition selected. **Retract rate** is the rate at which tool exits cutting and moves to the starting point of the next cutting path or moves for non-cutting operation. This parameter is also specified in inch per minute or mm per minute unit depending on the machine definition selected. **Plunge rate** is the rate at which tool perpendicularly enters in the workpiece along Z axis in case of lathe. The unit for this parameter is also inch per minute or mm per minute depending on the machine selected.

- Select the **Adjust feed on arc move** check box if you want to set the minimum feed rate for arc moves and specify desired value in the edit box below it.

Tool Clearance Options

- In the **Tool Clearance** area, specify the rapid feed rate and **Entry/Exit** tool movement speed in multiple of feed rate in the respective edit boxes.

Material Options

- Click on the **Select** button from the **Material** area of the **Machine Group Properties** dialog box to select a material. The **Material List** dialog box will be displayed as discussed in previous chapter.
- Select desired material and click on the **OK** button.
- To edit the properties of material related to lathe machining, click on the **Lathe Edit** button from the dialog box. The **Lathe Material Definition** dialog box will be displayed; refer to Figure-7.

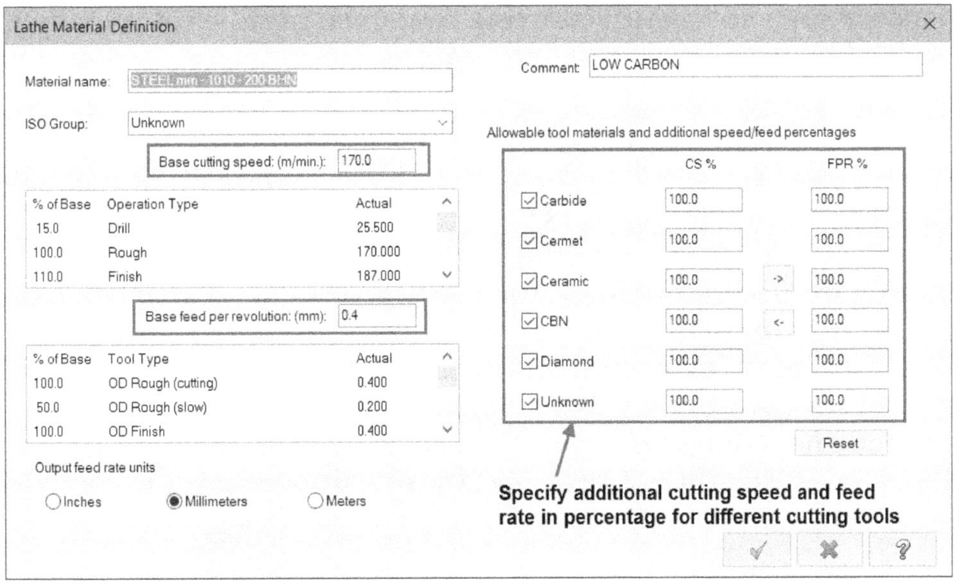

Figure-7. Lathe Material Definition dialog box

- Set desired base cutting speed and feed rate in the respective edit box of the dialog box. These parameters will be used while machining if you have selected the **From material** radio button in the **Feed Calculation** area of the **Machine Group Properties** dialog box.
- You can specify additional feed rate and cutting rates in percentage for different cutting tools in the **Allowable tool materials and additional speed/feed**

percentages area of the dialog box. Click on the **OK** button from the dialog box to apply the changes.

Toolpath Configuration Options

The options in the **Toolpath Configuration** area of the dialog box are used to set the parameters related toolpaths. These options are discussed next.

* Select the **Assign tool numbers sequentially** check box to assign tool numbers sequentially to the tools while generating various toolpaths. If this check box is deselected then the tool numbers manually assigned by you in **Tool Manager** will be used.
* Select the **Warn of duplication tool numbers** check box to display warning when same tool number is being used for two different tools.
* Select the **Warn on mill tool orientation conflict** check box to display warning if milling tool is not in correct orientation with respect to work piece while cutting material.
* Select the **Use tool's step, peck, coolant** check box to use tools parameters for tool step, peck and coolant control.
* Select the **Search tool library when entering tool number** check box to ask system to check tool library for matching tools with specified tool numbers.

Advanced options

* Select the **Override defaults with modal values** check box to override the clearance height, retract height and feed plane value specified in the toolpaths. Select the respective check boxes to apply the values specified in the model definition.

In the **Sequence #** area specify the starting sequence number of increment value for two consecutive line numbers on NC program.

Stock Setup on Lathe

* Click on the **Stock Setup** tab in the dialog box. The options in the dialog box will be displayed as shown in Figure-8.
* Select desired stock plane using the button in **Stock Plane** area.
* Select the spindle orientation from the **Stock** area of the dialog box using the **Left Spindle** or **Right Spindle** radio button.
* Click on the **Properties** button from the **Stock** area. The **Machine Component Manager - Stock** dialog box will be displayed as shown in Figure-9.
* Select desired shape for stock from the **Geometry** drop-down in the **Geometry** tab of the dialog box. Most of the time, **Cylinder** will be the right choice for turning machine.

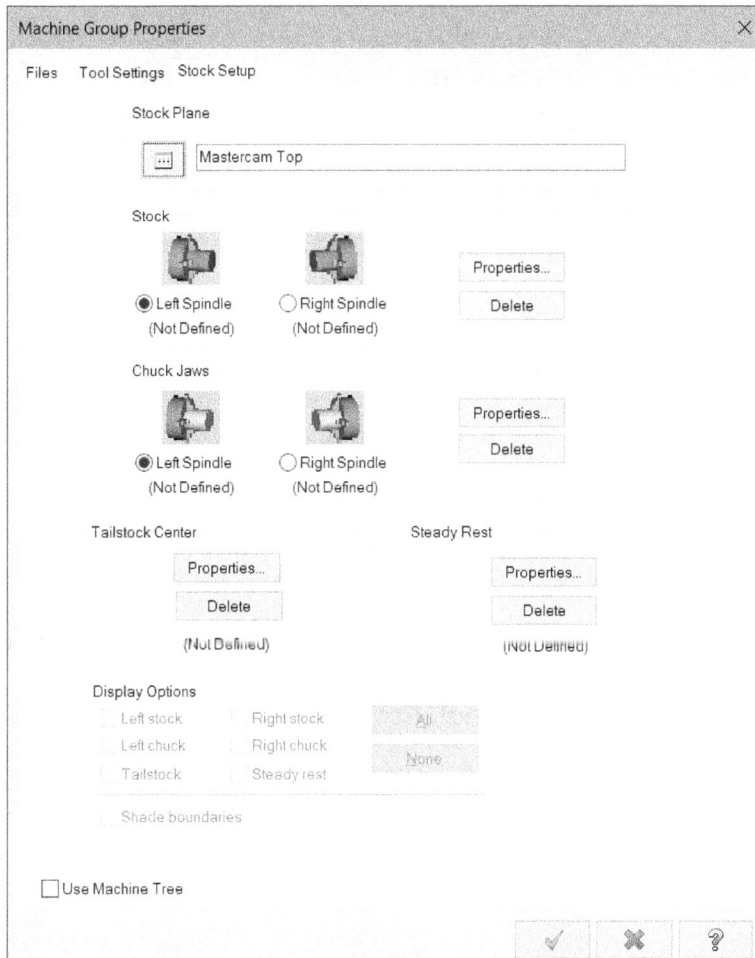

Figure-8. Stock Setup tab in Machine Group Properties dialog box

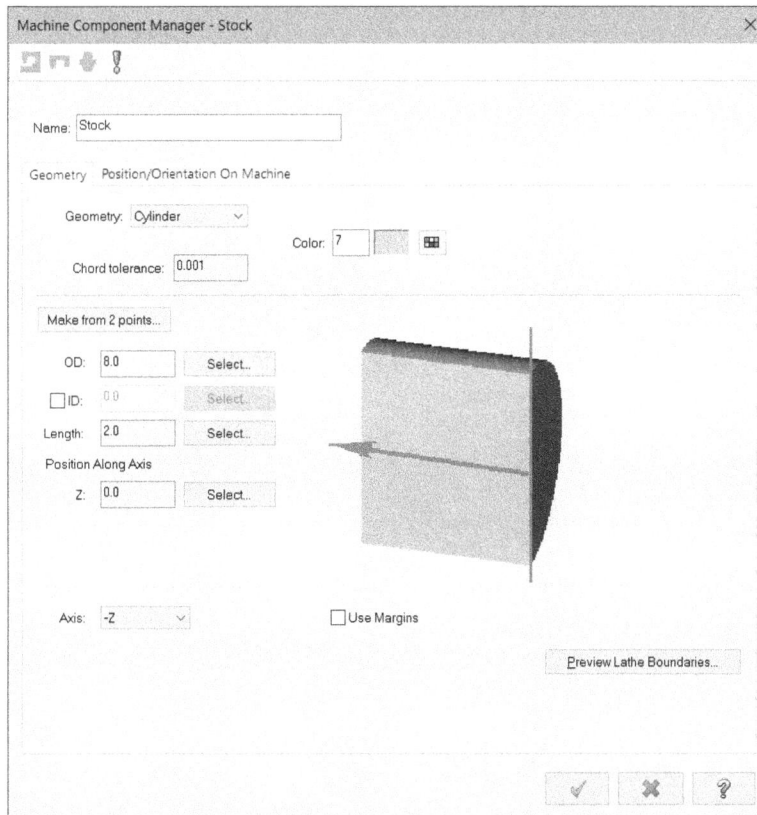

Figure-9. Machine Component Manager-Stock dialog box

- Click on the **Select** button from the next of **OD** edit box in the dialog box. The **Selection PropertyManager** will be displayed and you will be asked to select a reference to define outer diameter of stock. Select circular face or edge of the part whose stock is to be created; refer to Figure-10. Click on the **OK** button from the **PropertyManager**. The value of OD will be displayed in the **OD** edit box. Modify the value if the OD of stock piece is different.

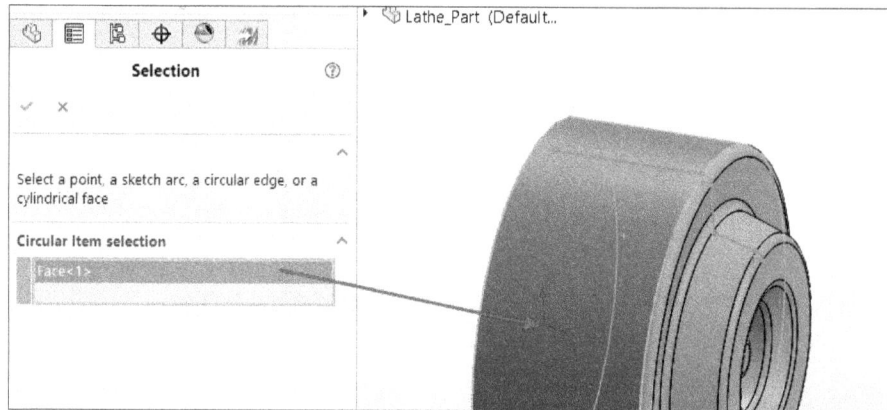

Figure-10. Round face selected for OD

- If your workpiece is hollow then select the **ID** check box and specify the internal diameter value in the same way.
- Click on the **Select** button next to **Length** edit box to define length of the stock. The **Selection PropertyManager** will be displayed and you will be asked to select two points to define length of stock. Select points or edges of the model to define length; refer to Figure-11 and click on the **OK** button from the **PropertyManager**.

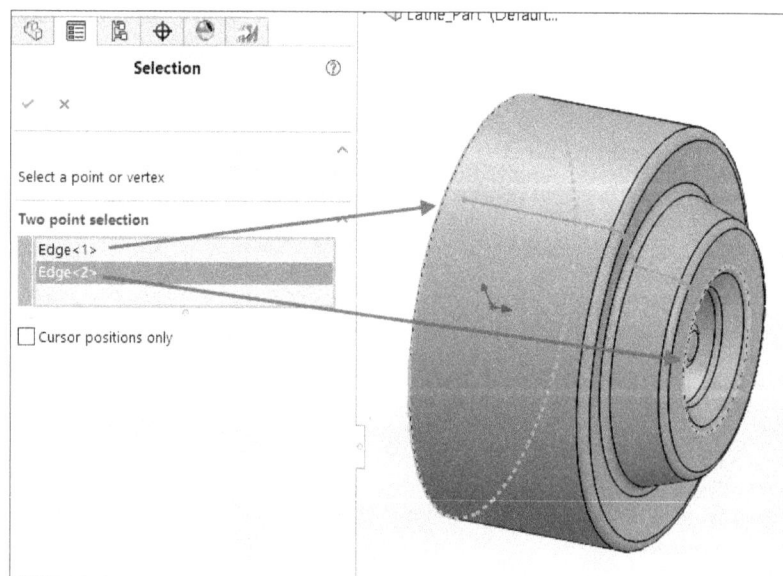

Figure-11. Edges selected to define length

- If you get the preview of stock in opposite direction then select the **+Z** option from the **Axis** drop-down; refer to Figure-12.

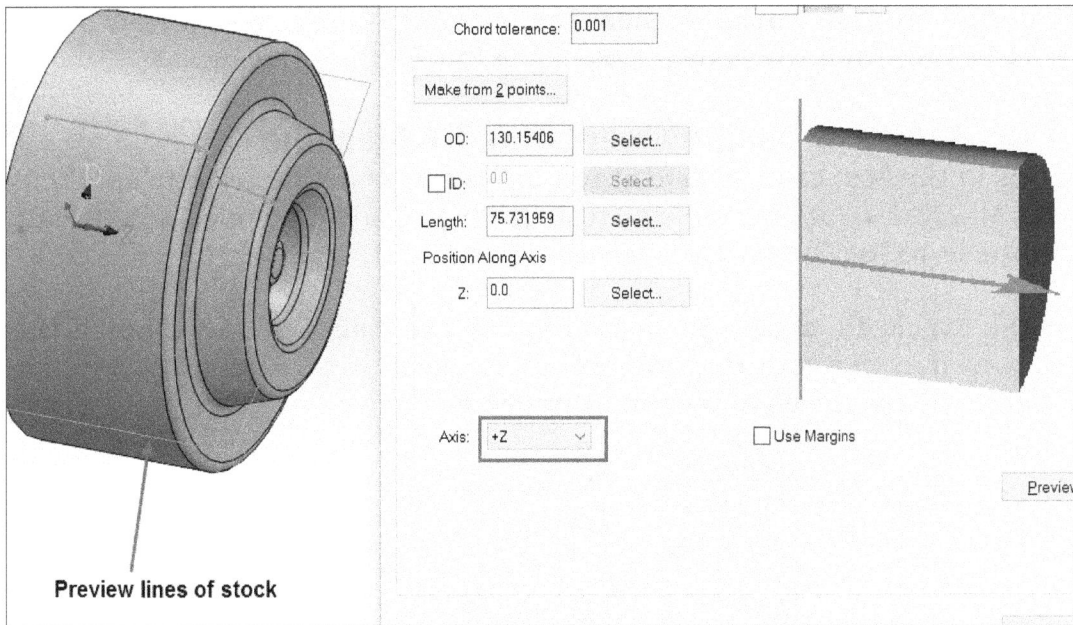

Figure-12. Axis option for stock direction

- Since we place part in chuck so some portion of part will not be machined in first side turning. To tell machine that 'x' length of stock will be held by chuck you need to specify the value in **Z** edit box of **Position Along Axis** area of the dialog box. You can also use the **Select** button next to **Z** edit box to specify the value.

Use Margins options

The **Use Margins** options are applicable when you have set the OD, ID and Length values exactly matching to the finished part. In that case, you can specify the extra material to be removed from stock by using the margin values. These margin values give dimensions of stock.

- Select the **Use Margins** check box from the dialog box. The options for specifying margins will be displayed in the dialog box.
- Specify desired values of margins, preview of the stock will be modified accordingly; refer to Figure-13.

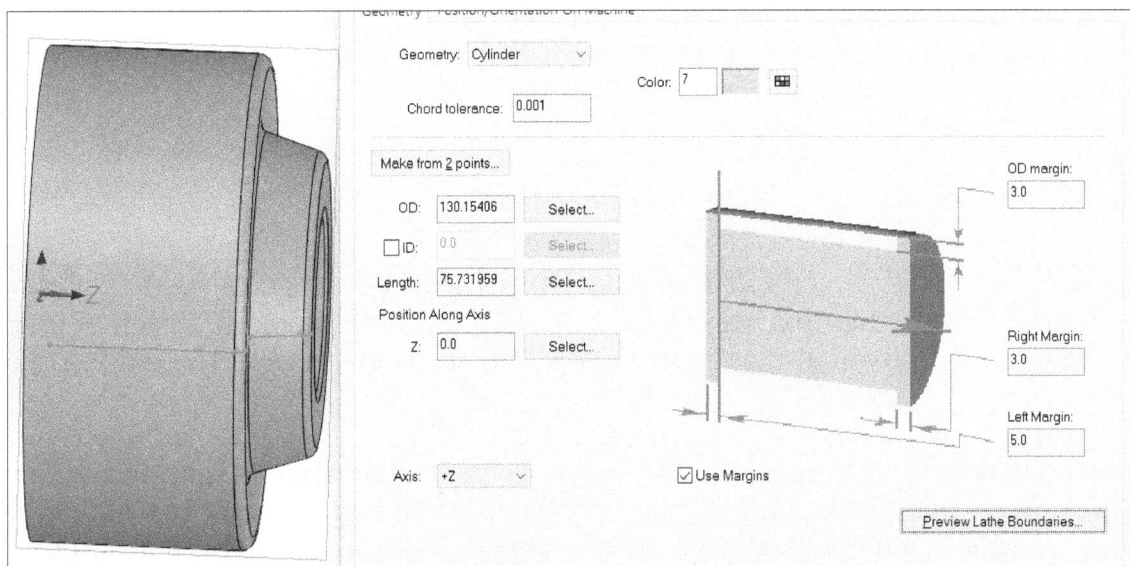

Figure-13. Specifying margin values

• Click on the **Preview Lathe Boundaries** button to check the preview of stock and click **OK** button from the **PropertyManager** to exit the preview.

Position/Orientation On Machine

The options in the **Position/Orientation On Machine** tab are used to set the WCS on Stock and Machine. Note that all the dimensions in the NC codes will be calculated based on these WCS. The options are discussed next.

• Clear the **Stock is drawn in position on the machine** check box if the origin is not placed as per your requirement in stock preview. The options in the dialog box will be displayed as shown in Figure-14.

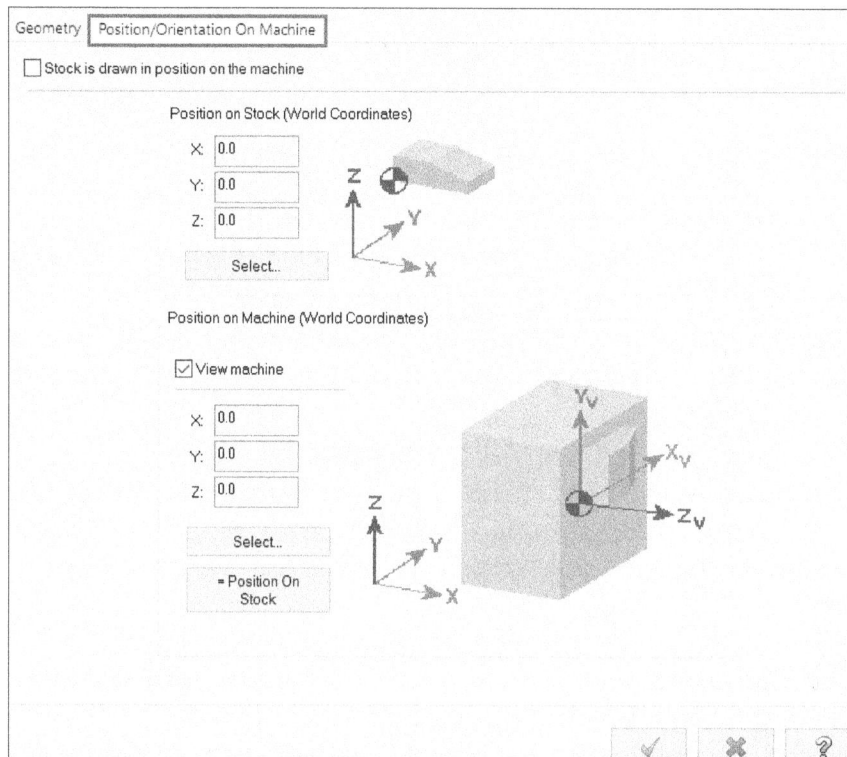

Figure-14. Position Orientation On Machine tab in dialog box

• Click on the **Select** button in the **Position on Stock (World Coordinates)** area of the dialog box. The **Selection PropertyManager** will be displayed and you will be asked to specify the position of WCS for stock.
• Select the point to be used as stock origin.
• Similarly, you can set WCS for machine.
• Click on the **OK** button to apply the parameters.

Chuck Jaw Positioning

After creating stock, the next step is to properly place the stock in chuck of lathe as it is to be done on physical machine. Follow the steps given next to position part in chuck.

• Select desired orientation of chuck using the **Left Spindle** or **Right Spindle** radio button from the **Chuck Jaws** area of the **Machine Group Properties** dialog box.
• Click on the **Properties** button in the **Chuck Jaws** area of the dialog box. The **Machine Component Manager - Chuck Jaws** dialog box will be displayed; refer to Figure-15.

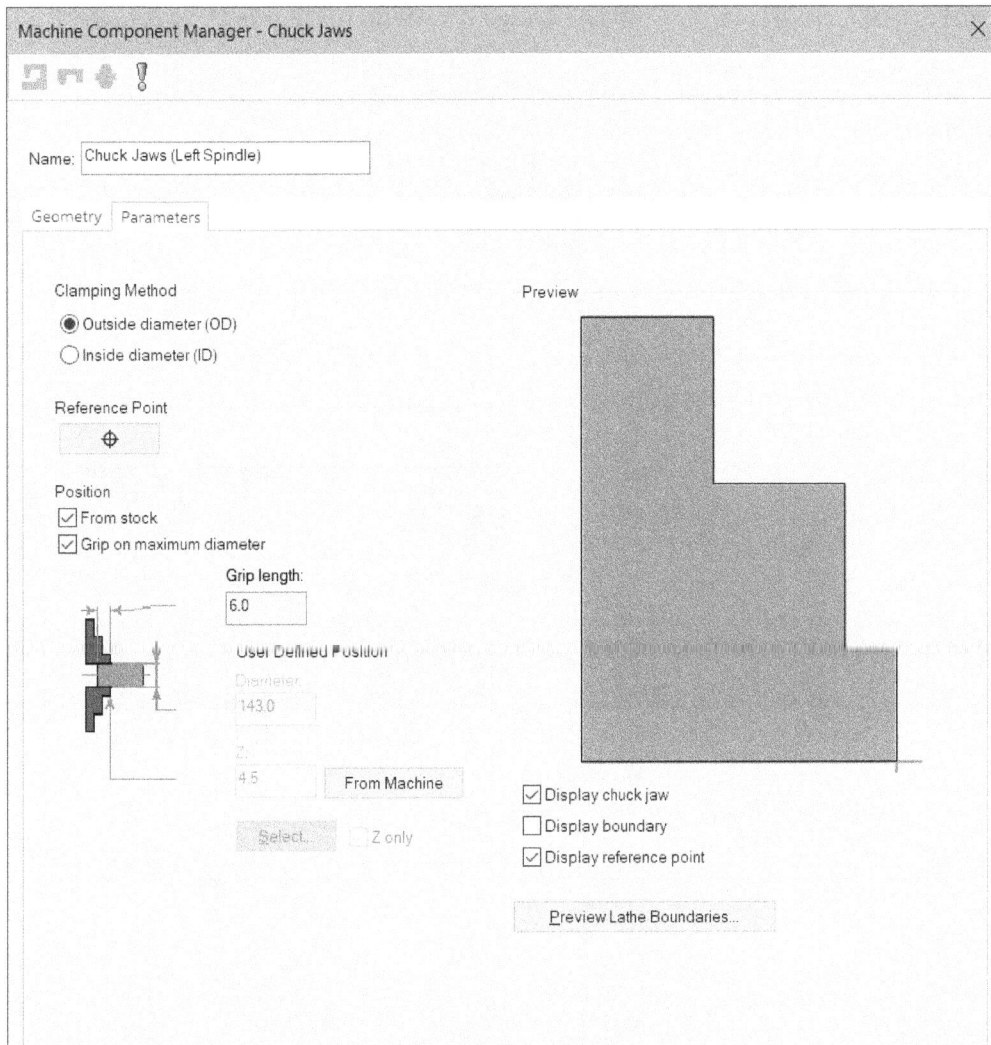

Figure-15. Machine Component Manager-Chuck Jaws dialog box

- Select desired radio button from the **Clamping method** area in the dialog box.
- Since we have setup the stock definition so select the **From stock** check box in the **Position** area of the dialog box to automatically position the chuck jaws on stock. Preview of chuck jaws will be displayed in graphics area; refer to Figure-16.

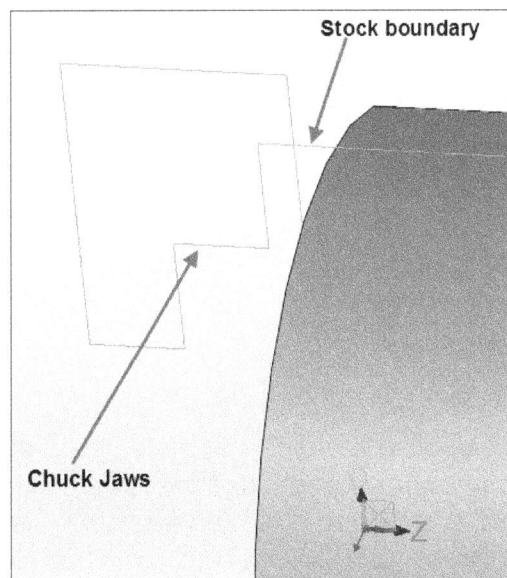

Figure-16. Chuck Jaws preview

- Set desired grip length in the **Grip length** edit box of **Position** area in the dialog box.
- Set the other parameters of chuck jaws as required and then click on the **OK** button from the dialog box.

Tail Stock Positioning

Tail stock is generally used to support long cylindrical parts that can bend during turning operation. The procedure to add tail stock in machining setup is given next.

- Click on the **Properties** button from the **Tailstock Center** area of the dialog box. The **Machine Component Manager - Center** dialog box will be displayed; refer to Figure-17.
- By default, the **Parametric** option is selected in the **Geometry** drop-down of the dialog box. Select the **Cylinder** or **Solid entity** option if you want different geometry of tail stock. The options for these parameters have already been discussed. (We will continue with **Parametric** option in the drop-down.)
- Specify the parameters of tail stock in the edit boxes in dialog box.
- Click on the **From Stock** button to automatically place the tail stock at the end of stock or specify desired position in the **Position Along Axis** edit box.
- Click on the **Preview Lathe Boundaries** button to check the preview of tail stock.
- Click on the **OK** button to apply the settings.

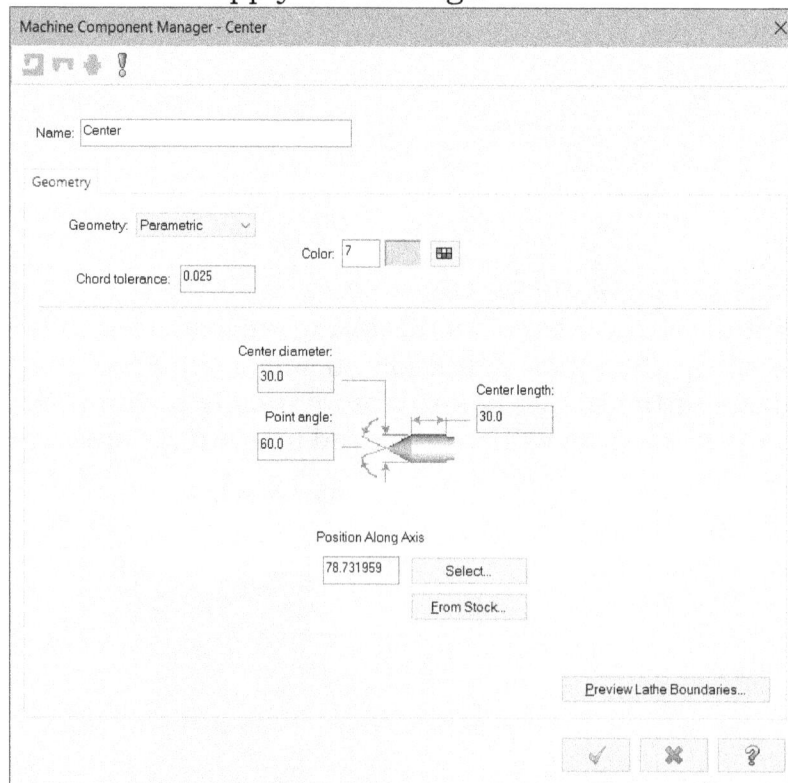

Figure-17. Machine Component Manager-Center dialog box

Steady Rest Positioning

- Click on the **Properties** button from the **Steady Rest** area of the dialog box. The **Machine Component Manager - Steady Rest** dialog box will be displayed; refer to Figure-18. Note that you must have a closed sketch in the graphics area to define shape of steady rest before using this option; refer to Figure-19.

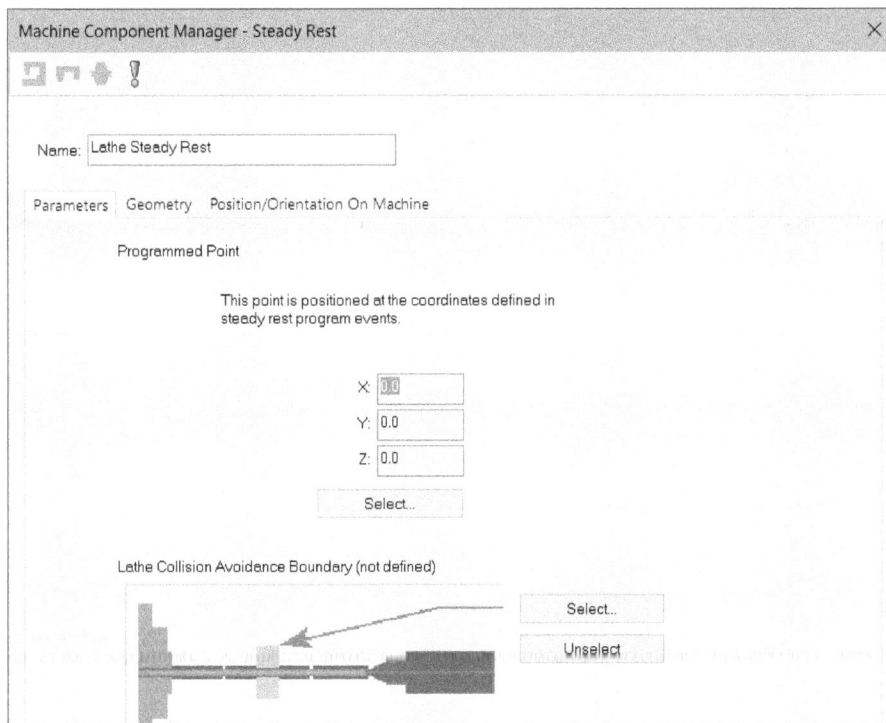

Figure-18. Machine Component Manager–Steady Rest dialog box

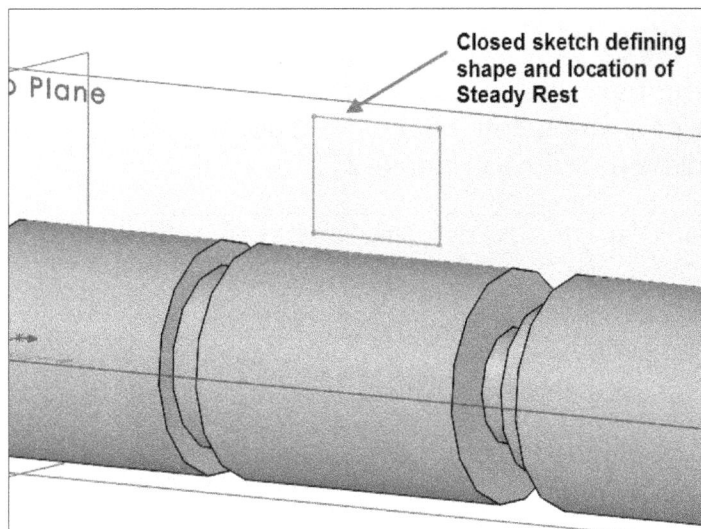

Figure-19. Sketch created for Steady Rest

- Click on the **Select** button from the **Lathe Collision Avoidance Boundary** area of the dialog box. The **Chain Manager** will be displayed and you will be asked to select the sketch defining boundary of Steady Rest.
- Select the sketch and click on the **OK** button from the **Chain Manager**. Click on the **OK** button from the **Machine Component Manager** dialog box. Preview of the Steady Rest component will be displayed; refer to Figure-20. Note that we have selected the **Shade boundaries** check box in the **Machine Group Properties** dialog box so that all the components are displayed as shaded.

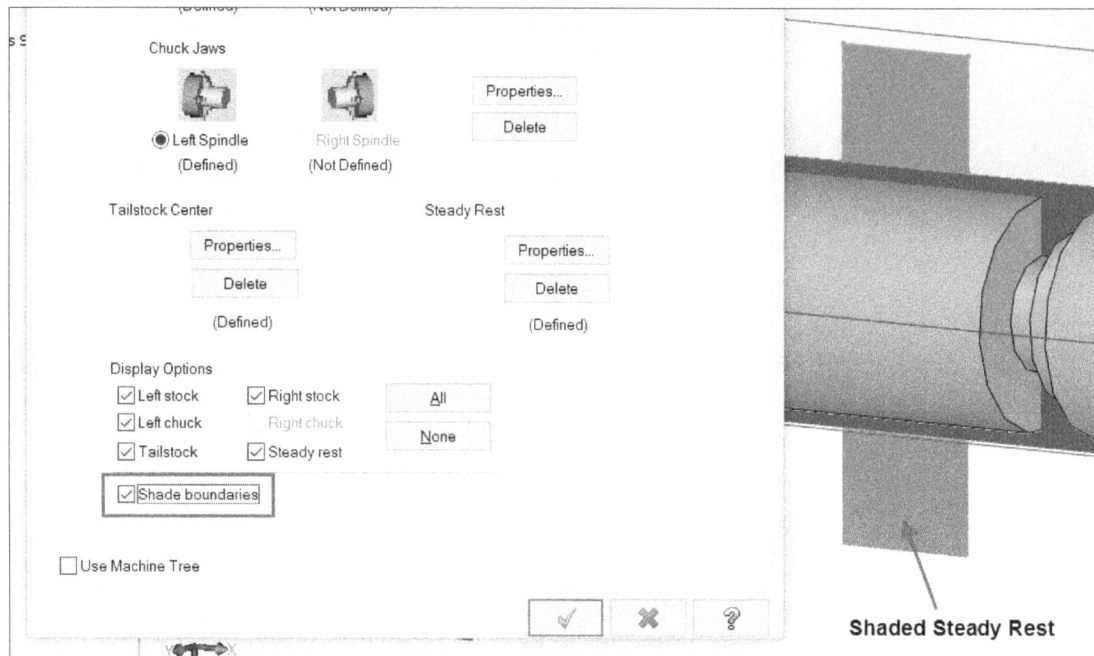

Figure-20. Preview of Steady Rest

• Click on the **OK** button from the **Machine Group Properties** dialog box to apply all the parameters.

CREATING ROUGH TOOLPATH ON LATHE

• Create stock of the part as discussed earlier.
• Click on the down arrow below **Lathe Toolpaths** button in the **Ribbon**. A list of tools will be displayed; refer to Figure-21.

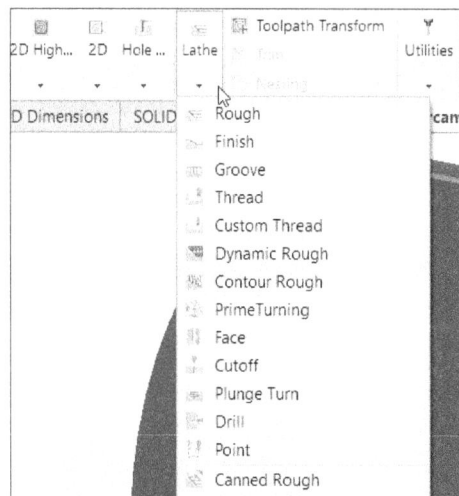

Figure-21. List of Lathe Toolpaths

• Click on the **Rough** tool from the menu. The **Chain Manager** will be displayed; refer to Figure-22. Also, you are asked to select the faces/edges/sketches for machining by roughing operation.
• You can create sketch of the trajectory for making toolpath; refer to Figure-23 or you can directly select the faces to be machined.
• Select the sketch for tool path or select the faces; refer to Figure-24 and click on the **OK** button from the **Chain Manager**.

Figure-22. Chain Manager

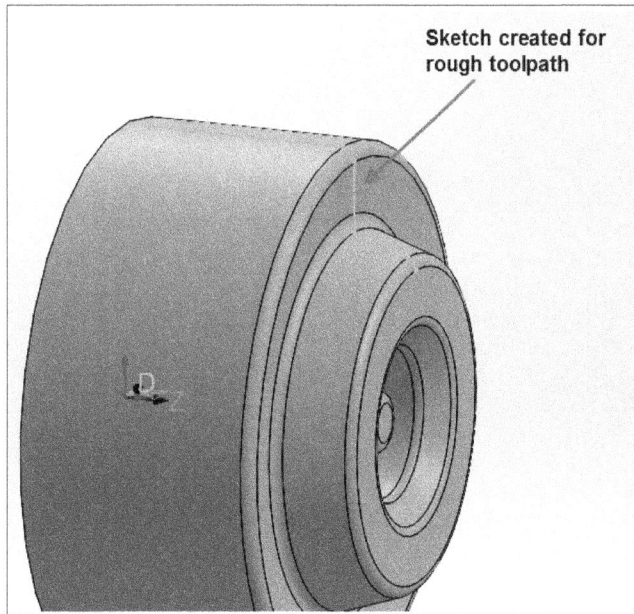
Figure-23. Sketch for rough toolpath

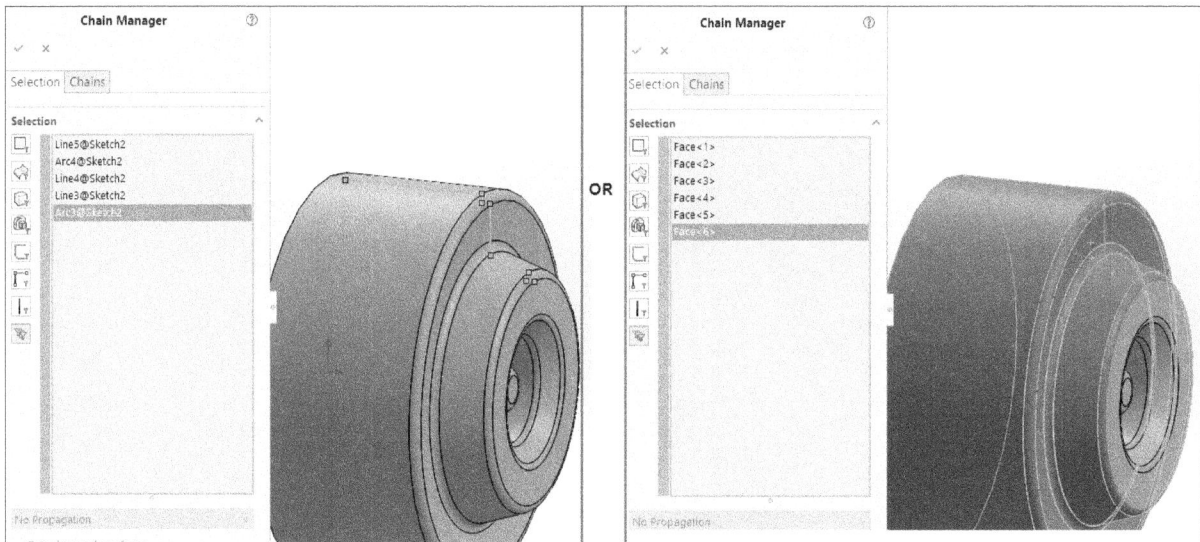
Figure-24. Selection for rough toolpath

- On clicking the **OK** button from the **Chain Manager**, the **Lathe Rough Properties** dialog box will be displayed; refer to Figure-25.

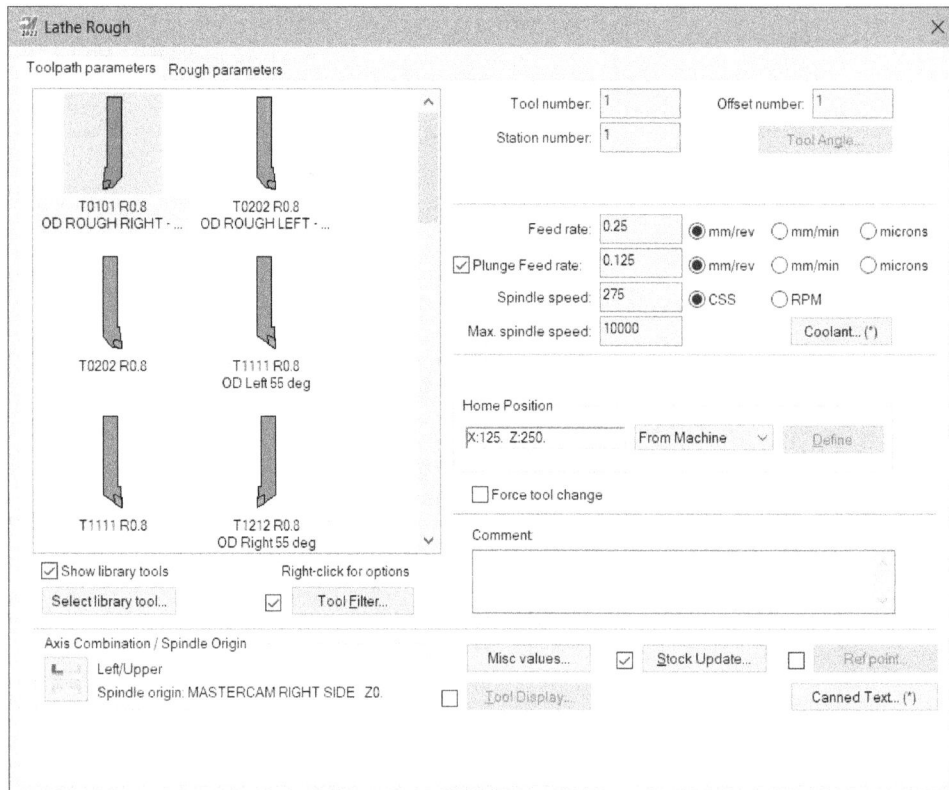

Figure-25. Lathe Rough Properties dialog box

Toolpath parameters

- Select desired tool from the list box in the left or click on the **Select library** tool; the **Tool Selection** dialog box will be displayed; refer to Figure-26.

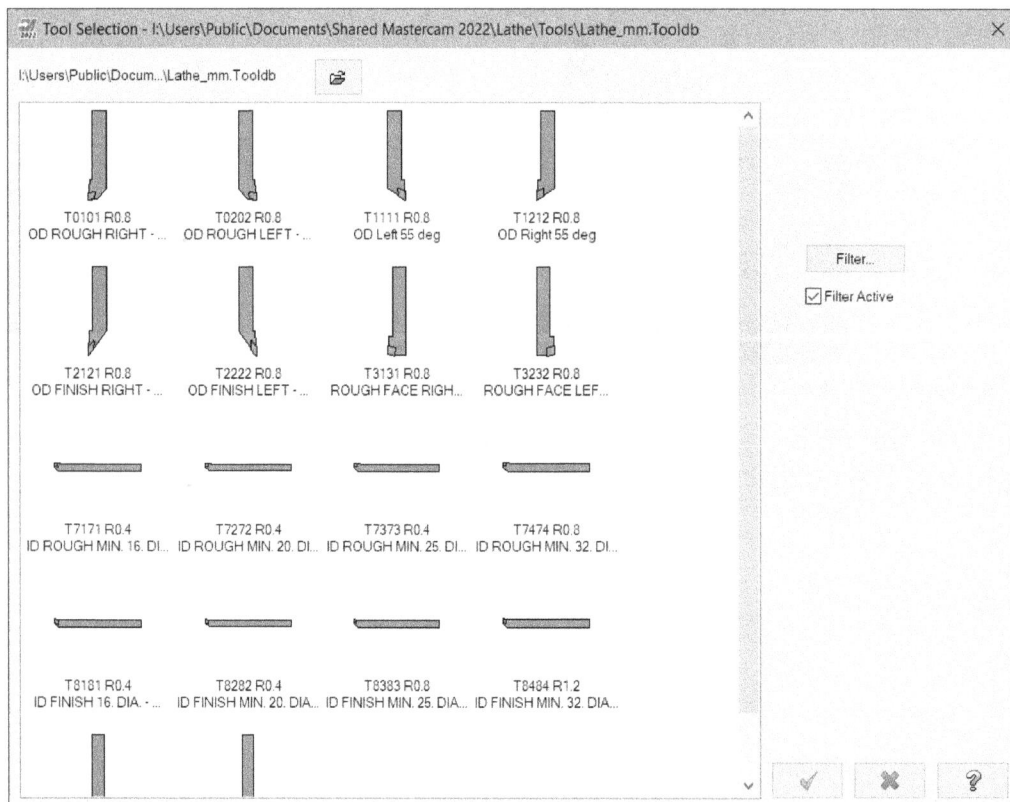

Figure-26. Tool Selection dialog box

- Select desired tool to add it in the list box.
- Now, select desired tool from the list box in the **Lathe Rough Properties** dialog box. The respective feed rate and speed parameters will be displayed in the right of the dialog box.
- Specify desired parameters in the right area of the dialog box. These parameters are discussed next.

Tool number, Offset Number, and Station Number

As the name suggests, these all the are the numbers that correspond to specific entities. The **Tool number** edit box of this dialog box is used to specify the tool number according to the turret. If this number is specified wrong then a wrong tool will be used in the machining as per the NC program.

The **Offset number** edit box is used to set an offset number for the tools that wear by same amount. For example, you have two tools that wear by same amount then to compensate for the wear, you need to specify the offset number for the tools. This offset number is linked to values in X and Z coordinates that positions the tool accordingly for machining.

The **Station number** edit box is used to set the station number of the tool. A machine can have multiple station of tools (turrets). For example, a machine can have a tools turret system instead of tail stock and a general tool turret. In such cases, we need to specify the station numbers.

Tool Angle

The **Tool Angle** button is used to change the angle of the tool for machining. This button is active in case of milling operations. For Lathe tools, the similar function regarding the angle is discussed later.

Feed Rate

The **Feed rate** options are used to specify the speed by which the tool will move while cutting material. Select desired radio button to specify the unit and specify the speed in edit box.

Plunge Feed rate

The **Plunge feed rate** options are used to speed by which the tool will be plunged in the workpiece. Select desired radio button to specify the unit and specify the speed in edit box.

Spindle Speed

The **Spindle Speed** option is used to specify the spindle rotation speed while performing cutting operations. It can be specified in constant surface speed(CSS) or rounds per minute (RPM).

Home Position

- Click on the **Define** button in the **Home Position** area of the dialog box to define the home position of the tool. The **Home Position - User Defined** dialog box will be displayed; refer to Figure-27.

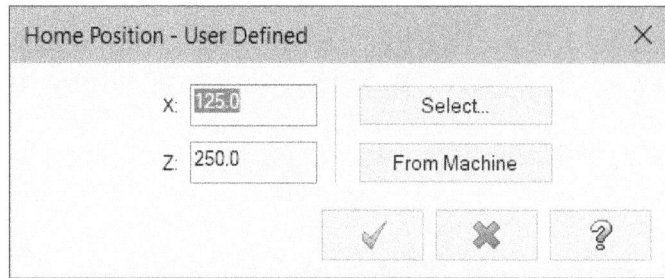

Figure-27. Home Position User Defined dialog box

- Specify desired values or use the **Select** button to specify the position.
- Click on the **OK** button from the **Home Position - User Defined** dialog box.
- To set the angle of tool in lathe, click on the **Tool Angle** button from the dialog box. The **Tool Angle** dialog box will be displayed; refer to Figure-28.
- Set desired direction of tool using the options and click on the **OK** button from the dialog box. Note that the angle specified in **Tool Orientation on Machine** area of the dialog box is used to specify the angle of tool about its own axis. You can use this option if your machine allows to place tool at rotated position.

Rough parameters

- Click on the **Rough parameters** tab to display the parameters related to cutting. Refer to Figure-29.

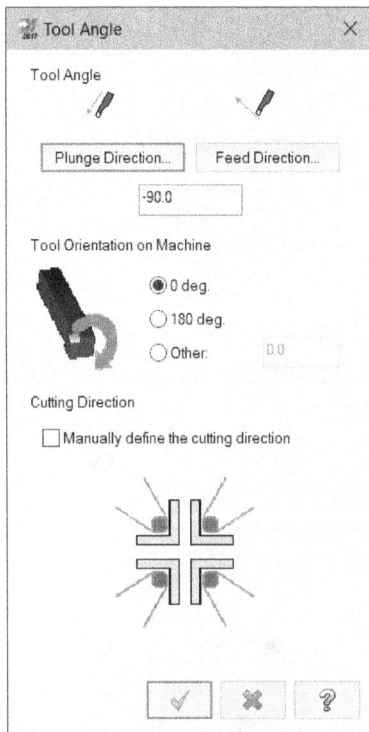

Figure-28. Tool Angle dialog box

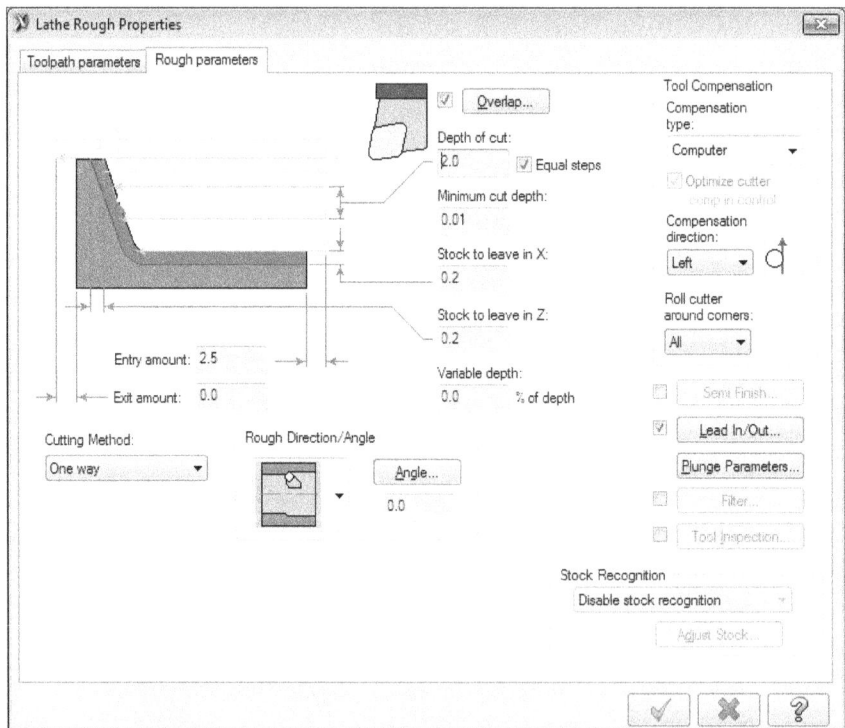

Figure-29. Rough parameters tab of Lathe Rough Properties dialog box

- Select desired cutting method from the **Cutting Method** drop-down. Figure-30 shows difference between various cutting methodologies available in the drop-down.
- Select the direction of roughing from the drop-down in the **Rough Direction/ Angle** area of the dialog box; refer to Figure-31. Set the tool cutting edge angle in the **Angle** edit box of this area.

- Specify the data related to cutting like; Depth of cut, minimum cut depth, cutting method and so on.

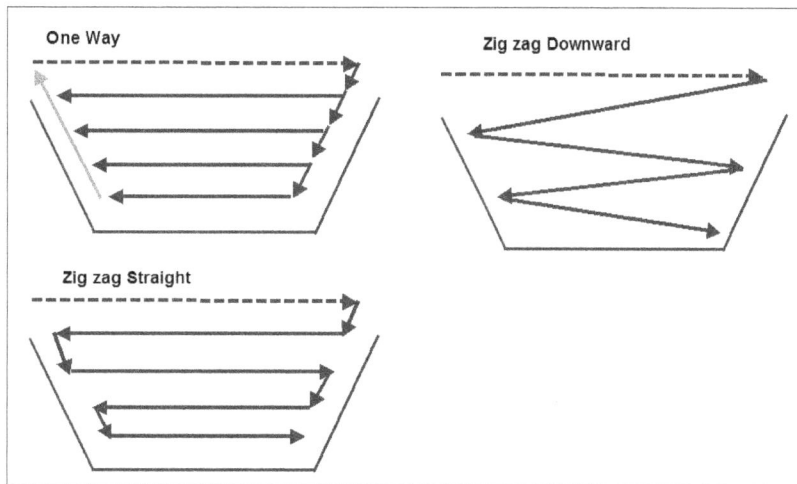

Figure-30. Cutting Methods on lathe

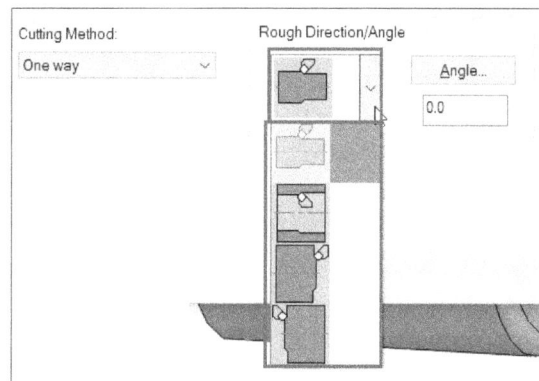

Figure-31. Rough Direction drop-down

Tool Compensation

- The options in the **Tool Compensation** area are used to adjust toolpath to allow for the cutting tool's radius. When we perform machining by a cutting tool, its cutting edge gets worn. To compensate for this wear, we apply tool compensation. Select desired option from the **Compensation type** drop-down in the dialog box. If you select the **Computer** option then tool compensation will be added in the coordinates of NC program by Mastercam but G41/G42 codes will not be generated and hence you will not be able to control compensation by machine panel. If you select the **Control** option then G41/G42 codes will be generated in the NC program and you will be able to control tool compensation using the machine panel. If you select the **Wear** option then Mastercam will calculate the compensated tool positions as if **Computer** is selected, but it will also outputs the G41/G42 codes. This lets the operator adjust for tool wear at the control panel also. If you select the **Reverse Wear** option then it will work in the same way as **Wear** option but you need to enter the positive compensation value in place of negative value. If you select the **Off** option then Mastercam will neither apply compensation in NC program nor it will create G41/G42 codes.

- Select the tool compensation direction from the **Compensation direction** drop-down in the **Tool Compensation** area of the dialog box. Select desired option from the **Roll cutter around corners** drop-down to define shape at corners.

Semi Finish Options

• Select the **Semi Finish** check box if you want to apply extra cutting pass for semi finishing the component during roughing. Click on the **Semi Finish** button after selecting the check box. The **Semi Finish Parameters** dialog box will be displayed; refer to Figure-32. Specify the parameters like number of semi-finishing passes, stepover, stock to be left in X direction, and stock to be left in Z direction. Click on the **OK** button from the dialog box to apply the parameters.

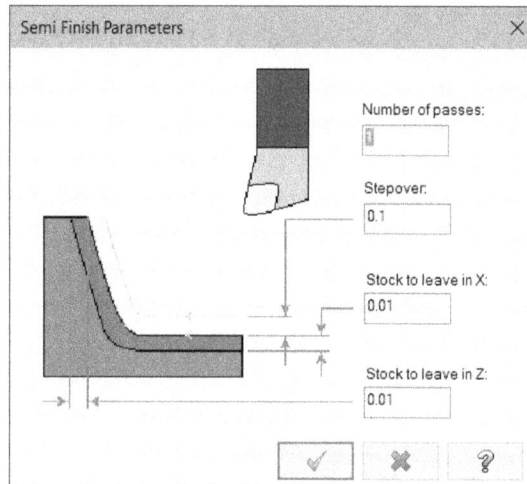

Figure-32. Semi Finish Parameters dialog box

Lead In/Out Options

• To specify the parameters related to entry and exit of tool in cutting toolpath, select the **Lead In/Out** check box and click on the **Lead In/Out** button. The **Lead In/Out** dialog box will be displayed; refer to Figure-33.

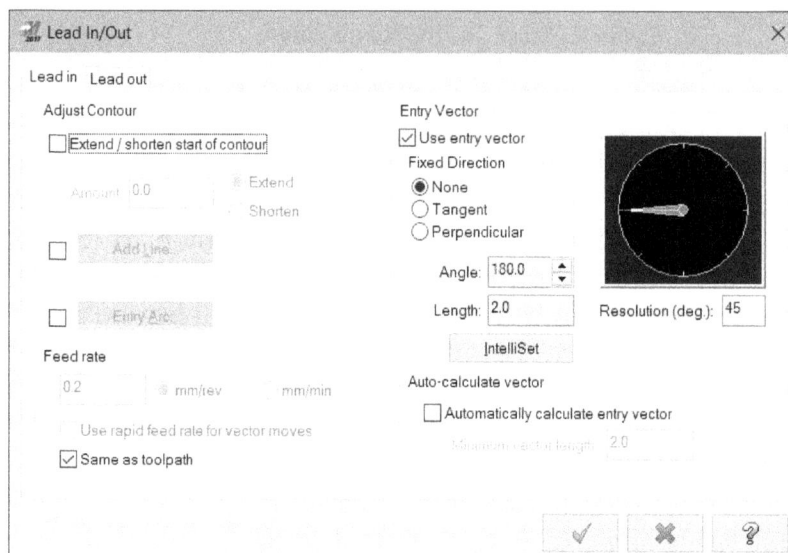

Figure-33. Lead In/Out dialog box

• Select the **Extend/shorten start of contour** check box if you want to specify desired value of extension or shortening in the contour when tool is entering the workpiece. After selecting check box, select the **Extend** or **Shorten** radio button and specify desired value.

- If you want to add a straight line to the cutting toolpath at beginning then select the check box before **Add Line** button and then click on the **Add Line** button. The **New Contour Line** dialog box will be displayed; refer to Figure-34. Specify the length of line and angle in the respective edit boxes and click on the **OK** button. Note that you can specify the same value in **Angle** and **Length** edit boxes of the **Entry Vector** area of the dialog box. If you want to use tangential or perpendicular direction of the toolpath to add Lead In then select the respective radio button from the **Entry Vector** area.
- Similarly, you can set arc lead in by selecting the **Entry Arc** check box and clicking on the **Entry Arc** button.
- You can let Mastercam decide the Lead In for you by selecting the **Automatically calculate entry vector** check box from the **Auto-Calculate vector** area of the dialog box. Specify the minimum vector length in the next edit box. Note that on selecting this option, Mastercam calculates entry or exit vectors based on the stock, chuck, and tailstock information, along with related parameter values, and calculates entry and/or exit vectors that begin outside the stock boundary and end at the start/finish of the toolpath.
- To specify the parameters for lead out, click on the **Lead Out** tab of the dialog box. The parameters in this tab are same as discussed for **Lead In** tab. Click on the **OK** button to apply lead in/out.

Plunge Parameters

Plunge parameters are used to define the method by which tool will plunge in the workpiece.

- Click on the **Plunge Parameters** button from the dialog box. The **Plunge Cut Parameters** dialog box will be displayed as shown in Figure-35.

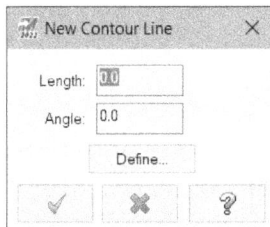

Figure-34. New Contour Line dialog box

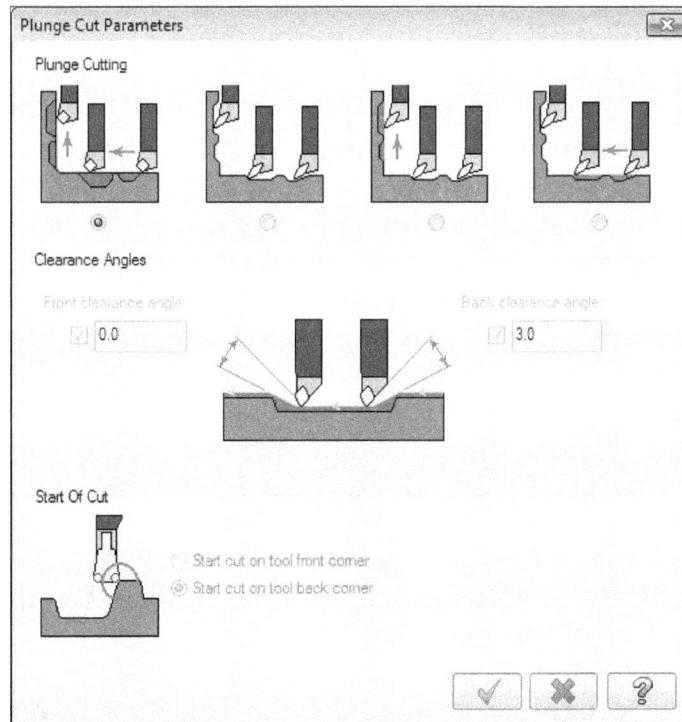

Figure-35. Plunge Cut Parameters dialog box

- Select desired radio button to specifying plunging criteria and set required values for them in respective edit boxes. Click on the **OK** button from the dialog box.
- Similarly, you can use **Filter**, **Tool Inspection** and **Chip Break** options.

Stock Recognition Options

Stock Recognition drop-down lets you choose from three options that control how the stock boundary is used for computing the toolpath. These options are available only if all of the following conditions are true:

- Only one contour is chained
- Stock for the current active spindle is defined in the Stock Setup
- The chained contour lies at least partially within the stock boundary.

There are three options which let you decide how Mastercam will use the stock boundary to compute the toolpath:

- **Use stock for outer boundary** - The stock is used as the outer boundary for roughing. If the ends of the chained contour lie within the stock, the chain is linearly extended to the stock using the parameters set on the **Adjust Stock toolbar** (Displayed on clicking **Adjust Stock** button; refer to Figure-36). This eliminates the need to create the (usually vertical) line added to the end of the inner-chained boundary to indicate the height of the stock.

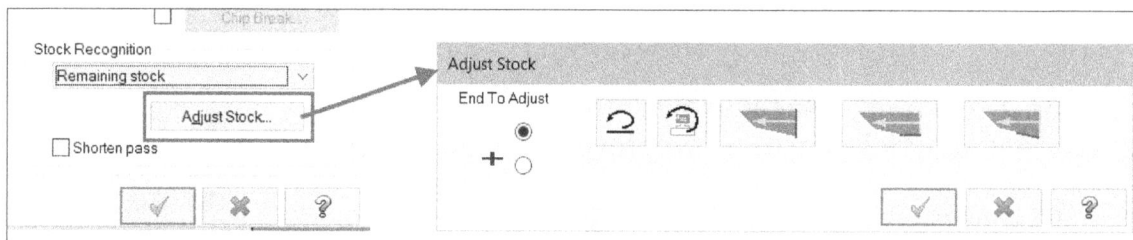

Figure-36. Adjust Stock toolbar

- **Extend contour to stock only** - The chained contour is linearly extended to the stock using the parameters set using the **Adjust Stock** toolbar if the endpoints are inside the stock boundary.
- **Remaining stock** - Uses only stock remaining from previous operations.
- **Disable stock recognition** - The roughing toolpath is computed as if the stock does not exist. You must chain the outer boundary for roughing between boundaries.

The first two options eliminate the need to chain the outer boundary for roughing between boundaries since Mastercam determines the stock to be removed.

You should select **Extend contour to stock** only for zigzag roughing of pocket/groove geometry. If you are using zigzag rough to remove stock, you should select **Use stock for outer boundary**.

When stock recognition is enabled, the following things happen:

- The chained contour is linearly extended to the stock boundary using the methods selected on the **Adjust Stock toolbar**.

- If no entry amount is set and no lead-in parameters are defined, the chained contour is extended past the stock boundary by the entry/exit tool clearance defined in **Stock Setup**. Mastercam also adds the tool nose radius to this amount if it is required.
- When roughing a depression, such as in zigzag roughing, and no lead-in parameters are defined and extend contour is disabled for the lead out, the chained contour is extended past the stock boundary by the entry/exit tool clearance defined in **Stock Setup**. The software also adds the tool nose radius to this amount if it is required.

Note: For finish toolpaths, canned rough and pattern repeat toolpaths, the only option is **Extend contour to stock**.

- Click on the **OK** button from the dialog box to apply the specified parameters. Preview of the tool will be displayed; refer to Figure-37. If the specified parameters are incorrect, then the **Click OK or Cancel to continue PropertyManager** will be displayed like if tool is colliding with jaws or stock directly. Apply the changes accordingly to rectify error

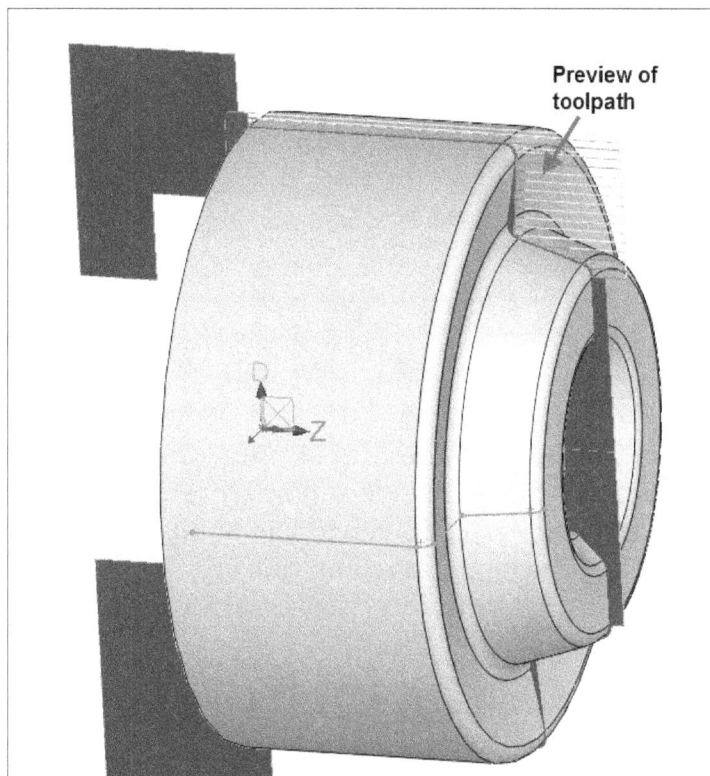

Preview of toolpath

Figure-37. Preview of rough toolpath

Before we move to the next toolpath, it is important to understand the function of **Chain Manager**.

CHAIN MANAGER

The **Chain Manager** (refer to Figure-38) is displayed when you click on any tool to create a toolpath and the toolpath requires multiple geometries to be selected as in case of Rough toolpath discussed earlier.

The options in the **Chain Manager** are discussed next.

Selection Tab Options

- Select desired filter from the left button bar in the **Selection** tab. By default, the **Clear Filters** button is selected and hence you can select any type of geometry for creating toolpath.
- Select the **Ignore Front face** and **Ignore Back Face** check boxes if you want to ignore the front face and back face of the model in selection.
- While selecting geometry, you can also use the standard SolidWorks options like you can right-click on a sketch entity in geometry and click on the **Select Chain** option from the shortcut menu; refer to Figure-39.

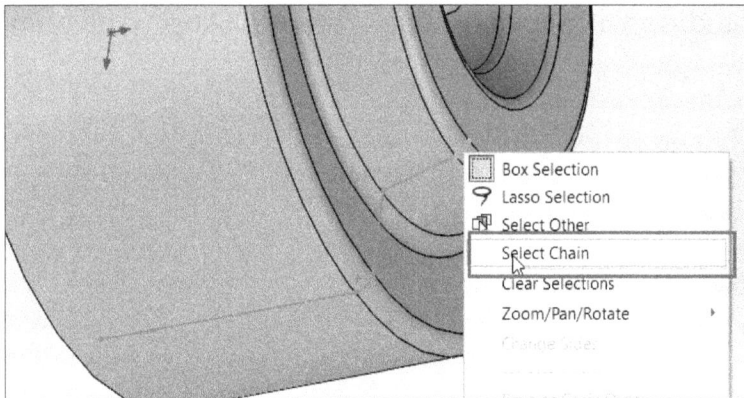

Figure-38. Chain Manager

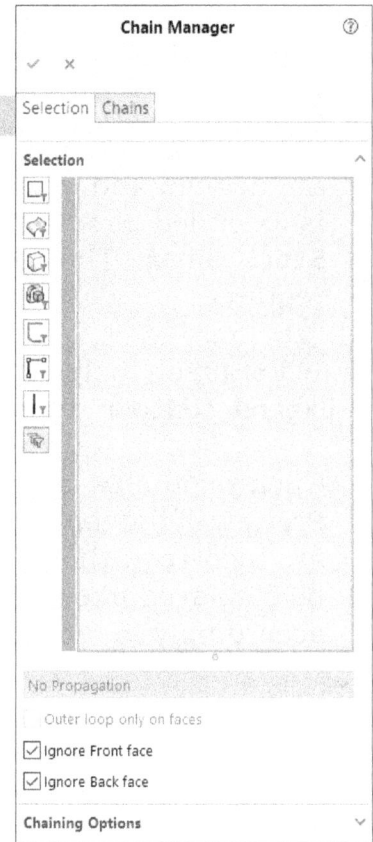

Figure-39. Select Chain option

- Expand the **Chaining Options** rollout and specify the profile creation method for solid part selected; refer to Figure-40. Note that the radio buttons for **Solid profile calculation method** will be active only when you have select the **Part** filter from the left button bar and select complete part as shown in Figure-40. For lathe parts, it is always better to select the **Spin** method as part spins on the lathe machine. Similarly, select the other options in the rollout. You will learn more about spin and slice parameters later in this chapter.

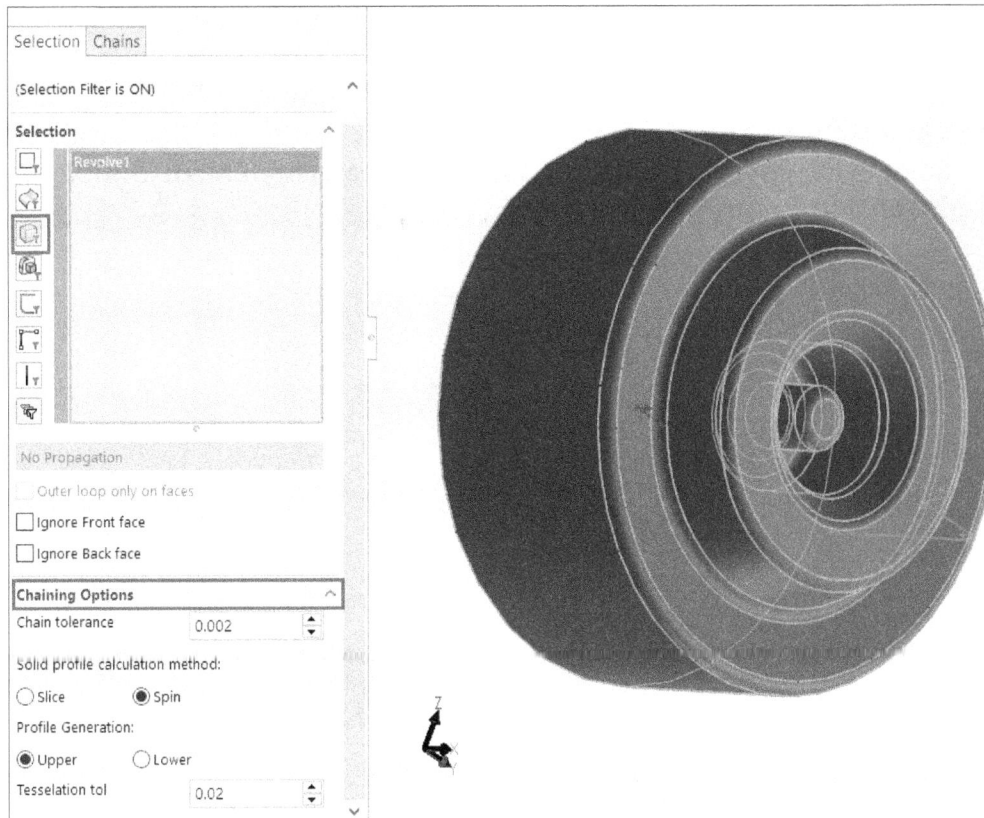

Figure-40. Part selected for profile calculation

Chains Tab Options

• Click on the **Chains** tab in the **Chain Manager** to modify options related to chain start and end. The number of chains is based on the selection done in previous tab. Select a chain to display its starting and end points; refer to Figure-41.

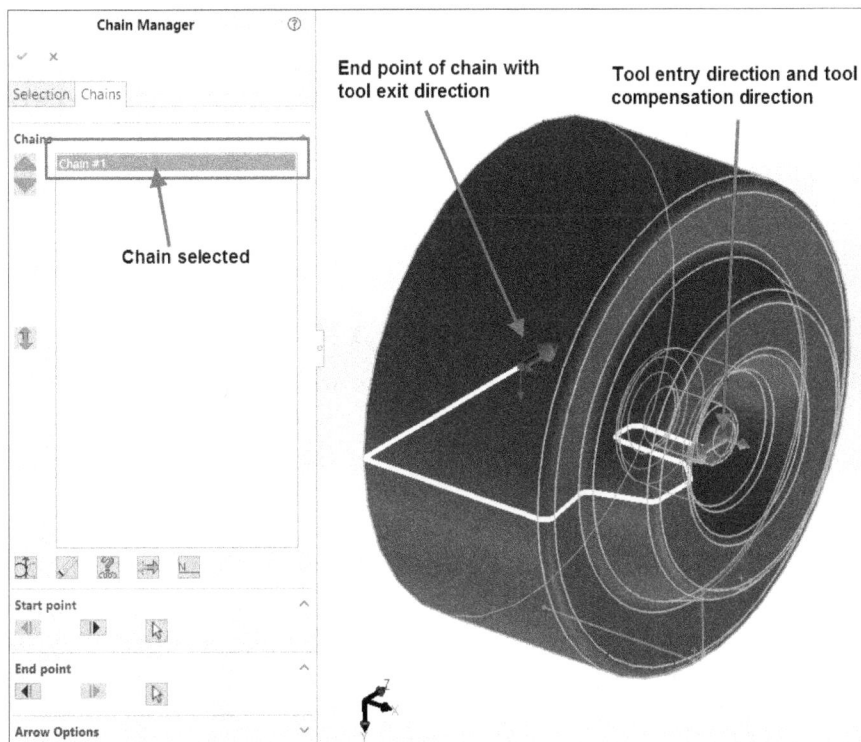

Figure-41. Chain selected in Chain Manager

- Click on the **Change sides** button [icon] below the chain box to change the direction of tool compensation.
- Click on the **Delete** button [icon] to delete selected chain.
- Click on the **Analyze Chain** button [icon] to analyze the selected chain and perform basic checks. On selecting this button, the **Analyze Chain** dialog box will be displayed; refer to Figure-42.

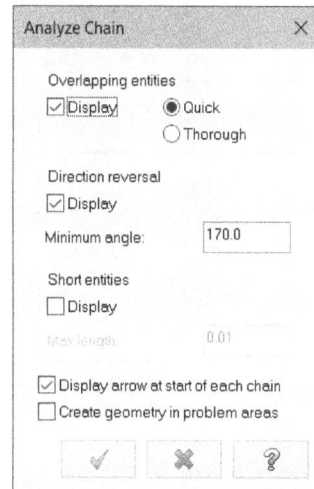

Figure-42. Analyze Chain dialog box

- Specify desired parameters for analyzing chain and click on the **OK** button from the dialog box. The result will be displayed with possible problem areas. Click **OK** to exit the result box.
- Click on the **Reverse chains order** button [icon] to reverse the starting and end point of the chain.
- Click on the **Rename** button [icon] to rename selected chain. On selecting this button a small input box will be displayed at the top. Specify desired name and press **ENTER**.
- Click on the **Backward** [icon] or **Forward** [icon] button from the **Start point** and **End point** rollouts to bring starting/end point backward or forward respectively along the chain. You can click on the **Select** [icon] button in these rollouts and directly select the start and end point of the chain.
- After modifying chain(s), click on the **OK** button from the **Chain Manager** to apply selection.

CREATING BOUNDARY OF PART

Creating boundary is very useful in case of lathe part. You can use these boundaries later to create toolpaths as you can select them as a sketch. The procedure to create boundary is given next.

- Click on the **Silhouette Boundary** tool from the **CAD** drop-down in the **Ribbon**; refer to Figure-43. The **Selection PropertyManager** will be displayed.
- Select the model whose boundaries are to be created and click on the **OK** button from the **PropertyManager**. The **Silhouette Boundary PropertyManager** will be displayed; refer to Figure-44.

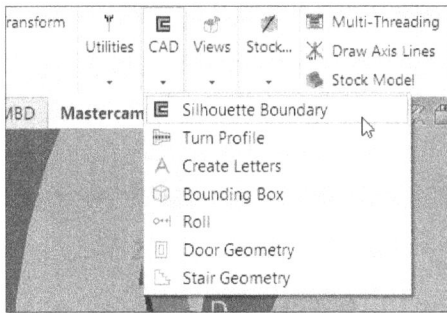
Figure-43. Silhouette Boundary tool

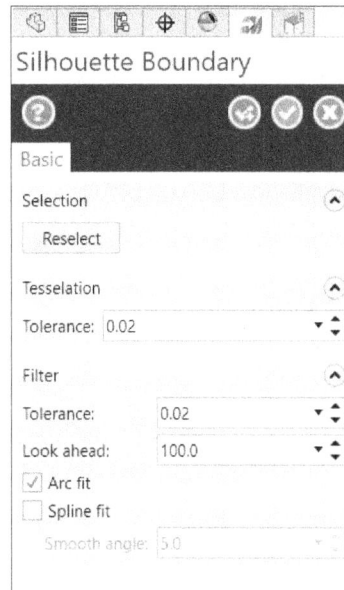
Figure-44. Silhouette Boundary PropertyManager

- The Silhouette Boundary projects a 2D sketch around the selected geometry onto the currently selected plane; refer to Figure-45. Select **Arc fit** check box to generate arcs for circular sections and select the **Spline fit** check box to generate splines for circular sections of the model.
- Set the other parameters and click on the **OK** button to create boundary.

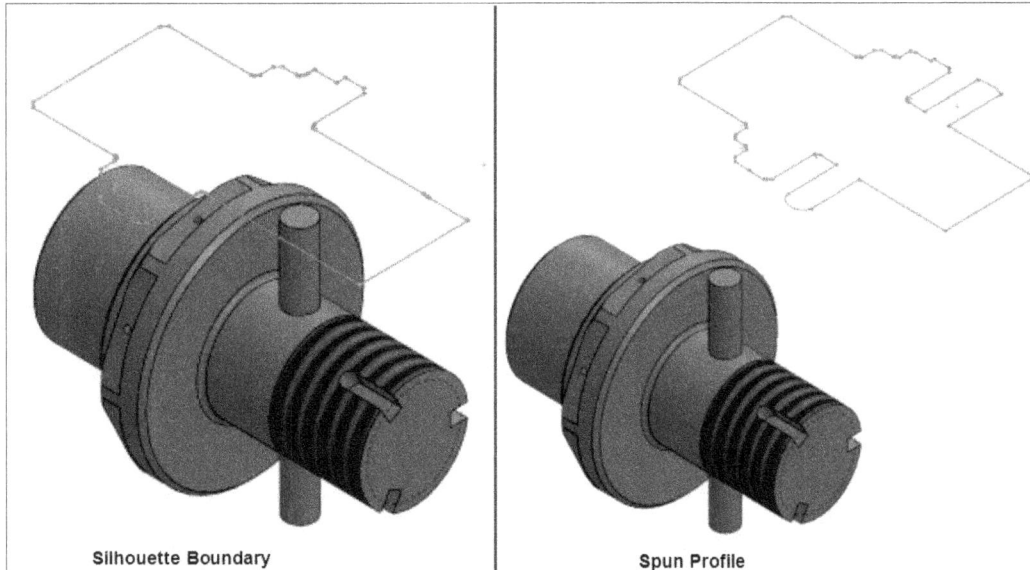

Silhouette Boundary   **Spun Profile**

Figure-45. Boundary types

Note that you can also use the **Convert Entities** tool in SolidWorks Sketch to perform the same action.

CREATING TURN PROFILE OF PART

The turn profile is used to define shape of part to be machined using turning toolpaths. For performing turning operations, we need wireframe curves to be followed by cutting tool. The procedure to create turning profile is given next.

• Click on the **Turn Profile** tool from the **CAD** drop-down in the **Ribbon**. The Selection PropertyManager will be displayed.
• Select the model whose profile is to be created and click on the **OK** button. The **Turn Profile Manager** will be displayed; refer to Figure-22.

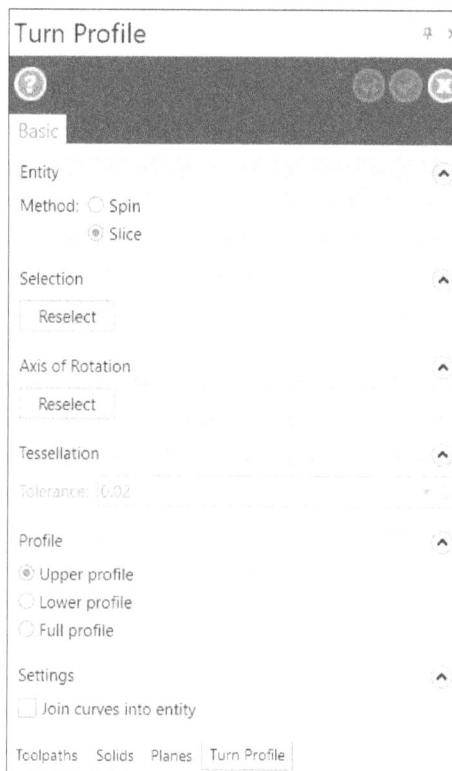

Figure-46. Turn Profile Manager

• Select the **Spin** radio button to create profile of solid body by revolving it about selected axis. Note that when this option is selected then undercuts are not included in the profile. Select the **Slice** radio button to create a profile by using section plane. The profile will be created at intersection of plane and part. Note that the plane used for slicing is currently selected as construction plane.
• Click on the **Reselect** button from **Axis of Rotation** rollout to define axis of rotation for spinning the model to generate profile.
• Select desired radio button from the **Profile** rollout to define which portion of part will be generated as profile. Select the **Upper profile** radio button to generate upper half of model as profile. Select the **Lower profile** radio button to generate lower half of model as profile. Select the **Full profile** radio button to generate both upper and lower profile of model with respect to construction plane. Generally, we generate lower profile for turning operations.
• Select the **Join curves into entity** check box to join all the individually generated curves in the profile as a single curve. Note that using this check box can cause irregular shaped splines, so make sure to check the preview.
• After setting desired parameters, click on the **OK** button. The turn profile will be generated; refer to Figure-23.

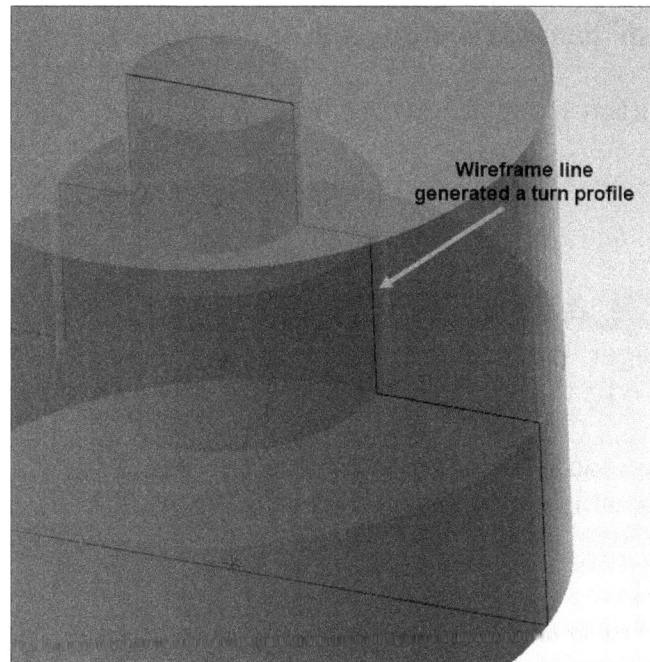

Figure-47. Turn profile generated

Note that we will use the turn profile for creating various turning operations.

FINISH TOOLPATH ON LATHE

The **Finish** tool in the **Lathe** drop-down is used to machine the stock left on the workpiece. The procedure to use this tool is given next.

- Click on the **Finish** tool from the **Lathe** drop-down. The **Chain Manager** will be displayed as discussed earlier.
- Select the sketch or faces that you selected for rough cutting and click on the **OK** button from the **Chain Manager**. The **Lathe Finish Properties** dialog box will be displayed similar to **Lathe Rough Properties** dialog box; refer to Figure-48.

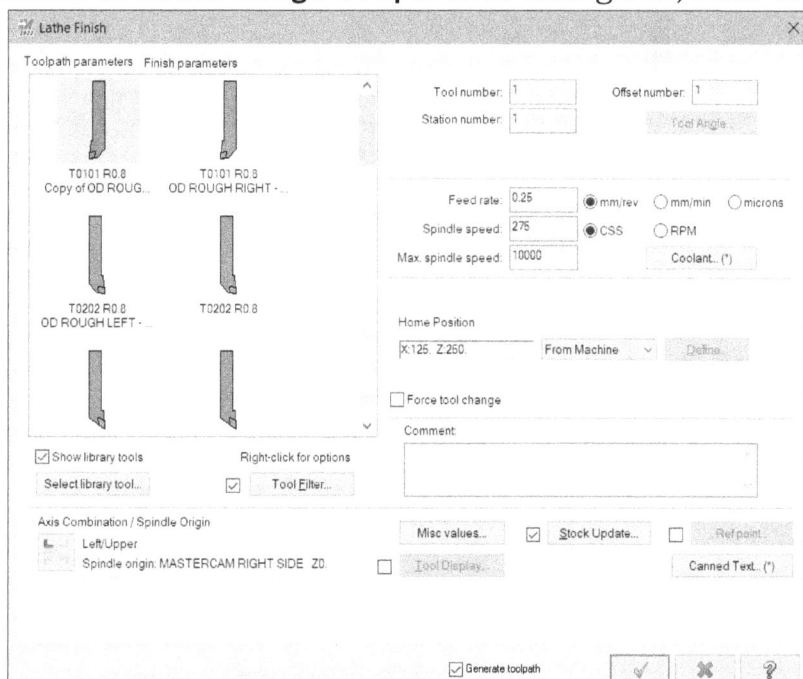

Figure-48. Lathe Finish Properties dialog box

- Select desired tool and specify the related parameters.
- Click on the **Finish parameters** tab and make sure the stock to leave in X and Z directions is **0**.
- Click on the **OK** button from the dialog box.

Before we move to the next toolpath, its important to understand how we set inserts for tool holders.

LATHE TOOL MANAGER

The **Lathe Tool Manager** is used to control the features of the tool used on lathe. This option allows to edit the tool properties. The procedure is given next.

- Click on the **Lathe Tool Manager** tool from the **Utilities** drop-down. The **Tool Manager** dialog box will be displayed; refer to Figure-49.

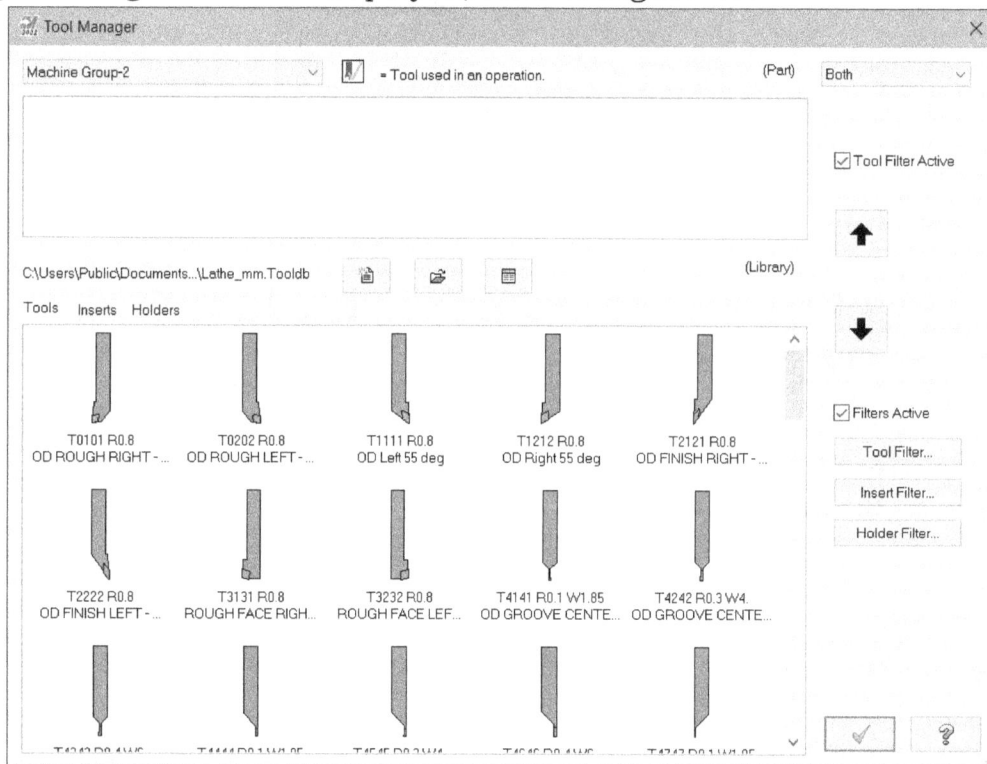

Figure–49. Tool Manager dialog box

- Double-click on the tool to be added in the machine group from the bottom list. The selected tool will be added to machine group and displayed in the upper list of dialog box.
- Now, double-click on the tool that you want to edit from the top box. The **Define Tool** dialog box will be displayed; refer to Figure-50.
- There are four tabs in the dialog box to edit 4 different sections of the tool. By default, the **Inserts** tab is selected in the dialog box. The options are as follows:

Inserts tab

- Click on the **Select Catalog** button and select desired catalog file by using the dialog box displayed. On selecting the catalog, the related inserts will be displayed in the **Define Tool** dialog box.

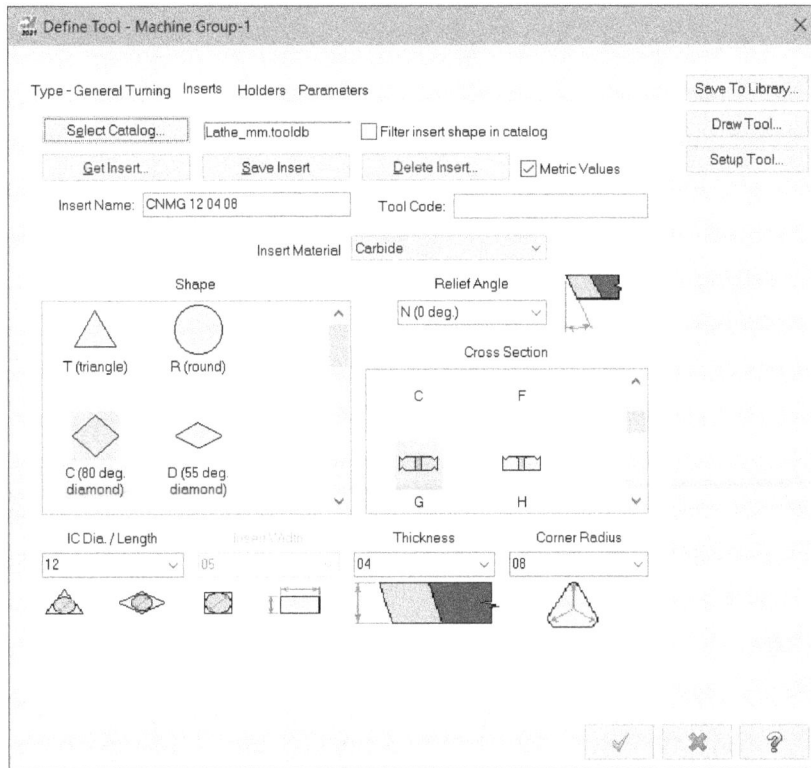

Figure-50. Define Tool dialog box

- Click on the **Get Insert** button to display the list of the tool inserts. The **General Turning/Boring Inserts** dialog box will be displayed; refer to Figure-51.

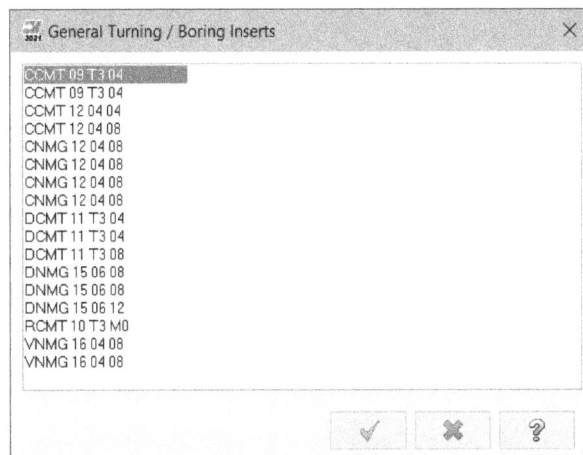

Figure-51. General Turning or Boring Inserts dialog box

- Select desired tool insert from the dialog box and click on the **OK** button. The shape and other parameters of the tool will be automatically decided on the basis of manufacturer's catalog.
- To change the parameters like **IC Diameter**, **Thickness**, and so on; click in the respective edit box and specify desired value.
- To specify the tool direction and machine assembly parameters, click on the **Setup Tool** button. The **Lathe Tool Setup** dialog box will be displayed; refer to Figure-52.

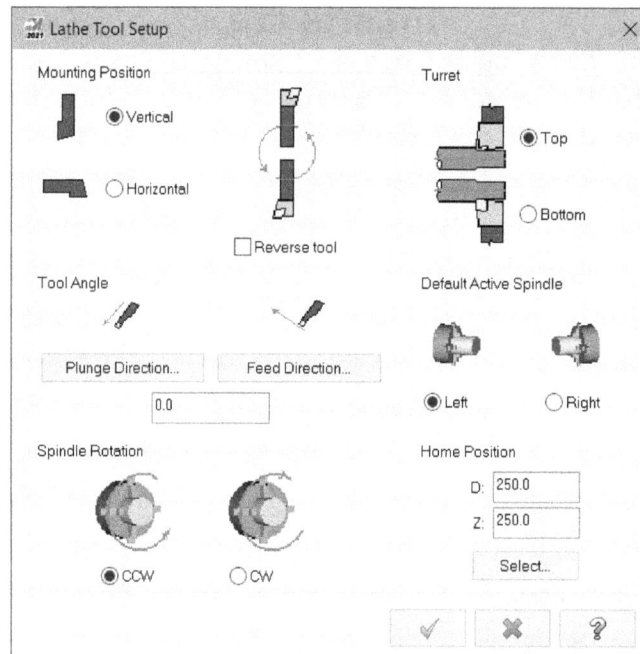

Figure-52. Lathe Tool Setup dialog box

- The radio buttons in the **Mounting Position** area are used to specify the direction in which the tool holder will be placed.
- The radio buttons in the **Turret** area are used to specify whether the tool insert will be facing downward or upward.
- Select the **Plunge Direction** or **Feed Direction** button from the **Tool Angle** area to specify the direction in which the tool will be aligned while cutting the workpiece.
- Similarly, select the radio buttons from the **Spindle Rotation** and **Default Active Spindle** areas as required.
- Click in the edit boxes in **Home Position** area and specify the home position of the tool tip. **Note that specifying correct home position of tool is important to avoid early collision of cutting tool with stock.** Sometimes, you may get warning at the start of toolpath telling you that there is a collision of cutting tool with stock, in those cases you need to change this value.
- Click on the **OK** button from the dialog box.

Type tab

- Click on the **Type** tab from the dialog box. The **Define Tool** dialog box will be displayed as shown in Figure-53.
- Select desired button to change the tool type if the tool you want to use is not displayed in the **Inserts** tab.
- On selecting desired button, the respective parameters will be displayed in the **Inserts** tab which have been discussed earlier.

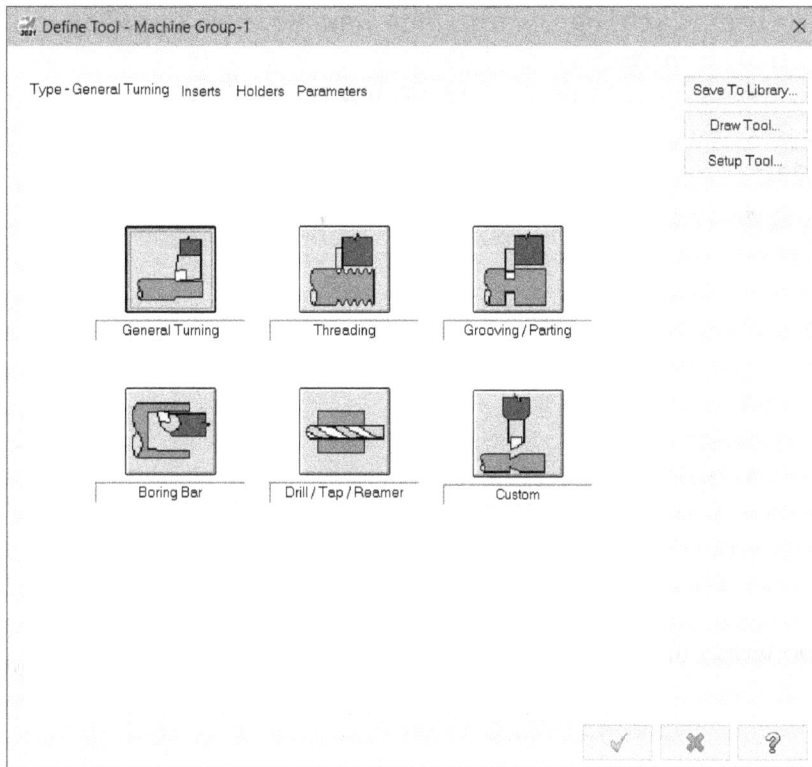

Figure-53. Define Tool dialog box with tool types

Holders tab

The options in the **Holders** tab are used to specify the size and shape of the tool holders. The steps to modify these options are given next.

• Click on the **Holders** tab in the **Define Tool** dialog box. The dialog box will be displayed as shown in Figure-54.

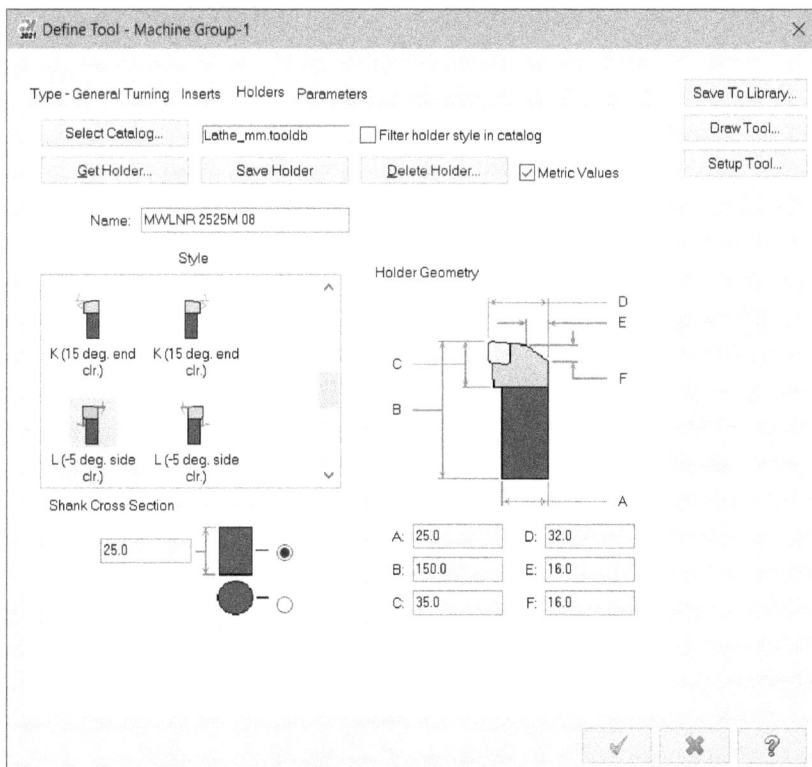

Figure-54. Define Tool dialog box with holder parameters

- Click on the **Select Catalog** button. The **New Holder Catalog** dialog box will be displayed; refer to Figure-55.
- Select desired catalog and click on the **Open** button from the dialog box.
- Click on the **Get Holder** button from the **Define Tool** dialog box. The **General Turning Holders** dialog box will be displayed; refer to Figure-56.

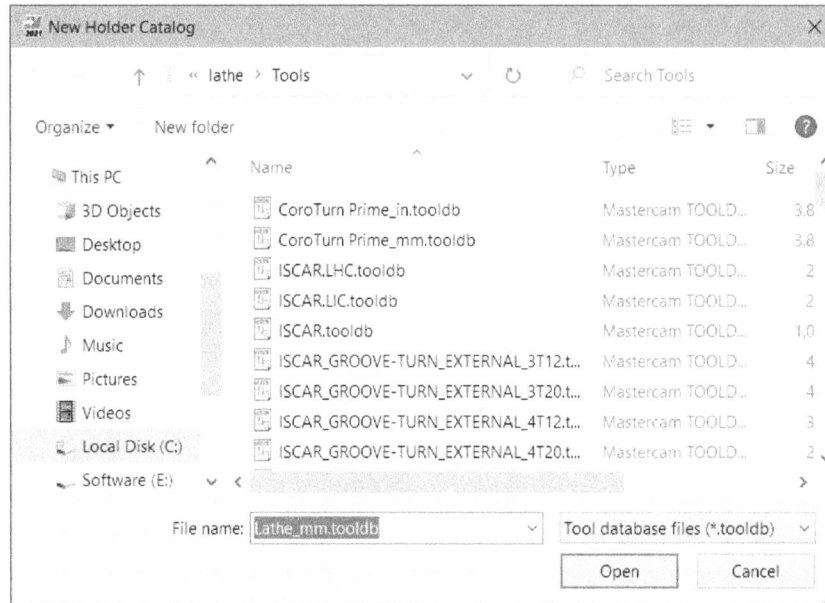

Figure-55. New Holder Catalog dialog box

Figure-56. General Turning Holders dialog box

- Select desired holder from the dialog box and click on the **OK** button. The parameters related to the selected holder will be displayed; refer to Figure-57.

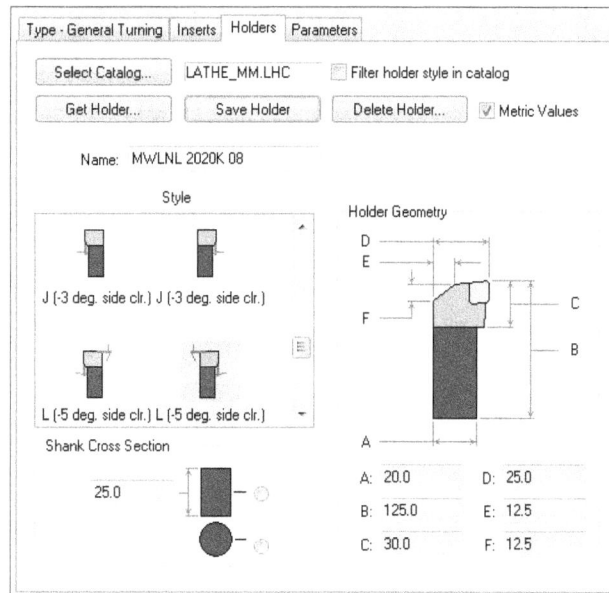
Figure-57. Parameters for selected tool holder

- Specify desired parameters in the dialog box to specify the shape and size of the tool holder. Note that you need to specify the parameters as per the insert used for cutting.

Parameters tab

- Click on the **Parameters** tab to specify the cutting parameters of the tool. The **Define Tool** dialog box will be displayed as shown in Figure-58.

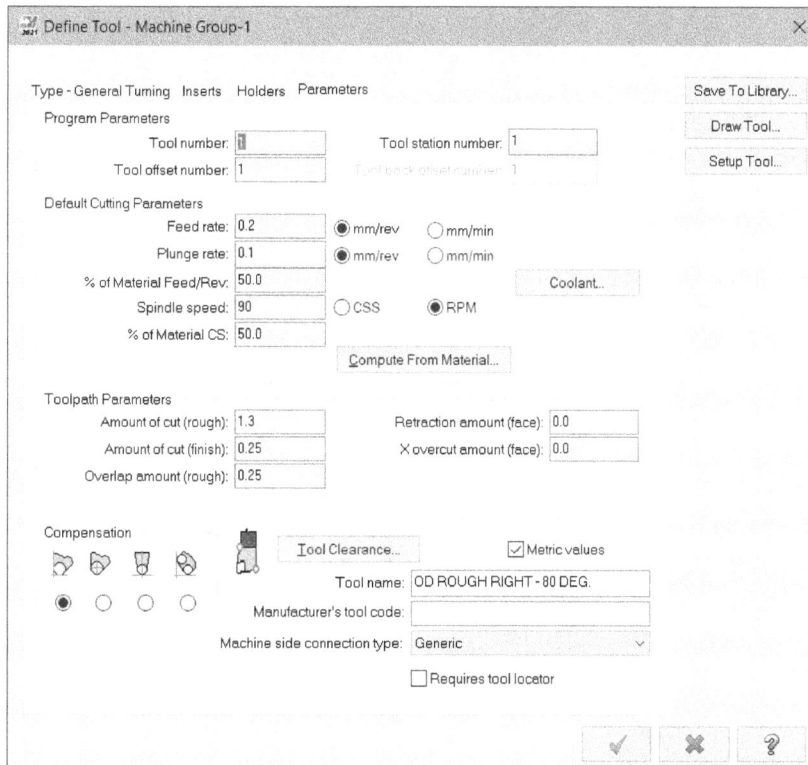
Figure-58. Define Tool dialog box with tool parameters

- Specify the parameters as per the requirement and click on the **OK** button from the dialog box. The **Tool Manager** dialog box will be displayed again.
- After adding desired tools, click on the **OK** button from the dialog box.

GROOVE TOOLPATH

The next toolpath that can be created on lathe is Groove. The **Groove** tool is used to create groove toolpaths in the workpiece. The procedure to use this tool is given next.

- Click on the **Groove** tool from the **Lathe** drop-down. The **Grooving Options** dialog box will be displayed; refer to Figure-59.

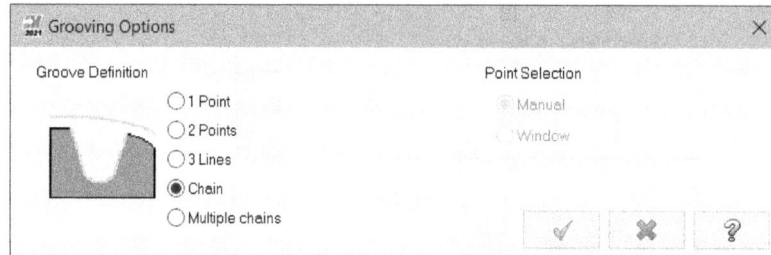

Figure-59. Grooving Options dialog box

- Select desired radio button from the dialog box and then click on the **OK** button from it. Note that if you select the **Multiple chains/Faces** radio button in the dialog box then you can select multiple grooves toolpath; refer to Figure-60.

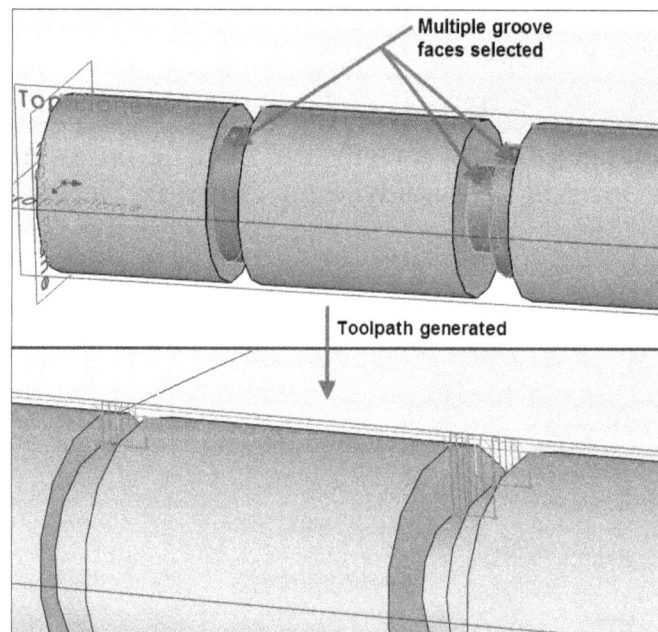

Figure-60. Multiple faces selected for groove toolpath

- The **Chain Manager** or **Selection Manager** will be displayed as discussed earlier.
- Select desired entities and then click on the **OK** button from the **Chain Manager**. The **Lathe Groove Properties** dialog box will be displayed, refer to Figure-61.

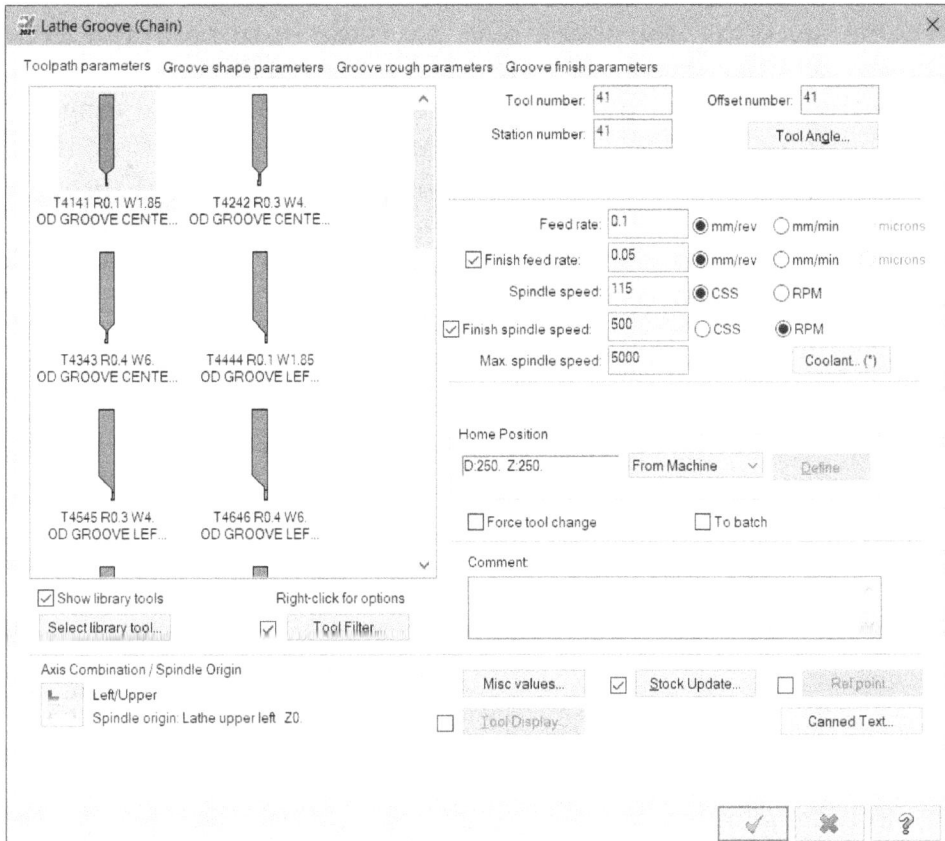

Figure-61. Lathe Groove Properties dialog box

- Select desired tool and specify the related parameters in the **Toolpath parameters** tab of the dialog box.

Groove shape parameters tab

- Click on the **Groove shape parameters** tab to specify parameters related to groove shape; refer to Figure-62.

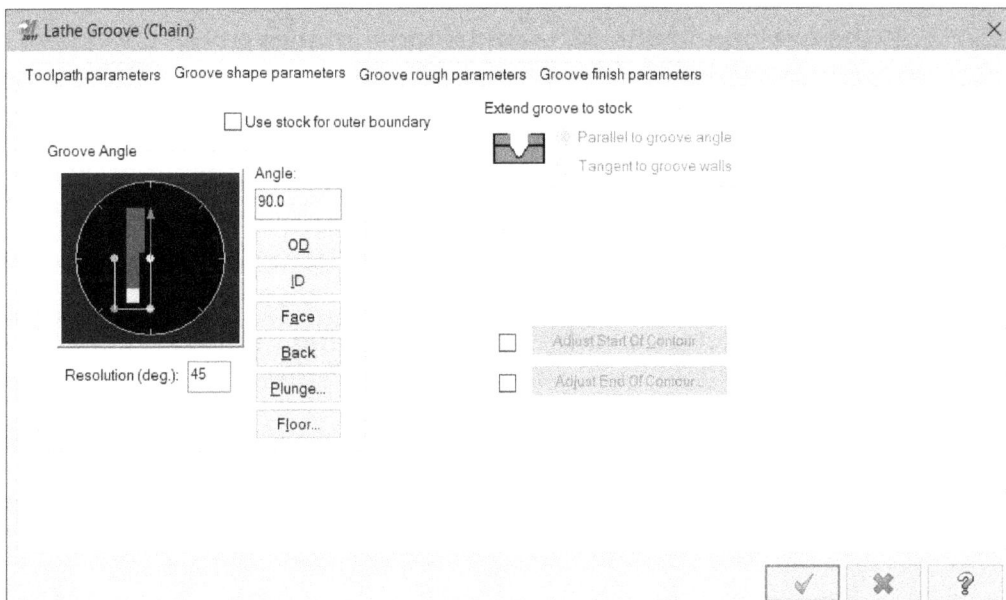

Figure-62. Groove shape parameters tab

- Select desired button to set angle of groove from the **Groove Angle** area.

- Select the **Use stock for outer boundary** check box to use stock as boundary of the groove. On selecting this check box, the radio buttons in the **Extend groove to stock** area will become active. Select the **Parallel to groove angle** radio button to cut the stock in parallel direction to groove. Select the **Tangent to groove walls** radio button to cut the stock tangent to groove walls. Note that preview of the groove is displayed with radio button on selecting desired radio button.
- If you want to adjust the starting or end of the toolpath contour then select the respective check box and click on the button. The **Adjust Contour** dialog box will be displayed; refer to Figure-63.

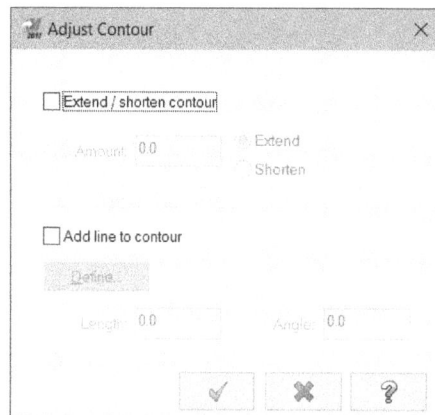

Figure-63. Adjust Contour dialog box

- Select the **Extend/shorten contour** check box, select **Extend** or **Shorten** radio button below it and specify desired value in the **Amount** edit box. Similarly, you can select the **Add line to contour** check and create a line to define extension of the contour. Click on the **OK** button from the dialog box to apply the parameters.

Groove rough parameters tab

The options in the **Groove rough parameters** tab are used to specify the parameters for roughing of the groove; refer to Figure-64.

- If you do not want to perform roughing operations for groove then clear the **Rough** check box, all the options in the tab will become inactive.
- To perform roughing, make sure Rough check box is selected. Select the **Finish each groove before roughing next** check box if you want to finish each groove before moving to the next groove for roughing. This option is useful when you are machining multiple grooves.
- Set the cutting direction from the **Cut Direction** drop-down. Chain direction is available only with inner/outer boundary grooves. For canned groove toolpaths, only positive and negative are available.
- Specify the cutting parameters as required for roughing in the respective edit boxes.

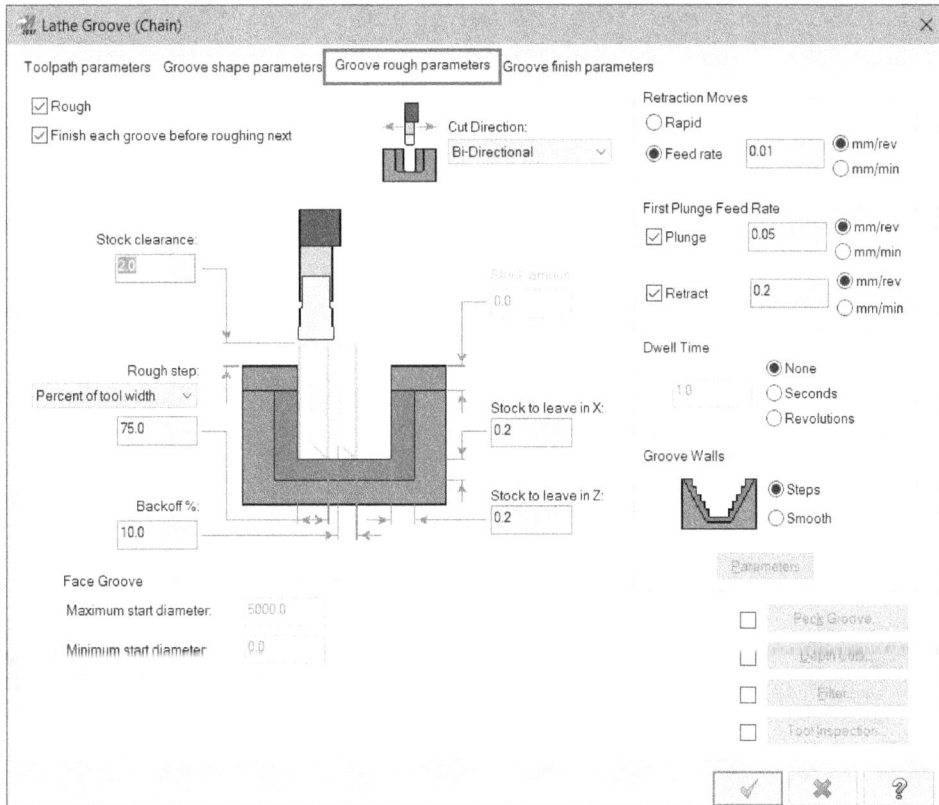

Figure-64. Groove rough parameters tab

Groove finish parameters tab

The options in the **Groove finish parameters** tab are used to specify parameters related to finishing groove; refer to Figure-65.

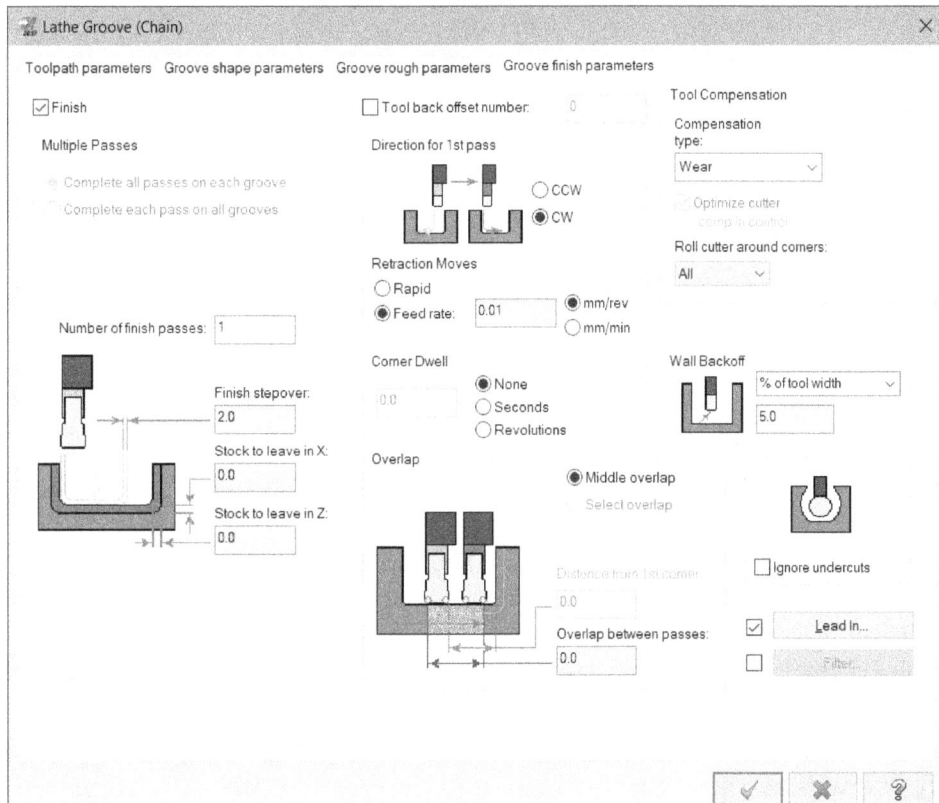

Figure-65. Groove finish parameters tab

- Clear the **Finish** check box from the tab if you do not want to perform finishing cut otherwise specify the finishing parameters as required in the respective edit boxes.
- Make sure you specify correct lead in for finishing pass using the **Lead In** button in the dialog box.
- Click on the **OK** button, the tool path will be created automatically.

THREAD TOOLPATH

The thread toolpath is used to create internal threads in a hole or external threads on a shaft. The steps to create thread toolpath are given next.

- Click on the **Thread** button from the **Lathe** drop-down. The **Lathe Thread** dialog box will be displayed; refer to Figure-66.

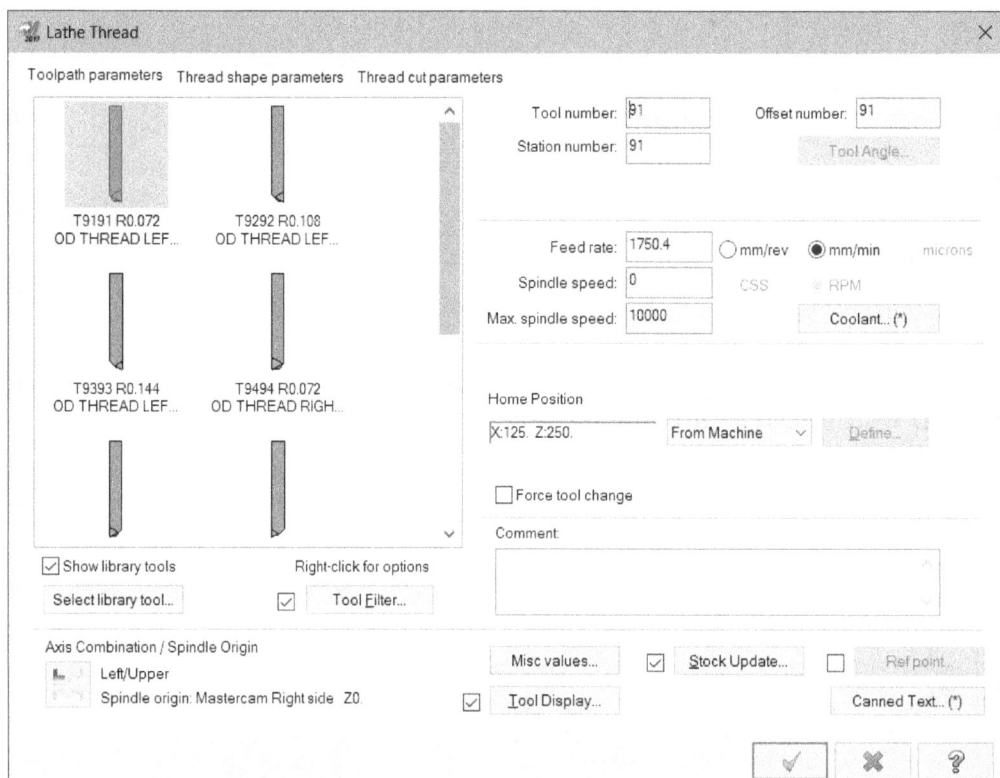

Figure-66. Lathe Thread dialog box

- Specify the parameters in the **Toolpath parameters** tab. Note that you need to specify either feed rate or spindle speed in this tab. The other value will be automatically decided based on the shape of thread specified in the **Thread shape parameters** tab.

Thread shape parameters tab

- Click on the **Thread shape parameters** tab. The dialog box will be displayed as shown in Figure-67.
- Select desired option from the **Thread Orientation** drop-down in the dialog box. Selecting **OD** option will create external threads, selecting **ID** option will create internal threads, and selecting the **Face/Back** option will create threads on the face. Here we will discuss the options for **OD**, you can create the other threads by using the respective options by yourself.

- Click on the **Start Position** button to specify the start point for the threading. The **Selection PropertyManager** will be displayed and you will be prompted to specify the start point.
- Click at desired point on the model to specify the start point; refer to Figure-68.

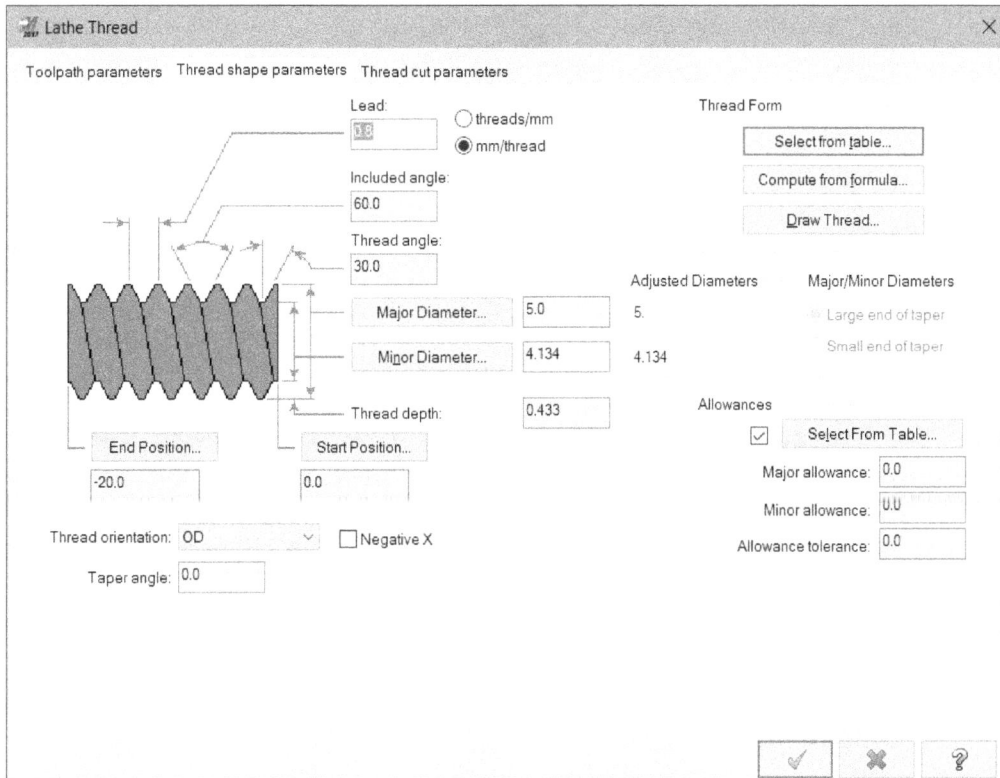

Figure-67. Thread shape parameters page in Lathe Thread dialog box

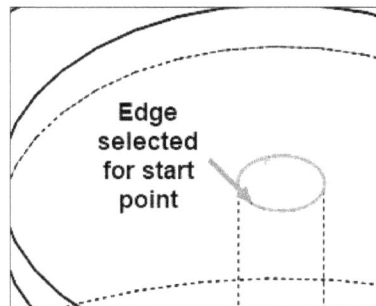

Figure-68. Start point for threading

- Click on the **OK** button from the **Selection PropertyManager**.
- Similarly, click on the **End Position** button from the dialog box and select the end position of the thread. Click on the **OK** button from the **Selection PropertyManager** to exit.
- Click on the **Major Diameter** button to define major diameter of thread. The Selection **PropertyManager** will be displayed and you will be asked to select a point defining the major diameter. Select a point on hole/shaft surface; refer to Figure-69 and click on the **OK** button. The value of major diameter will be displayed in the adjacent edit box.
- Similarly, set the value of minor diameter or enter the value directly in the edit box.

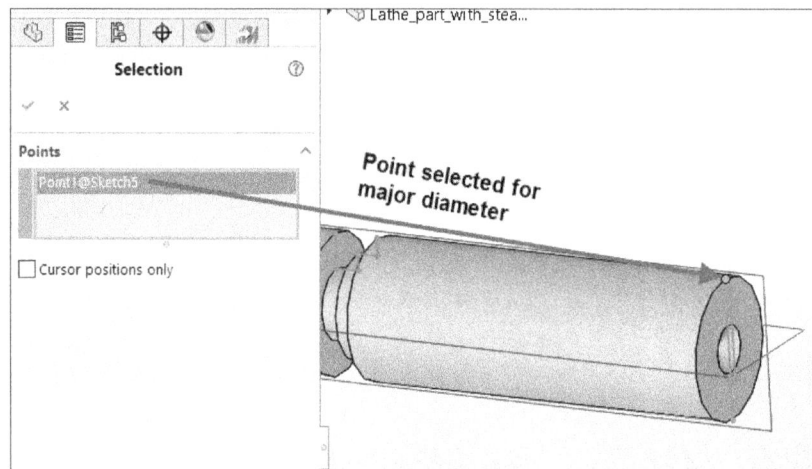

Figure-69. Point selected for major diameter

- If you want to create standard thread form then click on the **Select from table** button in the **Thread Form** area of the dialog box. The **Thread Table** dialog box will be displayed; refer to Figure-70. Select desired profile from the **Thread form** drop-down and double-click on desired thread from the table. The parameters of selected thread will automatically be reflected in the dialog box. Similarly, you can use the other options in the **Thread Form** area.

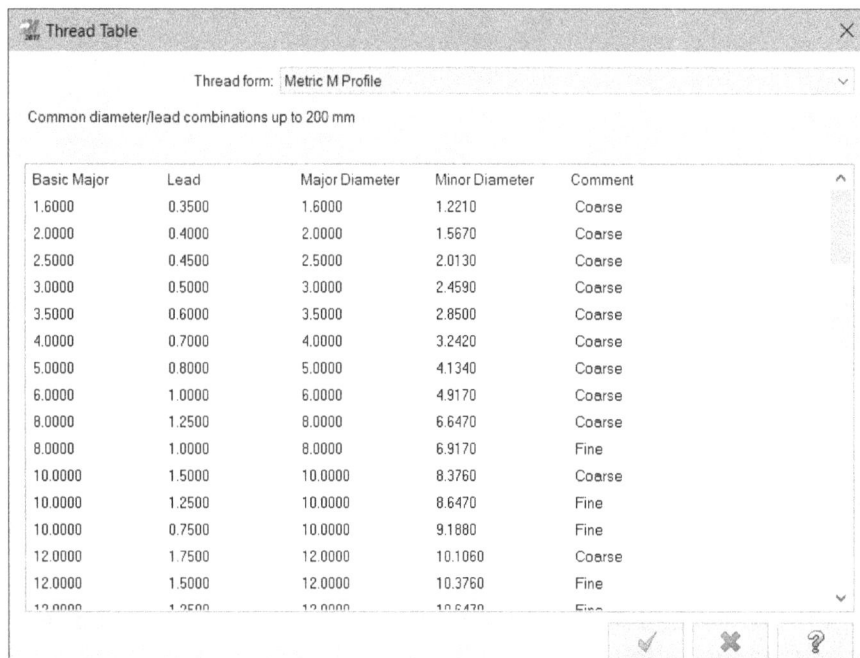

Figure-70. Thread Table dialog box

- Select the **Allowances** check box and specify desired values to set allowances in thread form.

Thread cut parameters

- Click on the **Thread cut parameters** tab in the dialog box. The dialog box will be displayed as shown in Figure-71.

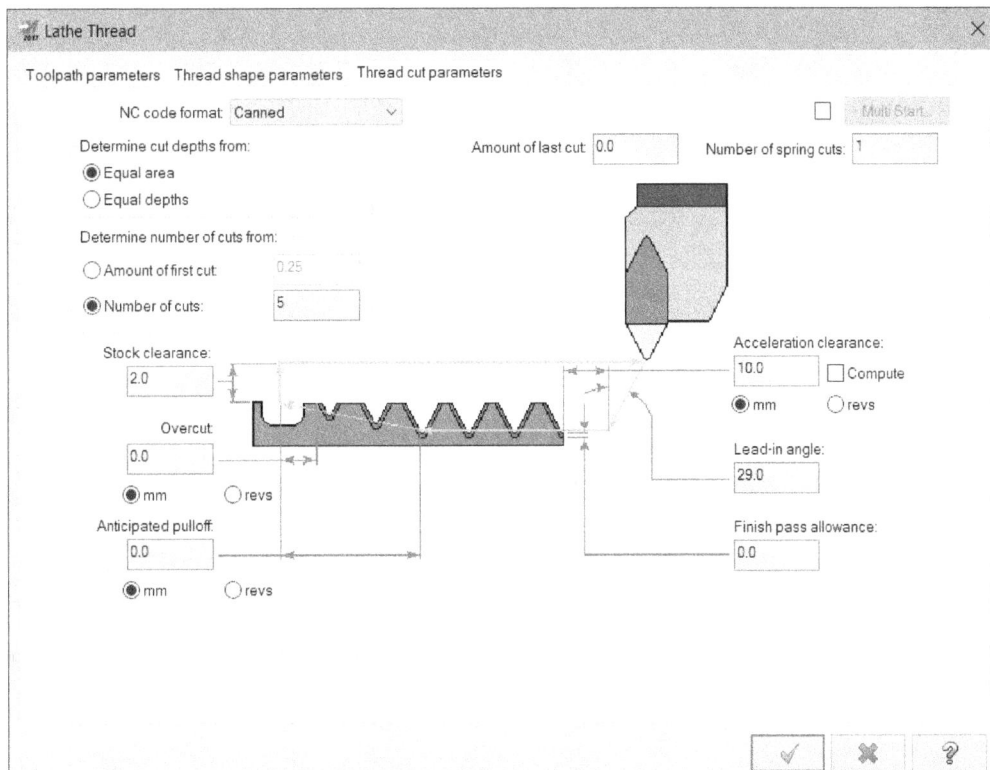

Figure-71. Thread cut parameters page in Lathe Thread Properties dialog box

- Specify the parameters of the thread cut as required by your manufacturing drawing.
- If you want to create multi start threads then select the check box adjacent to **Multi Start** button and click on the **Multi Start** button; refer to Figure-72.

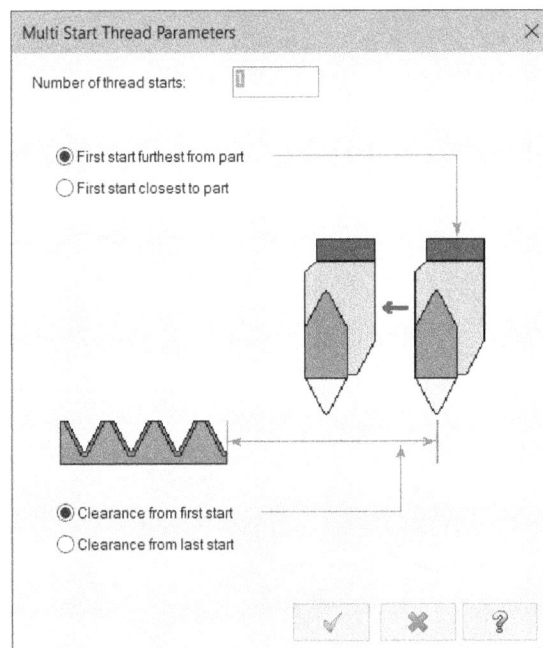

Figure-72. Multi Start Thread Parameters dialog box

- Specify the number of thread starts in the **Number of thread starts** edit box of the dialog box. Set the other parameters and click on the **OK** button.
- Click on the **OK** button from the dialog box. The tool path will be created automatically.

DYNAMIC ROUGH TOOLPATH

The dynamic rough toolpath is designed to cut hard materials with button inserts (i.e. radius or ball). The dynamic motion allows the toolpath to cut gradually, remain engaged in the material more effectively, and use maximum cutting surface of your insert, extending the tool life and increasing the cutting speed. Note that you should use the **Rough** tool and **Face** tool before using the **Dynamic Rough** tool if there are grooves in the dynamic rough toolpath so that tool do not collide with stock; refer to refer to Figure-73. The steps to create this toolpath are given next.

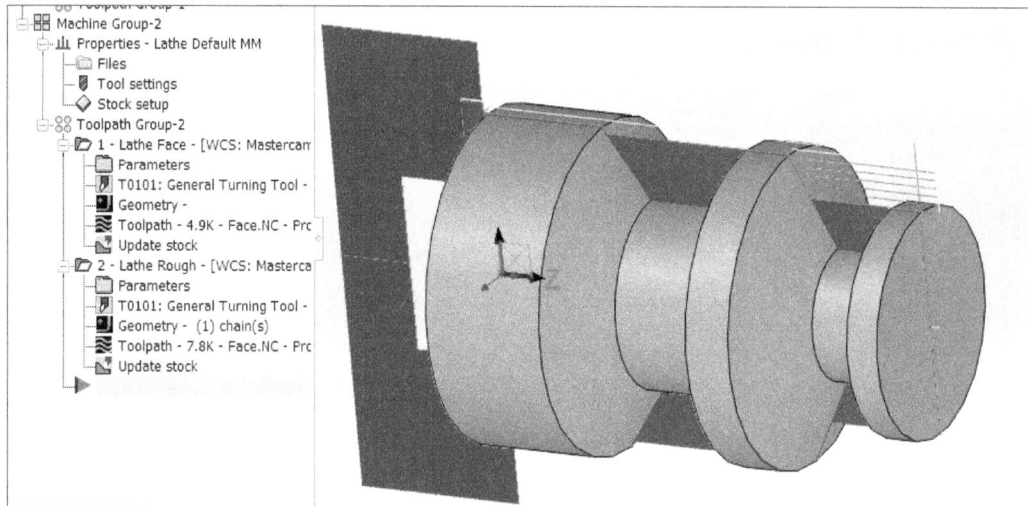

Figure-73. Part after facing and rough machining

- Click on the **Dynamic Rough** tool from the **Lathe** drop-down. The **Chain Manager** will be displayed as discussed earlier. Select the faces/lines/sketch for toolpath; refer to Figure-74.

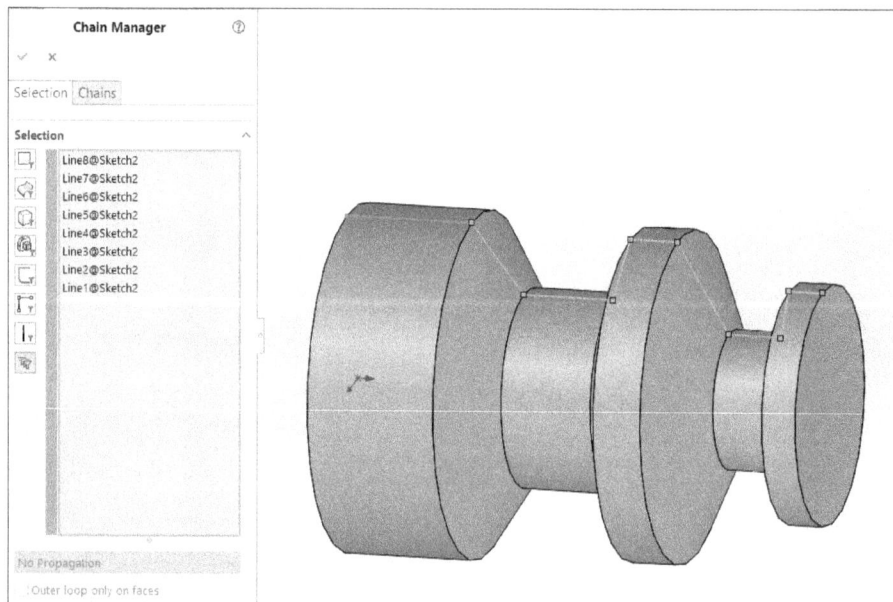

Figure-74. Sketch selected for dynamic rough machining

- Click on the **OK** button from the **Chain Manager**. The **Lathe Dynamic Rough** dialog box will be displayed; refer to Figure-75.

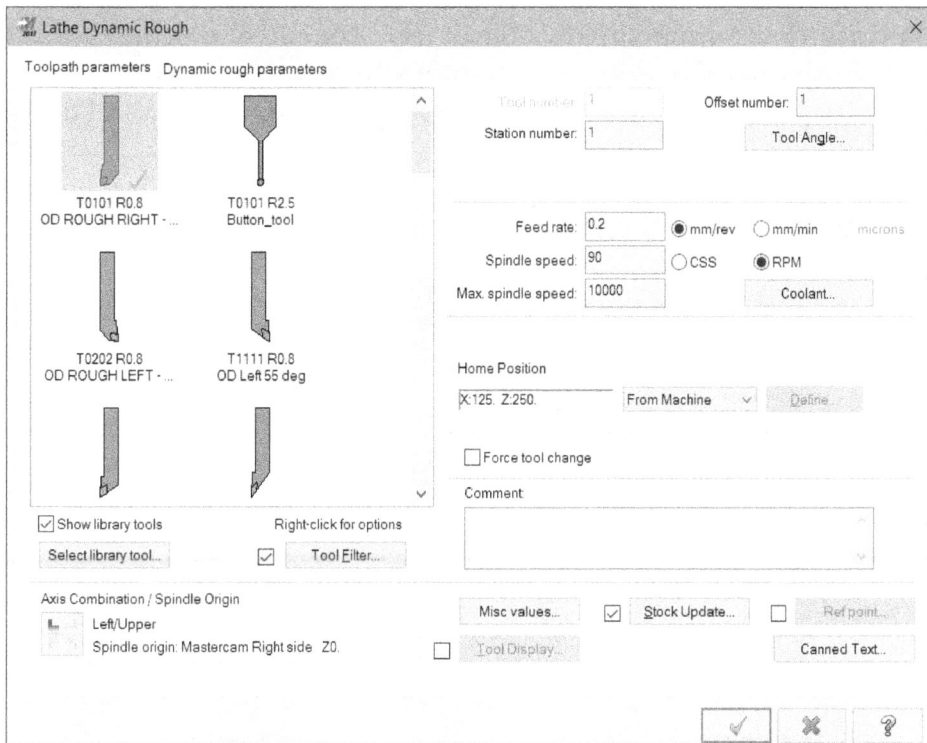

Figure-75. Lathe Dynamic Rough dialog box

- Select a tool that has Button insert (Round or Ball shaped) from the tool list. Specify the other parameters as discussed earlier in the **Toolpath parameters** tab of the dialog box.
- Click on the **Dynamic rough parameters** tab in the dialog box. The options in the dialog box will be displayed as shown in Figure-76.

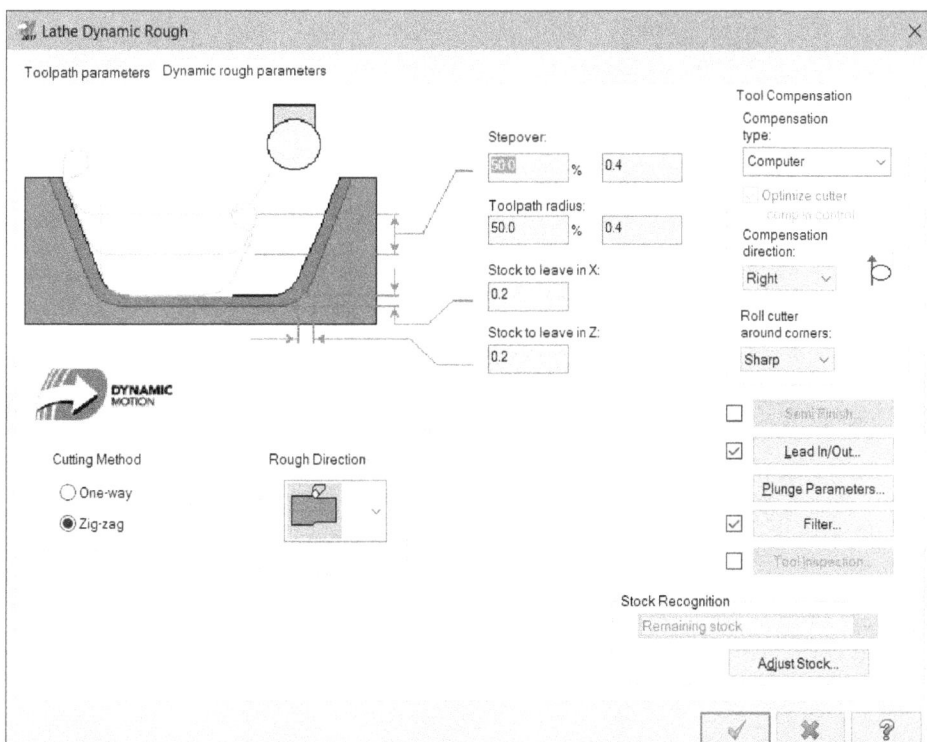

Figure-76. Dynamic rough parameters tab

- Specify the step-over percentage and toolpath radius with respect to tool tip radius in the **Stepover** and **Toolpath radius** edit boxes respectively.
- Set the cutting method and direction in the **Cutting Method** drop-down and **Rough Direction** drop-down respectively. Similarly, specify the other parameters. (OK! You are machinist set the other parameters by yourself!!)
- Click on the **OK** button from the dialog box to create toolpath. The toolpath will be displayed as shown in Figure-77.

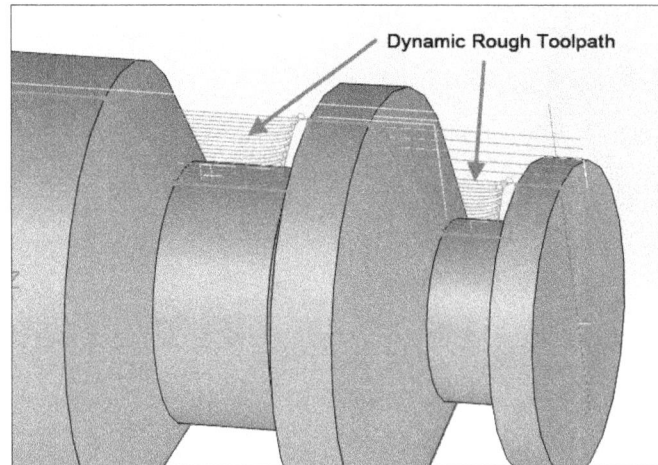
Figure-77. Dynamic rough toolpath created

CONTOUR ROUGH TOOLPATH

The Contour rough toolpath is used to create toolpath following the contour of workpiece. This toolpath is useful for parts where the initial stock shape is similar to the final part shape, such as using a casting for stock; refer to Figure-78. Note that this toolpath does not support collision detection on the holder so it is now your duty to verify path of tool holder. The procedure to create this toolpath is given next.

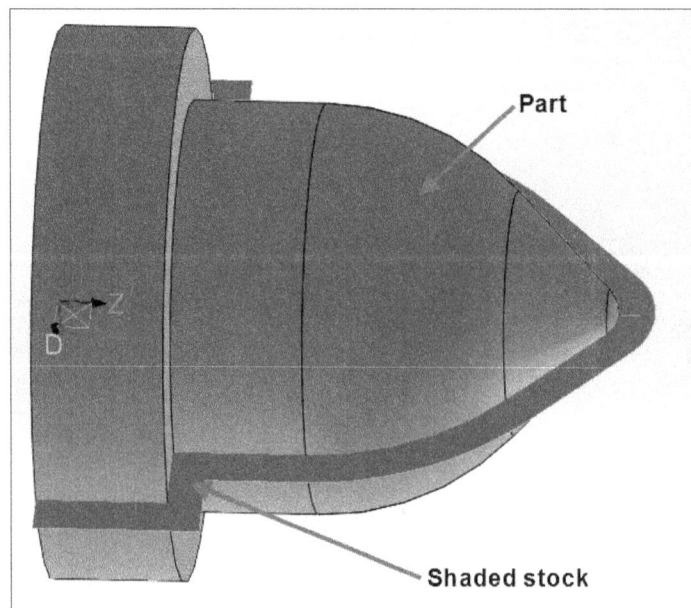
Figure-78. Part with stock of same shape

- Click on the **Contour Rough** tool from the **Lathe** drop-down. The **Chain Manager** will be displayed. Select the faces to be machined; refer to Figure-79. Click on the **OK** button from the **Chain Manager**. The **Enter new NC name** dialog box will be displayed if this is the first toolpath in the NC program.

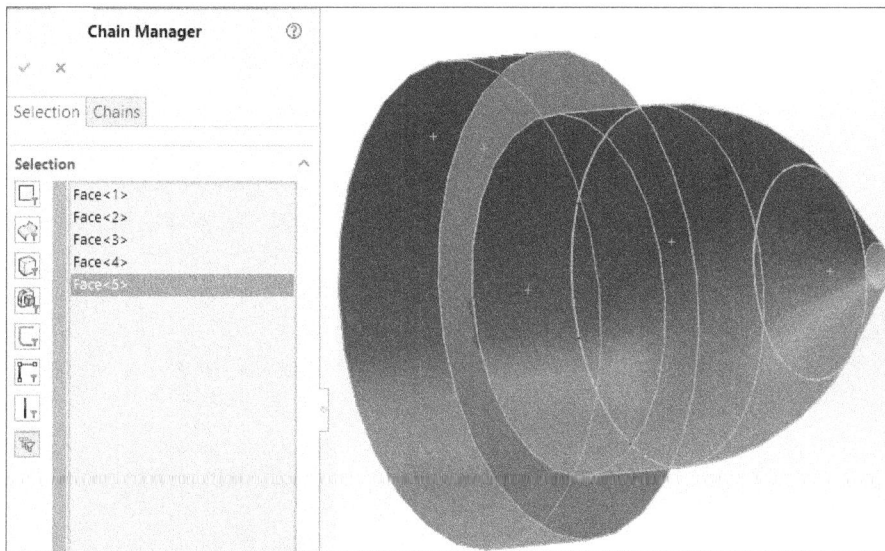

Figure-79. Faces selected for contour rough toolpath

- Specify desired name and click on the **OK** button from the dialog box. The **Lathe Contour Rough** dialog box will be displayed. Select any OD or ID roughing tool based on faces selected and set desired parameters in the dialog box.
- Click on the **Contour rough parameters** tab in the dialog box. The dialog box will be displayed as shown in Figure-80.

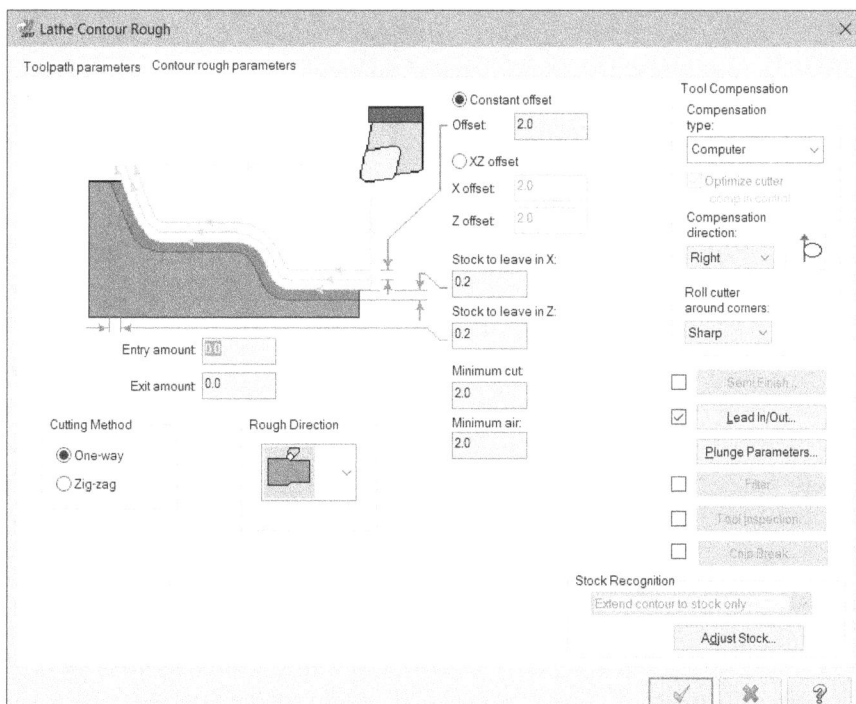

Figure-80. Lathe Contour Rough dialog box

- Set desired entry and exit distance values in the **Entry amount** and **Exit amount** edit boxes, respectively.
- Set the cutting method and cutting direction as required by using the **Cutting Method** area and **Rough Direction** drop-down.
- Select the **Constant offset** radio button and specify the constant depth of cut in **Offset** edit box or select the **XZ offset** radio button and specify different values of depth of cut in X offset and Z offset edit boxes.
- Set the stock to be left after roughing in the **Stock to leave in X** and **Stock to leave in Z** edit boxes.
- Define the value of smallest cut to be made in the **Minimum cut** edit box. Similarly, define the value of minimum distance required for a rapid move in the **Minimum air** edit box.
- Set the other parameters as discussed earlier. Make sure you specify correct **Lead In/Out** parameters to create accident free toolpath.
- After specifying desired parameters, click on the **OK** button to create toolpath; refer to Figure-81.

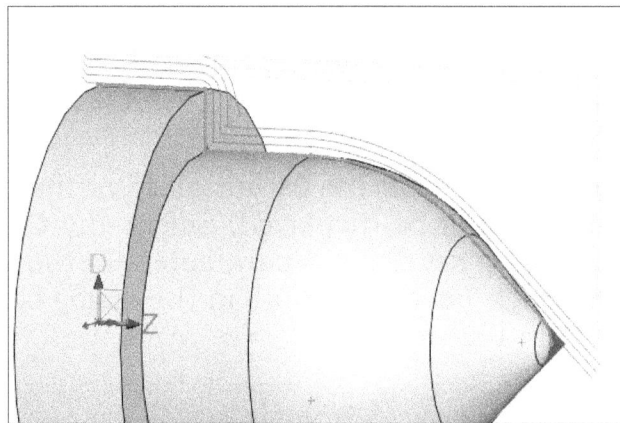

Figure-81. Lathe contour rough toolpath

PRIMETURNING TOOLPATH

The PrimeTurning toolpaths are generated specifically for Sandvik Coromant cutting tools. Mastercam has developed best possible cutting strategies for fast and accurate cutting operations using these tools. Note that this toolpath is not available in home learning edition of software. The procedure to generate this toolpath is given next.

- Click on the **PrimeTurning** tool from the **Lathe** panel in the **Ribbon**. The Chain Manager will be displayed.
- Select the geometry to be machined and click on the **OK** button from the **PropertyManager**. The **Lathe PrimeTurning** dialog box will be displayed; refer to Figure-82.
- Select desired cutting tool from the list. Note that you need to select **CoroTurn Prime** library to select related tools. To do so, click on the **Select library tool** button from the dialog box. The **Tool Selection** dialog box will be displayed. Click on the **Open** button at the top in the **Tool Selection** dialog box. The **Select tool library** dialog box will be displayed; refer to Figure-83. Select the **CoroTurn Prime mm.tooldb** file if you want to use tools in mm dimensions.

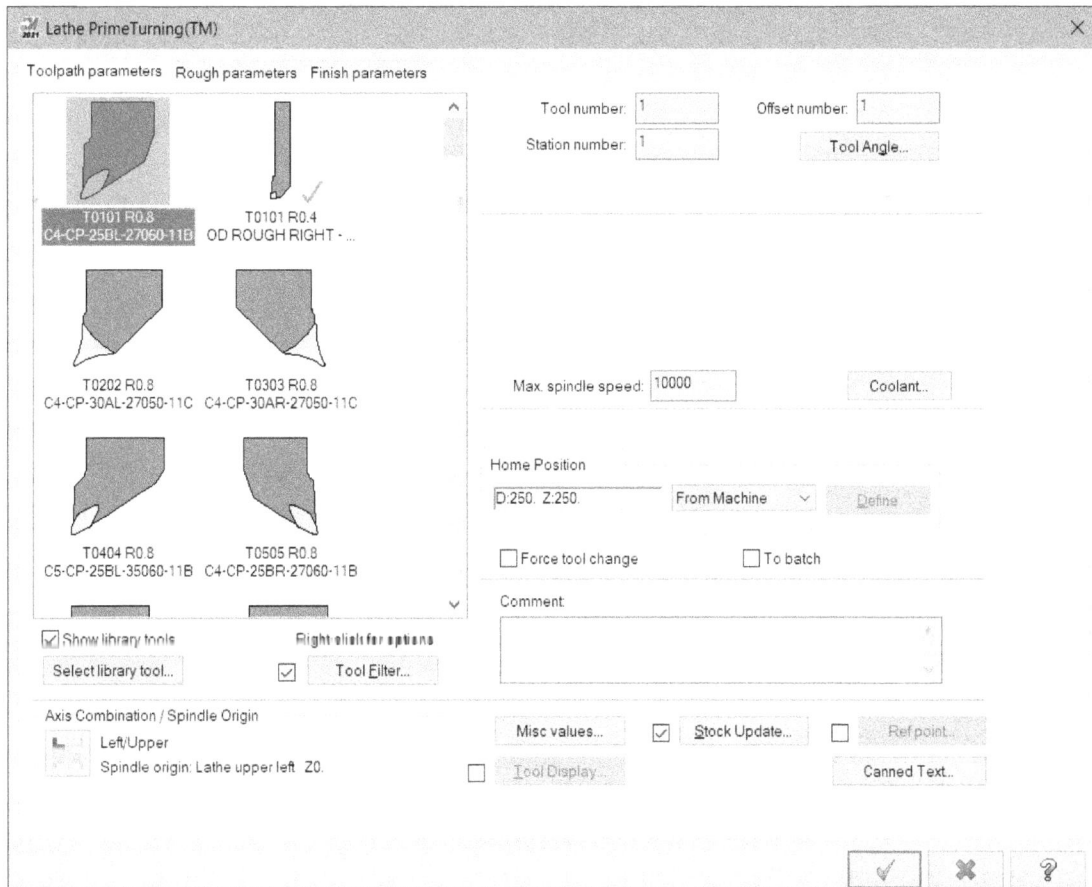

Figure-82. Lathe PrimeTurning dialog box

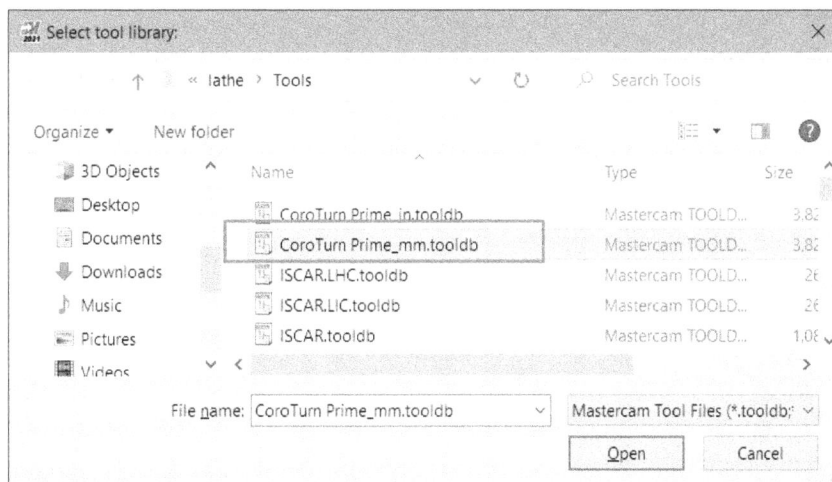

Figure-83. Select tool library dialog box

- After selecting the library file, click on the **Open** button from the dialog box. The tools related to selected library will be displayed in the **Tool Selection** dialog box; refer to Figure-84.
- Click on the **OK** button from the dialog box. The list of related tools will be displayed in the **Lathe PrimeTurning** dialog box.
- Select desired cutting tool and specify related parameters as discussed earlier.

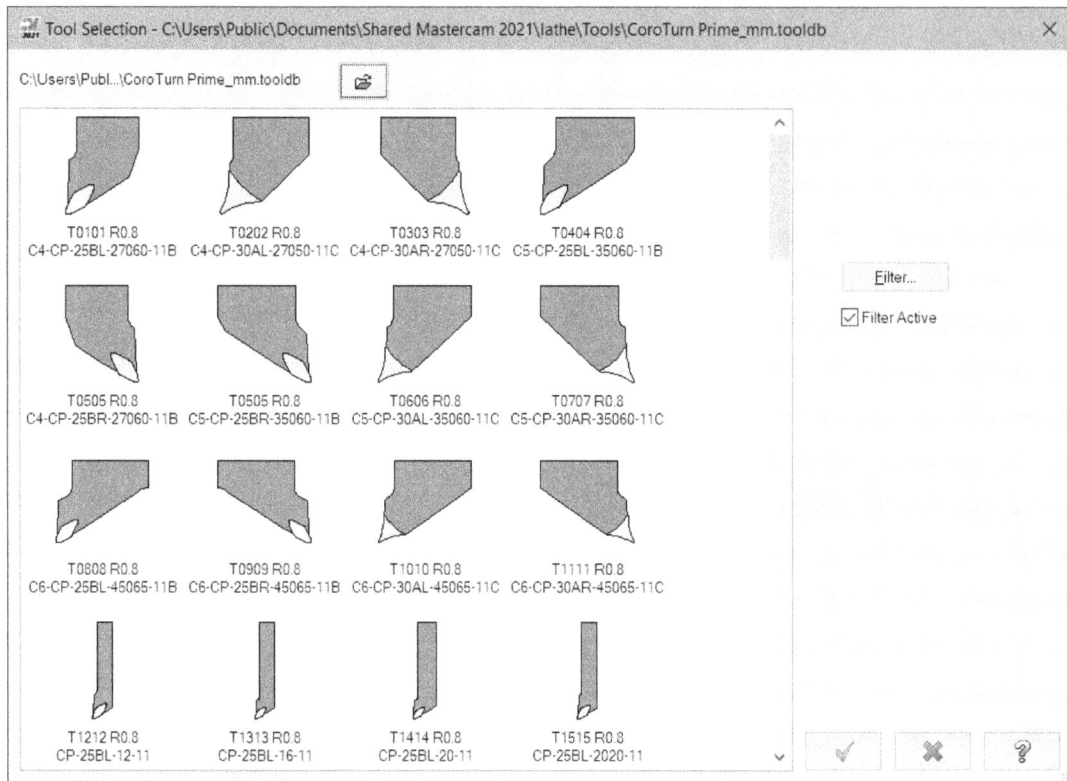

Figure-84. Tool Selection dialog box

Rough Parameters Tab

The options in the **Rough Parameters** tab are used to define strategy for removing large stock of material from the workpiece; refer to Figure-85.

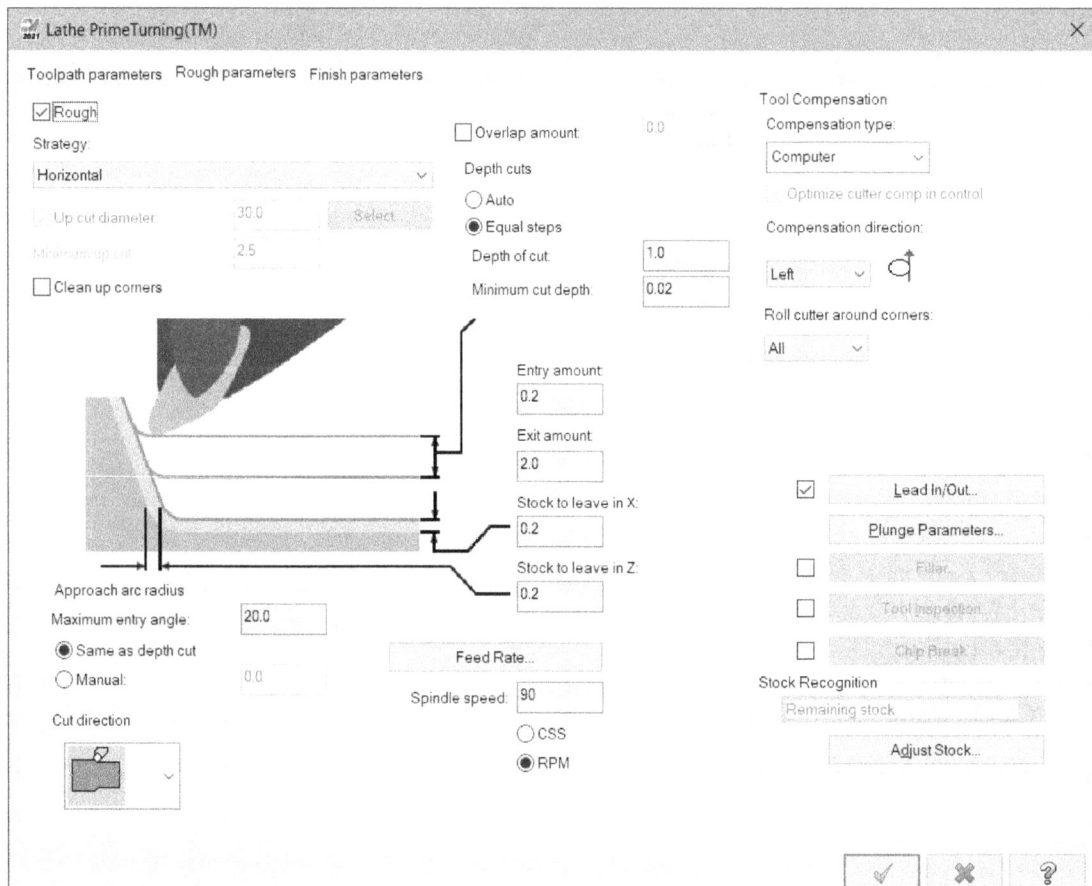

Figure-85. Rough parameters tab

- Clear the **Rough** check box if you do not want to perform roughing operation. On doing so, the options in this tab will get deactivated.
- Select the **Horizontal** option from the **Strategy** drop-down if you want to move cutting tool in horizontal direction while cutting. Select the **Vertical** option from the **Strategy** drop-down if you want to move cutting tool in vertical direction while cutting. Select the **Horizontal then vertical** option from the drop-down to move cutting tool in horizontal direction first and make all the necessary cuts, after move in vertical direction and remove rest of the material for better finish. Similarly, you can use the other strategies from the drop-down.
- If a strategy involving vertical direction is selected then **Up cut diameter** check box will become active. Select the **Up cut diameter** check box to specify upper diameter limit up to which the vertical cuts will be made. After selecting check box, specify the related value in edit box next to the check box.
- Specify desired value in **Minimum up cut** edit box to define the distance moved by cutting tool upward after making each cut.
- Select the **Clean up corners** check box to remove material from sharp edges of the workpiece.
- Specify desired parameters in the **Approach arc radius** area to define how cutting tool will approach the arc like shapes in workpiece.
- Click in the **Entry amount** edit box to specify the approach distance from where cutting tool will start cutting passes.
- Specify desired value in **Exit amount** edit box to specify the distance from workpiece up to which the cutting tool will move using cutting feed rate.
- The other parameters of this tab has been discussed earlier. Similarly, specify the parameters in **Finish parameters** tab of the dialog box and click on the **OK** button. The toolpath will be generated.

FACE TOOLPATH

The facing toolpaths are generated to remove material from the face of the workpiece. In any machining sequence, this is generally the first toolpath to be generated for flat head work-pieces. The steps to generate the face toolpath are given next.

- Click on the **Face** tool from the **Lathe** drop-down. The **Enter new NC name** dialog box will be displayed as discussed earlier if this the first toolpath.
- Specify desired name and click on the **OK** button from the dialog box. The **Lathe Face Properties** dialog box will be displayed; refer to Figure-86.

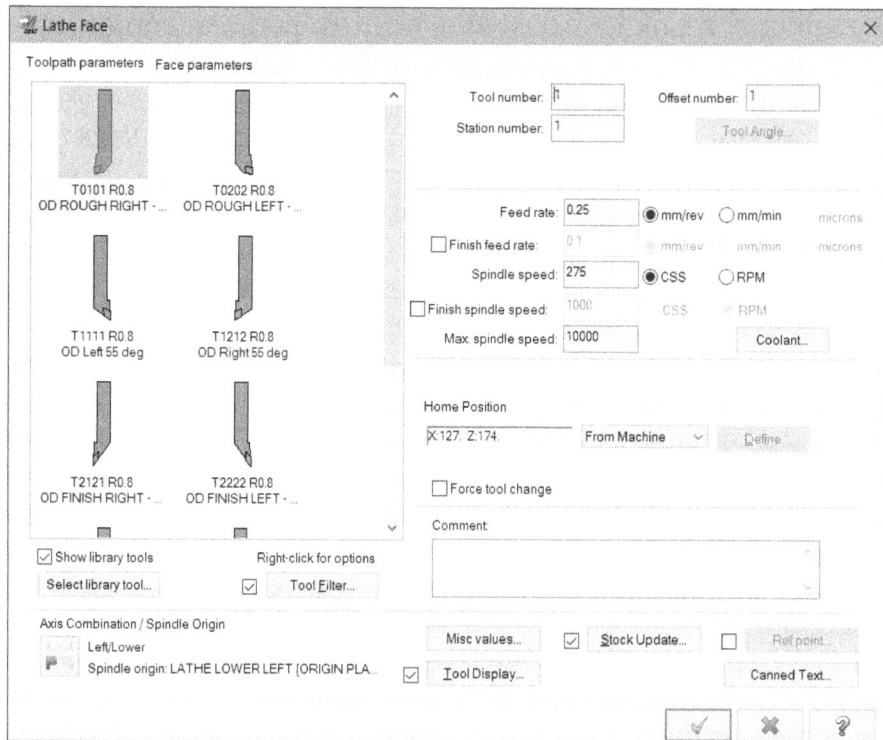

Figure-86. Lathe Face Properties dialog box

- Select the facing tool as per the requirement and specify the related parameters.
- Click on the **Face parameters** tab. The dialog box will be displayed as shown in Figure-87.

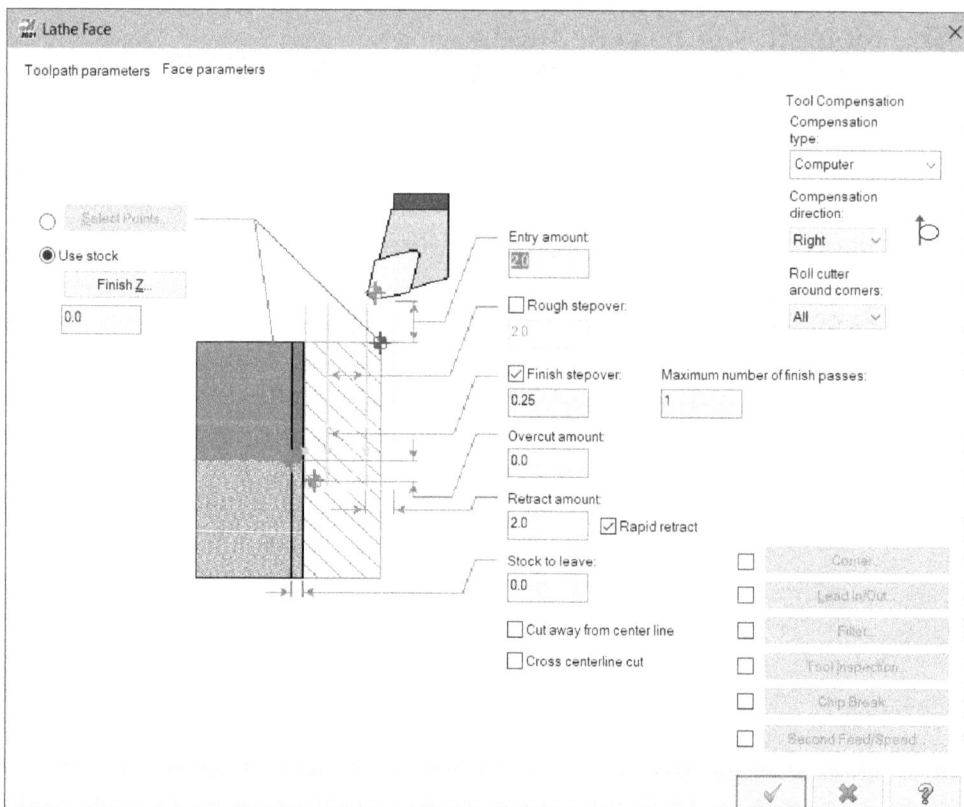

Figure-87. Face parameters page of Lathe Face Properties dialog box

- Specify the parameters related to material to be removed.
- Click on the **Finish Z** button to specify the depth to which the material is to be removed from the face. The **Selection PropertyManager** will be displayed; refer to Figure-88.
- Select the edge of the model up to which you want the material to be removed from the stock; refer to Figure-89.

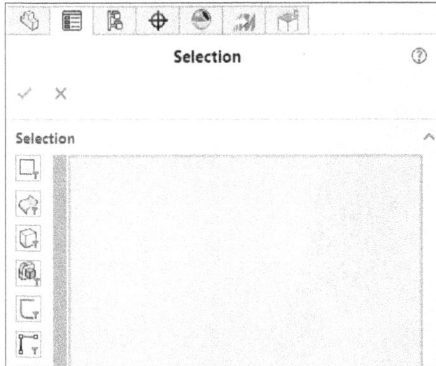

Figure-88. Selection PropertyManager

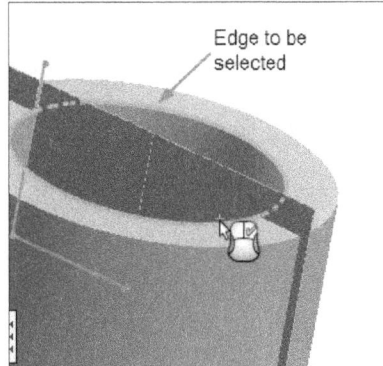

Figure-89. Edge to be selected

- Click on the **OK** button from the **PropertyManager**. The depth of finish Z will be displayed in the edit box below **Finish Z** button.
- Click on the **OK** button from the dialog box. Simulation of cutting will run and the tool path will be created.

LATHE CUTOFF TOOLPATH

We generate the Lathe Cutoff toolpath to cut the stock in two parts. Refer to Figure-90. The procedure to generate this tool path is given next.

Figure-90. Cut off operation.jpg

- Click on the **Cutoff** tool from the **Lathe** drop-down. The **Selection PropertyManager** will be displayed and you will be asked to select a point from where the cutoff operation is to be performed.
- Select the point for cutoff; refer to Figure-91.

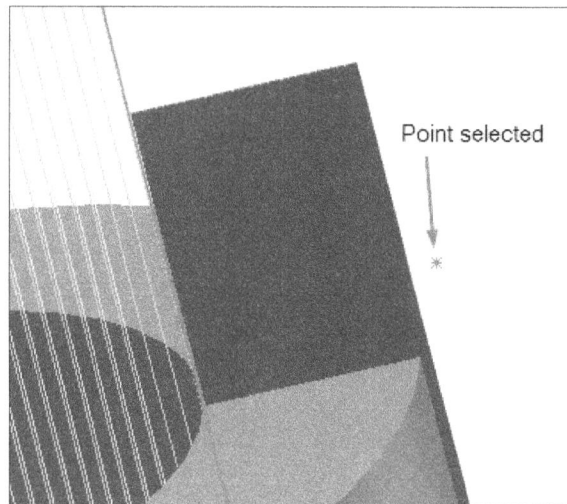

Figure-91. Point selected for cutoff

- Click on the **OK** button from the **Selection PropertyManager**. The **Lathe Cutoff Properties** dialog box will be displayed; refer to Figure-92.
- Select the cutoff tool (OD CUTOFF Tool in our case) from the tool list box and specify the related parameters in the dialog box.
- Set desired parameters in the **Cutoff parameters** tab of the dialog box like retract radius, corner geometry, tool compensation etc.
- Click on the **OK** button from the dialog box. The simulation of cutoff operation will run and the output will be displayed; refer to Figure-93.

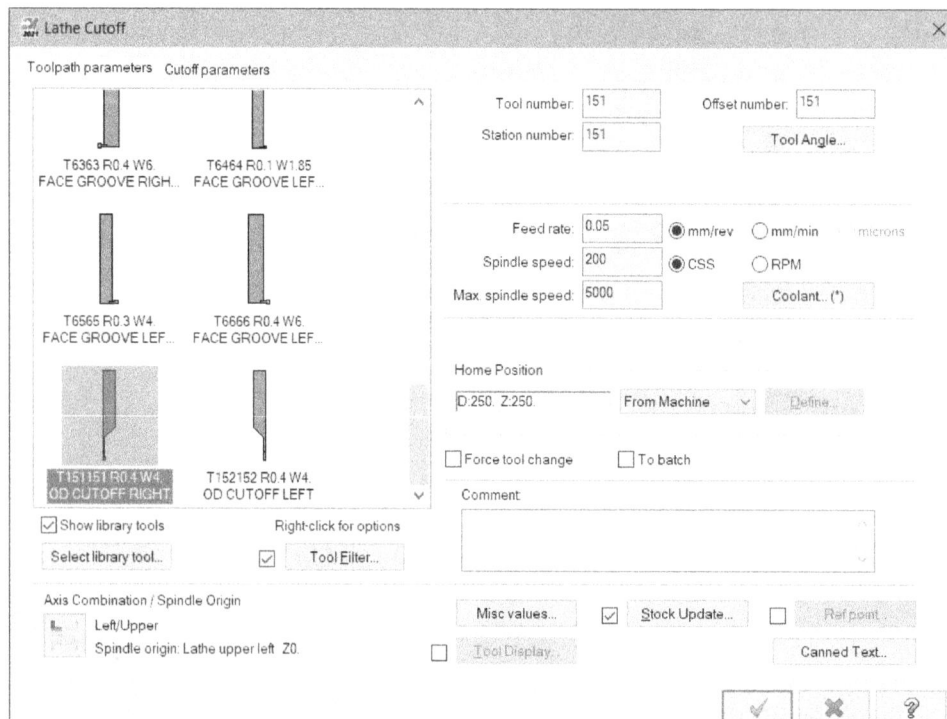

Figure-92. Lathe Cutoff Properties dialog box

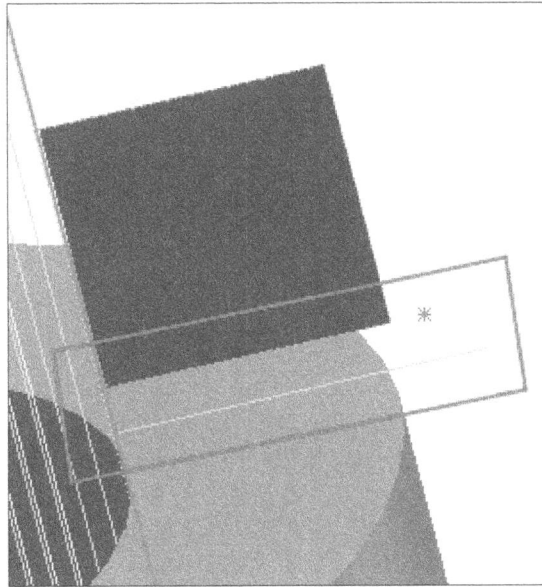

Figure-93. Output of cutoff operation

PLUNGE TURN TOOLPATH

The Plunge Turn toolpath is generated to remove the material from workpiece when there is no option left except plunging the tool in the workpiece. We say it as a last option because there are high changes of breaking tool or machine if this toolpath is not generated correctly. Make sure to select a tool specifically designed for this type of toolpath (Mastercam suggests using the ISCAR tools designed for plunge turn). Using a tool not designed for plunge turning could damage the tool and/or the part. In plunge turn toolpath, the tool plunges in the stock and moves laterally to cut material which is not in case of other toolpaths. Figure-94 shows a case where we need plunge turn toolpath. In this case, we have done facing, rough turning, and finish turning of part. Now, we need to remove material of the taper face and inner groove using the plunge turning. The procedure to generate this tool path is given next.

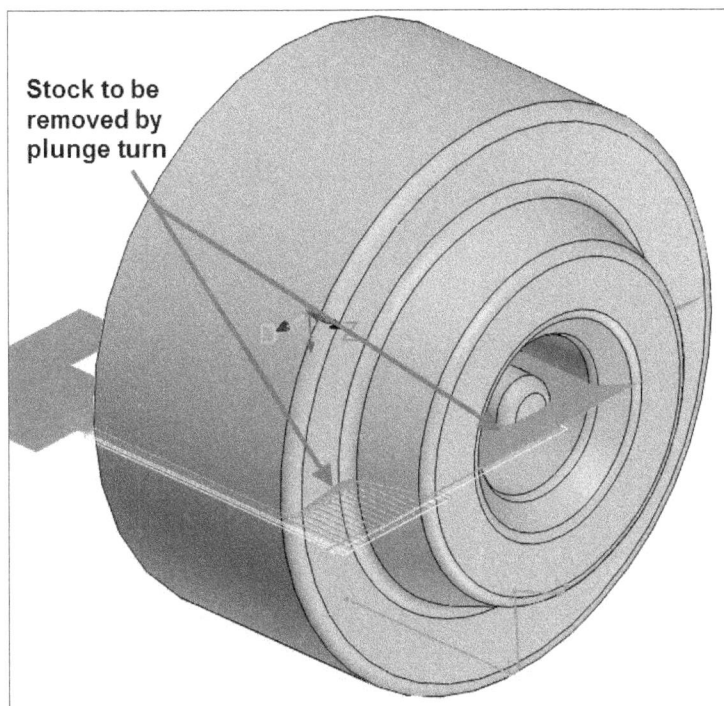

Stock to be
removed by
plunge turn

Figure-94. Case for plunge turn toolpath

- Click on the **Plunge Turn** tool from the expanded **General** panel of **Turning** tab in the **Ribbon**. The **Grooving Options** dialog box will be displayed; refer to Figure-95.

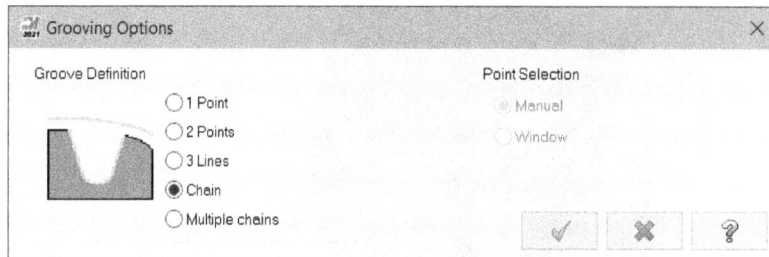

Figure-95. Grooving Options dialog box

- Select desired option and click on the **OK** button from the dialog box. (In our case, its **Multiple Chains** radio button). The **Wireframe Chaining** dialog box will be displayed and you will be prompted to select the chain of sketch entities.
- Select the sketch curves/faces for plunge turning; refer to Figure-96.

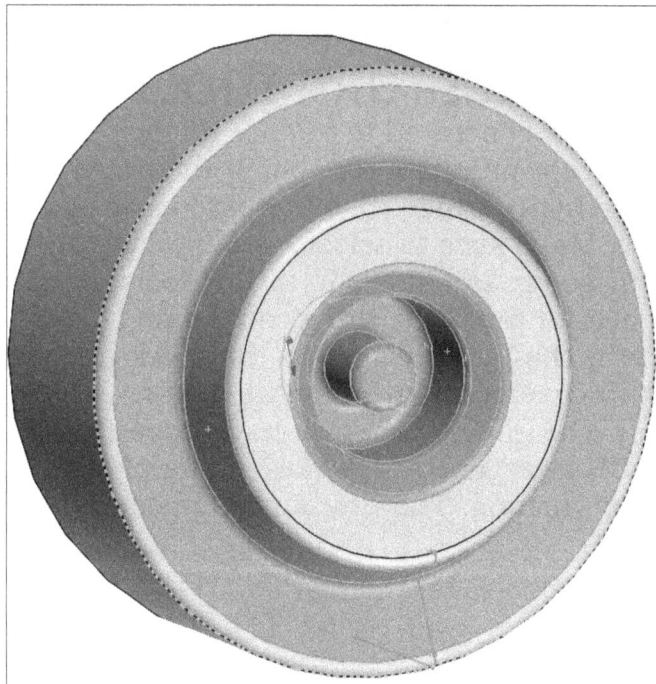

Figure-96. Faces selected for plunge turning

- Click on the **OK** button from the dialog box. The **Plunge Turn** dialog box will be displayed; refer to Figure-97.

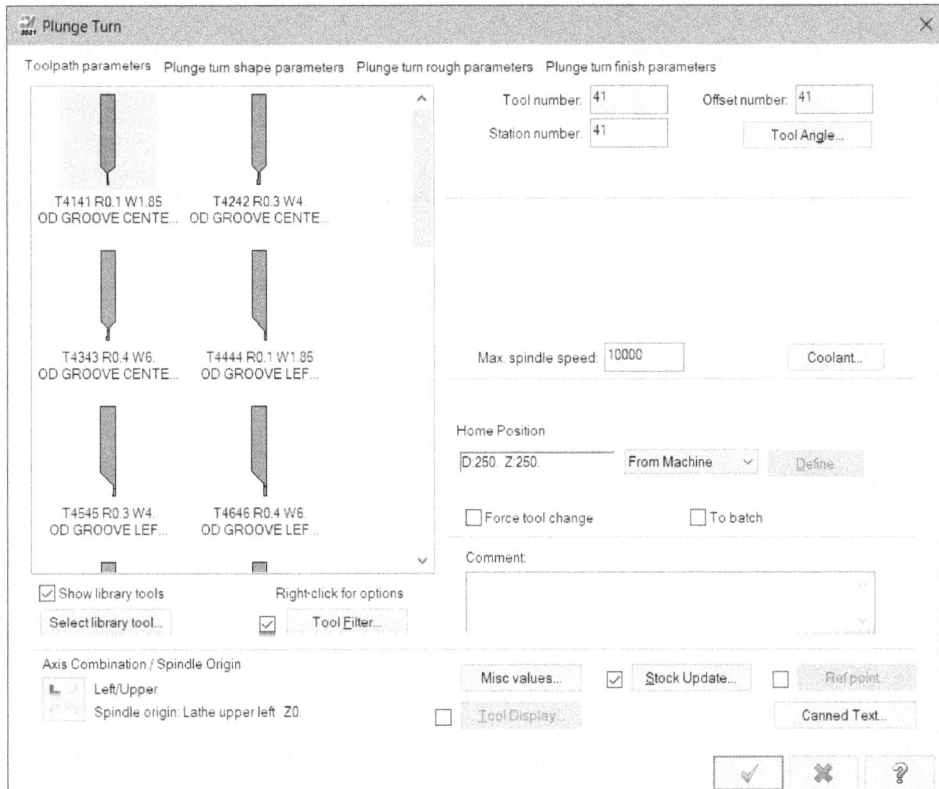

Figure-97. Plunge Turn dialog box

- Select a grooving tool with insert specially design for plunge turn machining.
- Click on the **Tool Angle** button and set the plunge direction normal to the stock face. Make sure the home position is set correctly in the **Home Position** edit box.

Plunge turn shape parameters tab

- Click on the **Plunge turn shape parameters** tab in the dialog box. The dialog box will be displayed as shown in Figure-98. Set desired angle if the groove is inclined otherwise groove will be created based on the plunge direction earlier set. Set the other options as discussed for groove toolpath.

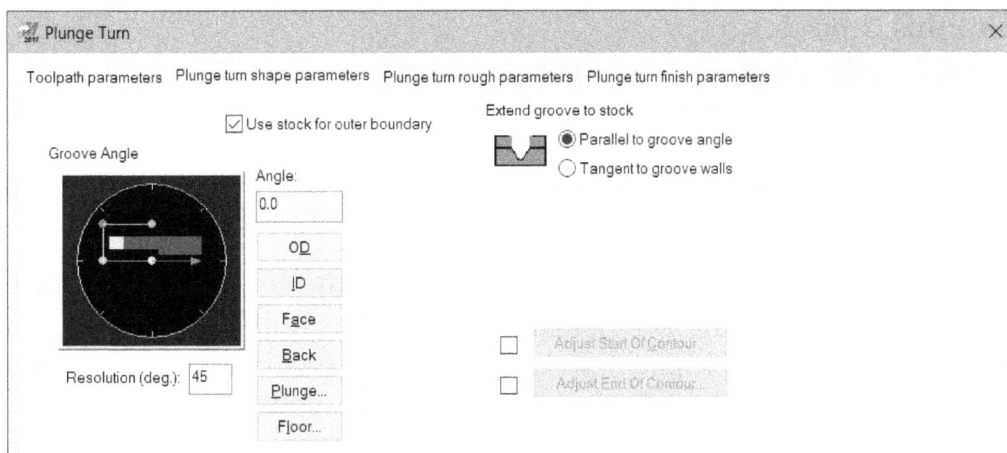

Figure-98. Plunge turn shape parameters tab

Plunge turn rough parameters tab

- Click on the **Plunge turn rough parameters** tab. The options in the dialog box will be displayed as shown in Figure-99.

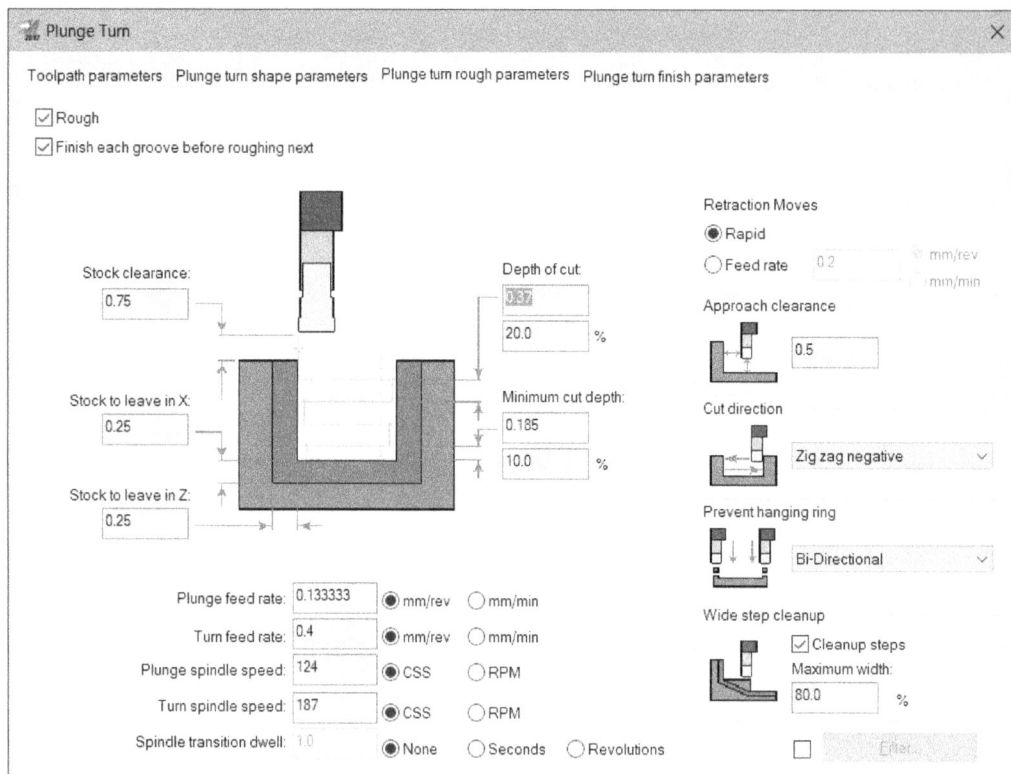

Figure-99. Plunge turn rough parameters tab

- Specify the **Stock clearance** value carefully (wrong value can cause collision). Specify the values in other edit boxes in the dialog box as required or allowed by tool specifications.
- If there are multiple grooves and there is a change of collision between tool and stock then make sure you select the **Finish each groove before roughing next** check box in this tab.
- Specify desired value in **Approach clearance** edit box to define the distance from stock, the tool will move before making the next cut in part. This option is similar to stock clearance for grooves.
- Select desired cutting direction in the drop-down of **Cut direction** area. Preview of the cutting direction is displayed in the **Cut direction** area on selecting an option in this drop-down.
- The options in the drop-down of **Prevent hanging ring** area are used to set the prevention of hanging rings generated after cutting step. A hanging ring occurs when the tool pushes off a small piece of material from the edge of the cut rather than cutting. With **Prevent hanging ring** enabled, the toolpath includes plunge moves to remove this extra material properly. Selecting :

  - Don't prevent - Hanging ring prevention is off.
  - Bi-Directional - Removes possible burrs from both sides of the cut.
  - Positive - Removes a possible burr from the positive side of the cut.
  - Negative - Removes a possible burr from the negative side of the cut.

- Select the **Cleanup steps** check box to turn on this function, which helps remove steps that are too large to remove on a finish cut. Such large steps can remain when the width of the cut is wider than the width of the insert. Wide step cleanup removes steps that are wider than a user definable width, based on the width of the tool's flat. Smoothing will take multiple passes if the extra width is more than the specified maximum percent of tool width.

- Similarly, specify the parameters as required in the **Plunge turn finish parameters** tab and click on the **OK** button from the dialog box. The simulation of machining will run and the toolpath will be generated; refer to Figure-100.

Figure-100. Toolpath generated by plunge turn

Lathe Drill

The **Drill** tool is used to generate toolpaths for drilling holes in the workpiece. The procedure to use this tool is given next.

- Click on the **Drill** tool from the **Lathe** drop-down. The **Lathe Drill** dialog box will be displayed; refer to Figure-101.
- Select desired drill and specify the related parameters for drilling.
- Click on the **Simple drill-no peck** tab. The dialog box will be displayed as shown in Figure-102.

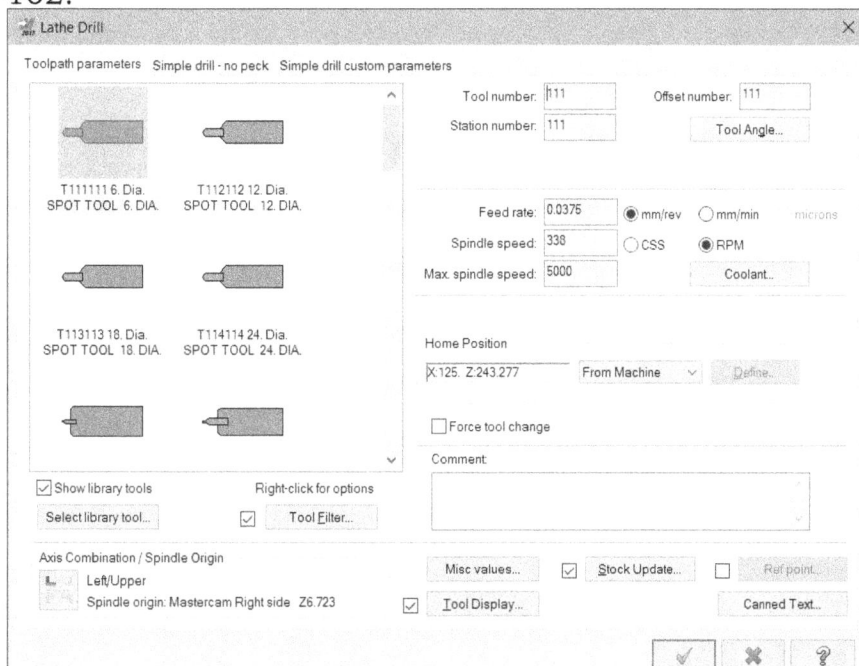

Figure-101. Lathe Drill dialog box

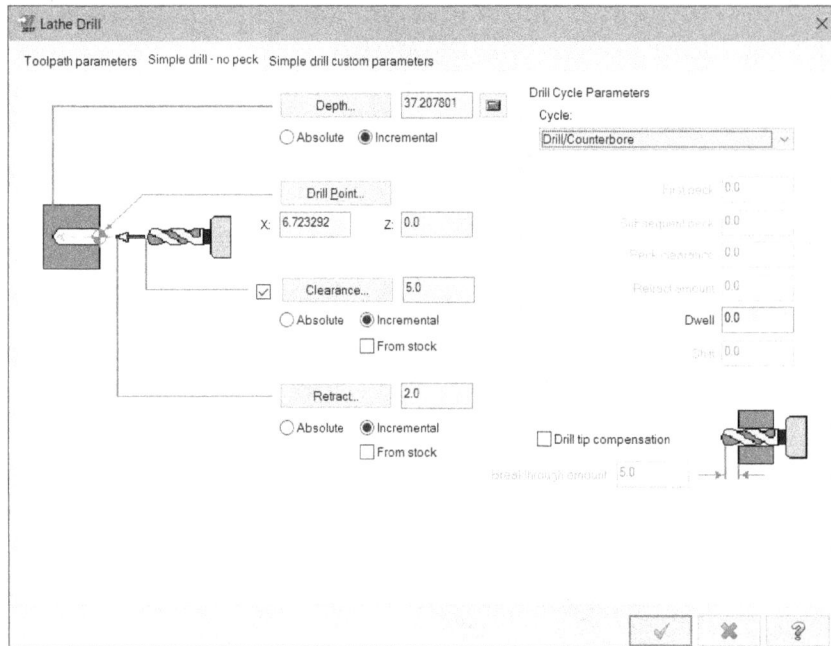

Figure-102. Simple drill page of Lathe Drill Properties dialog box

- Click on the **Drill Point** button and specify the starting point of the drill
- Click in the edit box next to **Depth** button and specify desired value of depth for the drilling or click on the **Depth** button and select a point up to which you want to drill the hole.
- Specify the tool clearance value in the **Clearance** edit box and retraction distance in the **Retract** edit box.
- If you want to create a peck drill cycle, chip break cycle or any other drilling cycle then click in the **Cycle** drop-down and select the respective option. The related parameters in the **Drill Cycle Parameters** area of the dialog box will become active. Set desired values in edit boxes of this area.
- If you want to specify any custom parameter for drill then click on click on the **Simple drill custom parameters** tab and specify the parameters. Note that the name of this tab will change based on the option selected in the **Cycle** drop-down of previous tab.
- Click on the **OK** button from the dialog box. The drilling toolpath will be created.

TOOLPATH THROUGH POINTS

The **Point** tool is used to create machining toolpaths with the help of points. Select the points as per their order of cutting and tool will move accordingly to generate the toolpath. The steps to perform the operation are given next.

- Click on the **Point** tool from the **Lathe** drop-down. The **Point Manager** will be displayed; refer to Figure-103.

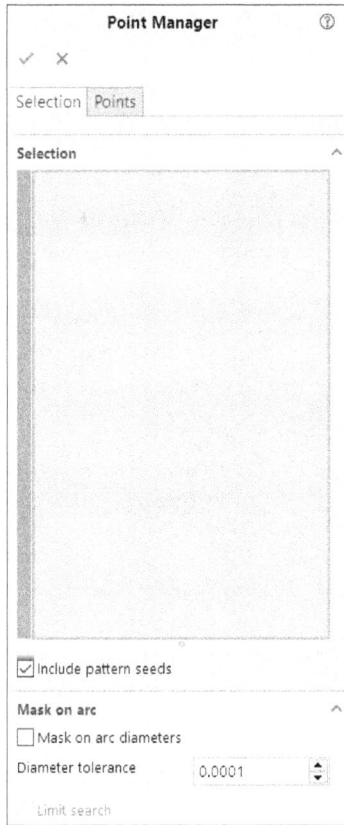

Figure-103. Point Manager

- Select the points of the toolpath one by one and then click on the **Points** tab in the **PropertyManager**. Selected points will be displayed. Click on the **Sorting Options** button in the **Sorting Options** rollout at the bottom of the **Propertymanager**. The **Sorting** dialog box will be displayed; refer to Figure-104.

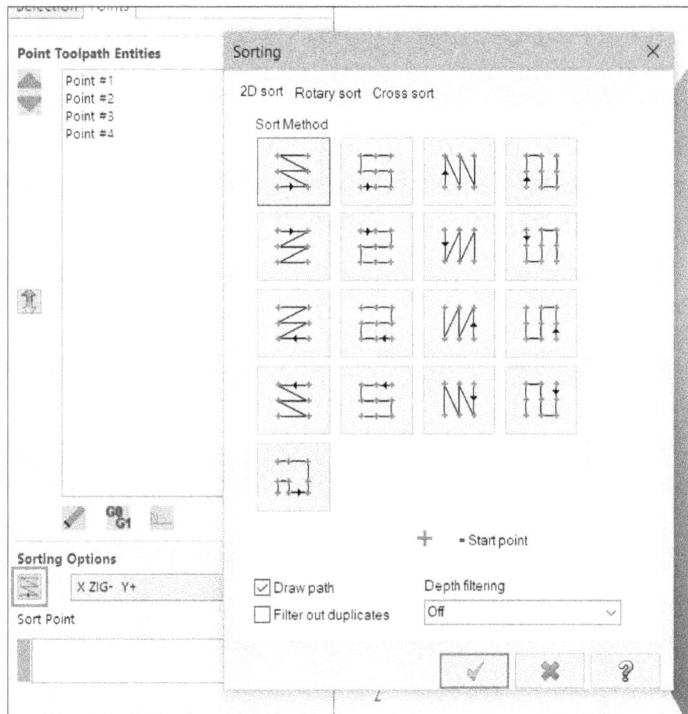

Figure-104. Sorting dialog box

- Select desired button from the dialog box to define cutting order for selected points and click on the **OK** button. The **Point Manager** will be displayed again.

- Click on the **OK** button from the **Point Manager**. The **Lathe Point Properties** dialog box will be displayed; refer to Figure-105.

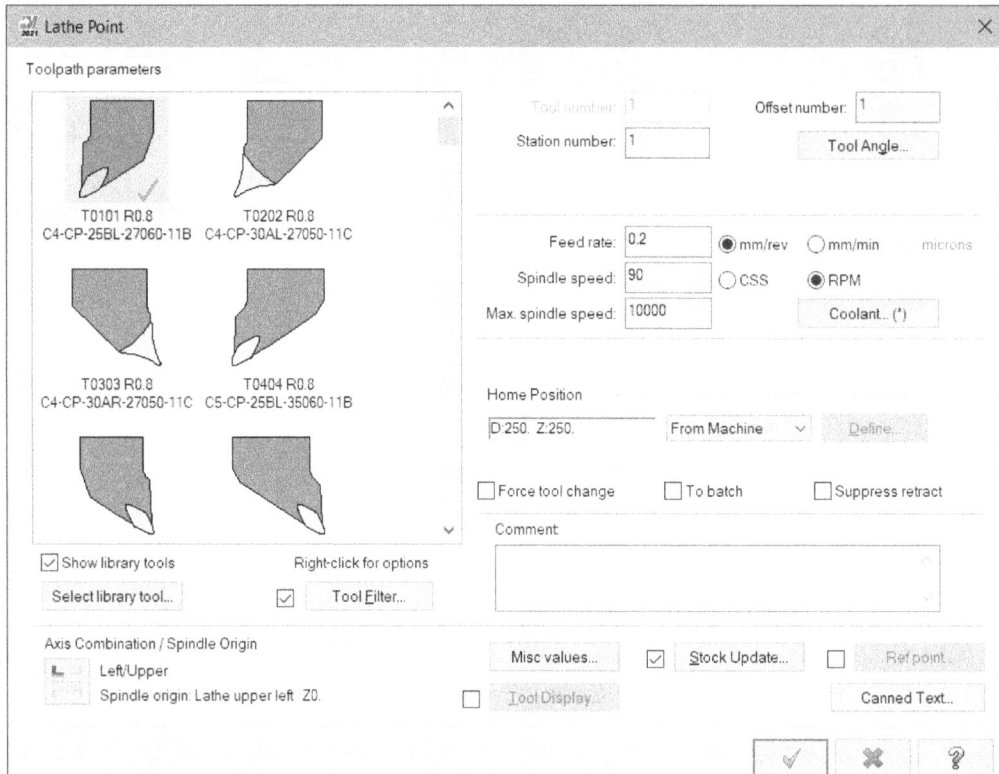

Figure-105. Lathe Point dialog box

- Specify desired parameters and then click on the **OK** button. The toolpath will be generated accordingly.

Canned Toolpaths

Some of you will be aware of canned cycles used in various machines for cutting workpiece. The canned cycles are repetitive cutting steps with specified depth increment. For example, if you want to remove 50 mm of stock by roughing operation then you need to use the canned roughing cycle to remove this much amount of material. For this cycle, you will specify start point, end point and number of steps required. There are four tools to generate canned cycles for various cutting operations:

Canned Rough Toolpath
Canned Finish Toolpath
Canned Groove Toolpath
Canned Pattern Repeat Toolpath

We will discuss the Canned Rough Toolpath and the other tools work in the same way.

Canned Rough Toolpath

The canned rough toolpath is generated in the same way as the rough toolpaths are used. The steps to create canned rough toolpaths are given next.

- Click on the **Canned Rough** tool from the **Lathe** drop-down. The **Chain Manager** will be displayed as discussed earlier.

- Select the sketched chain for toolpath and click on the **OK** button from the **Chain Manager**. The **Lathe Canned Rough Properties** dialog box will be displayed; refer to Figure-106.

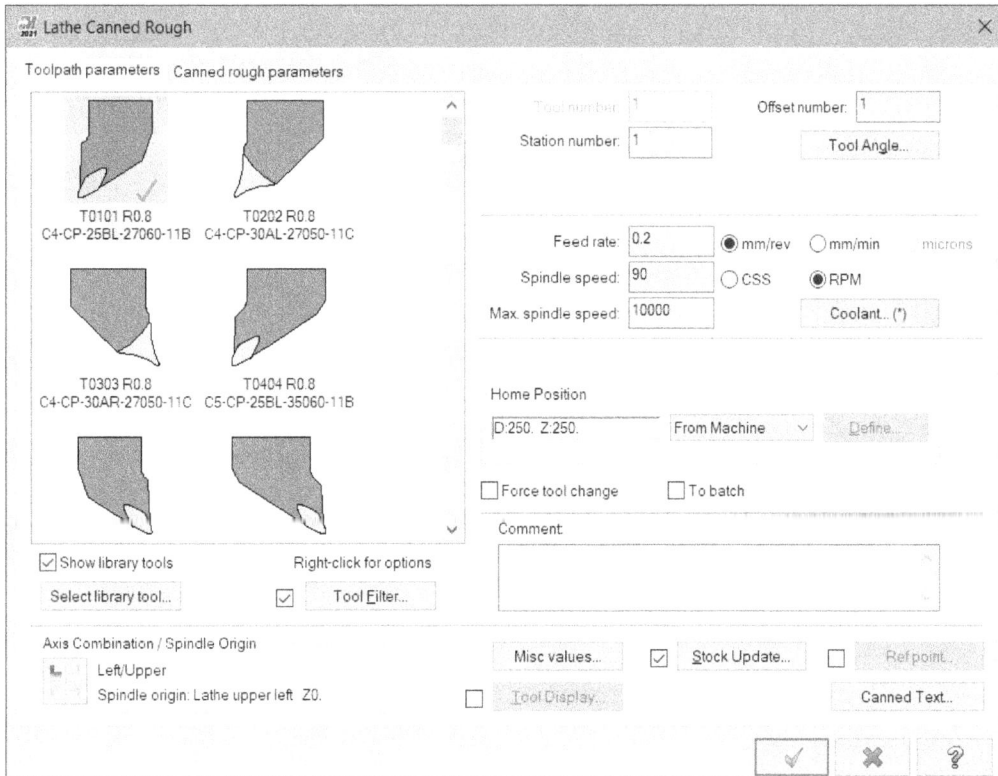

Figure-106. Lathe Canned Rough dialog box

- Select desired tool and specify the cutting parameters.
- Click on the **Canned Rough Parameters** tab in the dialog box to define toolpath parameters; refer to Figure-107.

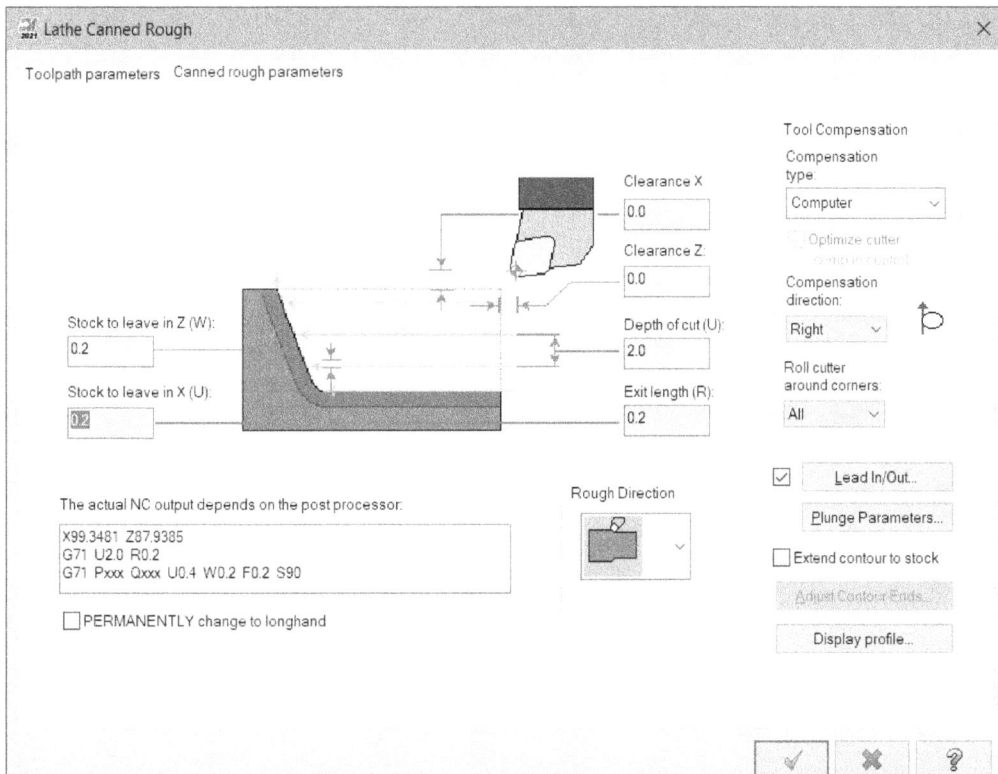

Figure-107. Canned rough parameters tab

- Specify the depth of cut and stock to be left in the respective edit boxes.
- Specify the other parameters as done in roughing operations and click on the **OK** button from the dialog box. The toolpath of canned roughing will be generated automatically.

In this chapter, we have worked with various toolpaths and tools related to Lathe machining. In the next chapter, we will work with some advanced lathe machining tools and toolpaths.

PRACTICAL 1

In this practical, we will create toolpaths for the given model; refer to Figure-108. The dimensions of stock are shown in Figure-109.

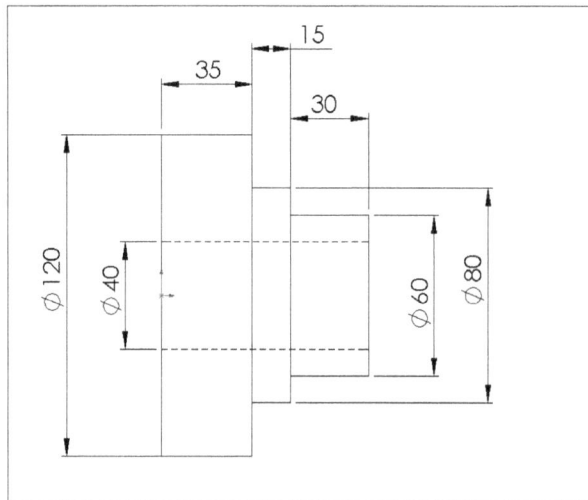

Figure-108. Practical 1

The first step before we start machining is to identify the operations that are required to machine the part. We can identify from the part that there is facing operation, one drilling operation, one boring operation to get good finish of hole and roughing operation. Note that these basic machining operations can be sub-divided into multiple operations based on the capabilities of the CAM software. We will start one by one for various operations to be performed.

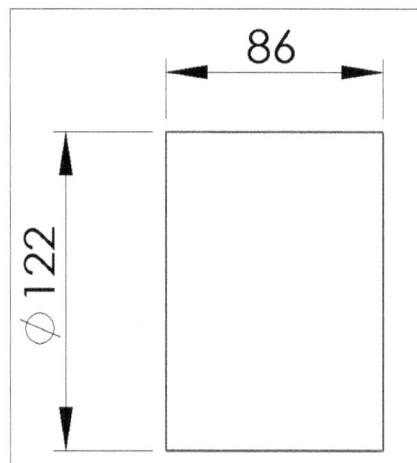

Figure-109. Stock dimension for Practical

Preparing the model

- Start SolidWorks and click on the **Open** button from the **Quick Access** toolbar displayed at top in the Application window. The **Open** dialog box will be displayed; refer to Figure-110.

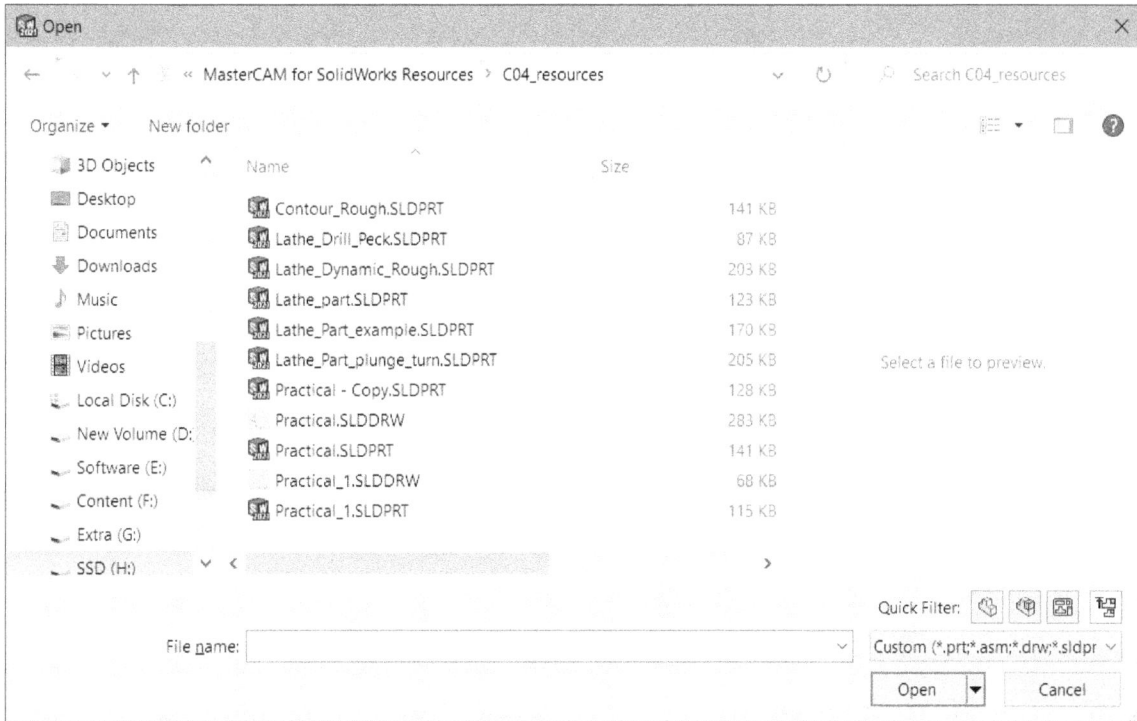

Figure-110. Open dialog box

- Select the model file provided for Practical of Chapter 4 in the Resources (ask at our Email address if do not have part files) and click on the **Open** button from the dialog box. The model will be displayed as shown in Figure-111.

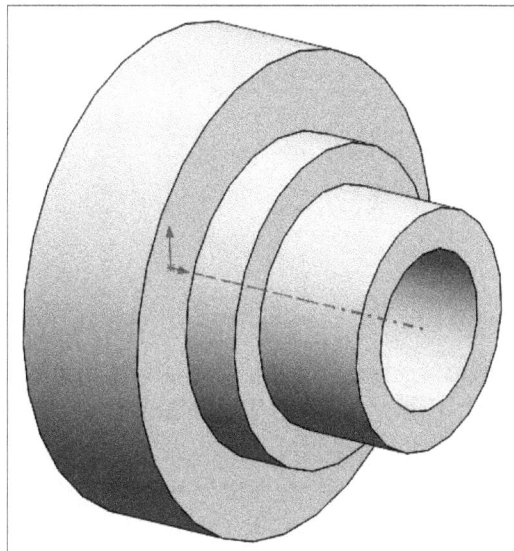

Figure-111. Model for Practical 1

- Add the **MasterCAM** add-in if not added till now.
- Select the **Lathe Default MM.MCAD-LMD** machine from **Tools > Mastercam2022 > Lathe Machines** menu. If not available then add it as discussed earlier.

Creating Stock

- Click on the **Planes Manager** tab from the **Design Tree**. The **Planes Manager** will be displayed to set the working plane.

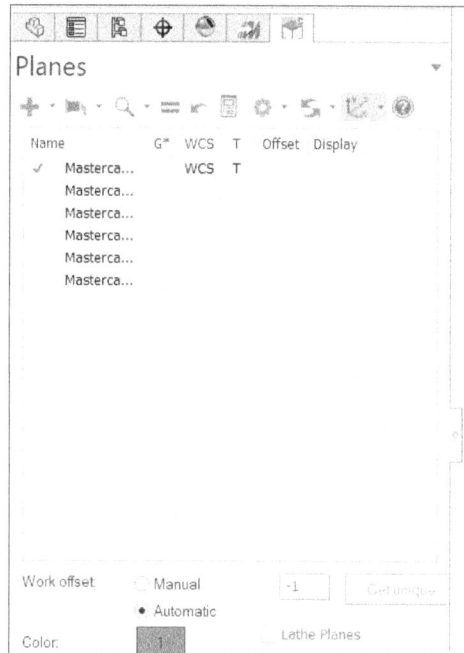

Figure-112. Plane Manager dialog box

- Make sure the **MASTERCAM TOP** plane is selected as tool plane as well as WCS. If not then set it as discussed earlier.
- Click on the **+** sign displayed before **Properties - Lathe Default** MM node. The node will expand and will be displayed as shown in Figure-113.

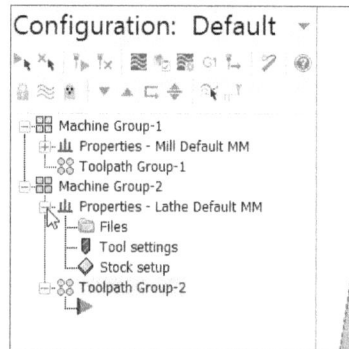

Figure-113. Expanded node of machine properties

- Click on the **Stock Setup** option from the **Properties** node in the **Mastercam Toolpath Manager**; refer to Figure-114. The **Machine Group Properties** dialog box will be displayed; refer to Figure-115.

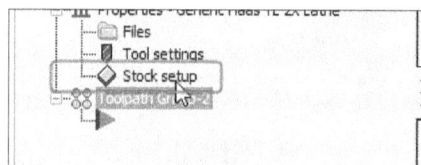

Figure-114. Stock setup option

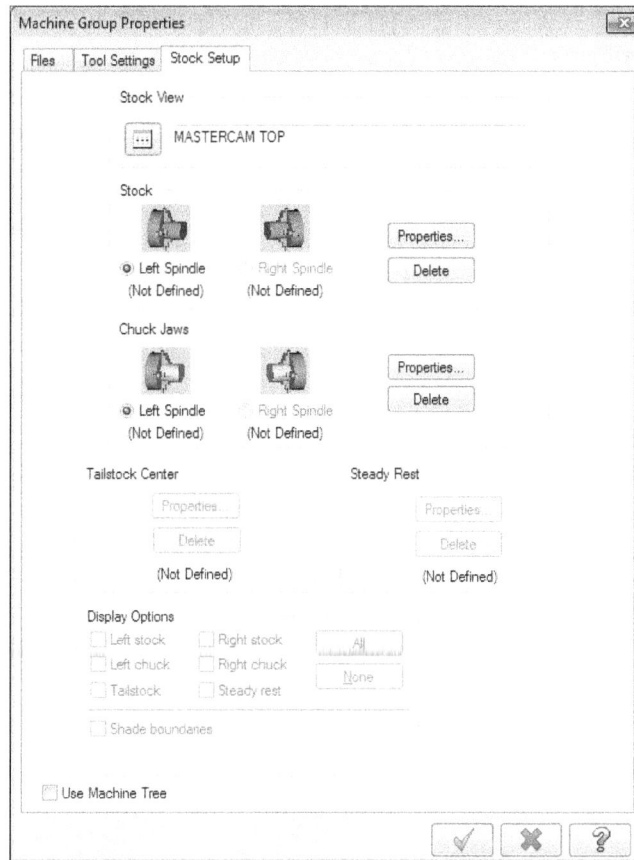

Figure-115. Machine Group Properties dialog box

- Make sure the **Left Spindle** radio buttons are selected in the dialog box and then click on the **Properties** button in the **Stock** area of the dialog box. The **Machine Component Manager-Stock** dialog box will be displayed. Specify the **OD** as **120** and **Length** as **80** in respective edit boxes. Select **+Z** option from the **Axis** drop-down and select the **Use Margins** check box.
- Specify **OD Margin** as **2.0**, **Right Margin** as **3.0**, and **Left Margin** as **3.0** in respective edit boxes; refer to Figure-116.
- Click on the **OK** button from the dialog box.

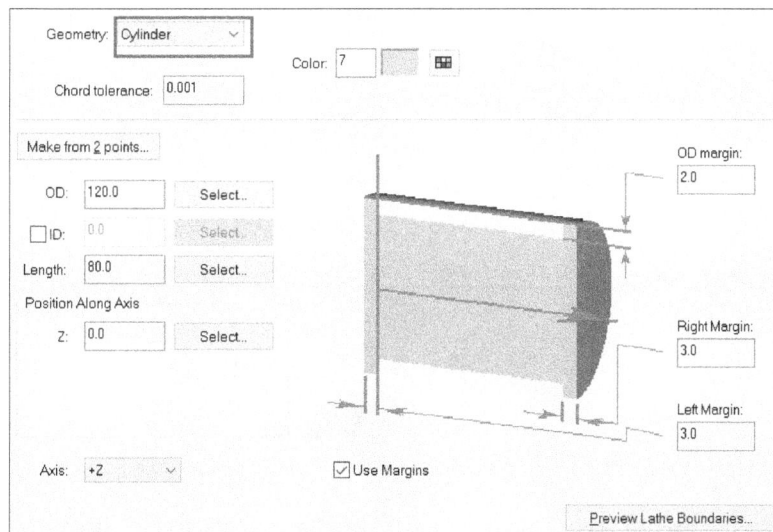

Figure-116. Margins and parameters for stock

Creating Chuck Jaws

- Click on the **Properties** button from the **Chuck Jaws** area of the dialog box. The **Machine Component Manager - Chuck Jaws** dialog box will be displayed.
- Select the **Outer diameter (OD)** check box from the **Clamping Method** area in the dialog box. Select the **From Stock** check box and **Grip on maximum diameter** check box from the **Position** area of the dialog box.
- Specify the **Grip length** as **6** in the edit box; refer to Figure-117.

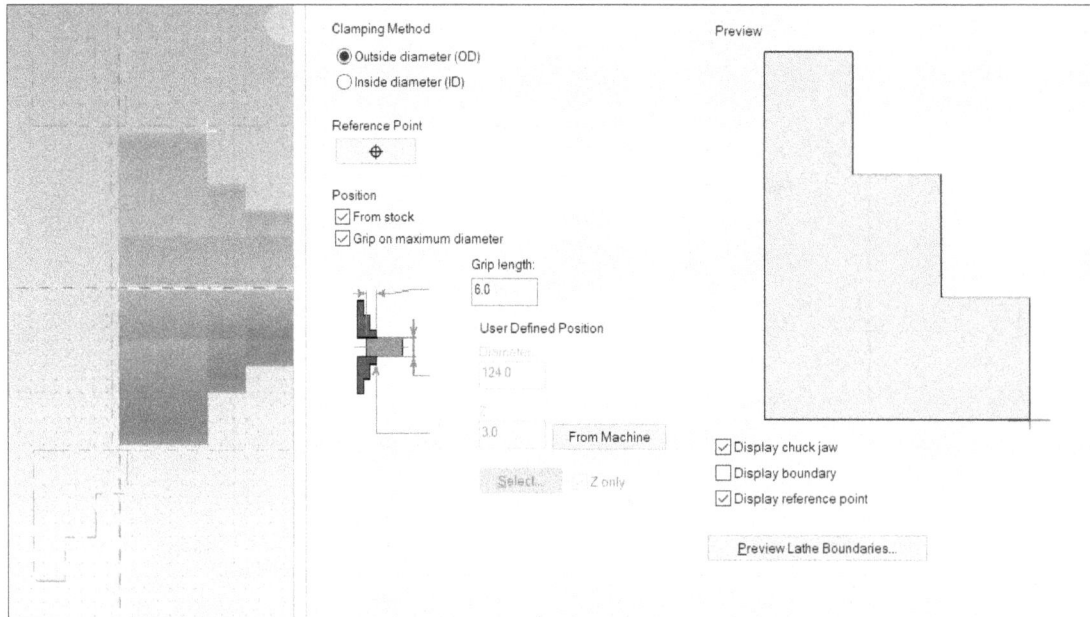

Figure-117. Parameters for chuck jaws

- Click on the **OK** button from the dialog box. The **Machine Group Properties** dialog box will be displayed again. Select the **Shade boundaries** check box and click on the **OK** button from the dialog box to exit.

Creating Sketch for Toolpaths

- Click on the **Sketch** button from the **Sketch CommandManager** in the **Ribbon** and create a sketch on the Top Plane as shown in Figure-118.

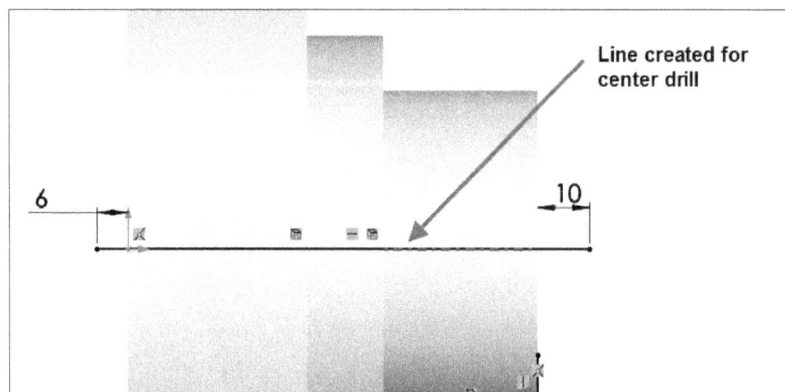

Figure-118. Sketch for toolpath of Practical 1

- Click on the **Exit Sketch** button from the left corner of the **Sketch CommandManager** to exit sketching.

Performing Facing Operation

- Click on the **Face** tool from the **Lathe** drop-down in the **Mastercam2022 CommandManager** of the **Ribbon**. The **Lathe Face** dialog box will be displayed.
- Select the **T3131 R0.8 ROUGH FACE RIGHT -80 DEG** tool from the tool list. The parameters as per the tool will be displayed on the right area in the dialog box.
- Select the **Finish feed rate** and **Finish spindle speed** check boxes from the right area.
- Set the values as per your machine and tool capabilities in the right area (We will continue with default values).
- Click on the **Face parameters** tab in the dialog box.
- Click on the **Finish Z** button in the dialog box. The **Selection PropertyManager** will be displayed and you will be asked to select a point.
- Select the edge of model as shown in Figure-119. Click on the **OK** button from the **PropertyManager**. The value of finish Z will be reflected automatically in the **Finish Z** edit box in the dialog box.

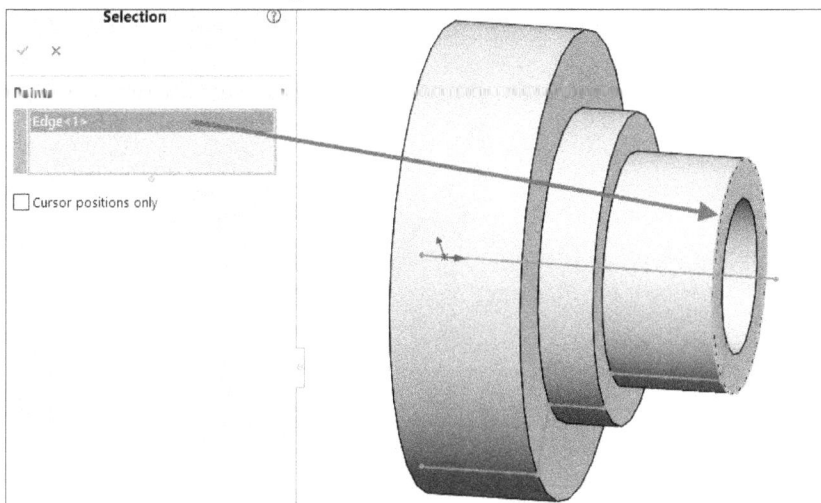

Figure-119. Edge selected for Finish Z

- Specify the other parameters as shown in Figure-120 and click on the **OK** button to generate toolpath.

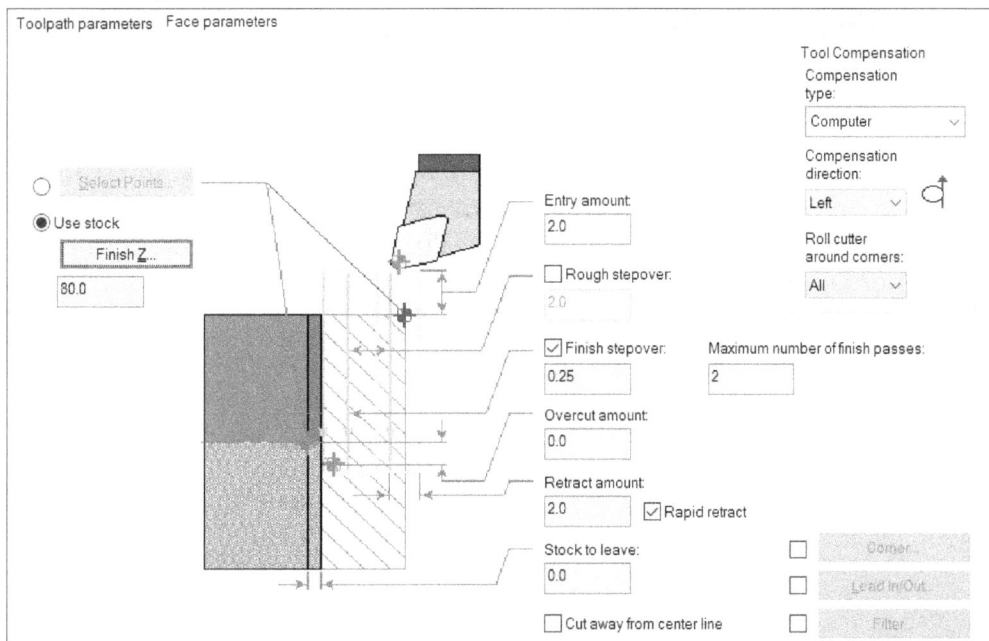

Figure-120. Parameters for facing

Performing OD Rough & OD FInish

- Click on the **Rough** tool from the **Lathe** drop-down in the **Ribbon**. The **Chain Manager** will be displayed asking you to select geometries for toolpath.
- Select the Outer faces of the model; refer to Figure-121 and select the **Upper** or **Lower** radio button for **Profile Generation** in the **Chaining Options** rollout as per your tool orientation. If the radio button selected is not correct then the tool can be in reverse direction. Click on the **Chains** tab and select the chain to check the tool entry and exit.

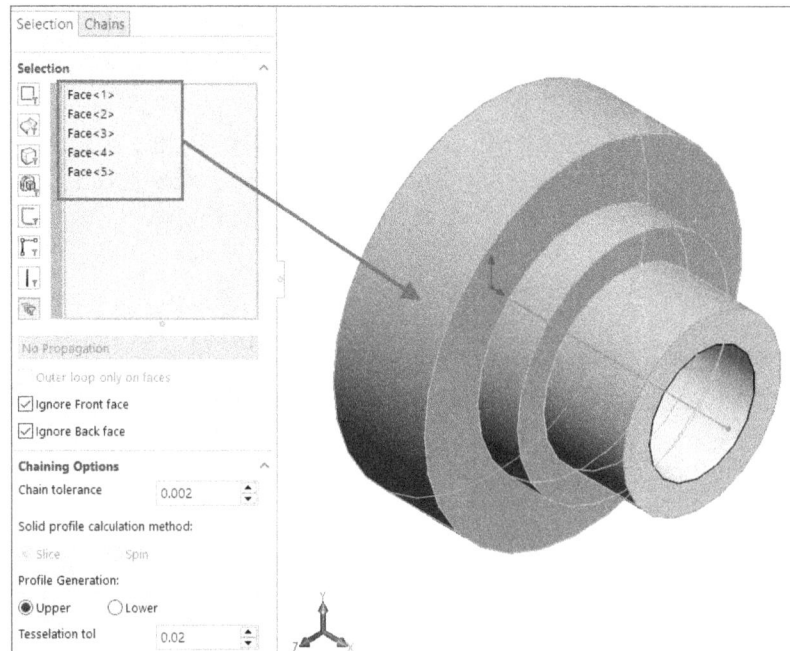

Figure–121. Sketch lines selected for OD rough

- Click on the **OK** button from the **Chain Manager**. The **Lathe Rough** dialog box will be displayed.
- Select the **T0101 R0.8 OD ROUGH RIGHT - 80 DEG** tool from the tool list box and specify desired feed rate parameters.
- Click on the **Rough parameters** tab and specify **0.5** in both **Stock to leave in X** and **Stock to leave in Z** edit boxes. If we run this toolpath then the tool is going to hit chuck and may kill the machine. We need to adjust the end contour of toolpath.
- Click on the **Lead In/Out** button and then click on the **Lead Out** tab in the **Lead In/Out** dialog box. Select the **Extend/shorten end of contour** check box from the **Adjust Contour** area of the dialog box. The related options will become active.
- Select the **Shorten** radio button and specify the amount as **4.0** in the edit box; refer to Figure-122.

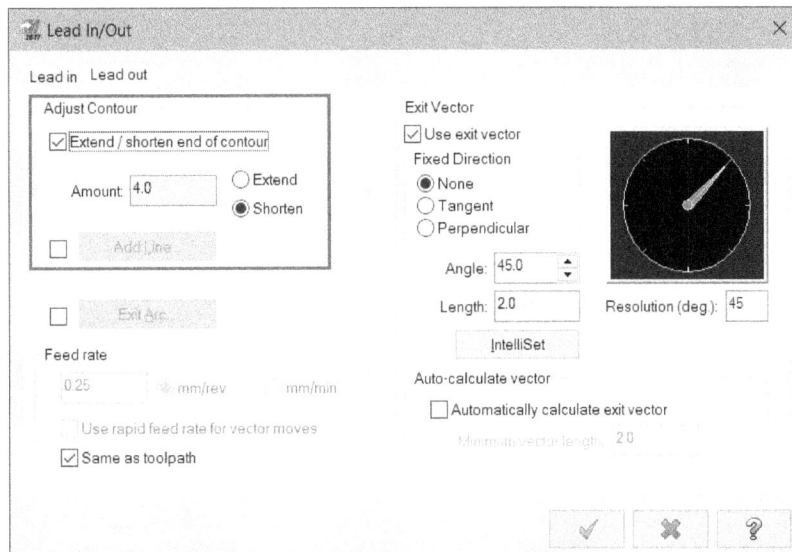

Figure-122. Adjusting contour to save tool and chuck

- Click on the **OK** button from the dialog box and then click on the **OK** button from the **Lathe Rough** dialog box.

Similarly, you perform the OD finish by using the **Finish** tool in the **Lathe** drop-down of **Ribbon**.

Performing Drilling

To get a better accuracy in center hole, we can plan to perform drilling with lesser diameter and then perform boring with exact diameter. The steps are given next.

- Click on the **Drill** tool from the **Lathe** drop-down in the **Ribbon**. The **Lathe Drill** dialog box will be displayed.
- Select the T129129 35 DIA DRILL from the tool list box. The related parameters will be displayed on the right in the dialog box. Again, set the parameters as per your machine and tool capabilities (We are continuing with default values).
- Sometimes, axis and spindle origin do not set automatically with default origins and can cause problem in specifying hole position of tool. Click on the button in **Axis Combination/Spindle Origin** area at the bottom left in the dialog box. Don't change anything and click on the **OK** button from the dialog box. This will align current tool plane with default WCS.
- Specify a suitable Home position by using the tools in the Home Position area as the default drill and drill holders in the application are bigger in size compared to other tools.
- Click on the **Simple drill - no peck** tab in the dialog box. The options related to drill depth and drill point will be displayed.
- Click on the **Depth** button and select the end point of centerline we created earlier in sketch; refer to Figure-123. Click on the **OK** button from the **Selection PropertyManager**. The **Lathe Drill** dialog box will be displayed again with depth value reflected in **Depth** edit box. Click on the **Depth Calculator** button next to the edit box and click on the **OK** button from the **Depth Calculator** dialog box displayed. This will add the drill tool tip length in the depth if you want to create a thorough hole.

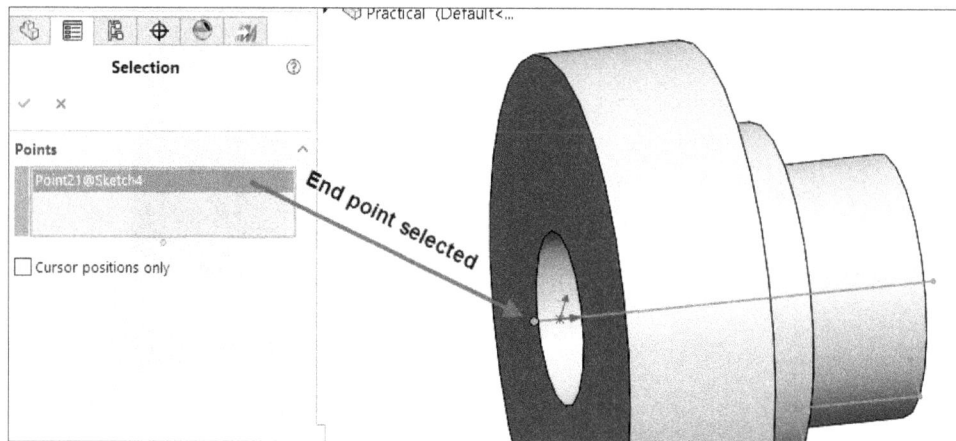

Figure-123. Drill depth point selected

- Click on the **Drill Point** button in the dialog box. The **Selection PropertyManager** will be displayed and you will be asked to select the starting point of drill toolpath. Select the other end point of the same line and click on the **OK** button from the **PropertyManager**.
- Click on the **OK** button from the dialog box to create the toolpath.

Performing Boring Operation

No! this operation is not that much 'boring (dull)'. This operation is performed to increases the diameter of hole in tight tolerance.

- Click on the **Finish** tool from the **Lathe** drop-down in the **Ribbon**. The **Chain Manager** will be displayed and you will be asked to select geometry for finishing operation.
- Select the inner round face of the model as shown in Figure-124 and click on the **OK** button from the **Chain Manager**. The **Lathe Finish** dialog box will be displayed.
- Select the **T8383 R0.8 ID FINISH MIN 25 DIA - 55 DEG** tool from the tool list box. Specify the feed parameters as per your machine and tool capabilities. Note that you need to modify or create new tools in Mastercam if you have different tools in your workshop. Otherwise, the NC program generated here will not be useful on your machine.

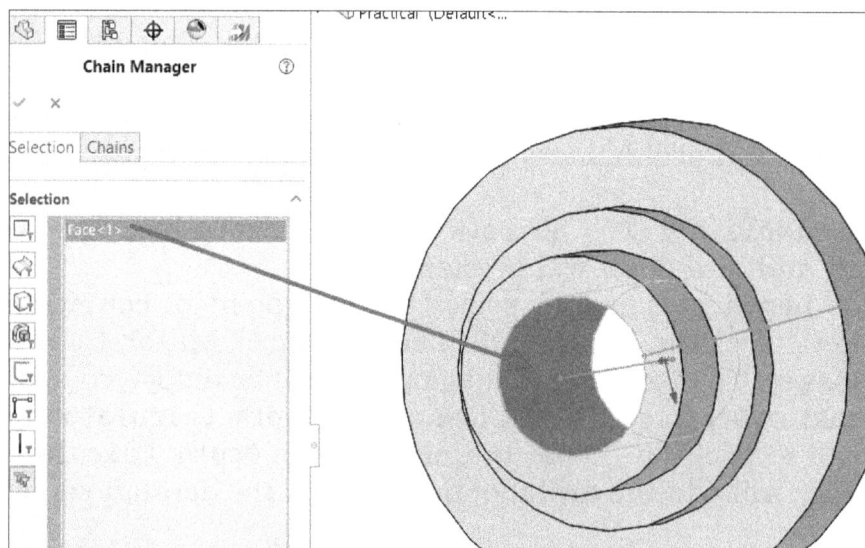

Figure-124. Round face selected for bore finishing

- Click on the **Finish parameters** tab in the dialog box. The options related to finish toolpath will be displayed.
- Set the **Finish stepover** value as **1** and **Number of finish passes** as 2 in respective edit boxes.
- If we click on **OK** button and create the toolpath now, then there will be some material left at the back face which need boring operation; refer to Figure-125. To solve this problem, we will set some extra lead out in the toolpath.

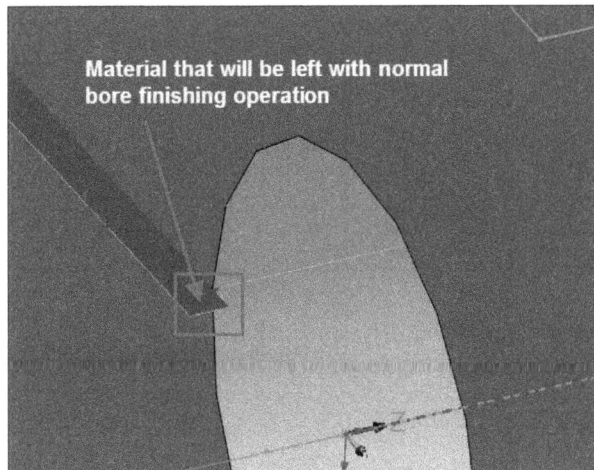

Figure-125. Material that could be left after boring

- Click on the **Lead In/Out** button from the dialog box. The **Lead In/Out** dialog box will be displayed.
- Click on the **Lead out** tab in the dialog box and extend the contour by **5**; refer to Figure-126(Options marked in boxes).

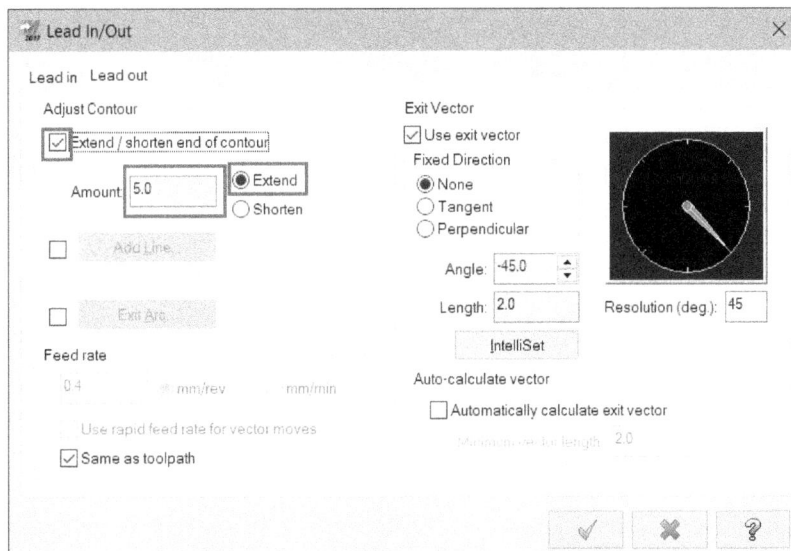

Figure-126. Options to extend contour

- Click on the **OK** button from the **Lead In/Out** dialog box and then click on the **OK** button from the **Lathe Finish** dialog box to create the toolpath.

Generating G-codes

- Select the specific operation like **Lathe Rough** from the **Mastercam Toolpath Manager** if you want to generate G codes for any specific operation; refer to Figure-127.

If you want to generate G codes for all the operations in current toolpath group then select the group from the **Mastercam Toolpath Manager**.

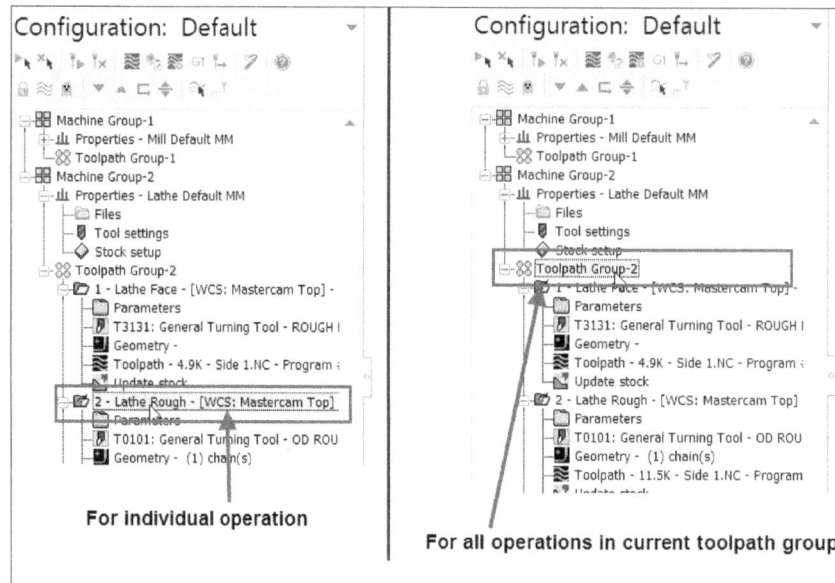

Figure-127. Operation to be selected

- Click on the **Post selected operations** button G1 from the toolbar in the **Mastercam Toolpath Manager**. The **Post processing** dialog box will be displayed; refer to Figure-128. (This option is not available in HLE Mastercam for SolidWorks.)

Figure-128. Post processing dialog box

- If you have a machine connected to the system then you can select the **Send to machine** check box and communicate the program directly to machine.
- Click on the **OK** button from the dialog box. Note that the post processor selected in Machine definition will be used to output G codes. The **Save As** dialog box will be displayed; refer to Figure-129.

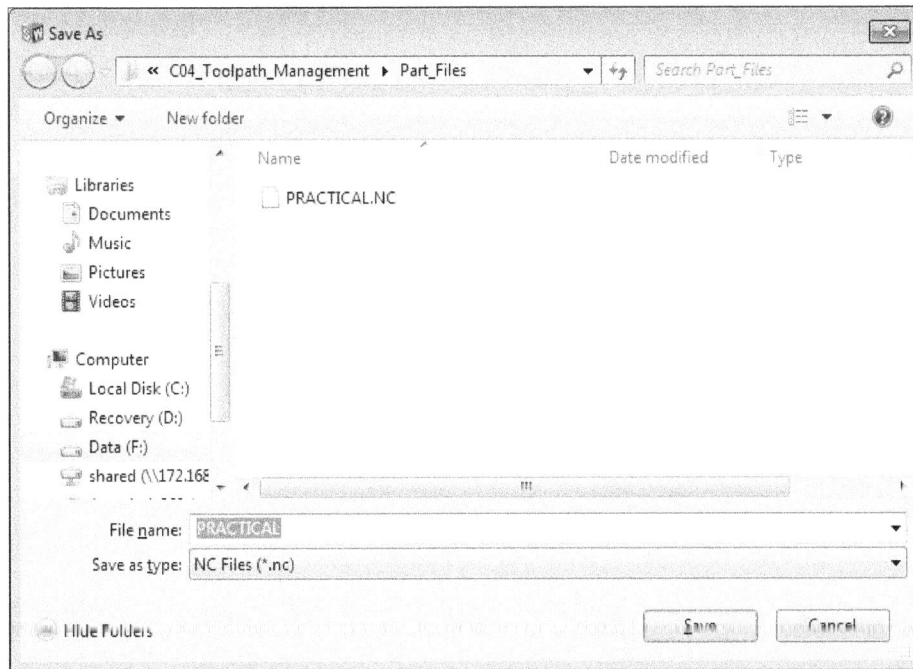

Figure-129. Save As dialog box

- Click on the **Save** button and save the file with desired name. The **Mastercam Code Expert** application will open. Verify the codes and close the application if you find the codes satisfactory. Otherwise you can edit the code file.

In this chapter, we covered the procedure to create toolpaths for lathe machines. We have worked on a real model for lathe machining. In the next chapter, we will cover some more lathe machining tools and practical exercises.

SELF-ASSESSMENT

Q1. Spindle speed is the rotation speed of spindle in RPM. (T/F)

Q2. Feed rate is the rate at which tool perpendicularly enters in the workpiece along Z axis in case of lathe. (T/F)

Q3. Plunge rate is the rate at which tool exits cutting and moves to the starting point of the next cutting pass or moves for non-cutting operation.

Q4. Retract rate is the rate at which tool perpendicularly enters in the workpiece along Z axis in case of lathe. (T/F)

Q5. To tell machine that 'x' length of stock will be held by chuck you need to specify the value in **Z** edit box of **Position Along Axis** area of the .... dialog box.

Q6. .... and .... generally used to support long cylindrical parts that can bend during turning operation.

Q7. Which of the following parameters is used to mark the tools that wear by same amount in same group?

a. Tool number   b. Station number
c. Offset number   d. Home Position

Q8. Which of the following tools/options is not used to generate machining profile of model for turning?

a. Chain Manager   b. Silhouette Boundary
c. Turn Profile   d. Selection PropertyManager

Q9. Which of the following cutting inserts can be used with Dynamic rough toolpaths for turning operations?

a. R inserts   b. V inserts
c. W inserts   d. D inserts

Q10. Which of the following cutting tools are used in conjunction with PrimeTurning toolpaths?

a. Sandvik Coromant cutting tools   b. ISCAR cutting tools
c. Kennametal cutting tools   d. Valenite cutting tools

Chapter 5

Advanced Lathe Tools and Toolpaths

Topics Covered

The major topics covered in this chapter are:

- *Introduction*
- *Stock Pickoff, Pull, and Cutoff Operations*
- *Stock Transfer*
- *Stock Flip*
- *Stock Advance*
- *Undoing Stock related operations*
- *Chuck and Tailstock movement code generation*
- *Face Contouring*
- *Cross Contouring*
- *Back Plot and Verifying Toolpaths*
- *Analyzing Toolpaths*
- *Manual Toolpath Code Entry*
- *Practical and Practice*

INTRODUCTION

In the previous chapter, you have learned about the basic operations that are done on simple 2 axis lathe. But the CNC turning machines are not confined to 2 axis lathe. The turning machine can be of as many as 4 axis. Note that we have left the back face of model un-machined in Practical of previous chapter. But in real machining, we will open the chuck and reverse the side of part to get back face machined. We will learn about such operations in this chapter. We will also learn about bar pulling or cutoff operations, C axis operations, and face contouring operations that can be performed on advanced lathe machines. In the end, we will simulate the process of machining in real-time.

PICKOFF/PULL/CUTOFF OPERATION

Although these are three different names of different operations but the mechanism behind them is same. Till now, we have worked with single chuck. If we have two chucks and other chuck can move back and forth then we can unclamp the first chuck and clamp the part in other chuck to perform back face machining, bar pulling, or cutoff operations. Note that `Pickoff/Pull/Cutoff` tool is not available in Home Learning Edition of software. The procedure to perform these operations is given next.

Preparing for Pickoff/Pull/Cutoff Operation

Preparation can have different meaning to different people but here by preparation we mean, select the machine or post processor which is capable of creating codes for such operations and creating secondary chuck which can move back and forth in machine. The steps to do so are given next.

- Click on the `Tools-> Mastercam2022-> Lathe Machines -> Lathe Multi Axis Mill -Turn 2-4-B machine` from the Menu bar. The new machine group will be added in the `Mastercam Toolpath Manager`. Note that if the machine is not available in the menu then add the machine as discussed in previous chapter or you can use default machine definition.
- Click on the `Stock setup` option from the `Properties` node of the newly added machine in `Mastercam Toolpath Manager`.
- Define stock and left spindle of chuck jaws for the part; refer to Figure-1.

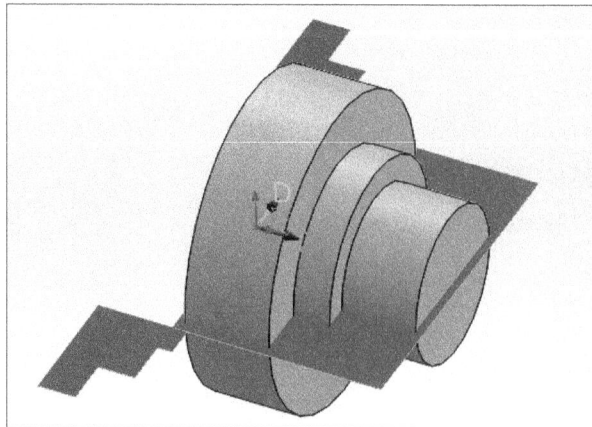

Figure-1. Left chuck and stock created

- Select the **Right Spindle** radio button from the **Chuck Jaws** area of the **Machine Group Properties** dialog box and click on the **Properties** button. The **Machine Component Manager - Chuck Jaws** dialog box will be displayed as discussed earlier.
- Select desired clamping method from the **Clamping Method** area of the dialog box.
- Specify the diameter of the shaft/boss which you want to clamp in jaws in the **Diameter** edit box and specify desired distance from the coordinate system in **Z** edit box to place the secondary chuck jaws. Preview of the chuck will be displayed; refer to Figure-2. Click on the **OK** button from the **Machine Component Manager - Chuck Jaws** dialog box and then from the **Machine Group Properties** dialog box to create chuck.

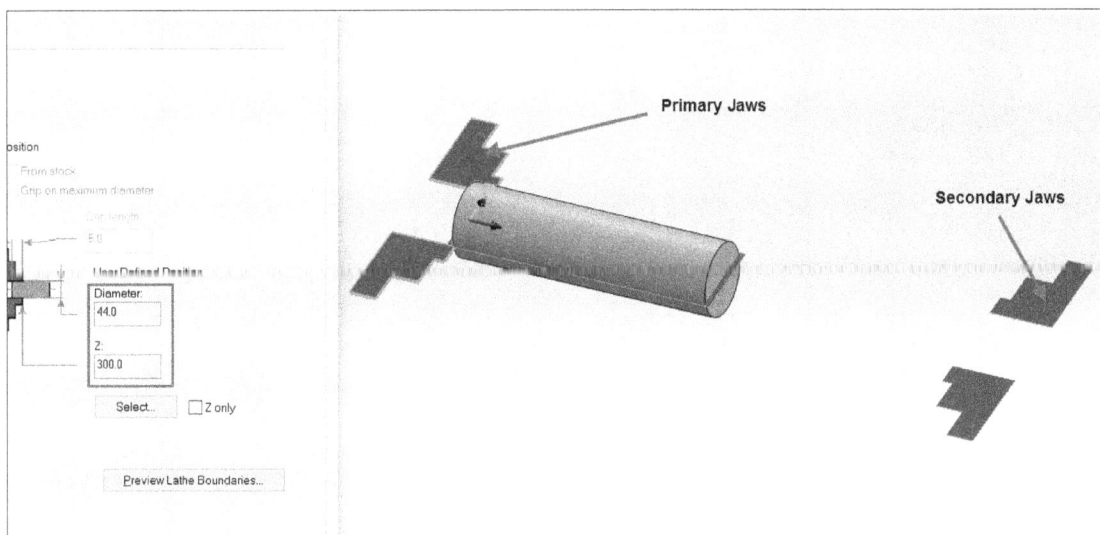

Figure-2. Preview of secondary jaws

Performing Pickoff, bar pull, cutoff Operation

- Click on the **Pickoff/Pull/Cutoff** tool from the **Lathe** drop-down in the **Ribbon**. The **Cutoff / Bar Pull / Pickoff** dialog box will be displayed; refer to Figure-3.
- Select the **Pickoff, bar pull, cutoff** option from the **Operations to program** box.
- Specify the parameters like length of finished part in the **Length of finished part** edit box. Once the part will cutoff, the specified length of part will remain in primary chuck.
- Specify desired value in the **Pickoff Z coordinate** edit box. The value specified here is the location at which secondary jaw will hold the cut off portion of part.
- Specify desired value in the **Cutoff width** edit box. This width will be lost in cutting. In place of specifying width manually, you can choose desired cut off tool by click on the button next to this edit box.
- Specify desired value of X coordinate for cut off location in the **Cutoff X coordinate** edit box. Generally, the diameter value is specified here.
- Specify the stock to be left on face of part in primary jaw in **Face stock** edit box and stock to be left on the back face of part in secondary jaw in **Back face stock** edit box.

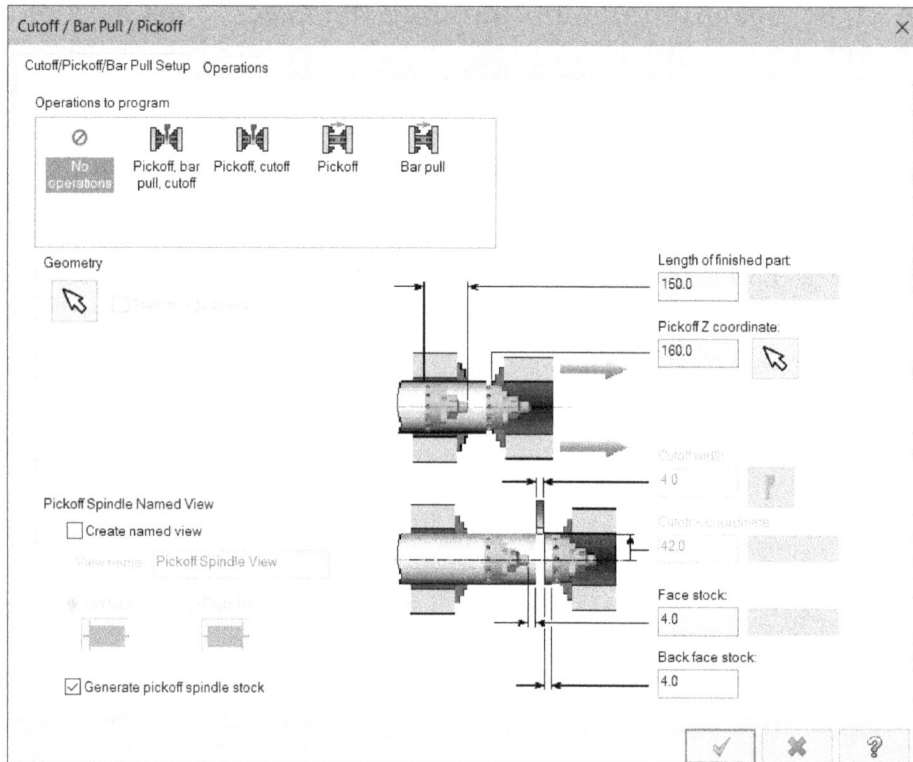

Figure-3. Cutoff / Bar Pull / Pickoff dialog box

- Click on the **Operations** tab in the dialog box. The parameters related to various operations will be displayed; refer to Figure-4.

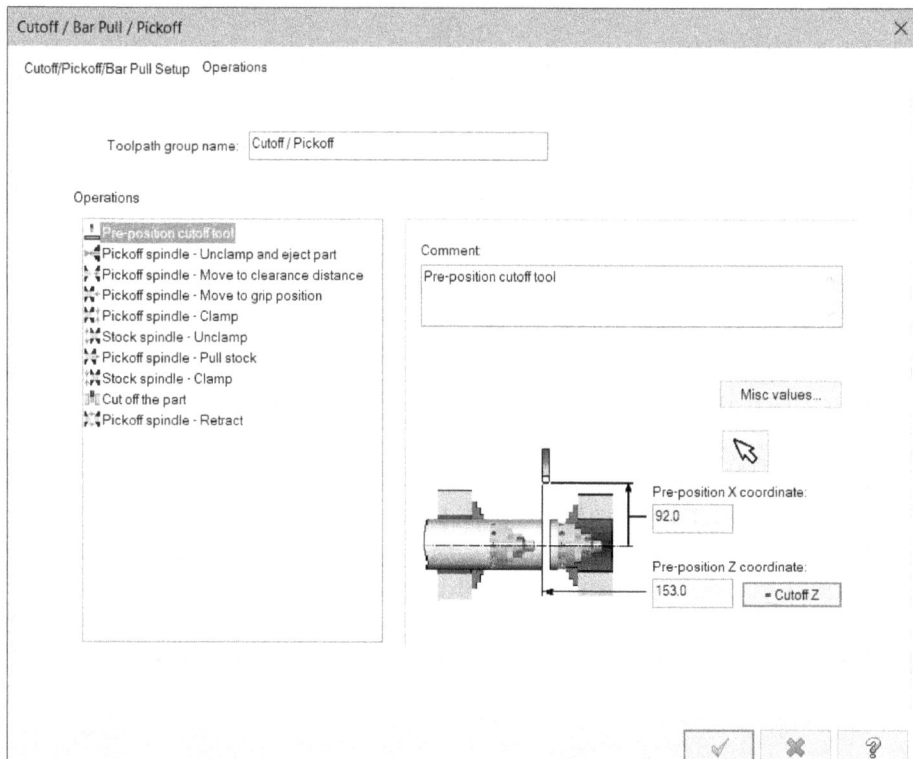

Figure-4. Operations tab

- Set desired parameters in the tab and click on the **OK** button. The process of cutting bar, pulling it out and taken away by secondary spindle will happen automatically. The related codes will be generated accordingly.

STOCK TRANSFER

The **Stock Transfer** tool is used to transfer stock from one chuck to another (sub-spindle). This tool is useful when you want to machine back face of part after machining the front side; refer to Figure-5. The procedure to use this tool is given next.

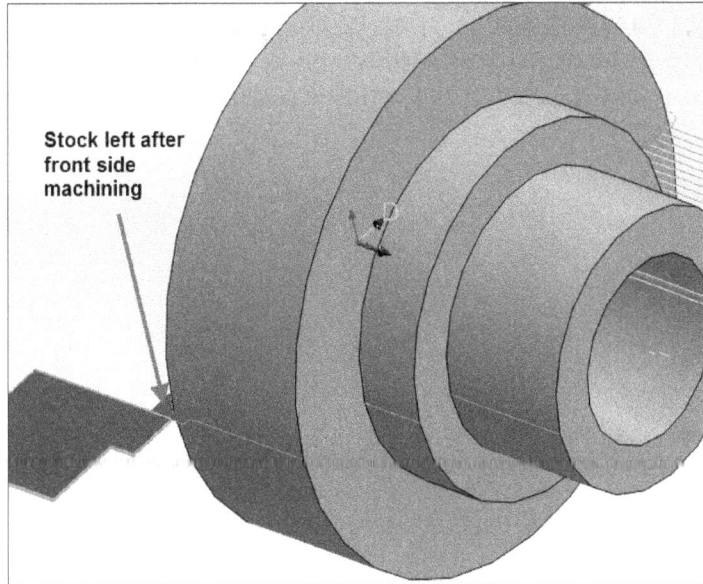

Figure-5. Stock left after front side machining

- Click on the **Stock Transfer** tool from the **Lathe** drop-down in the **Ribbon**. The **Lathe Stock Transfer** dialog box will be displayed; refer to Figure-6.

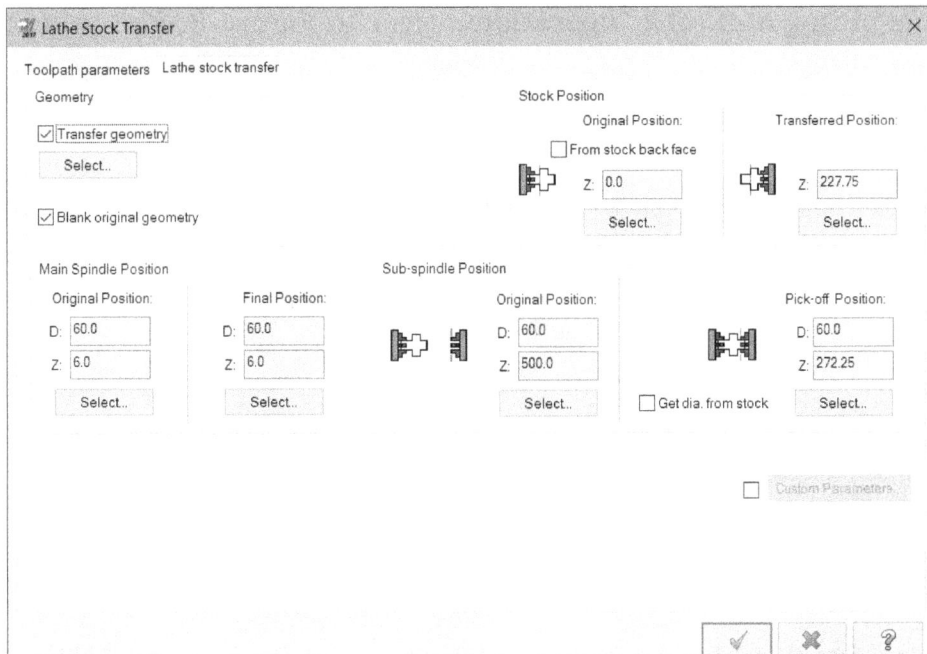

Figure-6. Lathe Stock Transfer dialog box

- Click on the **Select** button from the **Geometry** area of the dialog box. You will be asked to select the geometry to be transferred.
- Select the geometry to be transferred. The parameters in the dialog box will be modified according to geometry selected. In most of the cases, you can directly hit **OK** button to transfer the geometry but if you need to modify parameters then details are given next.

- Specify the distance of a reference point which you want to be shifted in the **Z** edit box of **Original Position** area. Like, you can specify the position of back face point.
- Specify the distance in **Z** edit box of **Transferred Position** area to which you want to move the back face point earlier selected.
- In the **Pick-off Position** area, specify the diameter and Z value at which you want to clamp the part in sub-spindle.
- After specifying desired parameters, click on the **OK** button to transfer stock; refer to Figure-7.

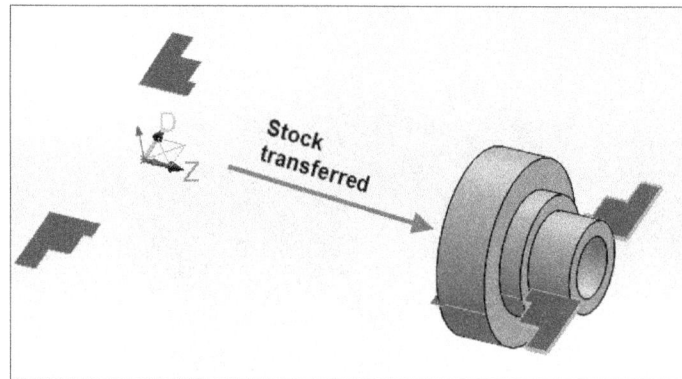

Figure-7. Stock transferred

STOCK FLIP

The **Stock Flip** tool is used to reverse the face of model so that you can perform back face machining and other operations; refer to Figure-8. The procedure to use this tool is given next.

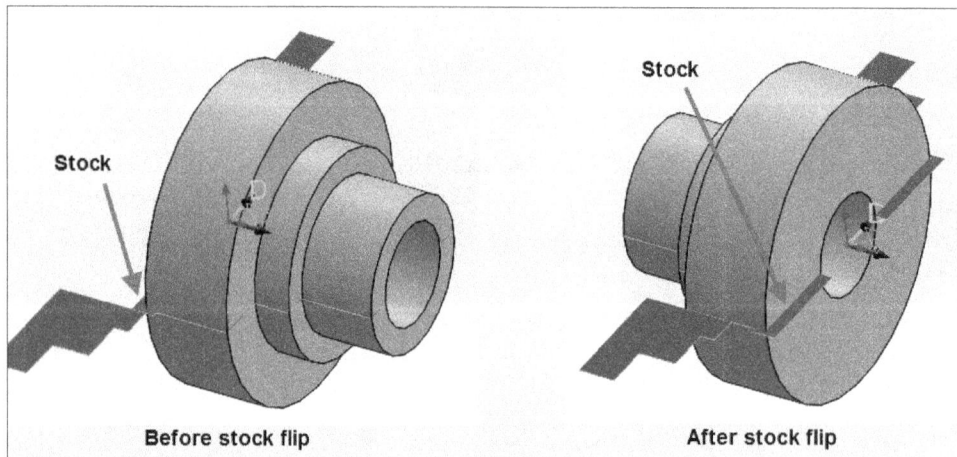

Figure-8. Model before and after stock flip

- Click on the **Stock Flip** tool from the **Lathe** drop-down in the **Ribbon**. The **Lathe Stock Flip** dialog box will be displayed; refer to Figure-9.

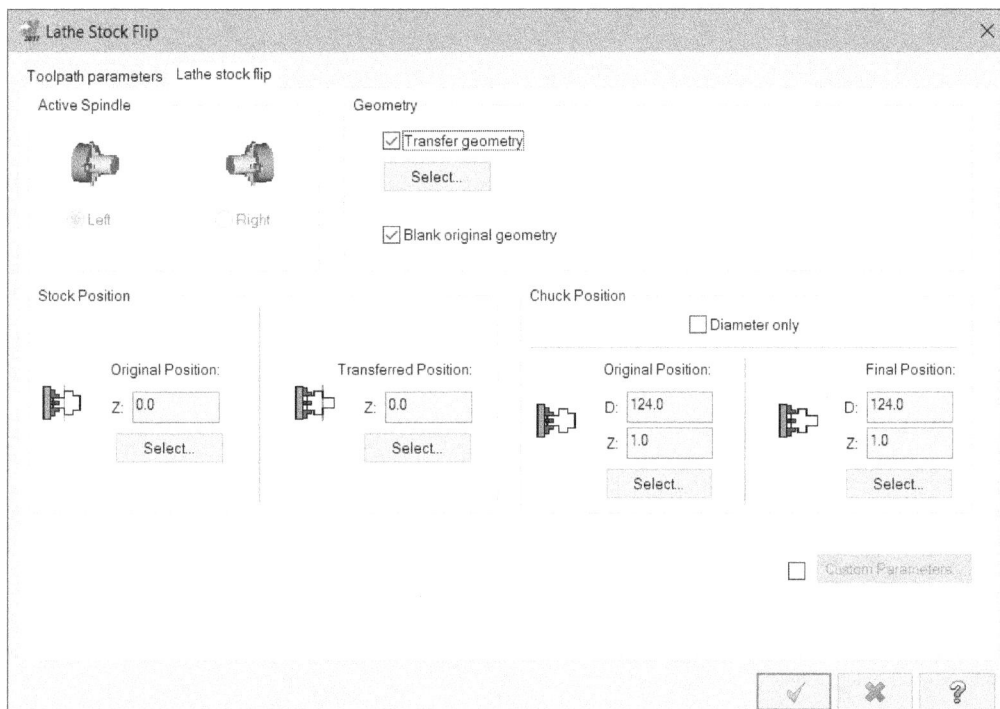

Figure-9. Lathe Stock Flip dialog box

- Click on the **Select** button from the **Geometry** area of the dialog box. You will be asked to model to be flipped.
- Select the model to be flipped and click on the **OK** button from the **Selection PropertyManager**.
- In the **Final Position** edit boxes, specify the diameter value and Z position where chuck should hold the part. Note that when part will be flipped, the back face of the model will be at new Z=0 position. So, you may need to specify a negative Z value hold the part; refer to Figure-10.

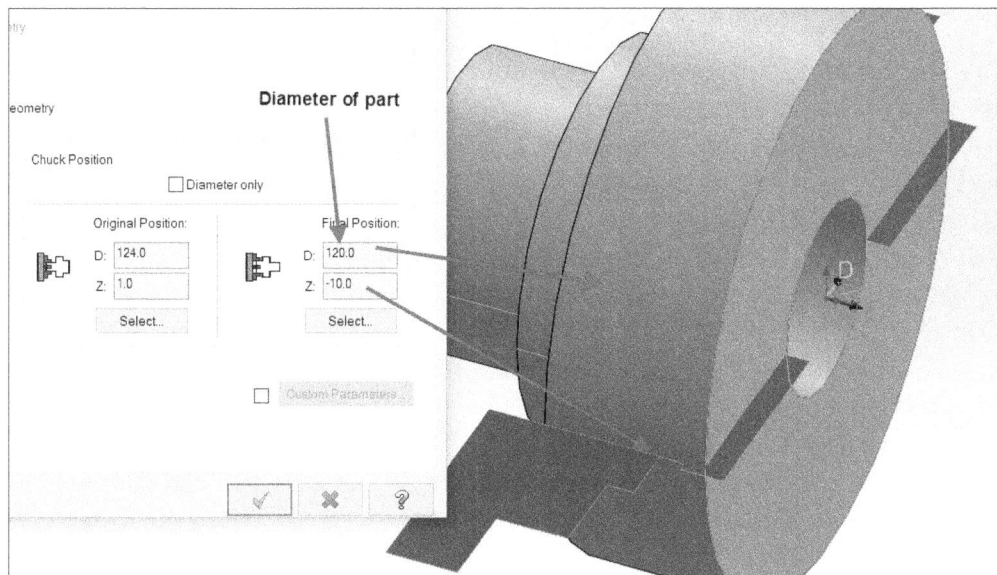

Figure-10. Model after stock flip

- After specifying desired parameters, click on the **OK** button to create stock flip. Now, you can perform machining on back face of the model.

STOCK ADVANCE

The **Stock Advance** tool is used to push or pull the stock. The procedure to use this tool is given next.

- Click on the **Stock Advance** tool from the **Lathe** drop-down in the **Ribbon**. The **Lathe Stock Advance** dialog box will be displayed; refer to Figure-11.

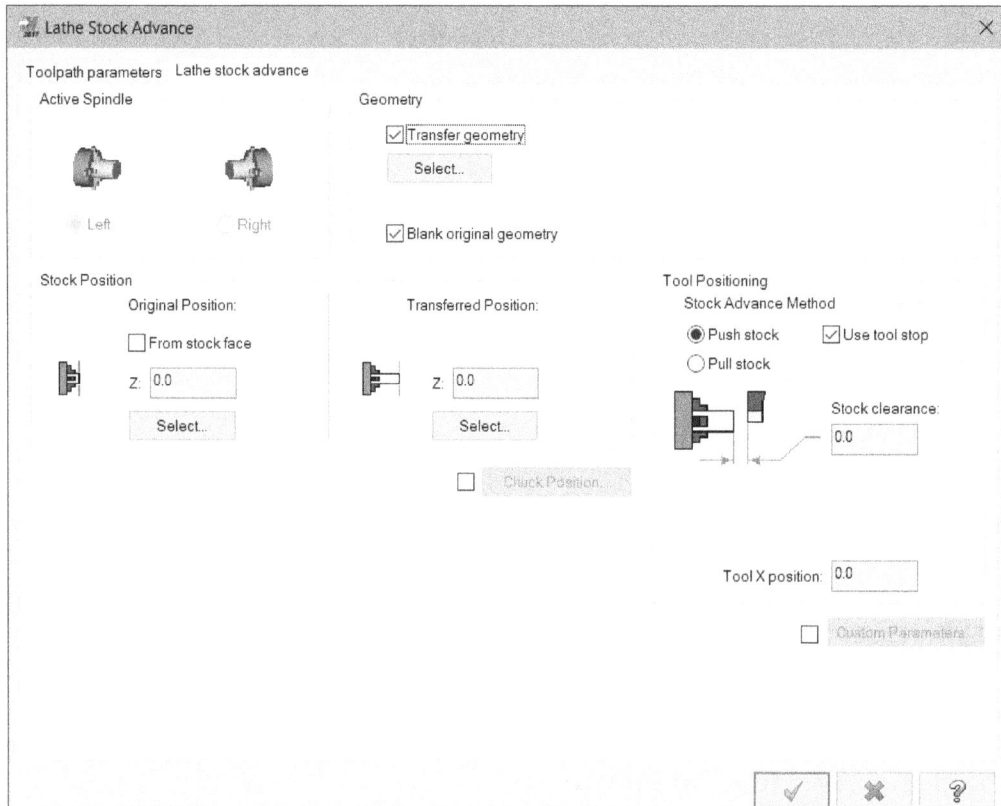

Figure-11. Lathe Stock Advance dialog box

- Click on the **Select** button from the **Geometry** area of the dialog box and select the part you want to move.
- If you select the **Transfer geometry** check box then the part will also move by specified distance along with stock. (We recommend you to select this check box.)
- If the **Blank original geometry** check box is selected then the original part will get hidden and only the moved part will be displayed.
- Specify the original position of stock in the **Original Position** edit box of **Stock Position** area in the dialog box or you can select the **From stock face** check box to automatically take the Z value of front face of stock.
- In the **Transferred Position** edit box, specify desired value of Z for front face of stock after moving it.
- Select the **Push stock** or **Pull stock** radio button to use the respective method for stock advance. Specify the related parameters in the edit boxes below the radio buttons.
- Click on the **OK** button from the dialog box to create the feature.

UNDOING STOCK RELATED OPERATIONS

This topic is more of a case rather than tool procedure. One simple question: What if you have performed stock advance and later you realize that you do not need it.

What will you do? One answer will come as delete the operation from the **Mastercam Toolpath Manager**. Figure-12 shows a case where we have done stock advance operation but now we want to reverse it. The steps are given next.

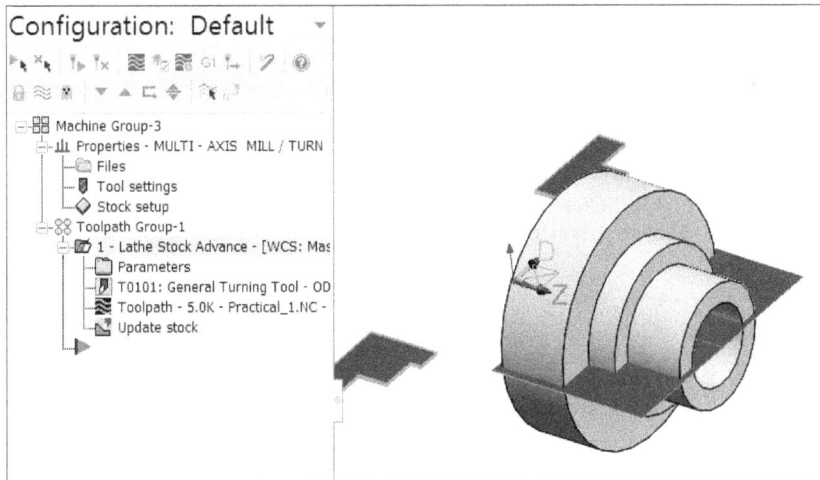

Figure-12. Lathe stock advance done on part

- Select the stock operation that you want to reverse from the **Mastercam Toolpath Manager** and right-click on it. A shortcut menu will be displayed.
- Select the **Delete** option from the shortcut menu; refer to Figure-13. The stock will move back to chuck but the model will not move back.

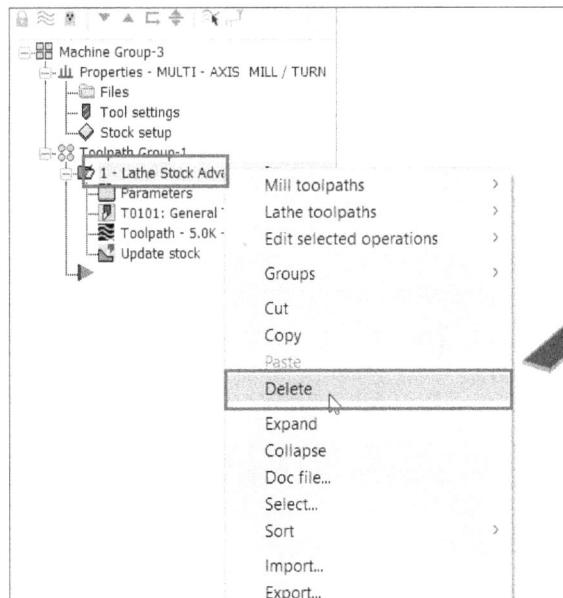

Figure-13. Delete option in shortcut menu

- Click on the **FeatureManager Design Tree** button from the left pane in SolidWorks. The model parameters will be displayed; refer to Figure-14.
- Click on the **Stock Advance 1** feature which was created due to **Stock Advance** operation and press **DELETE** button from the keyboard; refer to Figure-14. The **Confirm Delete** dialog box will be displayed. Click **Yes** button from the dialog box.
- Right-click on the original model and select the **Show** button from the mini-toolbar displayed; refer to Figure-15.

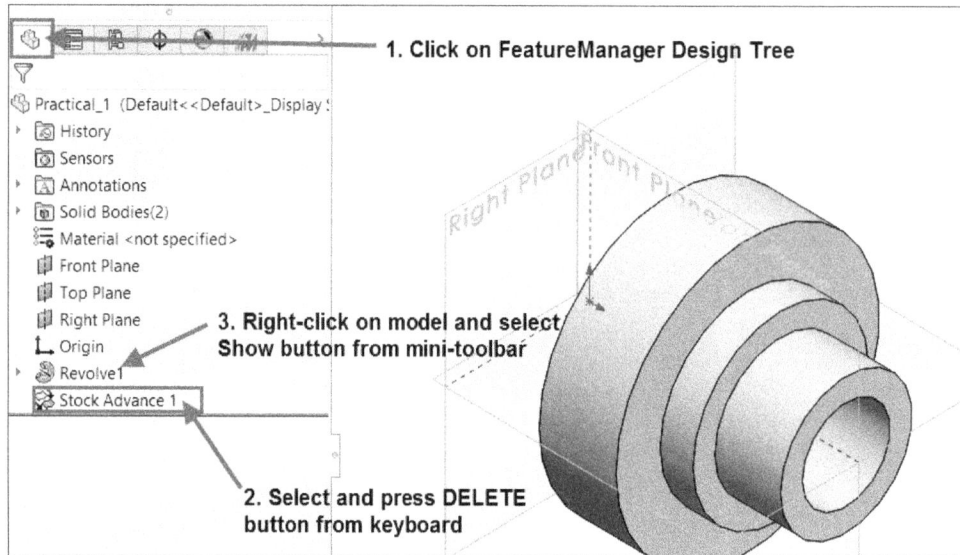

Figure-14. Displaying original model

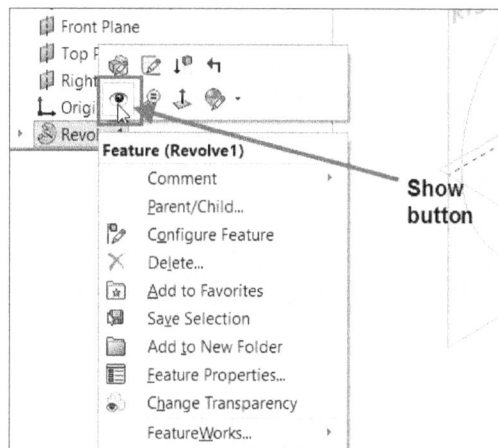

Figure-15. Show button to display original model

• Now, click on the **Mastercam Toolpath Manager** button from the left pane to return to Mastercam environment.

CHUCK

The **Chuck** tool is used to output M codes for clamping and de-clamping of the chuck. The procedure to use this tool is given next.

• Click on **Chuck** tool from the **Lathe** drop-down. The **Lathe Chuck** dialog box will be displayed; refer to Figure-16.

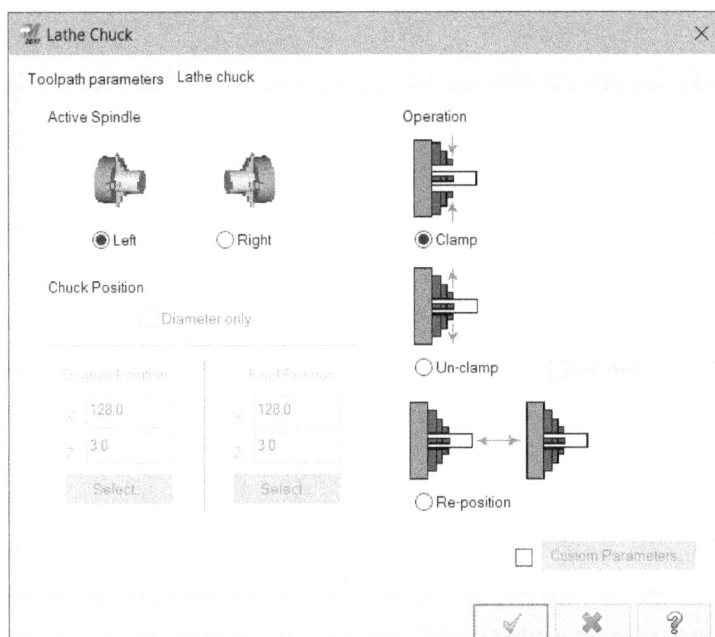

Figure-16. Lathe Chuck dialog box

- Select desired radio button from the **Active Spindle** area to perform operations on that spindle.
- Select desired operation like, clamp, un-clamp, or re-position from the **Operation** area of the dialog box and click on the **OK** button.

TAIL STOCK

The **Tail Stock** tool in **Lathe** drop-down is used to manage operations related to tail stock. To use this tool, you must have already defined tailstock in machine definition. The procedure to add tail stock has been discussed in previous chapter. The procedure to use **Tail Stock** tool is discussed next.

- Click on the **Tail Stock** tool from the **Lathe** drop-down in the **Ribbon**. The **Lathe Tailstock** dialog box will be displayed; refer to Figure-17.
- Select desired operation from the **Operation** area of the dialog box. If you want to move the tailstock towards the part then select the **Advance** radio button. If you want to move the tailstock away from part then select the **Retract** radio button.
- Specify the parameters related to selected operation in the **Tailstock Position** area of the dialog box.
- Click on the **OK** button to perform the operation and add codes in NC program.

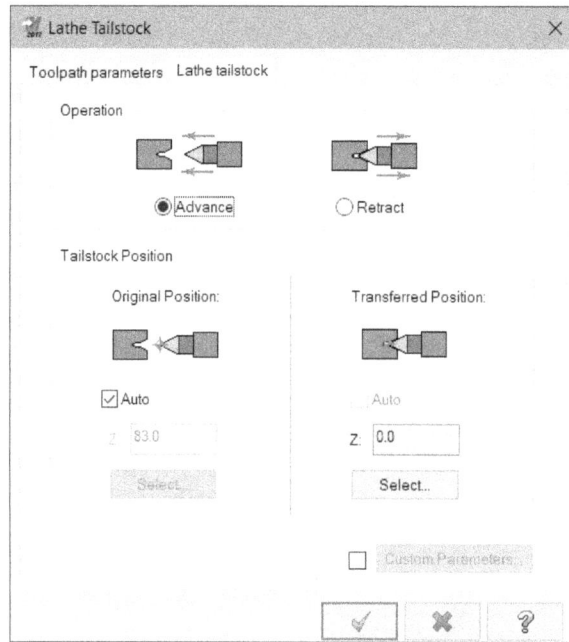

Figure-17. Lathe Tailstock dialog box

Similarly, you can use the **Steady Rest** tool to manage operations related to steady rest.

Note that the operations discussed in this chapter are not available in all the machines so you should check your machine manual before programming. Like the operations to flip stock or stock advance are not available in general turning machines. Now, we will learn about some toolpaths that are available for turning machines which have 3 or 4 motion axes.

FACE CONTOUR

The **Face Contour** tool is used to draw a toolpath for machining curves on face of part; refer to Figure-18.

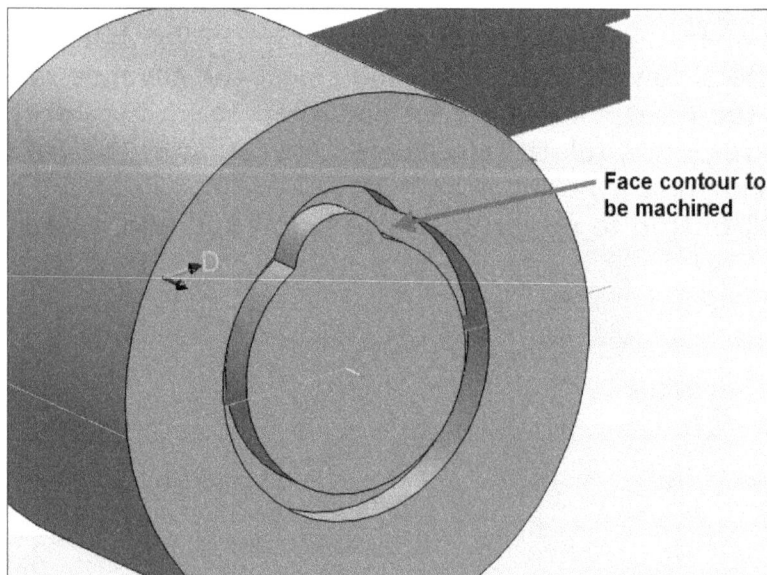

Figure-18. Face contour to be machined

Note that you should use this toolpath after you have done facing and roughing operation so that there is a low amount of stock left on the face. The procedure to create a face contour toolpath is given next.

- Click on the **Face Contour** tool from the **Lathe** drop-down in the **Ribbon**. The **Chain Manager** will be displayed.
- Select the faces on the front face of part to be machined; refer to Figure-19. Click on the **OK** button from the **Chain Manager** to apply selection. The **C-Axis Toolpath - C-Axis Face Contour** dialog box will be displayed; refer to Figure-20.

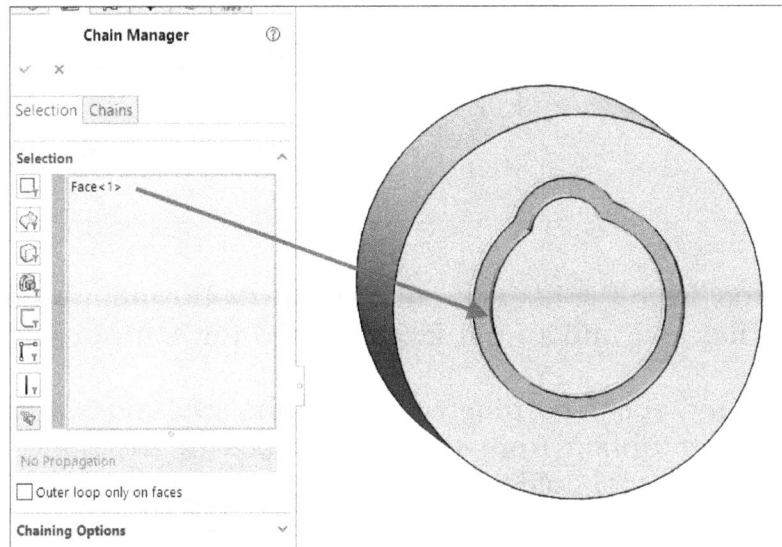

Figure-19. Face selected for face contouring

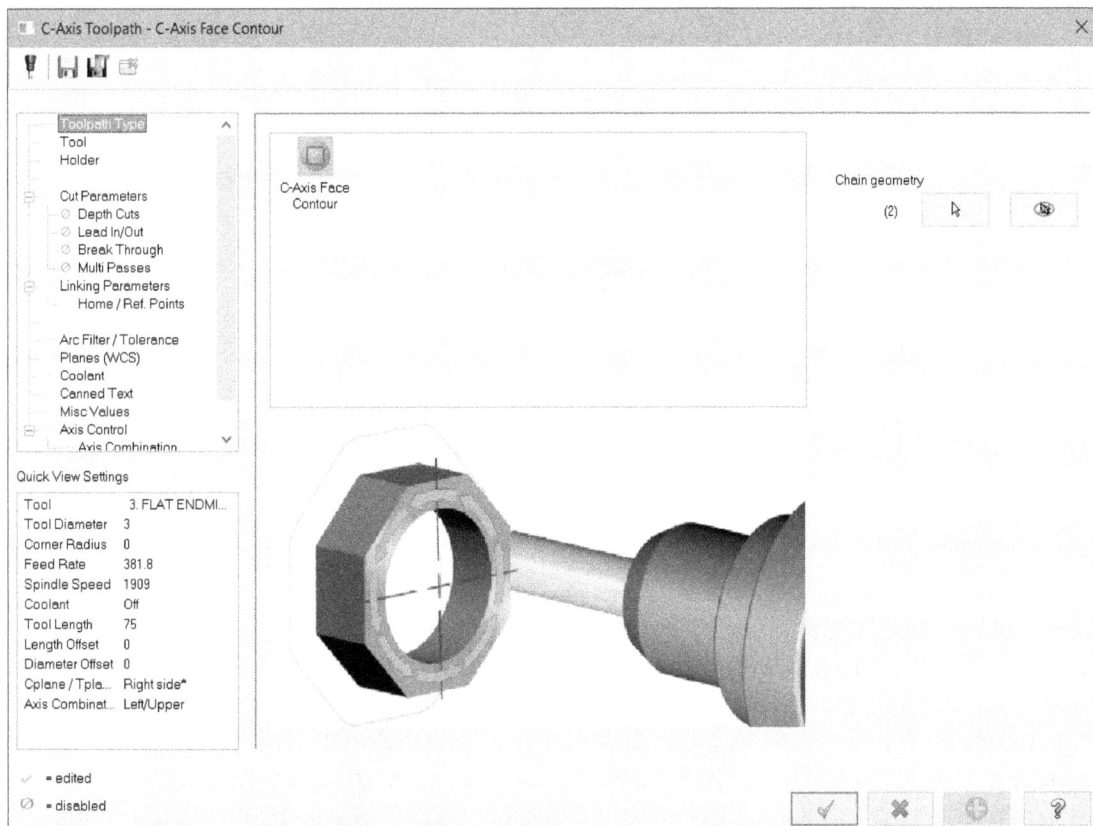

Figure-20. C-Axis Toolpath-C-Axis Face Contour dialog box

- Click on the **Tool** option from the top left box in the dialog box. The options related to tool selection will be displayed. Click on the **Select library** tool from the right area. The **Tool Selection** dialog box will be displayed; refer to Figure-21.

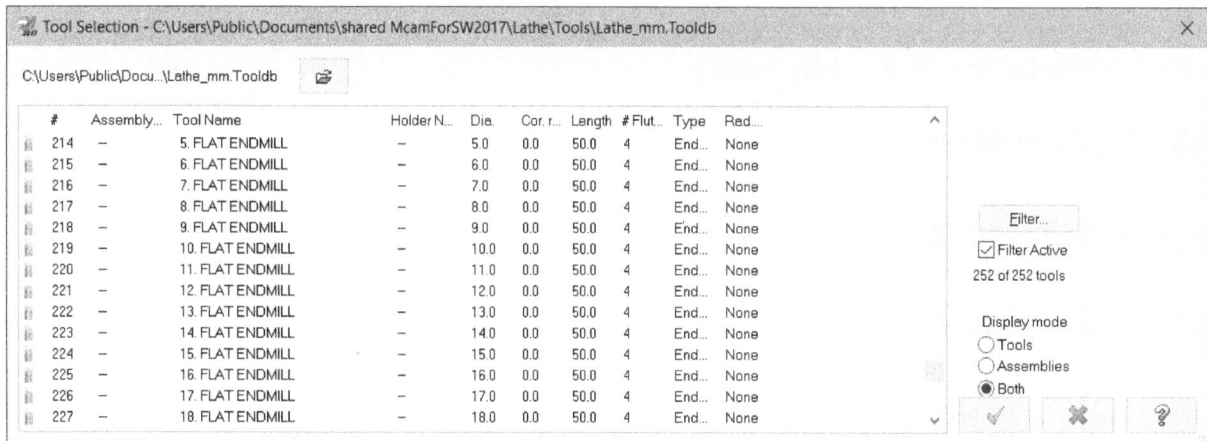

Figure-21. Tool Selection dialog box

- Select desired tool from the list and click on the **OK** button. (In our case, we have selected 3 mm flat end mill as the face selected for contour is flat and the slot width is small.
- Specify the feed and speed parameters as per the tool and material.
- Click on the **Holder** option from the top left box in the dialog box to set the parameters related to tool holder.
- Click on the **Cut Parameters** option from the top left box in the dialog box. The options will be displayed as shown in Figure-22.

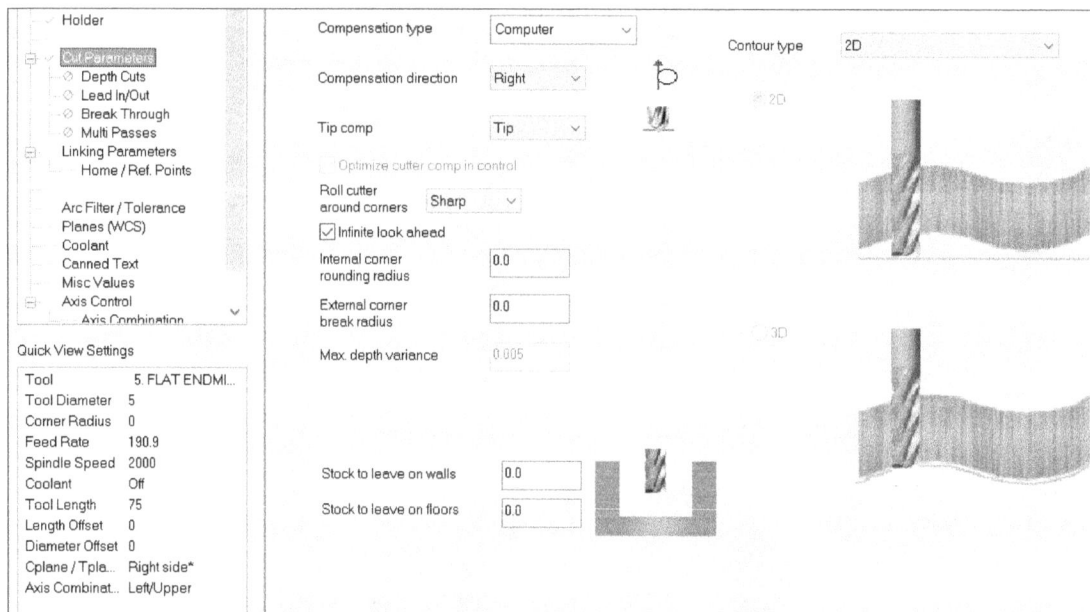

Figure-22. Cut Parameters page for face contour tool

- Select desired contour type from the **Contour type** drop-down. We need 2D contouring for our case. Set the other parameters as required. Note that some of the parameters are disabled by default in the dialog box as they are not useful for current toolpath. If you still need to specify any disable parameters then click on it from the top left box and select the related check box from the right area to activate the options.

- After specifying desired parameters, click on the **OK** button from the dialog box. The toolpath will be generated.

CROSS CONTOUR

The **Cross Contour** tool is used to create toolpath for creating contour along C axis; refer to Figure-23. The procedure to use this tool is given next.

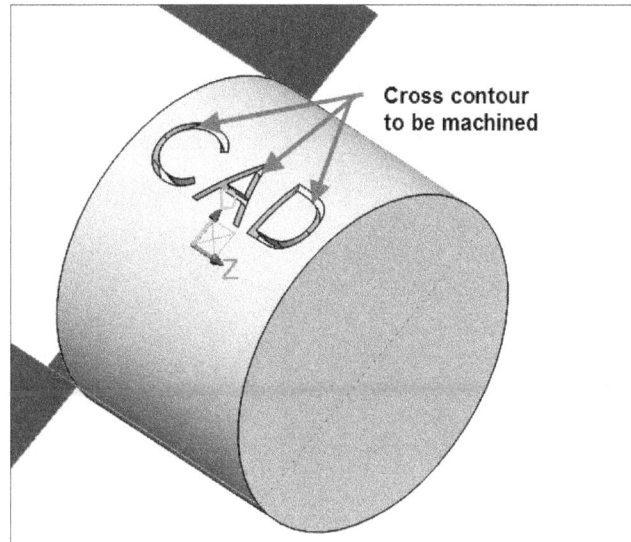

Figure-23. Cross contour to be machined

- Click on the **Cross Contour** tool from the **Lathe** drop-down in the **Ribbon**. The **Chain Manager** will be displayed.
- Select the faces that you want to be machined; refer to Figure-24. Click on the **OK** button from the **Chain Manager**. The **C-Axis Toolpath - C-Axis Cross Contour** dialog box will be displayed; refer to Figure-25.

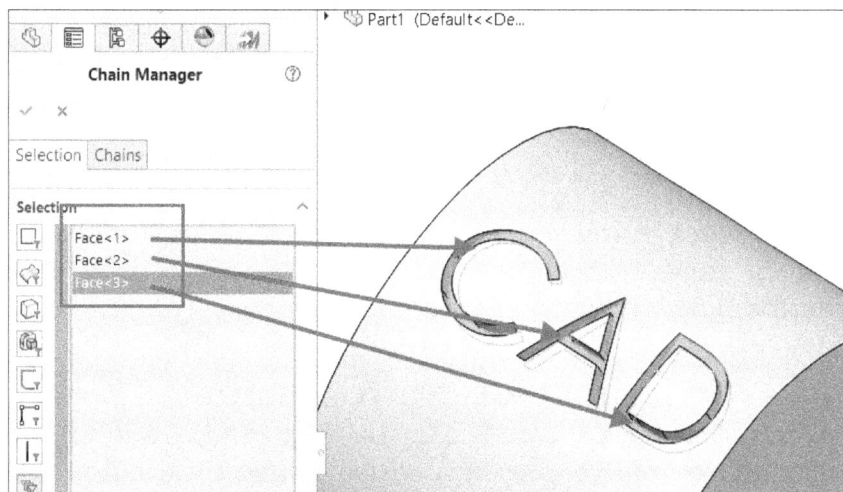

Figure-24. Faces selected for cross contour

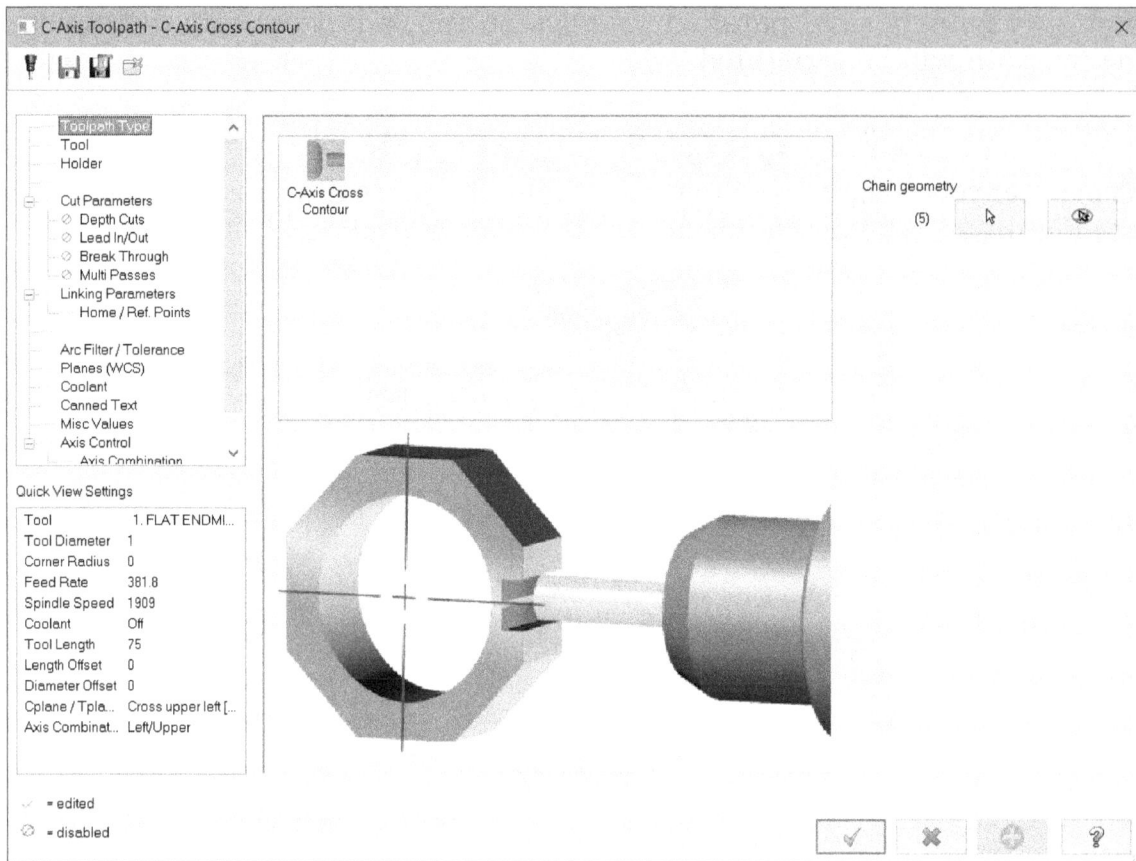

Figure-25. C-Axis Toolpath-C-Axis Cross Contour dialog box

- Select desired tool after clicking on the **Tool** option from the top left box in the dialog box as discussed in previous tool description.
- Specify the other parameters as required.
- Click on the **OK** button from the dialog box to create the toolpath. The toolpath will be displayed.

Similarly, you can use the other C-Axis toolpaths using **C-Axis Contour**, **Face Drill**, **Cross Drill**, and **C-Axis Drill** tools.

BACKPLOT AND VERIFYING TOOLPATH

Till now, we have created many toolpaths for CNC turning machines but we have not verified them in simulation environment. Here, we will discuss the options to graphically verify toolpaths created earlier.

Backplotting Toolpath

The **Backplot** button ≋ in **Mastercam Toolpath Manager** is used to check movement of the tool for selected toolpaths. The procedure to use this tool is given next.

- Select the toolpath group or toolpath which you want to check and click on the **Backplot selected operations** button ≋ from the **Mastercam Toolpath Manager**; refer to Figure-26. The **Backplot PropertyManager** will be displayed with the preview of tool and **Backplot VCR** bar; refer to Figure-27.

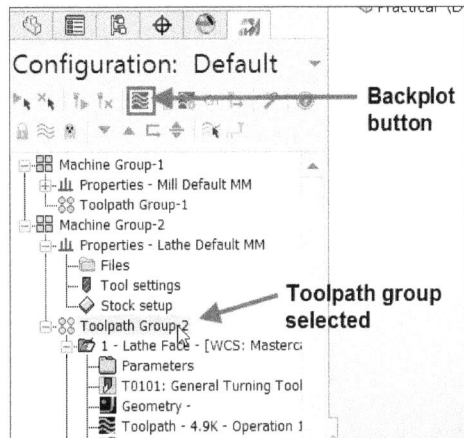

Figure-26. Backplotting toolpath group

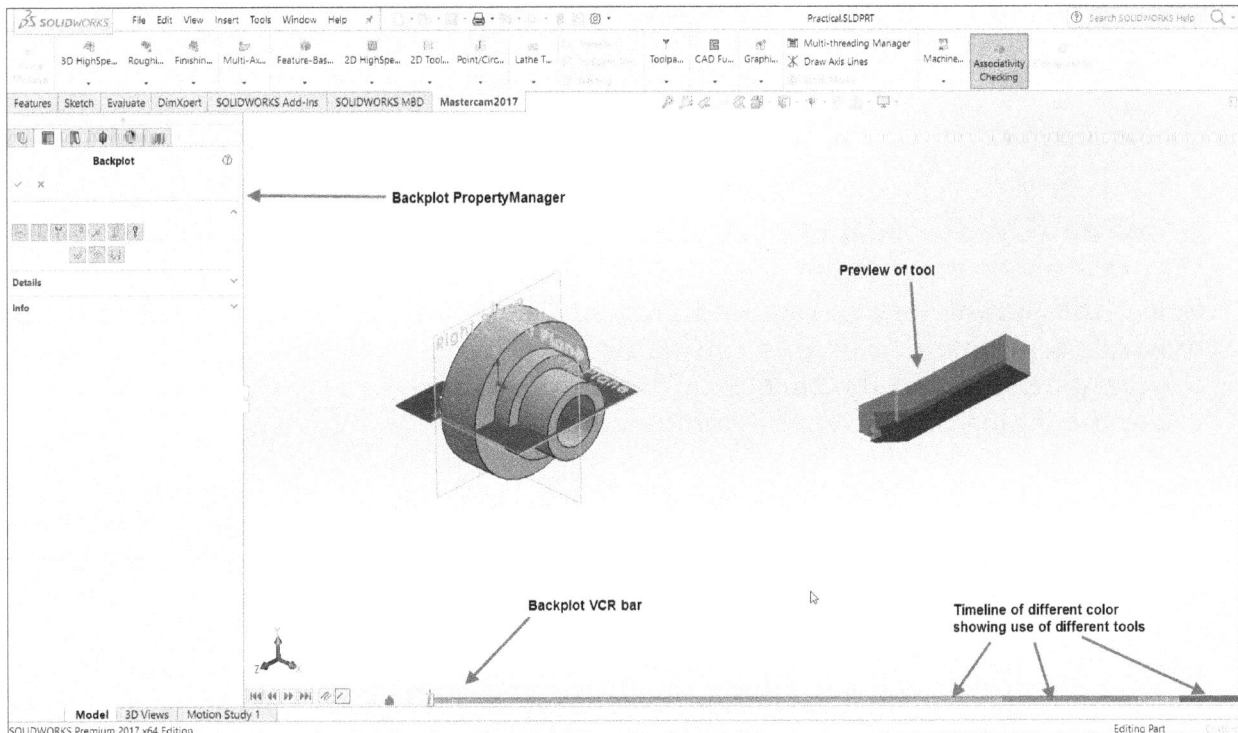

Figure-27. Backplot PropertyManager with Backplot VCR bar

- Click on the **Play** button from the **Backplot VCR bar** to play simulation of tool path generation. Select desired filters from the **Backplot PropertyManager** to display or hide elements in the simulation. Click on the **OK** button from the **PropertyManager** to exit the tool.

Verifying Toolpaths

The **Verify** tool is used to verify the toolpath in 3D dynamic environment. The procedure is given next.

- Select the operations to be verified and click on the **Verify selected operations** button from the **Mastercam Toolpath Manager**. The **Mastercam Simulation** window will be displayed; refer to Figure-28. You can also invoke this tool by right-clicking on the toolpath of operation in **Mastercam Toolpaths Manager**.

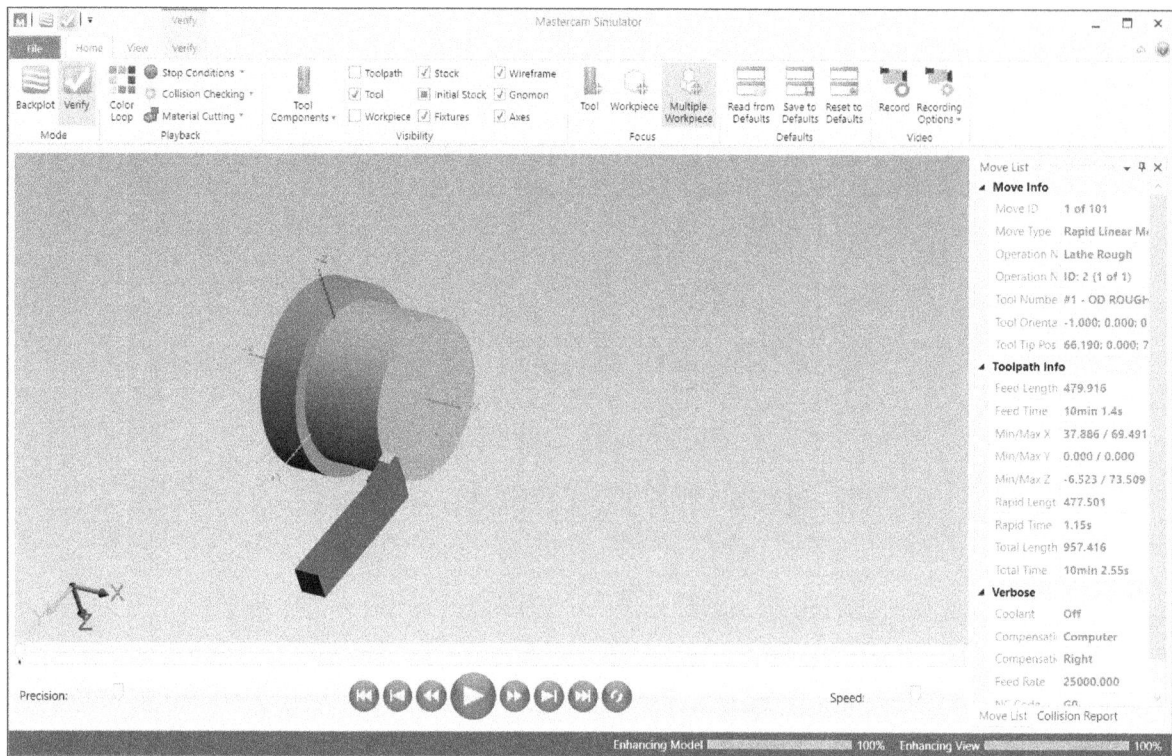

Figure-28. Mastercam Simulator window

- Select the parameters as required and then click on the **Play** button to run the dynamic simulation. Use this simulator to verify tool collisions. You can record the simulation using the **Record** button in the **Ribbon**.
- Close the dialog box after verifying toolpath.

ANALYZING TOOLPATH

The **Analyze Toolpath** tool is used to analyze toolpath at desired point. The procedure to use this tool is given next.

- Select the toolpath that you want to analyze from the **Mastercam Toolpath Manager** in the left pane and click on the **Only display selected toolpaths** button; refer to Figure-29. This will hide all the other toolpaths from the graphics window.
- Click on the **Analyze Toolpath** tool from the **Utilities** drop-down in the **Ribbon**. Now, if you move the cursor over the toolpath in graphics window, you will see the details of toolpath like feed rate, spindle speed, and so on; refer to Figure-30.

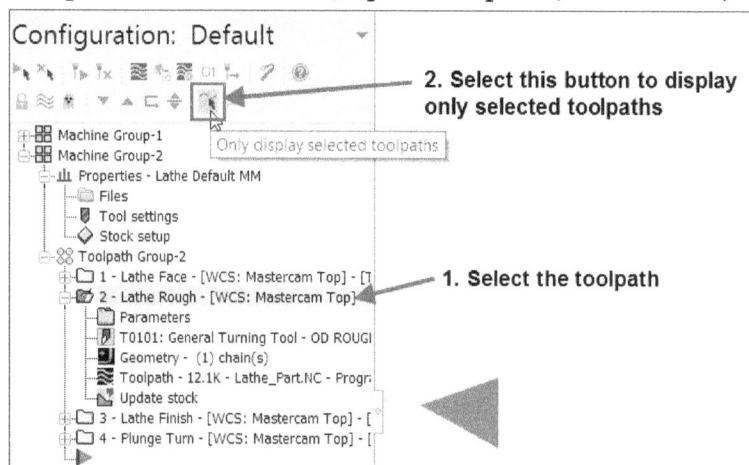

Figure-29. Displaying selected toolpaths

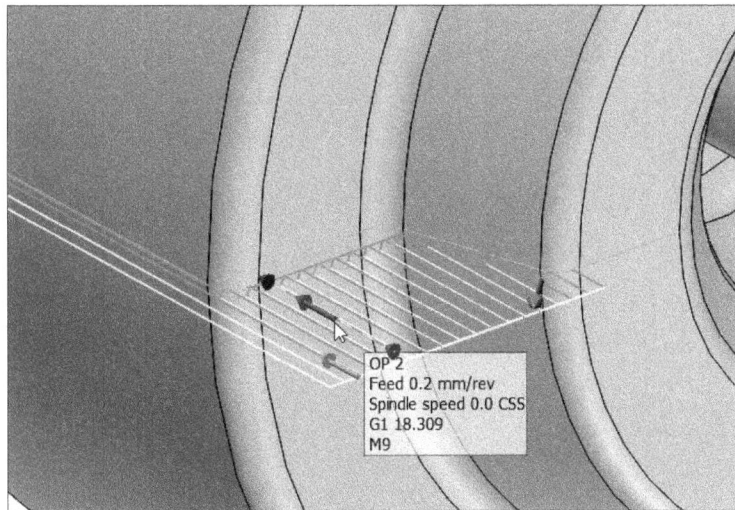

Figure-30. Details of toolpath

- Click the **OK** button from the **PropertyManager** after analyzing toolpaths.
- To display all the other toolpaths, select the toolpath group and click on the **Only display selected toolpaths** button from the **Mastercam Toolpath Manager**.

MANUAL TOOLPATH CODES ENTRY

If you have prepared NC codes manually or from any other cam software and want to use then in current Mastercam project then you can do so by using the **Manual Entry** tool in the **Utilities** drop-down. The procedure to use this tool is given next.

- Move the insertion point at desired location in the **Mastercam Toolpath Manager** and click on the **Manual Entry** tool from the **Utilities** drop-down in the **Ribbon**. The **Manual Entry** dialog box will be displayed; refer to Figure-31.

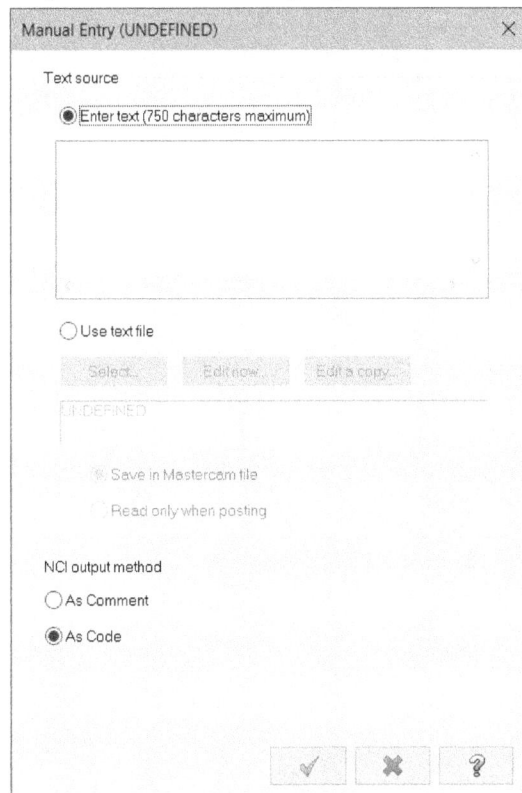

Figure-31. Manual Entry dialog box

- Enter the codes in the text box or click on the **Use text file** radio button and select desired text file of codes after clicking the **Select** button.
- Set the other options as required.
- Click on the **OK** button from the dialog box to add the file/text in NC codes.

PRACTICAL 1

Create an NC program for part shown in Figure-32. The stock dimensions are given in Figure-33. Note that the SolidWorks part file for model is available in the resources of this book.

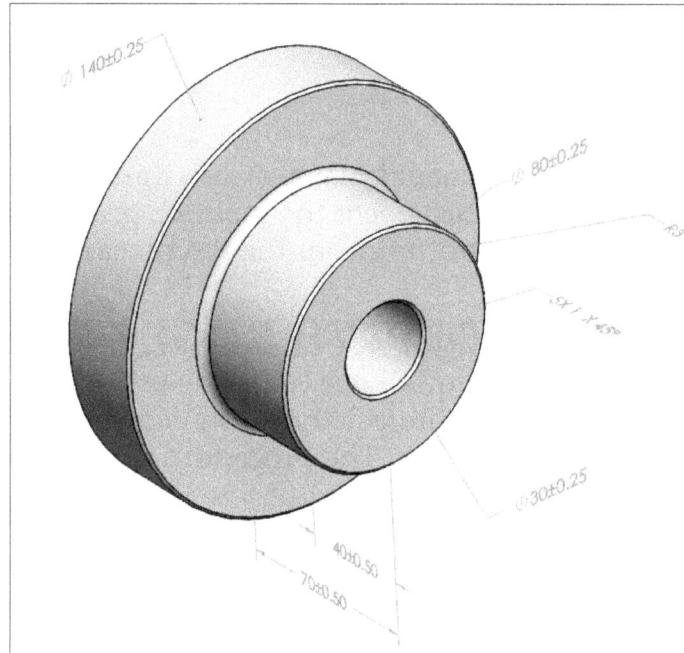

Figure-32. Model for Practical 1

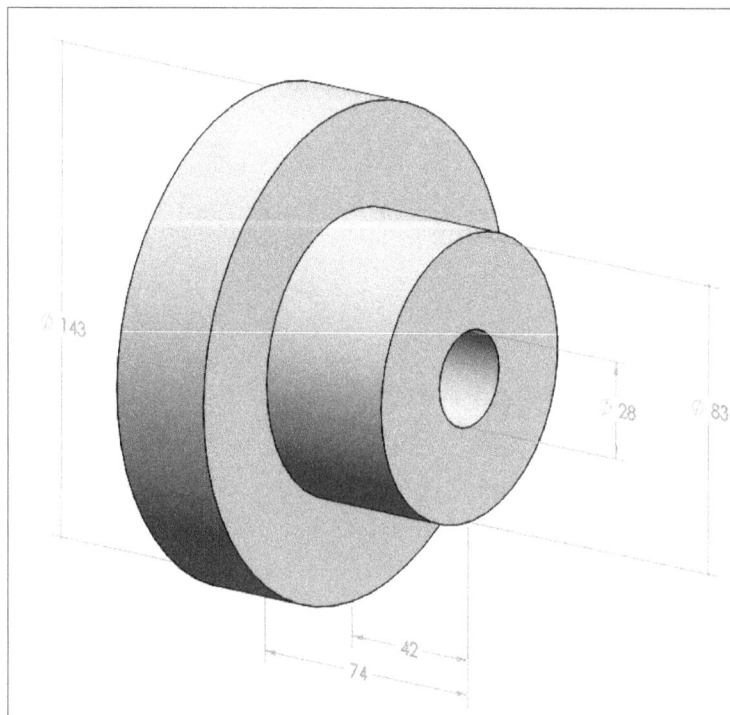

Figure-33. Dimensions of stock for gear blank

Opening Part and Setting Lathe Machine

- Start SolidWorks with Mastercam add-in.
- Open the **Gear blank for practical.sldprt** file from the resources folder of the book. The model will open.
- Click on the **Tools-> Mastercam2022-> Lathe Machines-> Default** option from the menu bar. A new machine group will be added with lathe machine settings; refer to Figure-34.

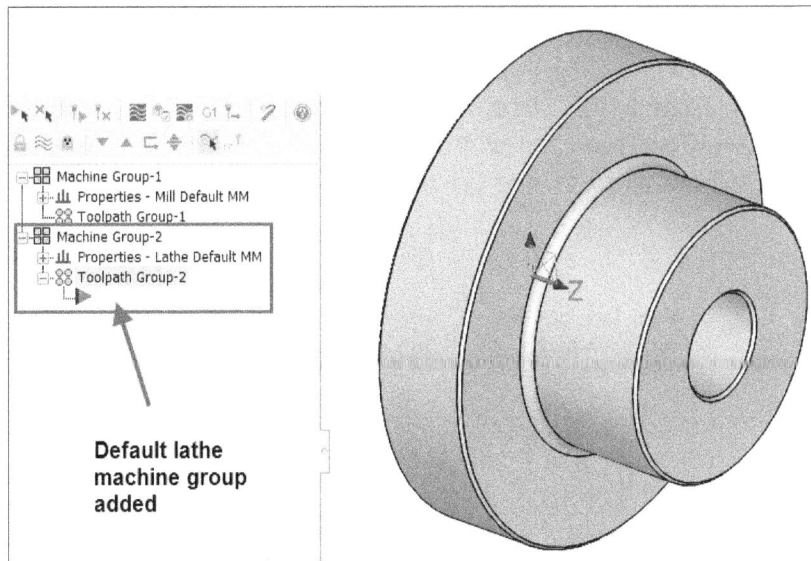

Figure-34. Default lathe machine group added

- Now, we will import a part to create stock. Click on the **Insert -> Features -> Imported** tool from the menu bar. The **Open** dialog box will be displayed.
- Select the **IGES** format from the **File Type** drop-down and select the stock for practical gear blank file from book resources; refer to Figure-35.

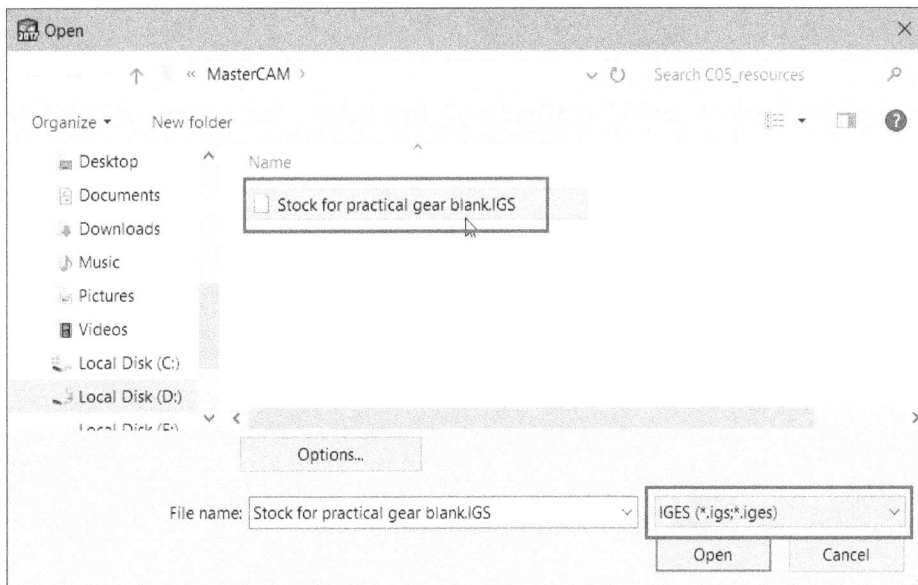

Figure-35. Open dialog box for imported files

- After selecting file, click on the **Open** button from the dialog box. The imported file will be in the drawing area.
- Select the **Front** view from the **View Orientation** drop-down in the **Heads-Up View Toolbar**; refer to Figure-36 or press **CTRL+1** to orient the model in front view.

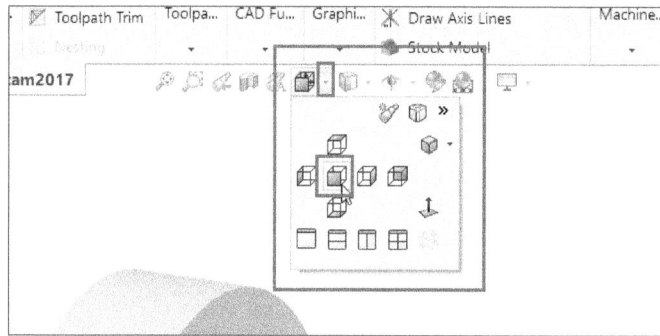

Figure-36. Selecting front view

- Select the **Hidden Lines Visible** option from the **Display Style** drop-down in the **Heads Up View Toolbar**; refer to Figure-37. You will find that the back face of stock model and main part are coincident; refer to Figure-37.

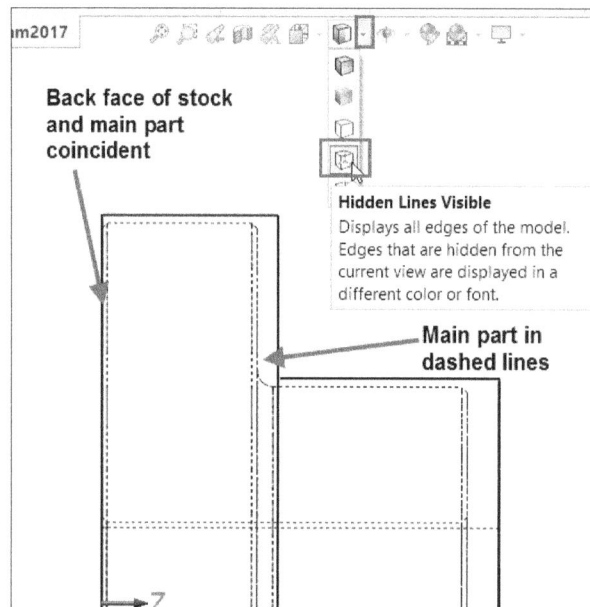

Figure-37. Placement of stock and main part

- Press the Middle mouse button and drag to rotate the model to check back faces in different orientation. Now, we need to move the back face of stock away from main part.

- Click on the **Insert -> Features -> Move/Copy** tool from the menu bar. The **Move/ Copy PropertyManager** will be displayed; refer to Figure-38 and you will be asked to select the body to be moved. Select the imported part.

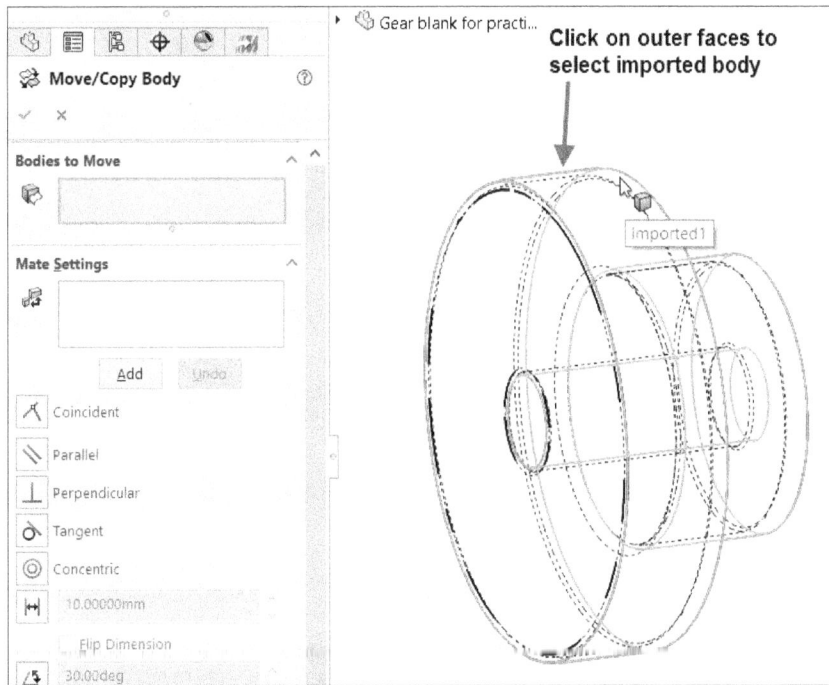

Figure-38. Move Copy Body PropertyManager

- Click on the **Distance** button from the **Mate Settings** rollout of the **PropertyManager** and specify the value of distance as **1.5** in the next edit box. Click on the **Flip** button to reverse the direction.
- Click in the **Entities to Mate** selection box and select the back faces of stock model and base part; refer to Figure-39.
- Click on the **OK** button from the **PropertyManager** to apply modifications. The model will be displayed as shown in Figure-40.
- Change the display style back to **Shaded With Edges** from the **Display Style** drop-down in the **Heads Up View Toolbar**.

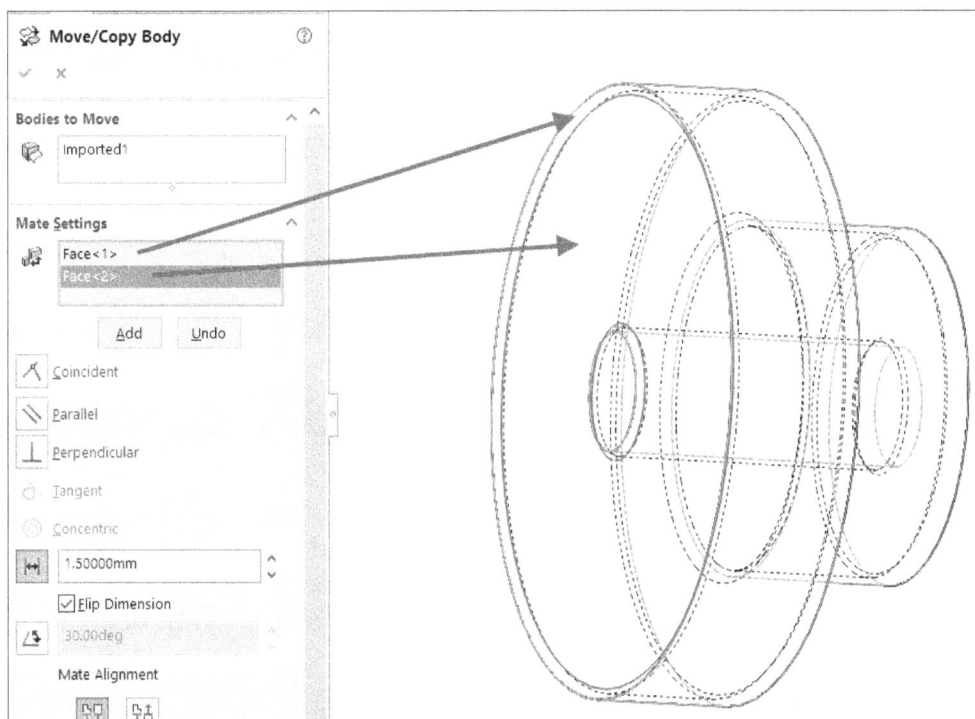

Figure-39. Selecting back faces of stock model and part

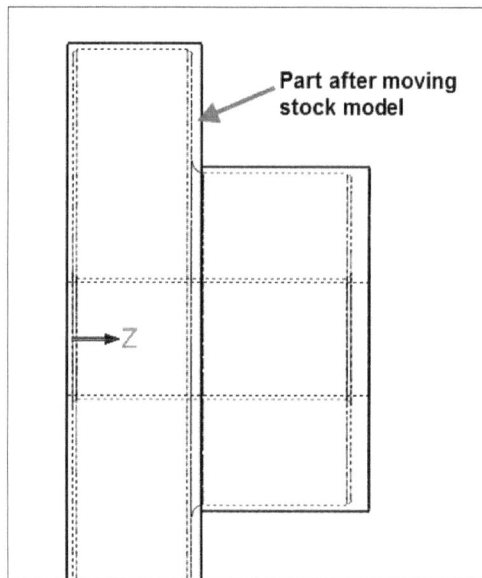

Figure-40. After moving stock

- Click on the face of imported part and select the **Hide** button from mini toolbar displayed; refer to Figure-41. The main part will be displayed again.

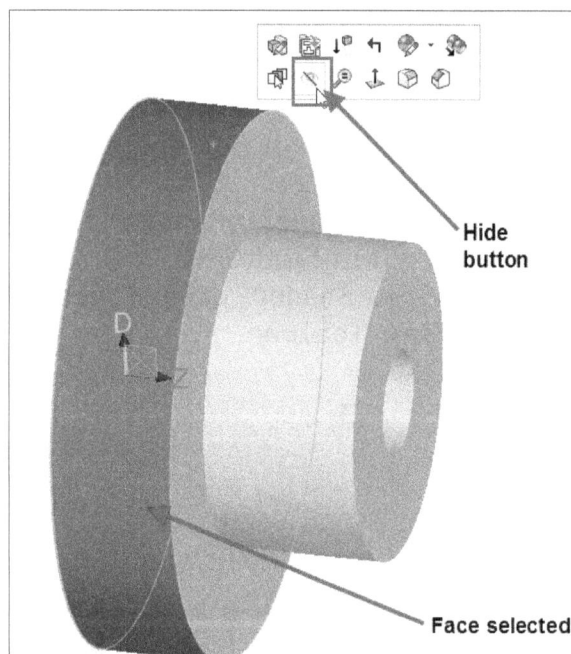

Figure-41. Hiding imported part

Defining Stock and Chuck Jaws

- Click on the **Stock setup** option from the **Properties** node in the **Mastercam Toolpath Manager**. The **Machine Group Properties** dialog box will be displayed with **Stock Setup** tab selected.
- Make sure the **Left Spindle** radio buttons are selected in **Stock** and **Chuck Jaws** areas of the dialog box and then click on the **Properties** button from the **Stock** area in the dialog box. The **Machine Component Manager - Stock** dialog box will be displayed.
- Select the **Solid entity** option from the **Geometry** drop-down in the dialog box. The options in the dialog box will be displayed as shown in Figure-42.

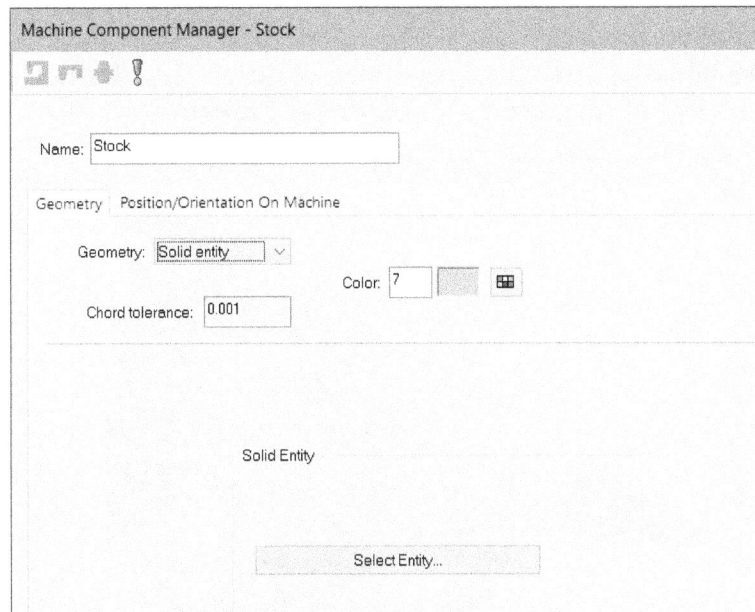

Figure-42. Machine Component Manager-Stock dialog box partial view

- Click on the **Select Entity** button in the **Solid Entity** area of the dialog box. You will be asked to select a body to be used as stock.
- Expand the **FeatureManager Design Tree** node and select the imported part; refer to Figure-43.

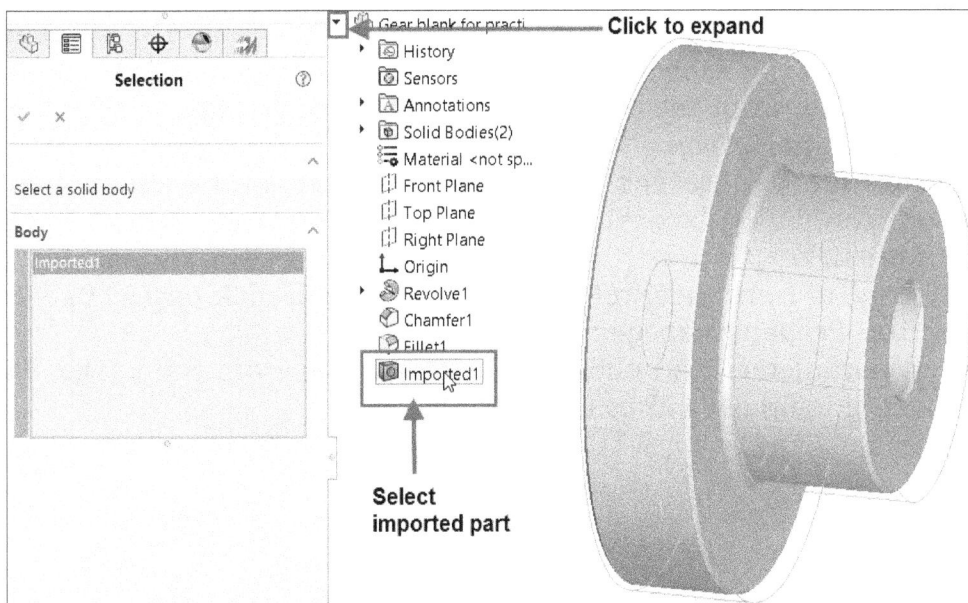

Figure-43. Selecting imported part for stock

- Click on the **OK** button from the **Selection PropertyManager** and then click on the **OK** button from the **Machine Component Manager-Stock** dialog box. The **Machine Group Properties** dialog box will be displayed again.
- Click on the **Properties** button from the **Chuck Jaws** area of the dialog box. The **Machine Component Manager-Chuck Jaws** dialog box will be displayed.
- Select the **From Stock** and **Grip on maximum diameter** check boxes in the **Position** area.
- Enter the value of grip length as **6** in the **Grip Length** edit box of the **Position** area of the dialog box.

- Click on the **OK** button from the dialog box. The **Machine Group Properties** dialog box will be displayed again.
- Select the **Shade boundaries** check box and click on the **OK** button from the dialog box. The model will be displayed as shown in Figure-44.

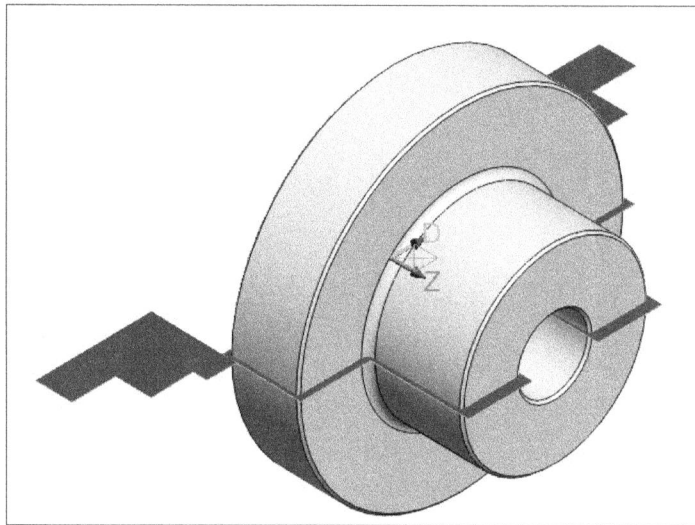

Figure-44. Model after applying stock and chuck jaws

Facing Toolpath

Facing is the first operation to be done on flat face parts. The steps are given next.

- Click on the **Face** tool from the **Lathe** drop-down in the **Mastercam2022 CommandManager** of the **Ribbon**. The **Lathe Face** dialog box will be displayed.
- Select desired tool from the tool list box and specify related parameters in the right area of the dialog box.
- Click on the **Face parameters** tab in the dialog box and click on the **Finish Z** button from the dialog box. You will be asked to select a point to specify the location of finish Z.
- Select the edge of front face; refer to Figure-45 and click on the **OK** button from the **Selection PropertyManager**.
- Set desired parameters and click on the **OK** button from the **Lathe Face** dialog box. The facing toolpath will be created.

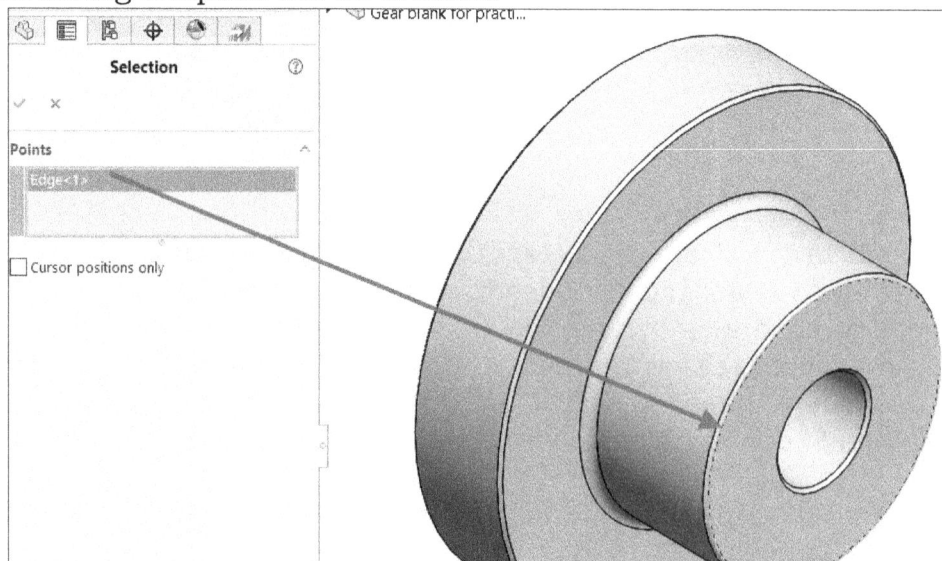

Figure-45. Edge selected for facing

ID Finish

- Click on the **Finish** tool from the **Lathe** drop-down. The **Chain Manager** will be displayed.
- Select the inner round face and chamfered face of the model; refer to Figure-46 and click on the **OK** button from the **Chain Manager**. The **Lathe Finish** dialog box will be displayed.

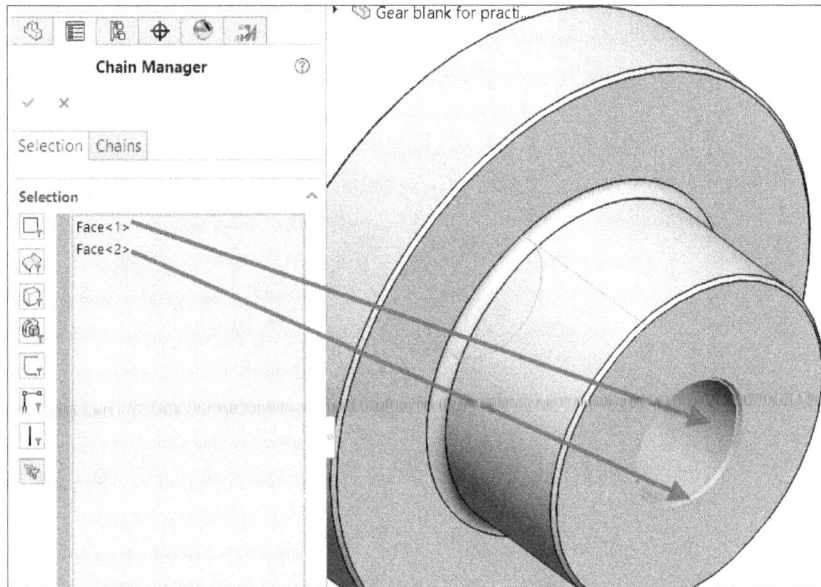

Figure-46. Inner face to be selected

- Select the **T8181 R0.4 ID FINISH 16 DIA- 55 DEG** tool from the tool list box and specify the related parameters in the right area of the dialog box.
- Click on the **Finish Parameters** tab in the dialog box and click on the **Lead In/ Out** button. The **Lead In/Out** dialog box will be displayed.
- Click on the **Lead Out** tab and select the **Extend/shorten end of contour** check box. The options below it will become active.
- Select the **Extend** radio button and specify the value **5** in the **Amount** edit box. (We are extending the contour so that no material of hole is left on the back face.)
- Click on the **OK** button from the dialog box. Specify desired parameters in the **Lathe Finish** dialog box and click on the **OK** button.

Roughing and Finishing Outer faces

- Click on the **Rough** tool from the **Lathe** drop-down in the **Ribbon**. The **Chain Manager** will be displayed.
- Select the outer faces of the model including chamfer and round; refer to Figure-47 and click on the **OK** button from the **Chain Manager**. The **Lathe Rough** dialog box will be displayed.

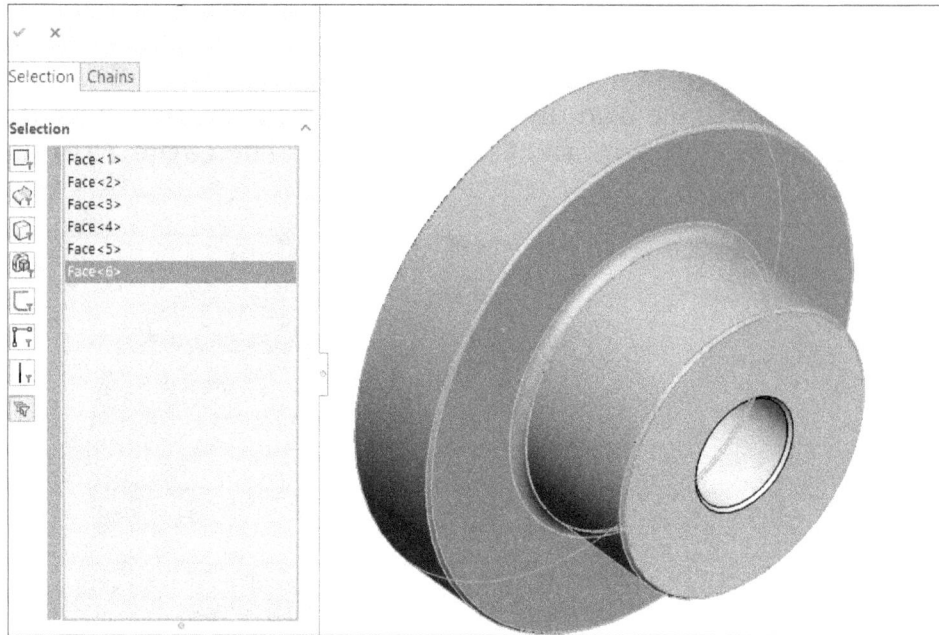

Figure-47. Faces selected for roughing

- Select the **T0101** tool from the tool list box and specify related parameters in the right area of the dialog box.
- Click on the **Rough parameters** tab in the dialog box and then click on the **Lead In/Out** button. The **Lead In/Out** dialog box will be displayed.
- Click on the **Lead Out** tab in the dialog box and shorten the end of contour by **7** mm using the options in the **Adjust Contour** area of the dialog box.
- Click on the **OK** button from the dialog box. The **Lathe Rough** dialog box will be displayed again.
- Select the **Remaining Stock** option from the drop-down in the **Stock Recognition** area of the dialog box and specify the other parameters as required.
- Click on the **OK** button to create toolpath.

Similarly, click on the **Finish** tool from the **Lathe** drop-down and create the finish toolpath using the same faces and **T2121** tool. Make sure you specify the same **Lead out** for finish tool path as you have specified for rough toolpath.

Performing Stock Flip

- Click on the **Stock Flip** tool from the **Lathe** drop-down in the **Ribbon**. The **Stock Flip** dialog box will be displayed.
- Make sure **Transfer geometry** and **Blank original geometry** check boxes are selected in the **Geometry** area of the dialog box and then click on the **Select** button. You will be asked to select the geometry to be flipped.
- Select the part from graphics area and click on the **OK** button from **Selection PropertyManager**.
- Specify the diameter as **80** and **Z** value as **-50** in the **D** and **Z** edit boxes of **Final Position** in the **Chuck Position** area of the dialog box, respectively.
- Click on the **OK** button from the dialog box. The part will flip with stock; refer to Figure-48.

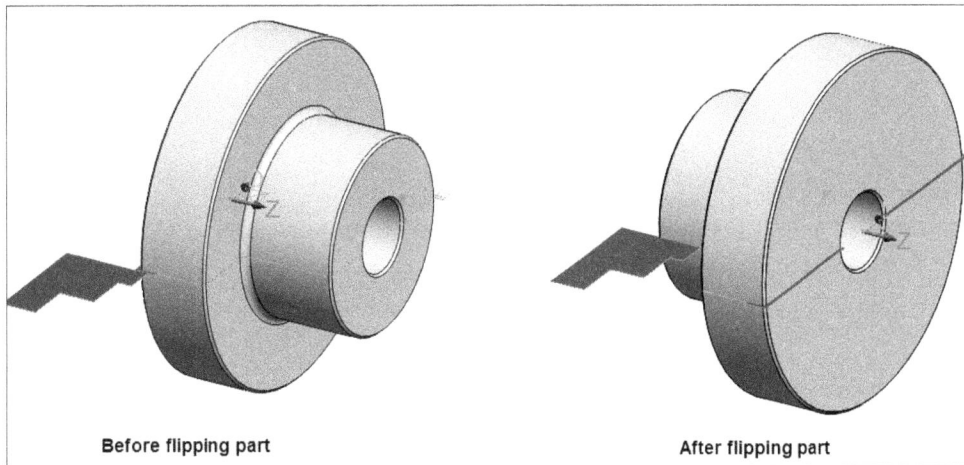

Figure-48. Flipping stock

Now, create the facing and finishing tool path for the remaining stock as discussed earlier. Save the file at desired location and generate NC file using the button in **Mastercam Toolpath Manager** as discussed earlier.

SELF ASSESSMENT

Q1. It is mandatory to have secondary chuck for performing cut off operation in cnc lathe. (T/F)

Q2. The **Stock Flip** tool is used to transfer stock from one chuck to another. (T/F)

Q3. Using **Stock Advance** tool, you can create operation to push the stock but you cannot pull stock by using this tool. (T/F)

Q4. Clamping and un-clamping of chuck can be controlled by M codes. (T/F)

Q5. The .... tool is used to draw a toolpath for machining curves on face of part in Mill-Turn machine.

Q6. By default, in CNC machines, C axis is rotation axis about

a. X axis   b. Y axis
c. Z axis   d. Center of model

FOR STUDENT NOTES

Chapter 6

Milling Toolpaths

Topics Covered

The major topics covered in this chapter are:

- *Basics of Milling.*
- *2D Milling Toolpaths*
- *Letter writing and engraving*
- *Sketched toolpaths*
- *2D High speed toolpaths*

MILLING TOOLPATHS

Milling toolpaths are generated for performing cutting operations on NC Milling machines. There are around 60 tools to generate different toolpaths for milling operations. These tools are one by one discussed in this chapter. But, before we move for milling tools, we need to setup the milling machine for doing the milling operations.

SETTING A MILLING MACHINE

*   Start SolidWorks with Mastercam add-in.
*   Open a model on which you want to performing milling operation.
*   Click on the **Tools -> Mill Machines -> Manage List** tool from the menu bar. The **Machine Definition Menu Management** dialog box will be displayed; refer to Figure-1.

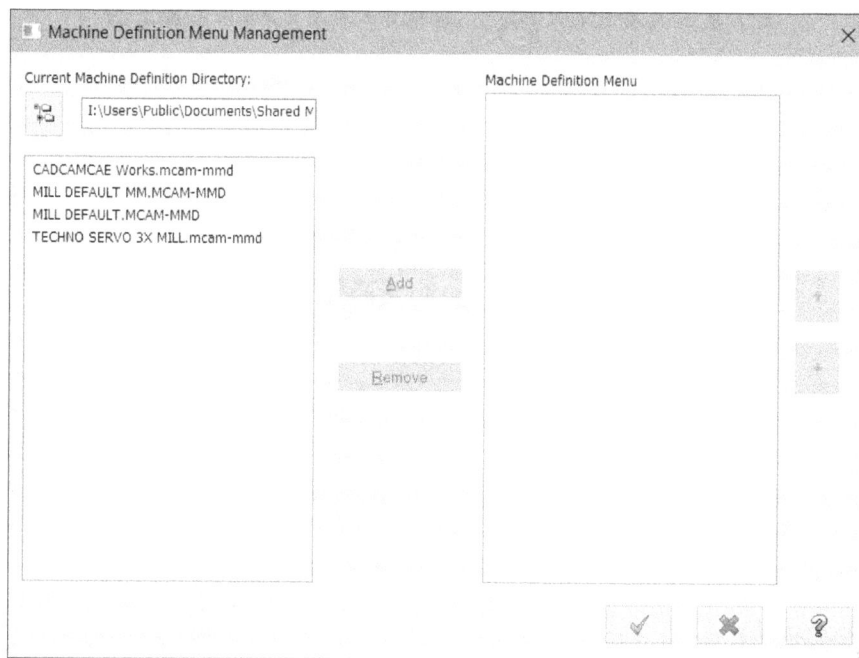

Figure-1. Machine Definition Menu Management dialog box for milling machines

*   Select all the machines that you will be using during the use of this software and click on the **Add** button. The machines will be added to the list in right side.
*   Click on the **OK** button from the dialog box and restart Mastercam add-in.
*   After restarting Mastercam add-in, select desired machine from the **Tools -> Mill Machines** cascading menu from the menu bar.

Or

*   Click on the **Mastercam Toolpath Manager** tab from the **PropertyManager** window; refer to Figure-2.

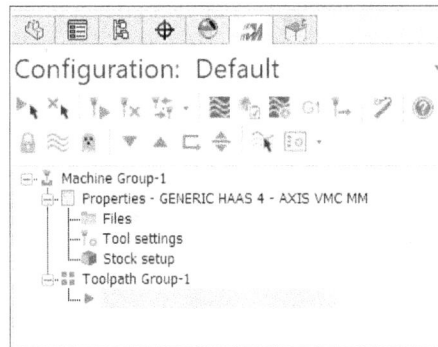

Figure-2. Mastercam Toolpath Manager

- Click on the **Files** option in the **Properties** node of the **Mastercam Toolpath Manager**. The **Machine Group Properties** dialog box will be displayed; refer to Figure-3.

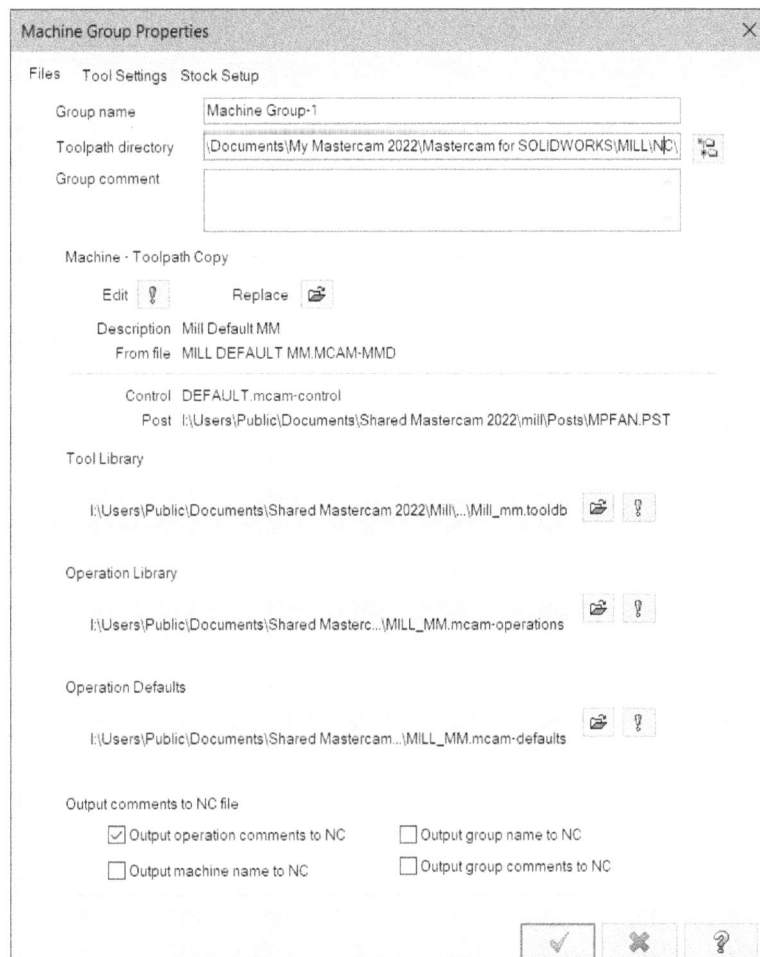

Figure-3. Machine Group Properties dialog box

- Click on the **Replace** button from the **Machine - Toolpath Copy** area. The **Open Machine Definition File** dialog box will be displayed; refer to Figure-4.
- Select desired machine file from the dialog box and click on the **Open** button. The machine will be loaded.
- Click on the **OK** button from the dialog box.

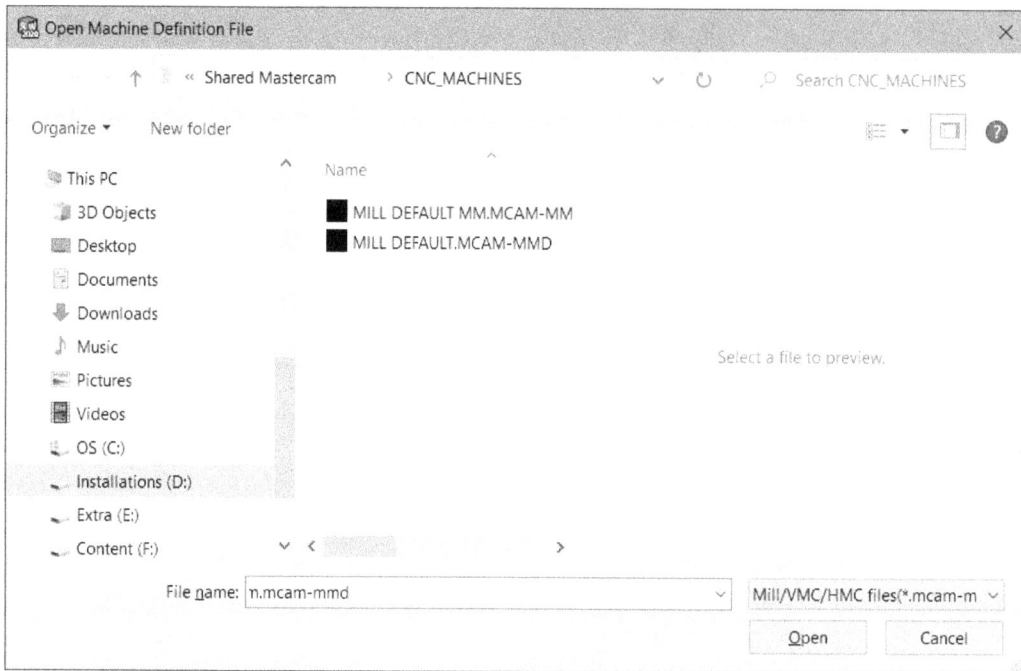

Figure-4. Open Machine Definition File dialog box

Defining Stock

- Click on the **Stock setup** option from the **Properties** node in the **Mastercam Toolpath Manager**. The **Machine Group Properties** dialog box will be displayed with the **Stock Setup** page; refer to Figure-5.
- Select the **Cylindrical** or **Rectangular** radio button.

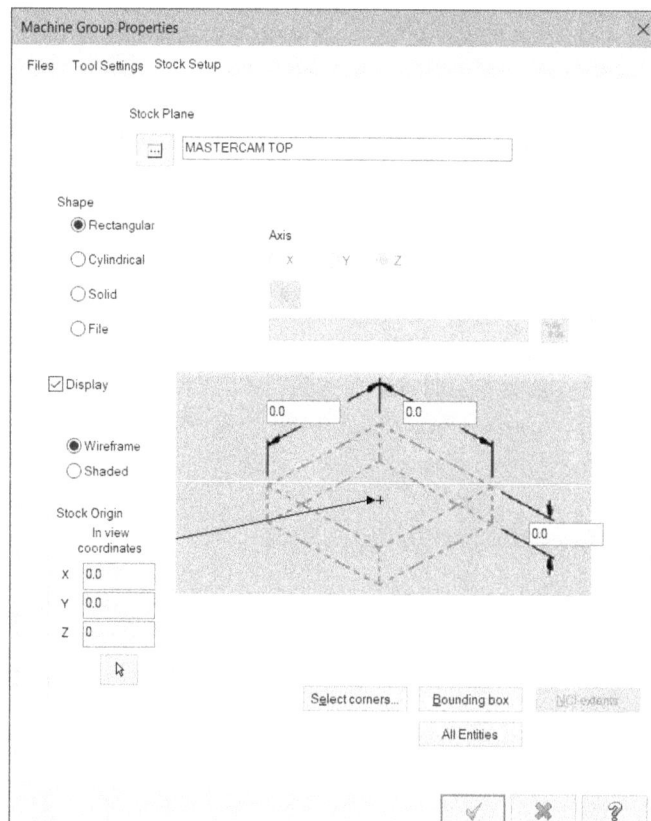

Figure-5. Stock setup page

- Click on the **Bounding box** button to create a bounding box of the stock around the workpiece.
- Adjust the stock and click on the **OK** button from the dialog box. The stock will be created. Note that you can use the other options for defining stock as discussed in previous chapters.

TYPES OF MILLING TOOLPATHS

There are various tools to create milling toolpaths categorized in different drop-downs in the **Ribbon**. In previous chapters, we have seen the **Lathe Toolpaths**. Now, we will work with different drop-downs of milling toolpaths, which are discussed next.

1. Hole making
2. 2D Toolpaths
3. 2D HighSpeed
4. Feature Based Toolpaths
5. Roughing Toolpaths
6. Finishing Toolpaths
7. 3D HighSpeed Toolpaths
8. Multi-Axis Toolpaths

HOLE MAKING

The tools in **Hole making** drop-down are used to create specialized toolpaths like creating holes, making threads, helix boring, and so on; refer to Figure-6. Various toolpaths in this drop-down are discussed next.

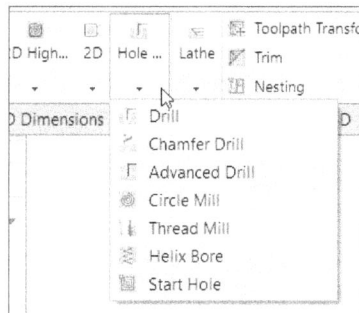

Figure-6. Hole making drop-down

Drill Toolpath

The **Drill** tool in the **Hole making** drop-down is used to generate toolpaths for drilling holes at selected points. The procedure to create the drill toolpath is given next.

- Click on the **Drill** tool from the **Hole making** drop-down. The **Point Manager PropertyManager** will be displayed as shown in Figure-7.

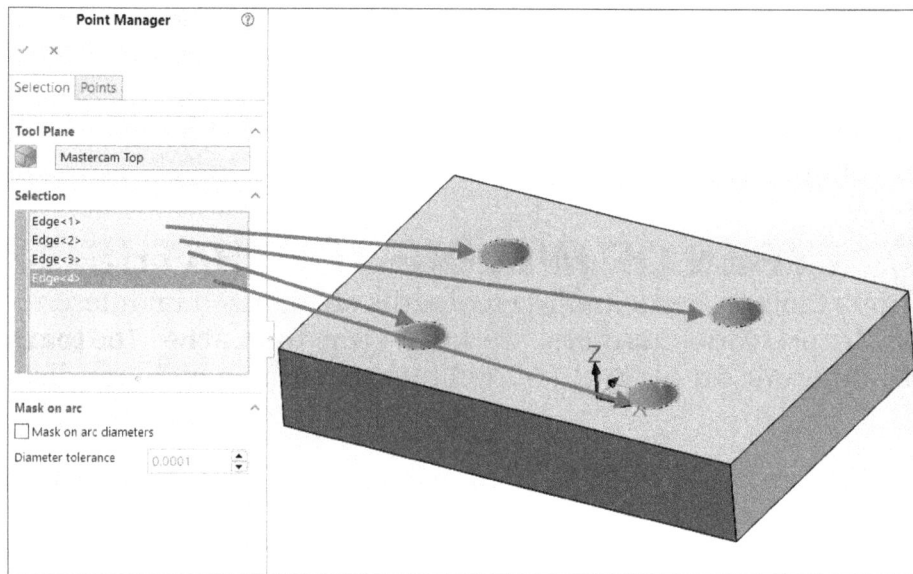

Figure-7. Point Manager PropertyManager

- Select desired points from the model where you want to drill the holes or you can select the edges of holes in your model; refer to Figure-7.
- Click on the **Points** tab in the **Point Manager PropertyManager**. The order of drilling holes will be displayed; refer to Figure-8. If you want to change the order then click on the **Sorting** button at the bottom in the **PropertyManager**. The **Sorting** dialog box will be displayed; refer to Figure-9.

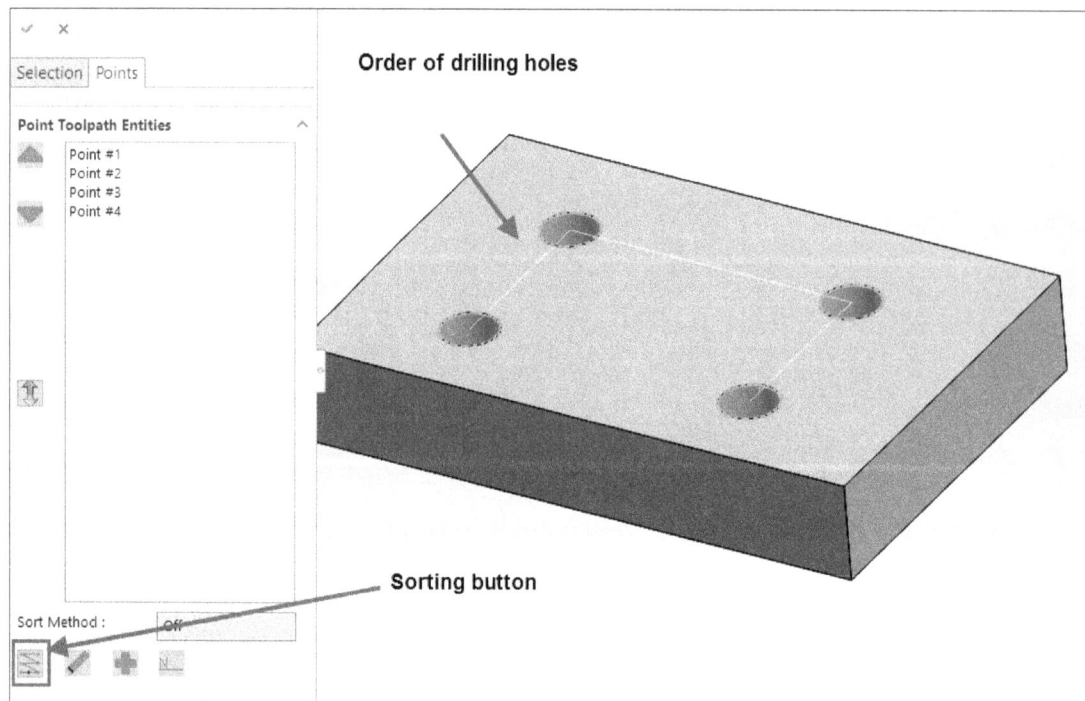

Figure-8. Order of drilling multiple holes

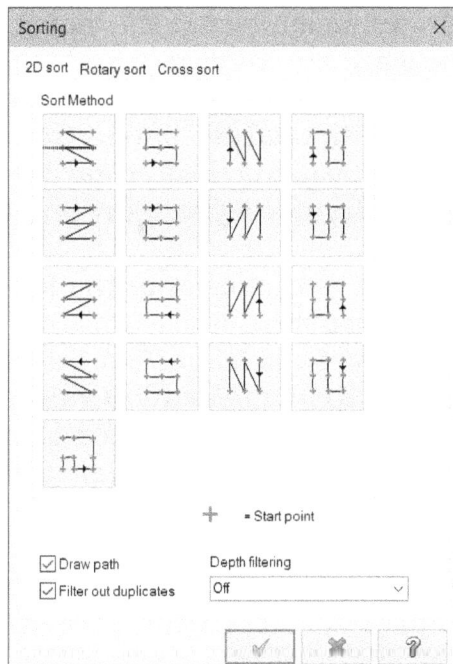
Figure-9. Sorting dialog box

- Select desired sorting method from the using the options in the dialog box and click on the **OK** button.
- If you want to manually change the order of drilling then you can use the **Move Up** and **Move Down** buttons in Points tab of the **Point Manager PropertyManager** after select a point.
- Click on the **OK** button from the **PropertyManager**. The **2D Toolpaths - Drill/ Circles Simple drill** dialog box will be displayed; refer to Figure-10.

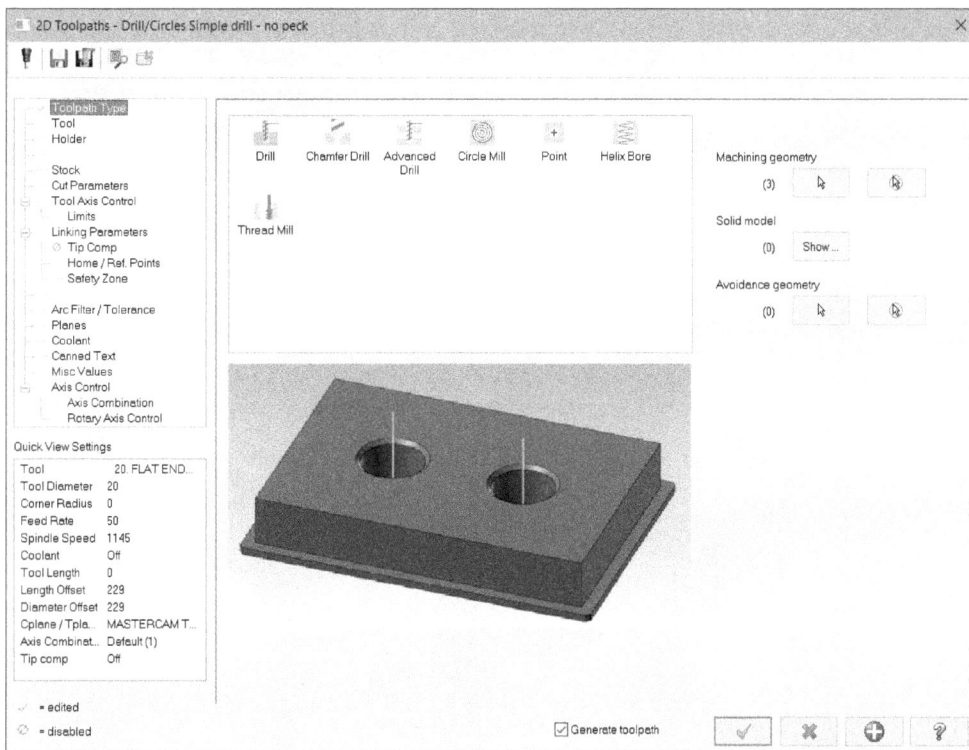
Figure-10. 2D Toolpaths-Drill/Circles Simple drill dialog box

- Click on the **Tool** option from the left box in the dialog box and select desired drill from the tool library. If tools are not displayed in the box then click on the **Select library** tool button and select desired drill from the dialog box.
- Select desired holder by using the **Holder** option from the left box and similarly, specify the cutting parameters by using the **Cut Parameters** option. You can create a desired type of cutting cycle from the **Cycle** drop-down in **Cut Parameters** page of the dialog box.
- Click on the **Linking Parameters** option from the left box and specify the depth of the drill in the edit box for **Depth**.
- Click on the **Tip Comp** option from the left, select the **Tip Comp** check box from the page and specify desired drill tip compensation value if required.
- The options in **Tool Axis Control** page are used to convert a toolpath to 5 axis toolpath. You will learn about these options later in the book.
- Select the **Generate toolpath** check box to generate toolpath for the operation once you click on the **OK** button from the dialog box.
- Click on the **Preview Toolpath** tool from the top in the dialog box to check preview of the toolpath. If you find problems in the preview then you can modify the toolpath parameters.
- Click on the **OK** button from the dialog box to create the toolpath. You can check the toolpath by using the **Backplot** and **Verify** buttons in the **Mastercam Toolpath Manager**.

Circle mill Toolpaths

The Circle mill toolpaths are used to remove material in circular pattern with respect to the specified point. After specifying point, you need to specify diameter and depth of cut to be made. The procedure to create circular milling toolpaths is given next.

- Click on the **Circle mill Toolpaths** tool from the **Hole making** drop-down. The **Point Manager PropertyManager** will be displayed as discussed earlier.
- Select the points that you want to include of circular milling; refer to Figure-11.

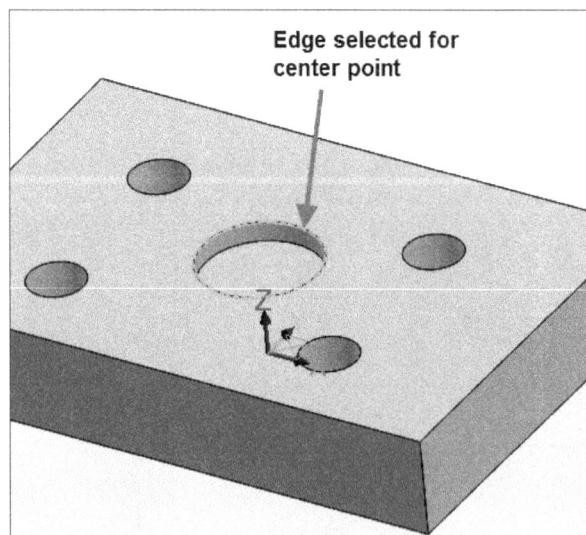

Figure-11. Points selected for circular milling

- Click on the **OK** button from the **PropertyManager**. The **2D Toolpaths-Circle Mill** dialog box will be displayed; refer to Figure-12.

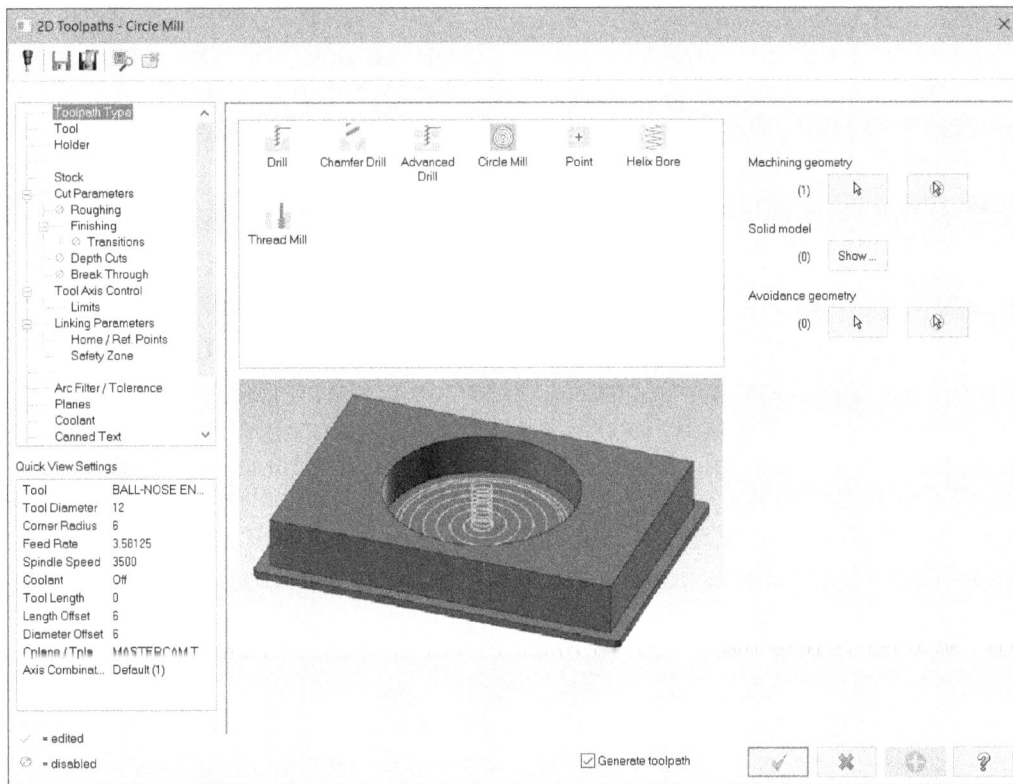

Figure-12. 2D Toolpaths–Circle Mill dialog box

- Click on the **Tool** option from the left box and specify desired tool by using the **Select library tool** button. Generally, end mill tool (with/without radius) is selected for this operation.
- Similarly, specify the holder by using the **Holder** option.
- Click on the **Cut Parameters** option from the left box. The **Cut Parameters** page of the dialog box will be displayed; refer to Figure-13.

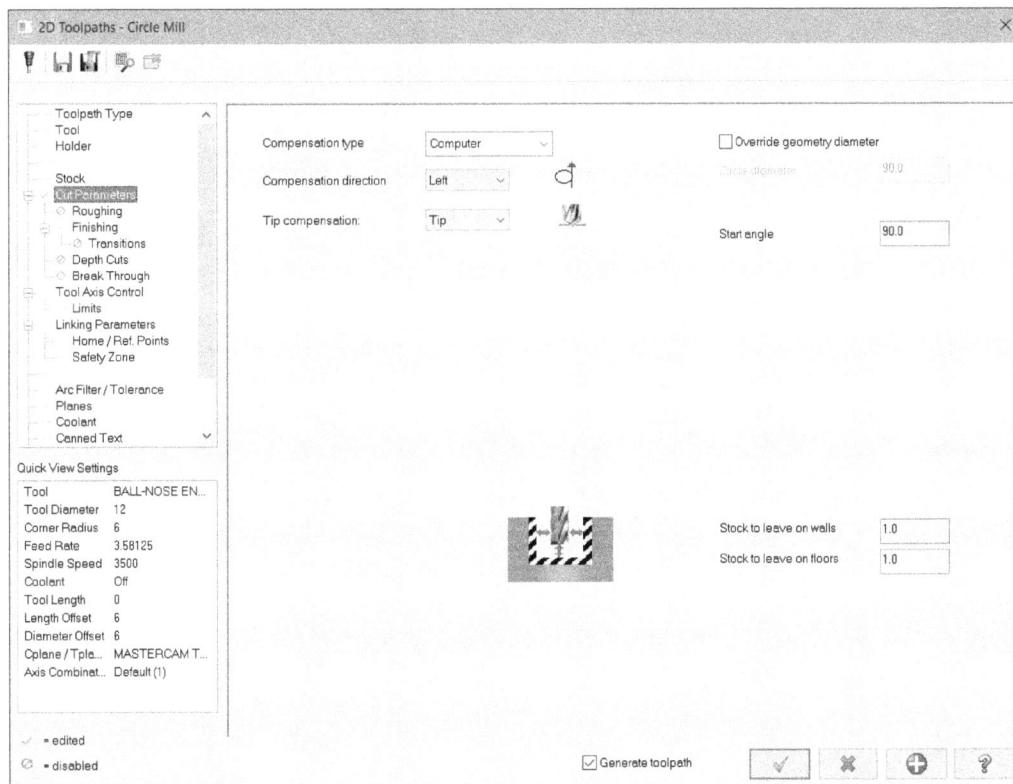

Figure-13. Cut Parameters page

- Click in the **Circle diameter** edit box to specify the diameter of the circular cuts being performed on the selected points. Note that if you have selected circular edge then this edit box will be locked with the value of diameter equal to selected round edge.
- Specify the stock to be left after milling in the **Stock to leave on walls** and **Stock to leave on floors** edit boxes. Set the compensation type and direction in the respective fields of this page.
- Click on the **Linking Parameters** option from the left. The options to define vertical movement of tool will be displayed.
- Specify the value of **Clearance**, **Retract plane**, and **Feed plane** as desired in the respective edit box. Note that clearance is the distance of tool from workpiece to avoid collision, retract plane is the location at which tool moves back after one cut operation, and feed plane is the location from where tool moves for cutting workpiece.
- Click on the **Top of stock** button and select top point of the workpiece. Be careful to specify the value manually if you have not done facing operation on the part before using this toolpath.
- Click on the **Depth** button and select a vertex/round edge to define total depth of cut.
- If we click on the **OK** button now then a circular cut will be made but inner material will remain intact; refer to Figure-14. This type of toolpath is useful when we need to remove large cylindrical stock from the workpiece and there is a thorough hole in the part in place of small depth cut. But this is not the requirement in our example.

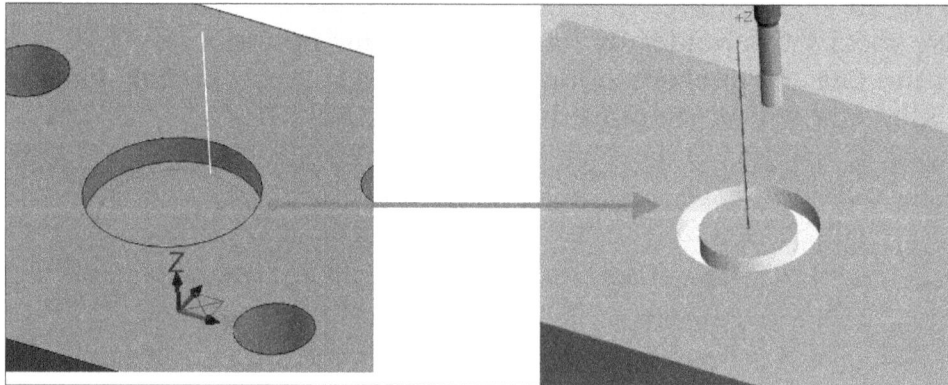

Figure-14. Circmill toolpath without roughing and finishing steps

- Click on the **Roughing** option from the left in the **2D Toolpaths - Circle Mill** dialog box. The options will be displayed as shown in Figure-15.
- Select the **Roughing** check box from the right area of the dialog box to activate roughing options.
- Specify desired values of parameters. Note that on click in each edit box, function of that edit box value in toolpath is displayed in preview box. It is better to specify the value of **Stepover**, **Minimum radius**, and **Maximum radius** in percentage of your tool diameter.
- Similarly, specify the **Finishing** parameters if required.

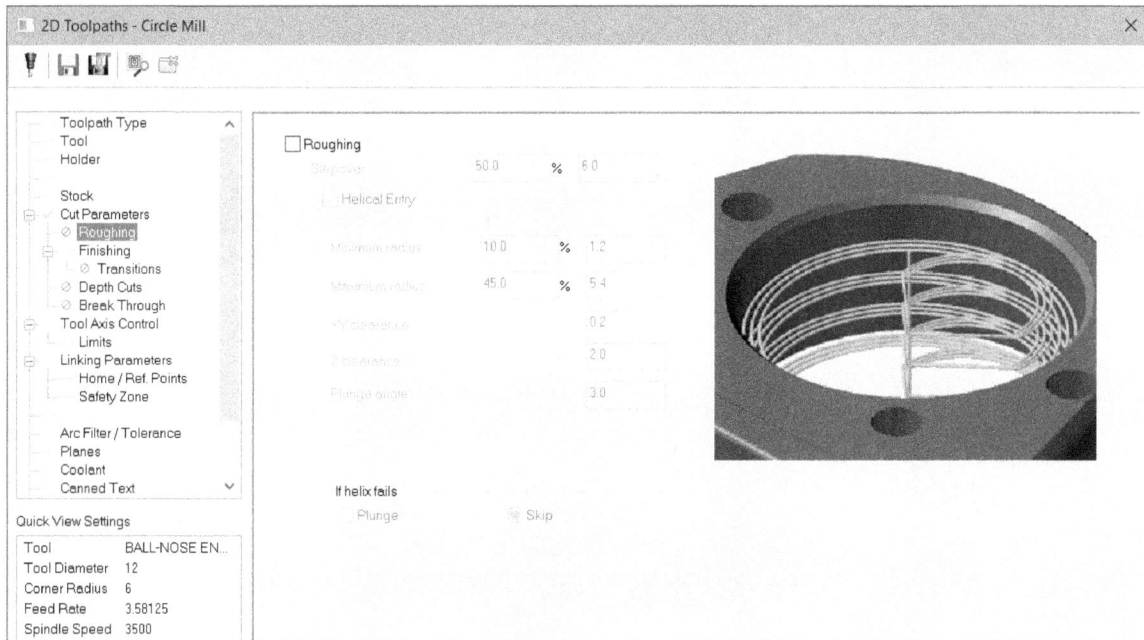

Figure-15. Options for roughing

- Specify the other related parameters like **Depth Cuts**, **Home/Ref. Points** etc. and click on the **OK** button from the dialog box. The toolpaths will be created as shown in Figure-16.

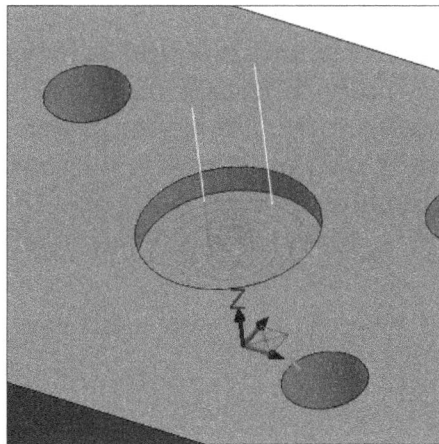

Figure-16. Circular mill toolpath generated

Note that you can cut multiple locations of same depth by using this toolpath.

Threadmill Toolpath

The **Threadmill Toolpath** tool is used to create threads in the hole while milling. The procedure to create the thread mill toolpath is given next.

- Click on the **Threadmill Toolpaths** tool from the **Hole making** drop-down. The **Point Manager PropertyManager** will be displayed as discussed earlier.
- Select the points at which you want to perform thread milling (you can also select edges of holes in which you want to form threads) and click on the **OK** button from the **PropertyManager**. The **2D Toolpaths - Thread Mill** dialog box will be displayed; refer to Figure-17.

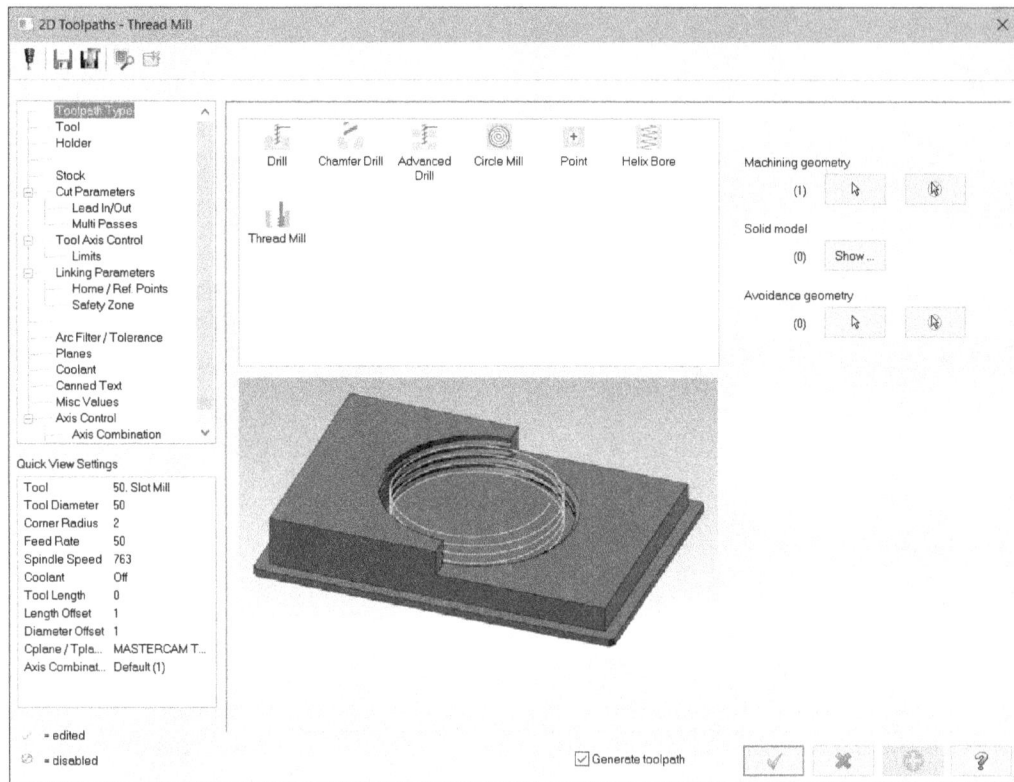

Figure-17. 2D Toolpaths-Thread Mill dialog box

- Click on the **Tool** option from the left box and select the tap of desired size. Note that in market the tool is sold with the name thread mill cutter.
- Click on the **Holder** option from the left box and select desired holder.
- Click on the **Cut Parameters** option from the left box. The **Cut Parameters** page of the dialog box will be displayed; refer to Figure-18.

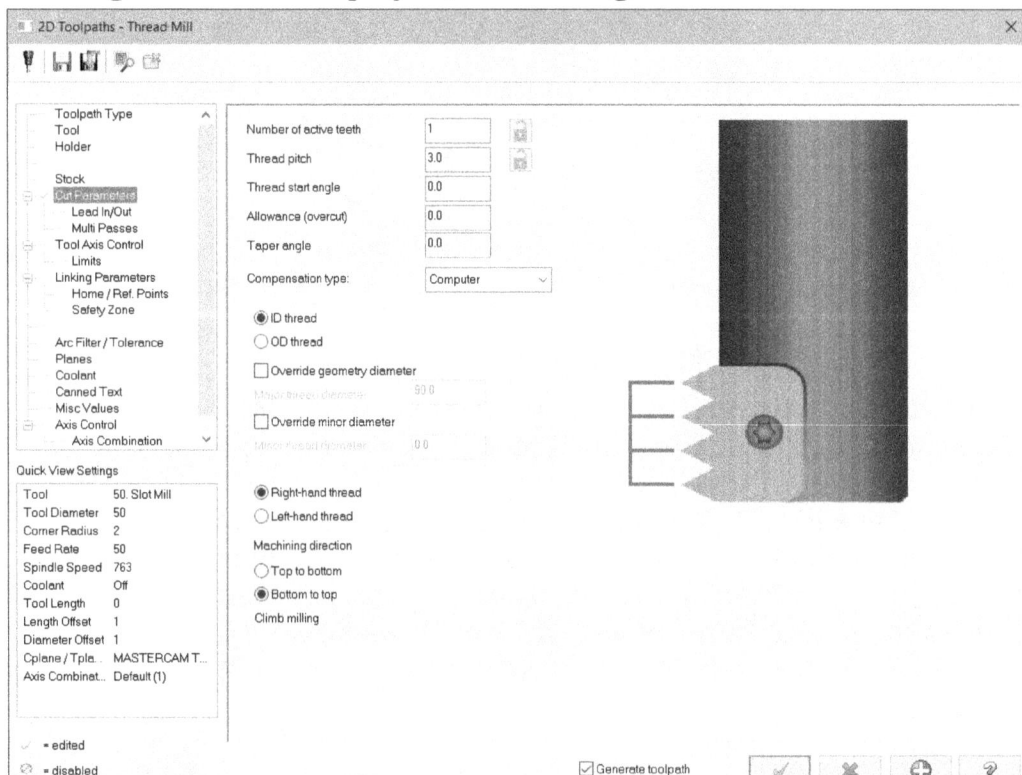

Figure-18. Thread cutting parameters page

- Specify desired threading parameters in this page.

- Click on the **Linking Parameters** option from the left box and specify the depth needed.
- Click on the **OK** button from the dialog box. The toolpath for thread milling will be created; refer to Figure-19.

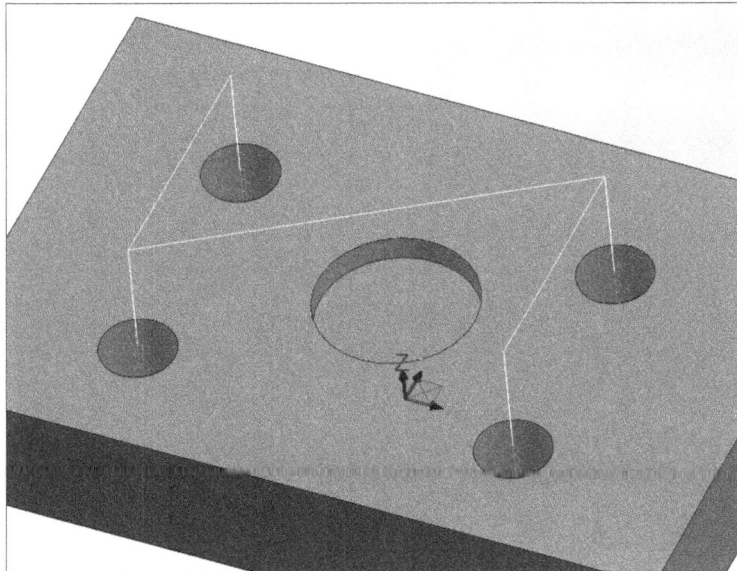

Figure-19. Toolpath for threadmilling

Helix Bore Toolpaths

The Helix Bore toolpaths are used to bore a hole in a helical path. The procedure to create helix bore toolpath is given next.

- Click on the **Helix Bore** tool from the **Hole making** drop-down in the **Ribbon** after applying drill. The **Selection PropertyManager** will be displayed and you will be asked to select the points to perform helical bore.
- Select the points at which you want to perform helical bore (you can also select edges of holes) and click on the **OK** button from the **PropertyManager**. The **2D Toolpaths - Helix Bore** dialog box will be displayed; refer to Figure-20.
- Click on the **Tools** option from the left and select desired end mill tool. Similarly, select tool holder after clicking the **Holder** option from the left.
- Click on the **Cut Parameters** option from the left. The options to specify cutting parameters will be displayed; refer to Figure-21.

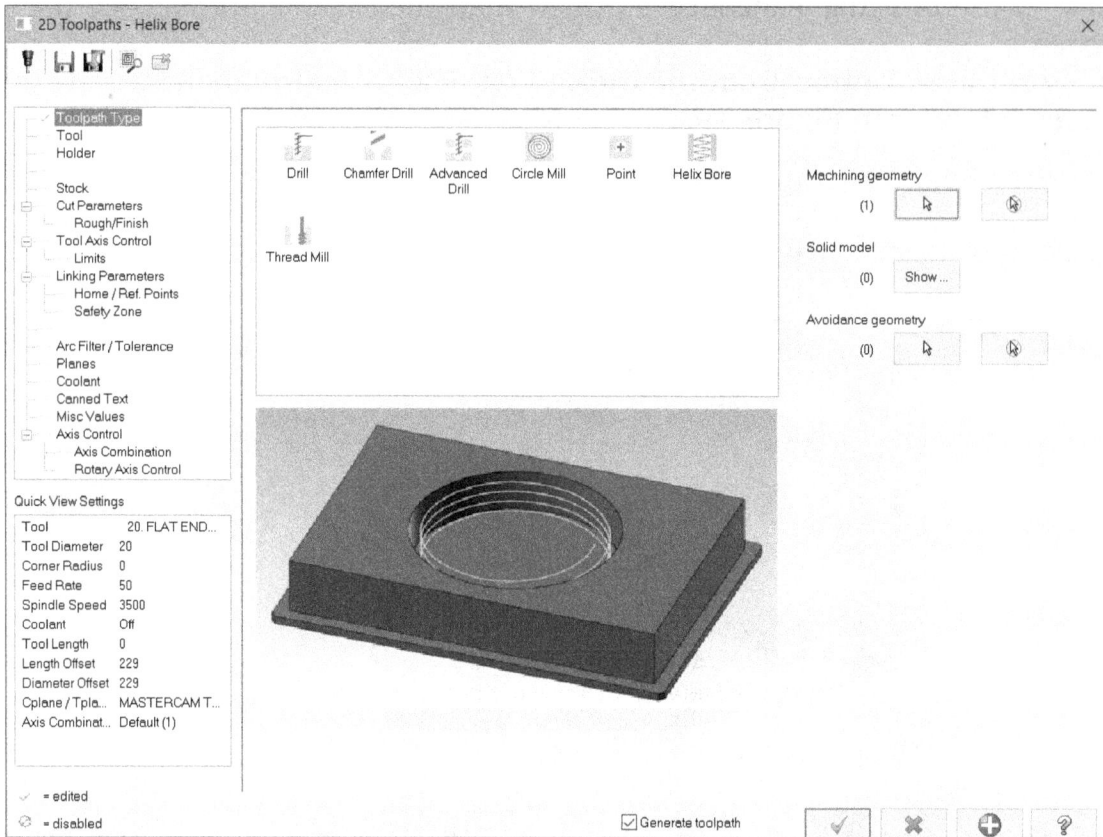

Figure-20. 2D Toolpaths-Helix Bore dialog box

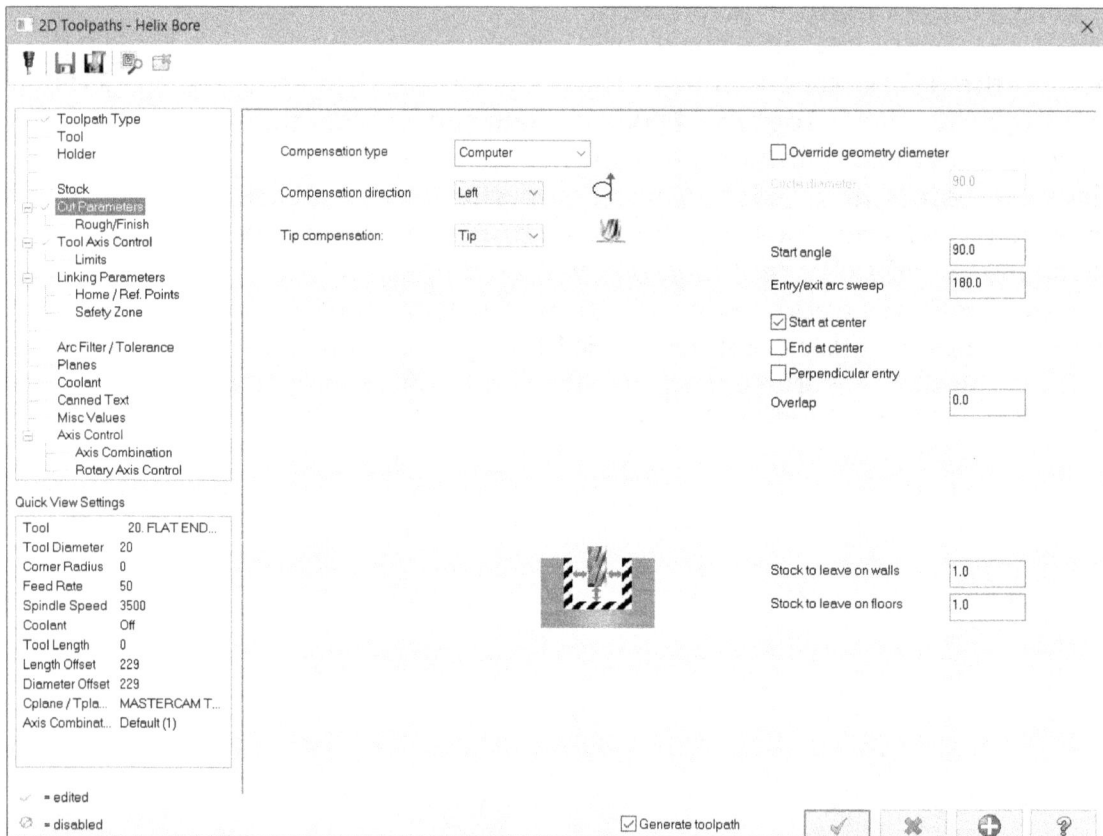

Figure-21. Cut Parameters page for helix bore toolpath

- Specify desired cutter compensation type, direction and tool tip compensation in the respective fields of the dialog box.

- Specify desired angle value in the **Entry/exit arc sweep** edit box to define entry and exit path of tool while cutting starts and ends. If the entry/exit arc sweep is less than **180** degrees, the system applies an entry/exit line.
- Specify desired starting angle value in the **Start angle** edit box. The tool will move at this specified angle while moving for the cutting helix toolpath. If you specify 90 degree value then the tool will move perpendicularly till the starting of helix toolpath.
- Select the **Start at center** check box if you want to start the toolpath from the center of the hole. Similarly, selecting the **Perpendicular entry** check box will make toolpath start with a perpendicular line.
- Specify desired amount in the **Overlap** edit box to set how far the tool goes past the end of the toolpath before exiting for a cleaner finish
- Similarly, specify the amount of material to be left after this operation in the **Stock to leave on walls** and **Stock to leave on floors** edit boxes.
- Click on the **Rough/Finish** option from the left box. The options related to roughing and finishing will be displayed in the right area of the dialog box. Specify desired parameters. Note that the preview of variable will be display as you click in an edit box of this page.
- Specify linking parameters as done in previous toolpaths by using the **Linking Parameters** page. Set the other options as required and click on the **OK** button from the dialog box. The toolpath will be created; refer to Figure-22.

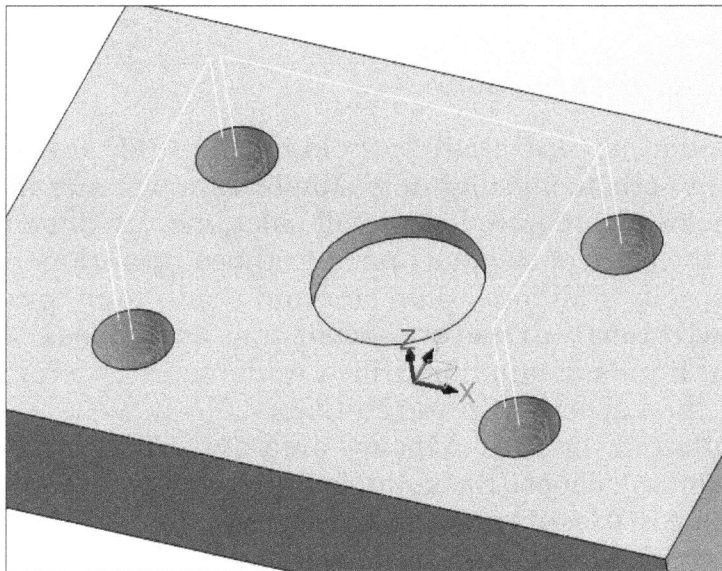

Figure-22. Helix bore toolpath created

Start Hole Toolpath

The Start Hole toolpath is used to create a hole by drilling before you perform any roughing operation on large stock. This option is used when your tool is delicate or there is not location to start toolpath due to part geometry. The Start Holes feature works with all toolpath types, but is especially effective when used together with the align plunge entries for start holes feature found in Surface Rough Pocket. This feature organizes all of the plunge points so that one pre-drilled hole can serve as the plunge position for multiple depth cuts. The procedure to use this tool is given next.

- Make sure you have earlier performed any milling operation before using this tool. Click on the **Start Hole** tool from the **Hole making** drop-down in the **Ribbon**. The **Drill Start Holes** dialog box will be displayed; refer to Figure-23.

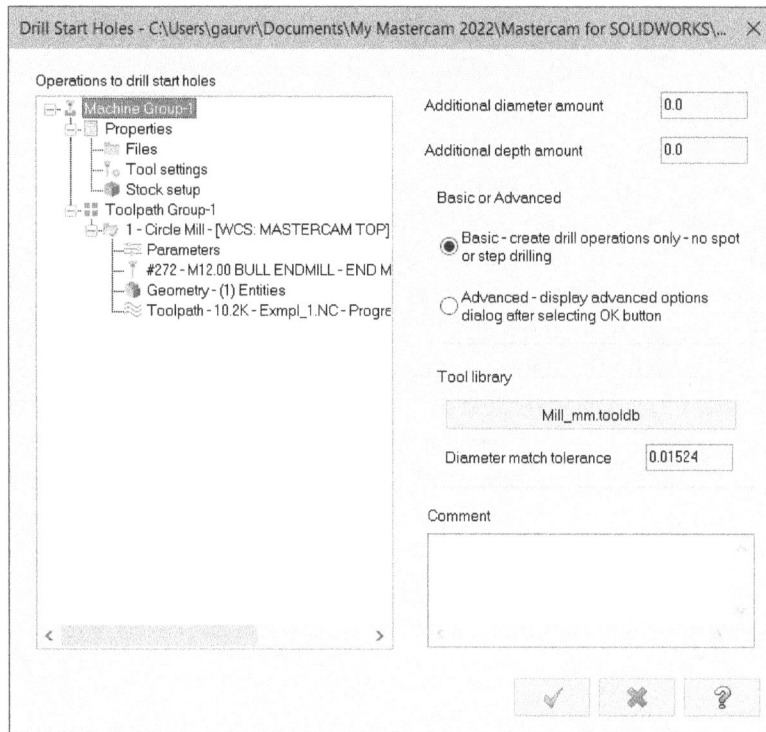

Figure-23. Drill Start Holes dialog box

- Select the operation (or operation by holding the **CTRL** key while selecting) for which you want to create starting hole. Mastercam will automatically figure out where plunge holes are required, and will calculate the dimensions of the start holes based on the sizes of the tools used in those operations.
- If you want to specify additional diameter and depth then specify the respective values in the **Additional diameter amount** and **Additional depth amount** edit boxes. Note that if mastercam has drilled a diameter of 5 mm but you want to drill 6 diameter then specify 1 in the edit box.
- Click on the button in the **Tool library** area of the dialog box to select desired library. Mastercam will choose the suitable drill from the select tool library. Specify the tolerance value in **Diameter match tolerance** edit box. Mastercam will search in the range of specified tolerance for the drill size.
- Click on **OK** button to apply settings.
- If you want to specify advanced options related drilling like spot drilling or drilling cycle then select the **Advanced** radio button from the **Basic or Advanced** area of the dialog box and click on the **OK** button. Mastercam will start analyzing the toolpaths and in a few moments the **Automatic Arc Drilling** dialog box will be displayed; refer to Figure-24.

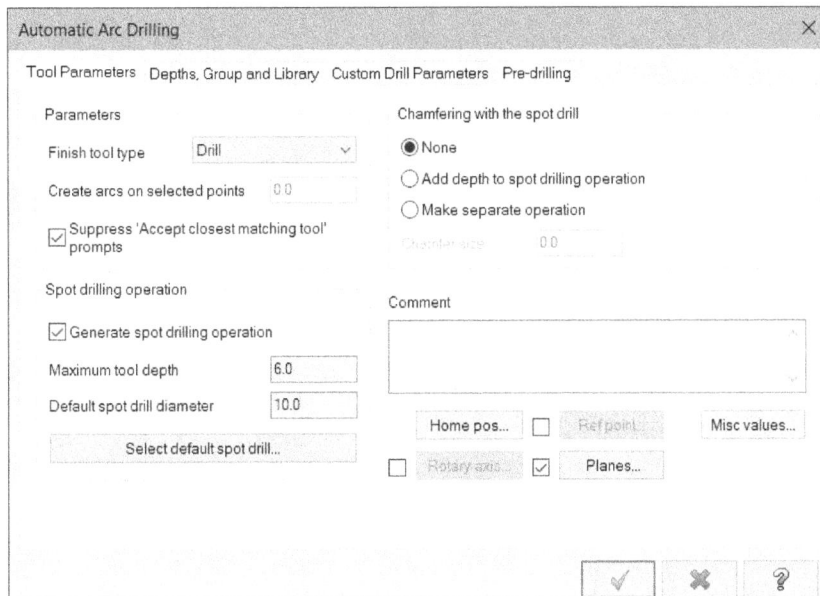

Figure-24. Automatic Arc Drilling dialog box

- Select desired finishing tool type for drill from the **Finish tool type** drop-down. To spot drill, select the **Generate spot drilling operation** check box. The options below it will become active.
- Specify the tool depth, default size or default spot drill as required.
- Click on the **Depths, Group and Library** tab in the dialog box. The **Automatic Arc Drilling** dialog box will be displayed; refer to Figure-25.

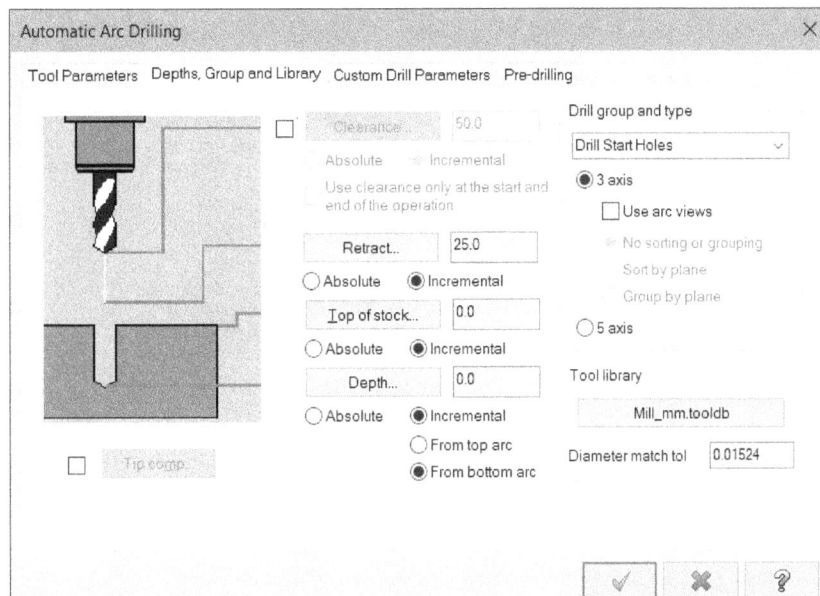

Figure-25. Depths, Group and Library tab

- Specify the retraction location, top of stock location and depth point for drilling. Similarly, parameters in the other tabs and click on the **OK** button from the dialog box. The toolpaths will be created for drilling but they might be inserted after main operation so you need to manually drag the drill toolpaths before the main operations; refer to Figure-26.

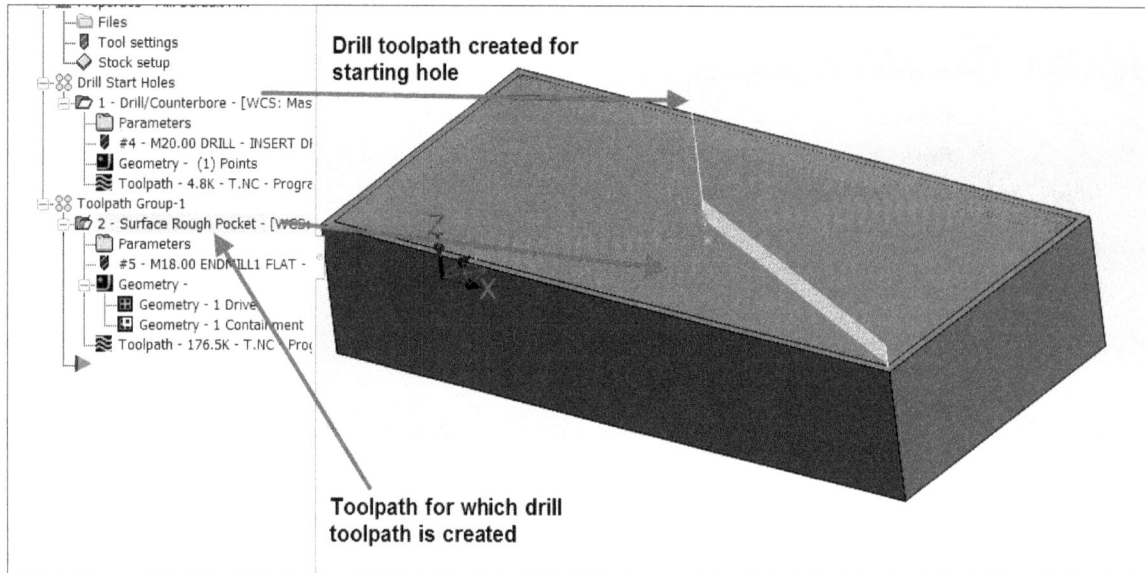

Figure-26. Start hole toolpath

Point Toolpaths

The Point toolpaths make the tool to follow the toolpaths formed by the selected points. The procedure to use this tool is given next.

* Click on the **Point Toolpaths** tool from the **Hole making** drop-down in the **Ribbon**. You are asked to select the points.
* Select the points and click on the **OK** button from the **PropertyManager**.
* The **2D Toolpaths-Point** dialog box will be displayed; refer to Figure-27.

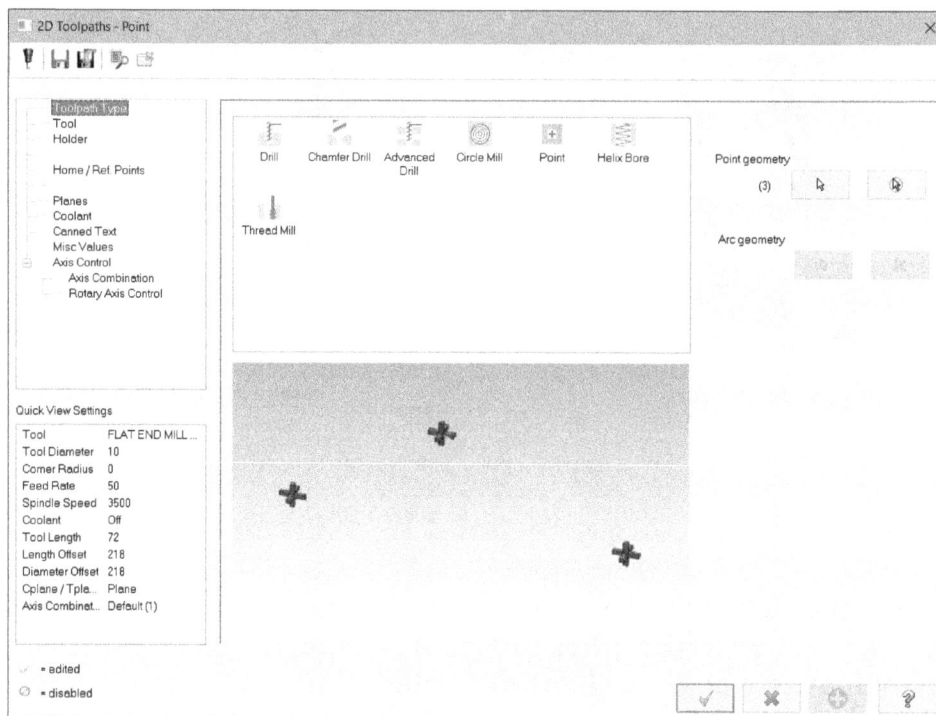

Figure-27. 2D Toolpaths-Point dialog box

- Specify the parameters as discussed earlier.
- Click on the **OK** button from the dialog box to create the toolpaths. The tool will move straight to the selected points while cutting material.

Creating Chamfer Drill Toolpath

The **Chamfer Drill** tool is used to create chamfer on the drilled holes. The procedure to use this tool is given next.

- Click on the **Chamfer Drill** tool from the **Home making** drop-down of the **Ribbon**. The **Point Manager** will be displayed
- Select desired holes with chamfers to be machined from the model and specify the parameters as discussed earlier.
- Click on the **OK** button from the **Manager**. The **2D Toolpaths - Chamfer Drill** dialog box will be displayed; refer to Figure-28.

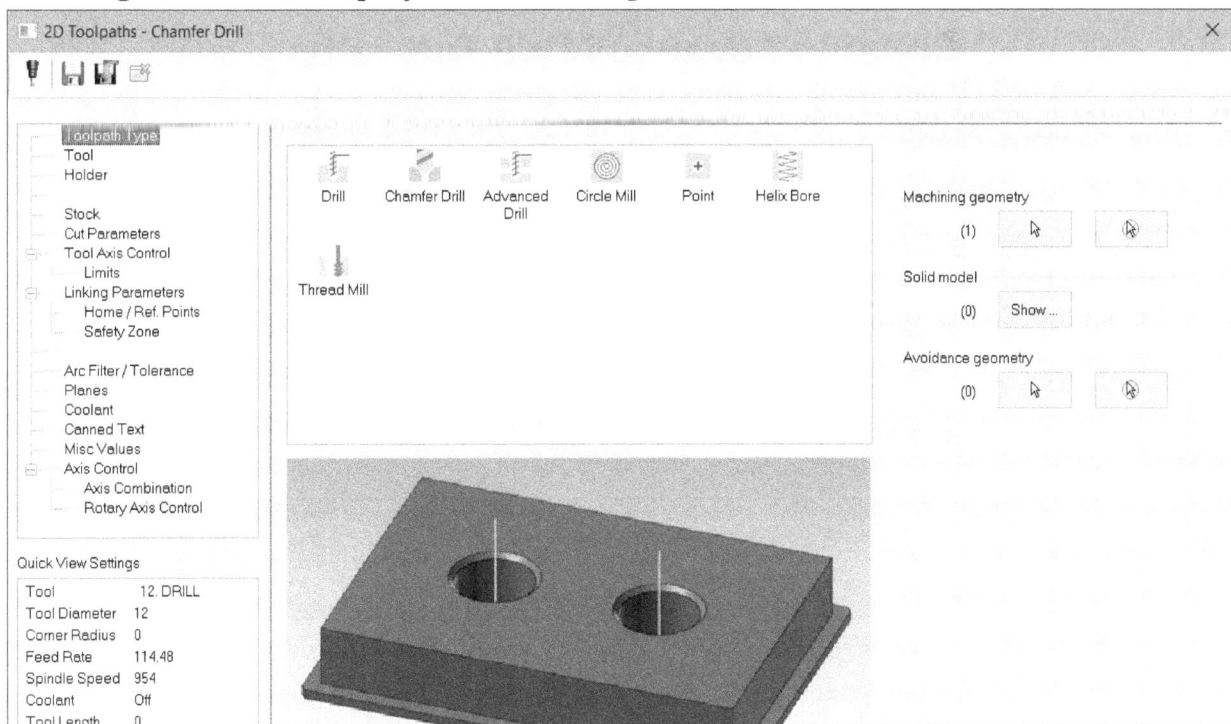

Figure-28. 2D Toolpaths-Chamfer Drill dialog box

- Select desired **Chamfer mill** tool and related tool holder from the dialog box.
- Click on the **Stock** option from the left area of the dialog box and select the stock model to be used for performing chamfer milling. Note that the stock left after performing various operations are available in the **Stock Model** drop-down of **Define stock from** area of the dialog box.
- Click on the **Cut Parameters** option from the left area of the dialog box and specify the chamfer cut depth/width.
- Set the other parameters as discussed earlier and click on the **OK** button from the dialog box. The toolpaths will be generated; refer to Figure-29.

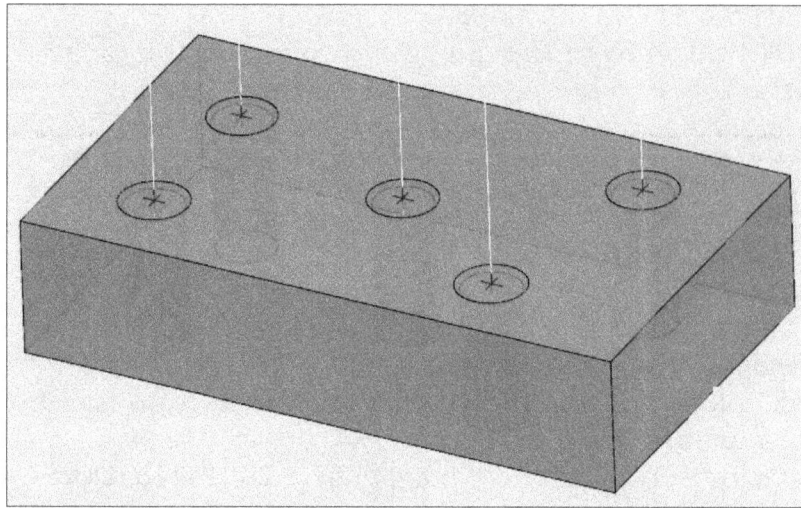
Figure-29. Chamfer drill toolpath

Creating Advanced Drill Toolpaths

The **Advanced Drill** tool is used to create advanced toolpaths with different behaviors for each toolpath segment. The procedure to use this tool is given next.

• Click on the **Advanced Drill** tool from **Hole making** drop-down of the **Ribbon**. The **Point Manager** will be displayed as discussed earlier.
• Select desired holes and click on the **OK** button from the **Manager**. The **2D Toolpaths - Advanced Drill** dialog box will be displayed.
• Select the drill tool and tool holder as discussed earlier.
• Click on the **Cut Parameters** option from the left area of the dialog box to define cutting parameters for each hole. The options will be displayed as shown in Figure-30.

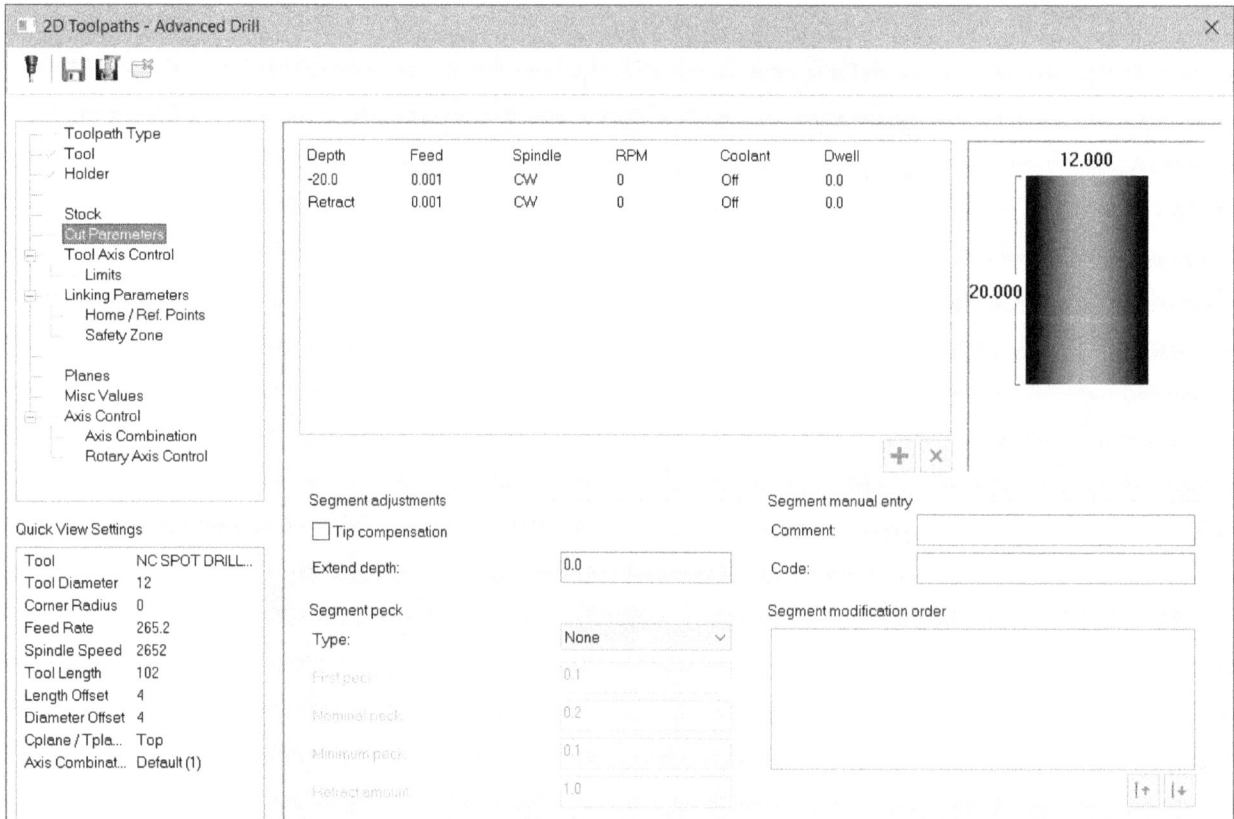
Figure-30. Cut Parameters for advanced drill operation

- Set desired parameters in the table to define parameters like depth of hole, rpm of tool, coolant, dwell time, and so on.
- Select desired option from the **Type** drop-down in the **Segment peck** area of the dialog box to define pecking parameters for chip breaking and cooling of drill bit. Select the **Full Segment** option from the drop-down if you want to create drill peck cycle for full segment. Using this option will stop the drill for a moment at specified depth intervals. Select the **Chip Break** option from the drop-down if you want to create a peck cycle in which tool retracts by specified distance value after each cutting peck.
- Set the other parameters as discussed earlier.
- You can add more cutting parameters by using the **+** button from the **Cut Parameters** page and click on the **OK** button. The toolpath will be generated.

2D TOOLPATHS

In 2D Toolpaths, tool moves on one plane and once cutting at that plane is complete, it moves down by specified cutting depth to cut another plane. Now, we will start with the 2D toolpaths and explain the procedure to create 2D toolpaths. There are five tools in this category:

- Contour
- Pocket
- Face
- Slotmill Toolpaths
- Engrave

The generation to these toolpaths are given next.

CONTOUR

The Contour toolpaths are used to remove the material by following the contour of the model. This toolpath is used to remove material from the outer sides of part; refer to Figure-31. The procedure to create contour toolpaths is given next.

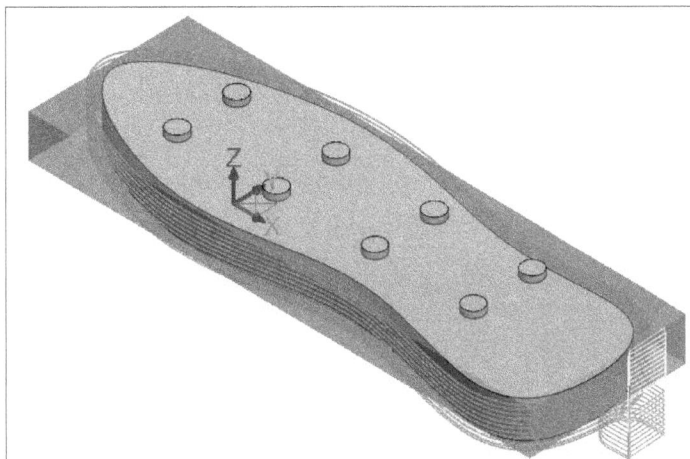

Figure-31. Contour toolpath example

- Click on the **Contour** tool from the **2D** drop-down; refer to Figure-32. The **Chain Manager** will be displayed as discussed earlier.

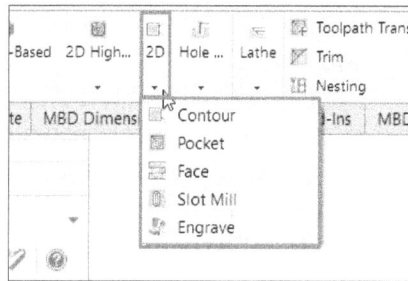

Figure-32. 2D drop-down

- Select the faces that you want to machine using the contour milling toolpaths. Note that you need to select the outer side faces of the part to remove material at the boundaries. Click on the **OK** button from the **Chain Manager**. The **2D Toolpaths-Contour** dialog box will be displayed; refer to Figure-33.

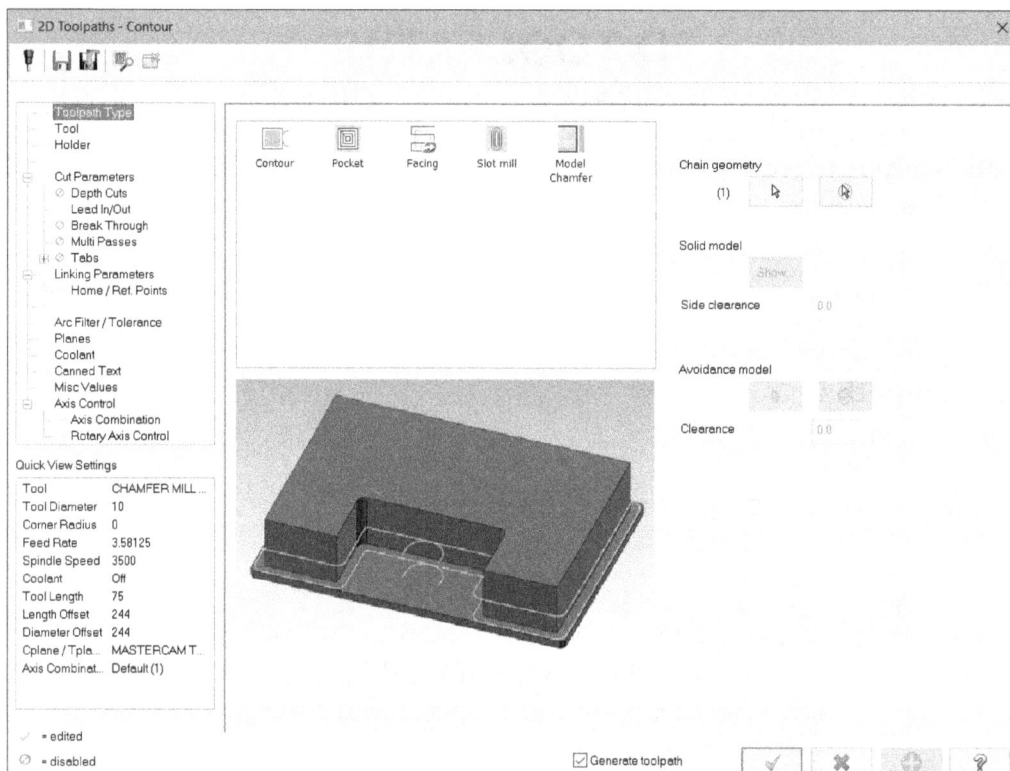

Figure-33. 2D Toolpaths-Contour dialog box

- Click on the **Tool** option from the left box of the dialog box. The **2D Toolpaths-Contour** dialog box will be displayed as shown in Figure-34.

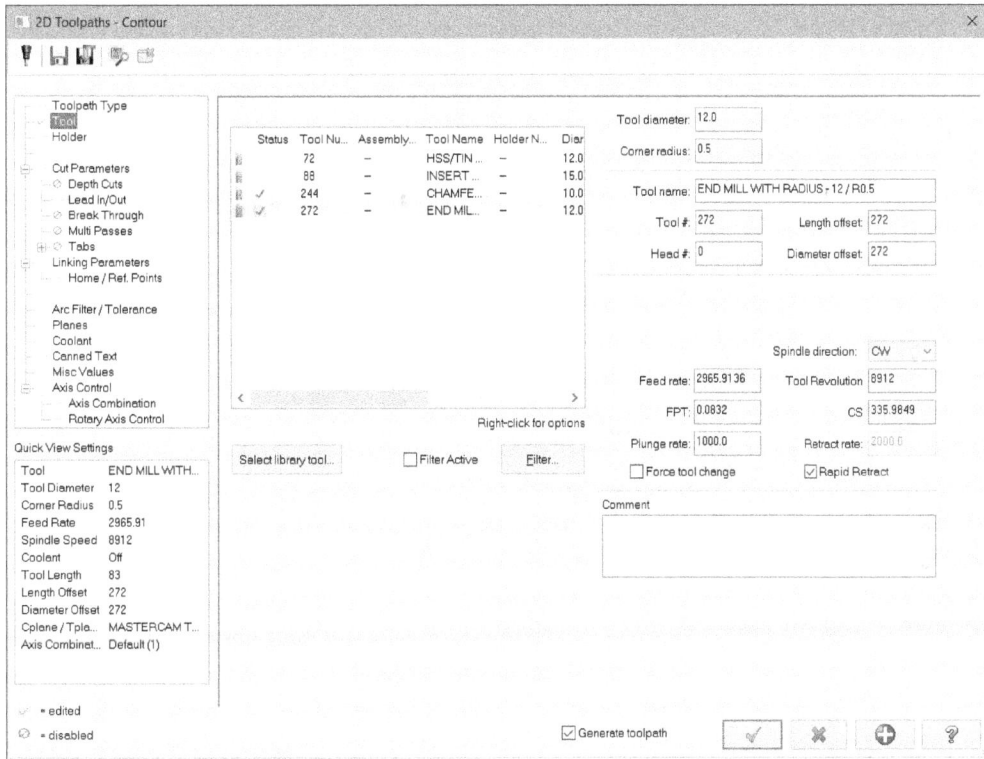

Figure-34. Tool page for 2D Toolpaths

- Click on the **Select library** tool button from the dialog box. The **Tool Selection** dialog box will be displayed; refer to Figure-35.
- Select desired tool from the dialog box and click on the **OK** button from the dialog box. The tool will be added for the current operation and the related parameters will be displayed in the right of the dialog box.
- Specify desired parameters and click on the **Holder** option from the left box. The dialog box will be displayed as shown in Figure-36.

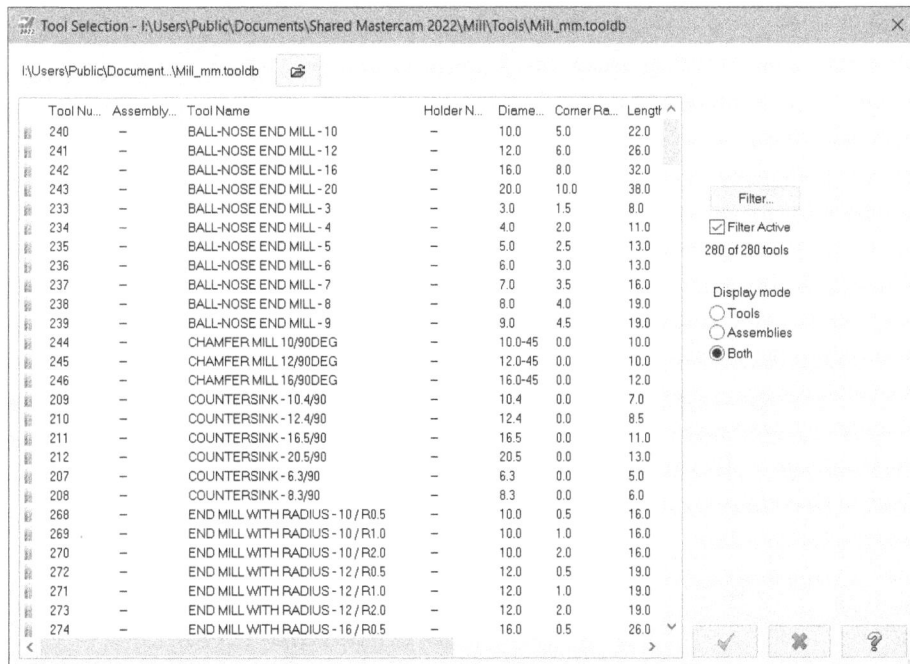

Figure-35. Tool Selection dialog box

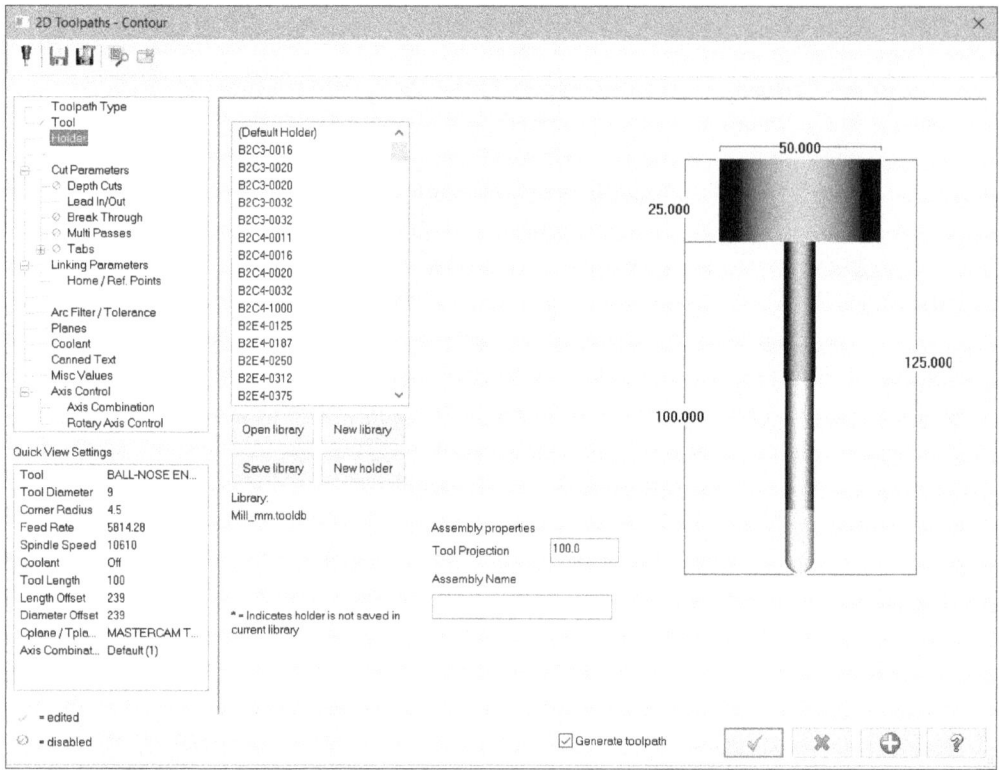

Figure-36. Holder page of 2D Toolpaths dialog box

- Click on the **Open library** button from the dialog box. The **Open** dialog box will be displayed; refer to Figure-37.

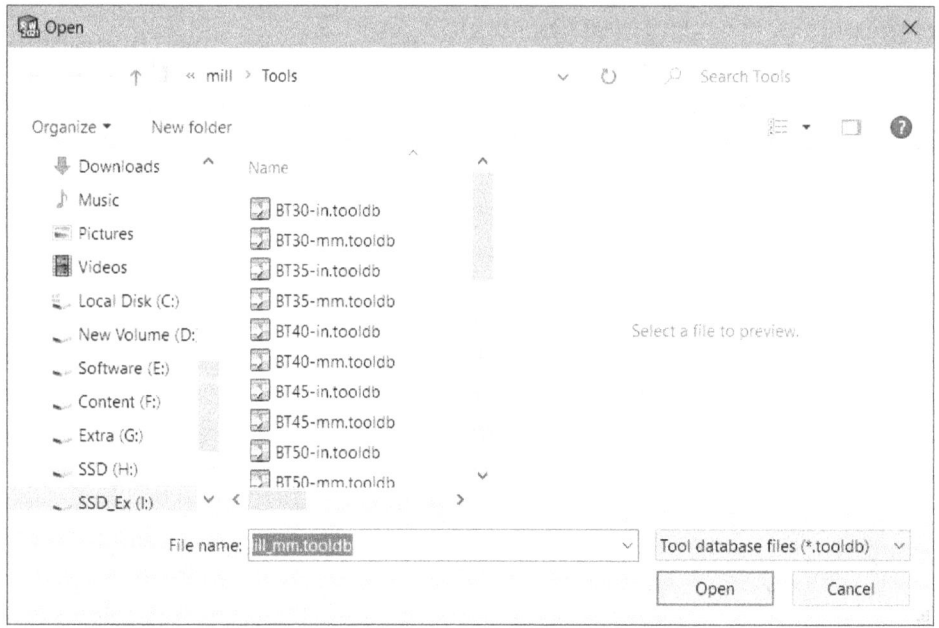

Figure-37. Open dialog box

- Select desired holder library from the dialog box and click on the **Open** button from the dialog box. The holder list will be displayed in the dialog box.
- Select desired holder and specify the related parameters.
- Click on the **Cut Parameters** option from the left and carefully define compensation direction.

- Select desired contour type from the **Contour Type** drop-down in this page. Select the **2D chamfer** option if you want to apply chamfer at the outer edges of the part. Note that you must have earlier selected a tool according to the chamfering option to creating chamfering toolpath. Select the **Ramp** option from the drop-down if you want to use a continuous ramp to transition smoothly between depth cuts, instead of individual plunge cuts. This technique is especially useful in high speed machining. Contour ramping is only available for 2D contour toolpaths. You can ramp by a set angle, by a set depth, or to plunge directly between depth cuts. Select the **Remachining** option from the drop-down if you want to remove material left by previous machining operations. Select the **Oscillate** option from the drop-down if you want to create an oscillating motion while cutting along the contour. Note that you need to select the depth of cut and other parameters according to the selected contour type in other pages.

- Similarly, click on desired parameters like **Depth Cuts**, **Break Through**, **Multi Passes**, and so on from the left box and specify the parameters. Make sure to adjust **Lead In/Out** correctly otherwise tool may collide with stock at the beginning or end.

- Select the **Tabs** option from the left area in the dialog box to define location of tabs holding the workpiece on table. The options in dialog box will be displayed as shown in Figure-38.

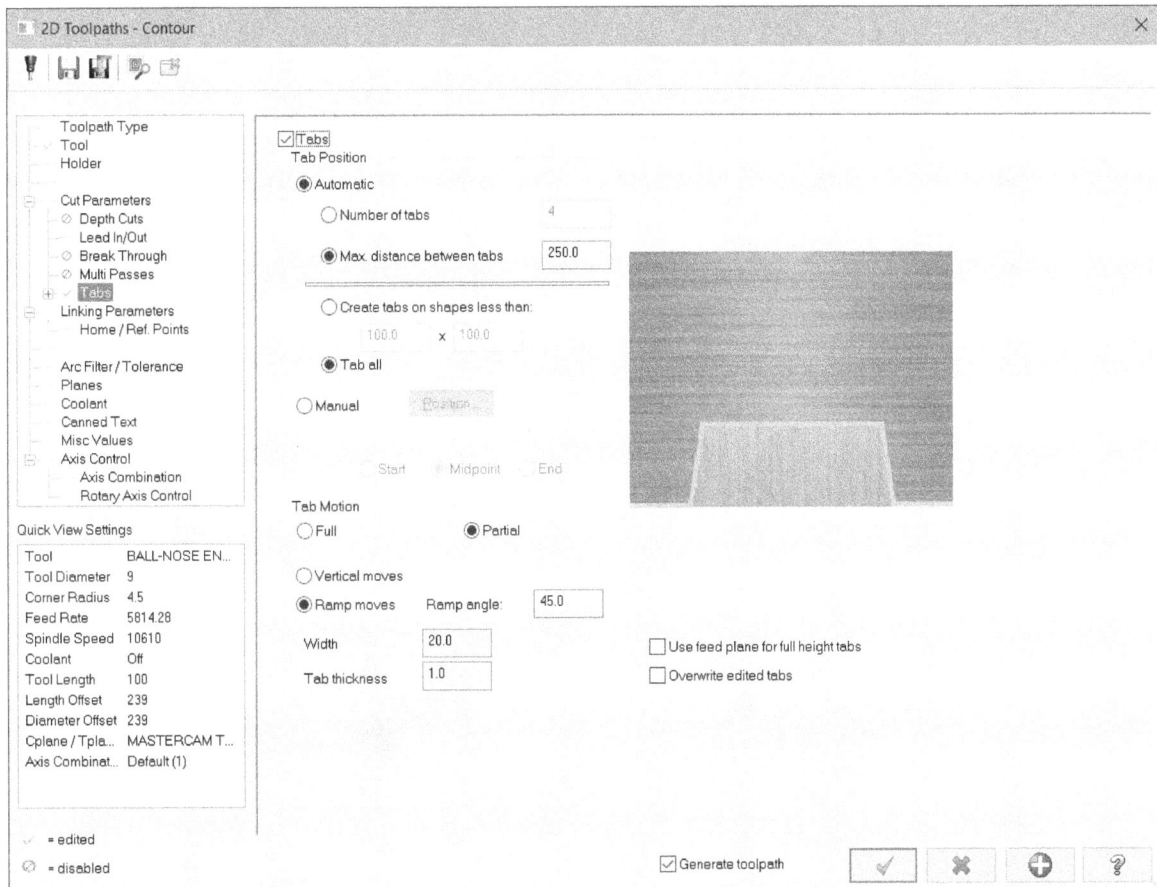

Figure-38. Tabs options

- Select the **Tabs** check box to activate options in this page. Select the **Automatic** radio button to place tabs at specified distance if **Max. distance between tabs** radio button is selected or you can specify total number of tabs placed along the contour of model if **Number of tabs** radio button is selected.

- Select the **Manual** radio button to manually place each tab at desired location on the contour.
- In the **Tab Motion** area of this page, you can define how cutting tool will avoid hitting tabs while cutting the workpiece.
- Expand the **Tabs** node in the left area of dialog box and select **Tab Cutoff** option to define how tabs will be machined after contour operation.
- After specifying the parameters, click on the **OK** button from the dialog box. The toolpath will be generated; refer to Figure-39.

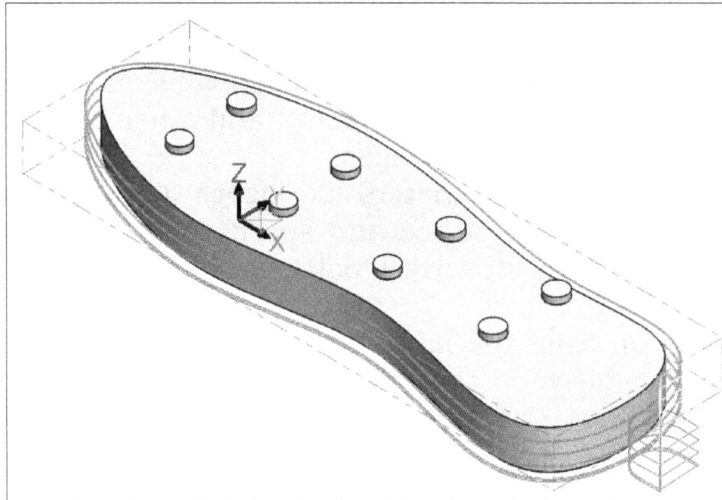

Figure-39. Toolpath generated

- To display the simulation of the material removal, click on the **Backplot selected operations** button ▧ from **Mastercam Toolpath Manager**. The **Backplot PropertyManager** will be displayed with the prototype of tool and model; refer to Figure-40.

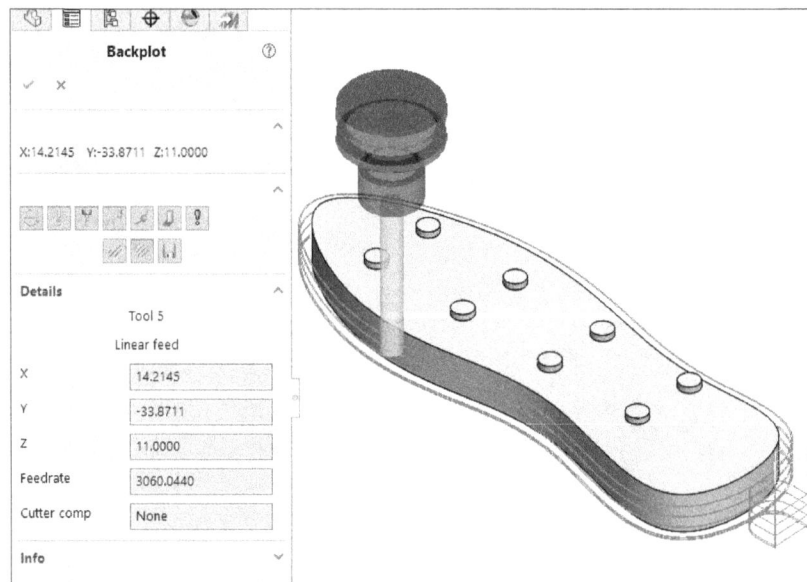

Figure-40. Backplot PropertyManager

POCKET

The Pocket toolpaths are used to remove material from the pocket in the model. The procedure to create pocket toolpaths is given next.

- Click on the **Pocket** tool from the **2D** drop-down. The **Chain Manager** will be displayed as discussed earlier.
- Select the bottom face of the pocket in the model; refer to Figure-41.

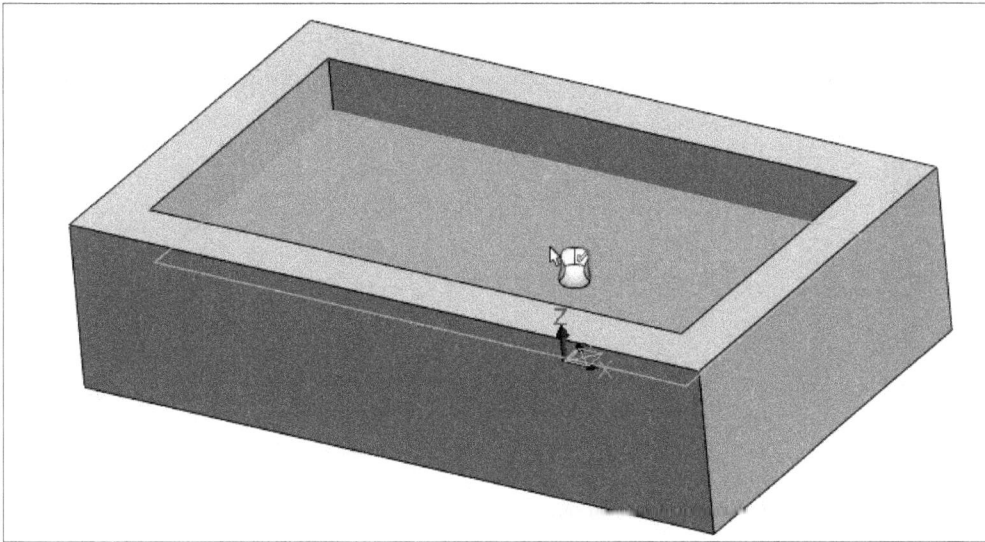

Figure-41. Pocket in the model

- Click on the **OK** button from the **PropertyManager**. The **2D Toolpaths-Pocket** dialog box will be displayed; refer to Figure-42.

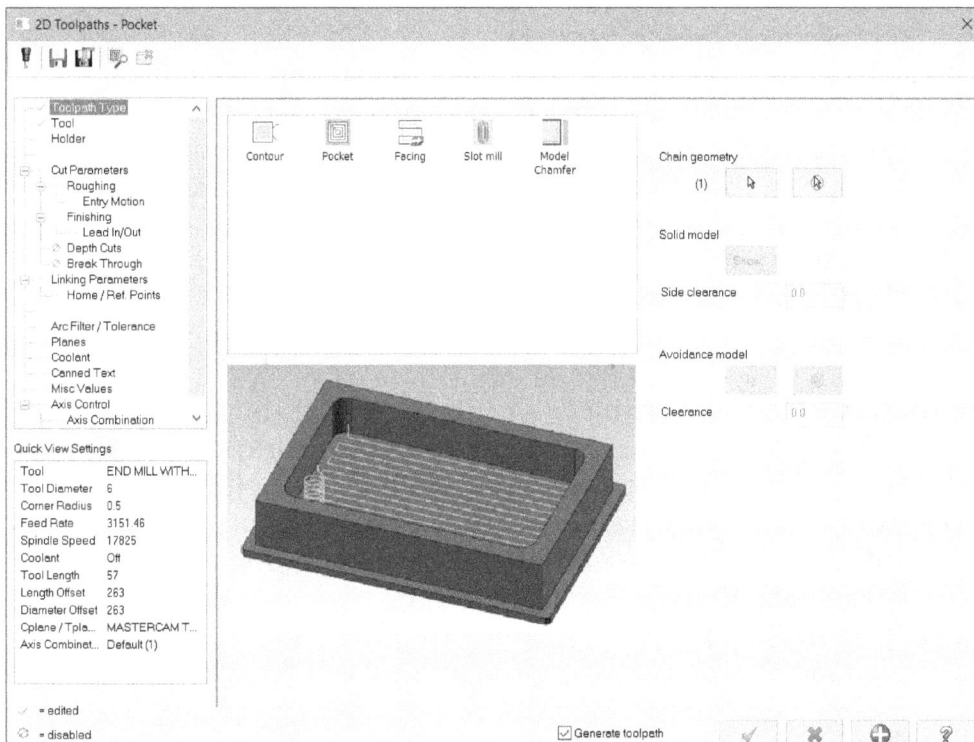

Figure-42. 2D Toolpaths–Pocket dialog box

- Click on the **Tool** option from the left box of the dialog box. The **Tool** page of the dialog box will be displayed as discussed earlier.
- Click on the **Select library** tool from the dialog box and select desired tool from the **Tool Selection** dialog box displayed. Most of the time a flat endmill or ball end mill cutter is used for this type of toolpath.
- Click on the **OK** button from the dialog box. The tool will be added in the list. Select to allot the tool for the operation.

- Click on the **Holder** option from the left box in the **2D Toolpaths - Pocket** dialog box. The **Holder** page of the dialog box will display.
- Set the holder as discussed for Contour toolpaths.
- Click on the **Cut Parameters** option from the left area. The options in the dialog box will be displayed; refer to Figure-43.

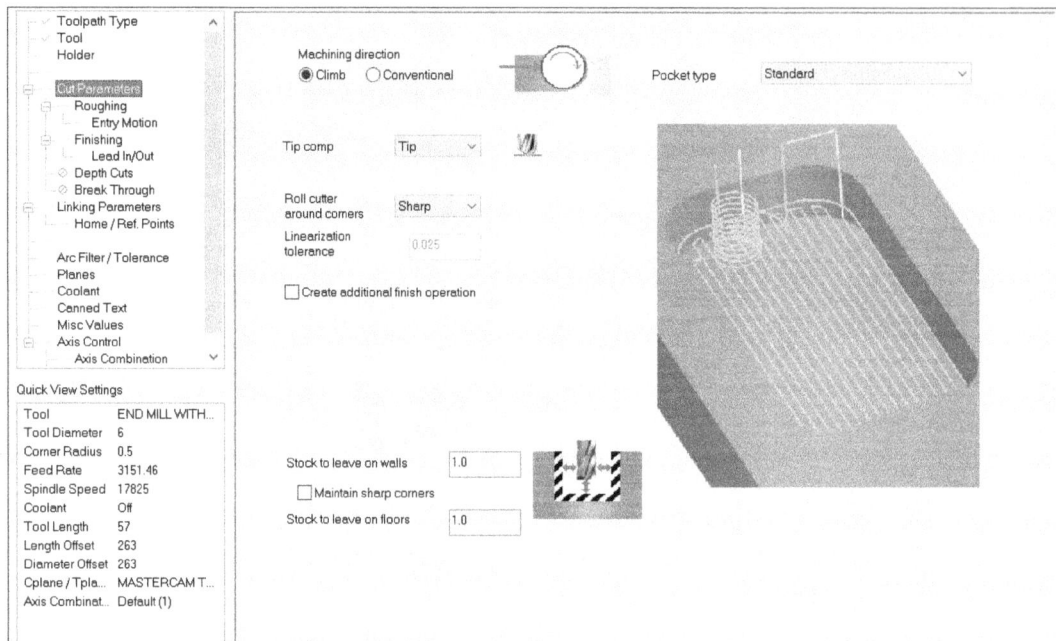

Figure-43. Cut Parameters options for pocket toolpath

- Select desired machining direction from the **Machining direction** are in the page. If the **Climb** radio button is selected then the tool will rotate in opposite direction to the tool movement direction. If the **Conventional** radio button is selected then the tool will rotate in the same direction to the tool movement direction.

Note: Conventional Milling V/S Climb Milling
Characteristics of Conventional Milling:
- The width of the chip starts from zero and increases as the cutter finishes slicing.
- The tooth meets the workpiece at the bottom of the cut.
- Upward forces are created that tend to lift the workpiece during face milling.
- More power is required to conventional mill than climb mill.
- Surface finish is worse because chips are carried upward by teeth and dropped in front of cutter. There's a lot of chip recutting. Flood cooling can help!
- Tools wear faster than with climb milling.
- Conventional milling is preferred for rough surfaces.
- Tool deflection during Conventional milling will tend to be parallel to the cut.

Characteristics of climb milling:
- The width of the chip starts at maximum and decreases.
- The tooth meets the workpiece at the top of the cut.
- Chips are dropped behind the cutter--less recutting.
- Less wear, with tools lasting up to 50% longer.
- Improved surface finish because of less recutting.
- Less power required.
- Climb milling exerts a down force during face milling, which makes work-

holding and fixtures simpler. The down force may also help reduce chatter in thin floors because it helps brace them against the surface beneath.
- Climb milling reduces work hardening.
- It can, however, cause chipping when milling hot rolled materials due to the hardened layer on the surface.
- Tool deflection during Climb milling will tend to be perpendicular to the cut, so it may increase or decrease the width of cut and affect accuracy.
- There is a problem with climb milling, which is that it can get into trouble with backlash if cutter forces are great enough. The issue is that the table will tend to be pulled into the cutter when climb milling. If there is any backlash, this allows leeway for the pulling, in the amount of the backlash. If there is enough backlash, and the cutter is operating at capacity, this can lead to breakage and potentially injury due to flying shrapnel.

Some worthwhile rules of thumb:

- When cutting half the cutter diameter or less, you should definitely climb mill (assuming your machine has low or no backlash and it is safe to do so!).

- Up to 3/4 of the cutter diameter, it doesn't matter which way you cut.

- When cutting from 3/4 to 1x the cutter diameter, you should prefer conventional milling.

- Specify the tool compensation parameters as required and set the value of stock to be left after machining in respective edit boxes.
- Select desired pocket type from the **Pocket type** drop-down in this page and set the related parameters as required. Note that on select the pocket type, preview of the selected pocket is also displayed on this page.
- Click on the **Roughing** option from the left area in the dialog box and specify the cutting speed and other parameters. Select the cutting method carefully from the Cutting method box in this page; refer to Figure-44.

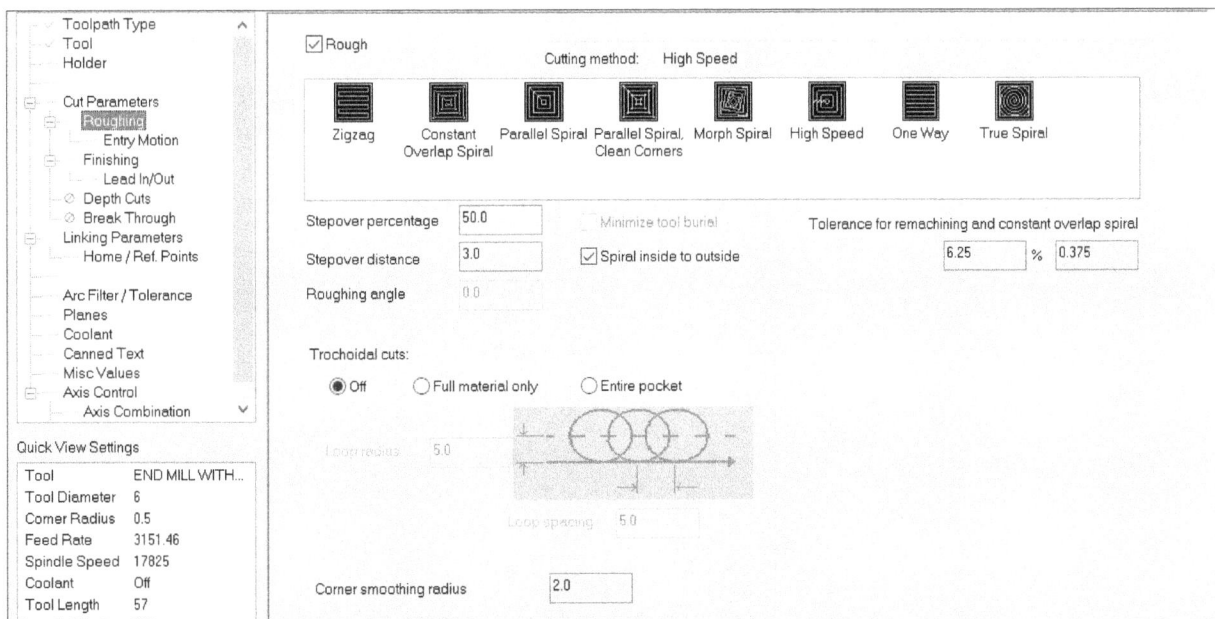

Figure-44. Roughing page in 2D Toolpath-Pocket dialog box

- For **High Speed** cutting method, you can also specify the parameters related to trochoidal cuts.
- Click on the **Entry Motion** option from the left and specify how the tool will enter while cutting material from the stock.
- Similarly, specify the other parameters as discussed in previous toolpaths like lead in/out, finishing, depth of cuts etc.
- Click on the **OK** button from the dialog box. The toolpaths will be generated; refer to Figure-45.

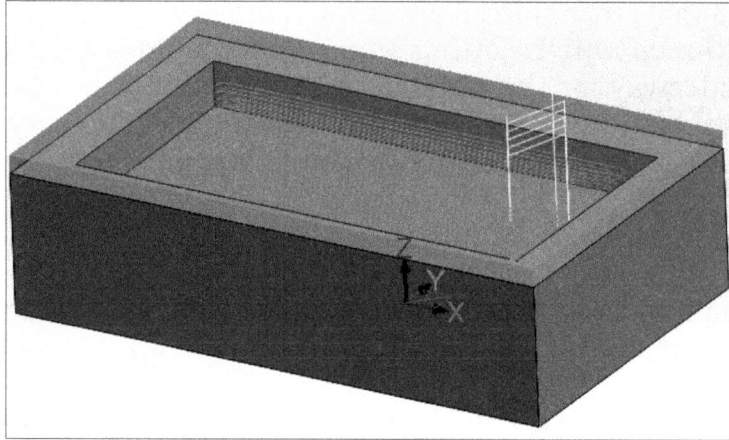

Figure-45. Pocket toolpaths generated

- You can also check the simulation of the tool by using the backplot option as discussed earlier. Note that sometimes Mastercam can move the tool in stock in rapid rate to avoid this make sure to specify adequate lead in and lead out or create start hole.

FACE

The Facing toolpaths are used to remove the material from the top faces in the model. This toolpath should be used before every other toolpath if the top surface of part is flat. The procedure to create facing toolpaths is given next.

- Click on the **Facing** tool from the **2D** drop-down. The **Chain Manager** will be displayed as discussed earlier.
- Select the top face of the model on which you want to perform the facing operation. Make sure that you select the ▣ button from the **Chain Manager** before selecting the face to filter face selection. Also, there should be only one chain in **Chain Manager** after selecting face.
- Click on the **OK** button from the **Chain Manager**. The **2D Toolpaths - Facing** dialog box will be displayed; refer to Figure-46.

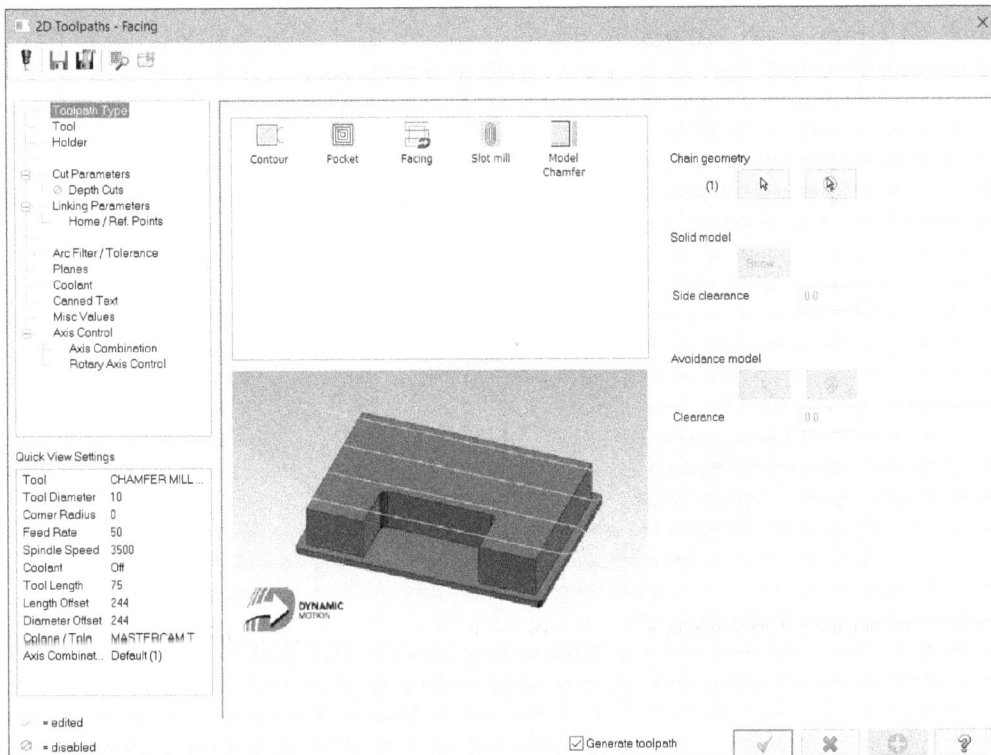

Figure-46. 2D Toolpaths-Facing dialog box

- Select the facing tool by clicking on the **Tool** option from the left box in the dialog box and similarly select desired Holder.
- Specify the other parameters and click on the **OK** button from the dialog box. The toolpath will be generated; refer to Figure-47.

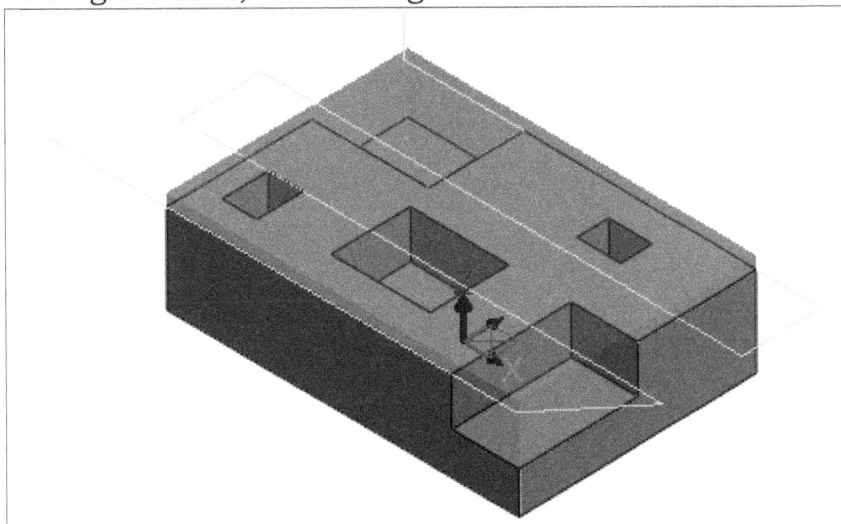

Figure-47. Toolpath for facing

- Backplot the toolpath to verify.

SLOT MILL TOOLPATHS

The Slot mill toolpaths are used to remove the material from the slots in the model. Note that this slot milling is governed in 2D plane so the depth of cut need to be specified explicitly. The procedure to create slot milling toolpaths is given next.

- Click on the **Slotmill** tool from the **2D** drop-down. The **Chain Manager** will be displayed as discussed earlier.
- Select the flat face of the slot; refer to Figure-48.

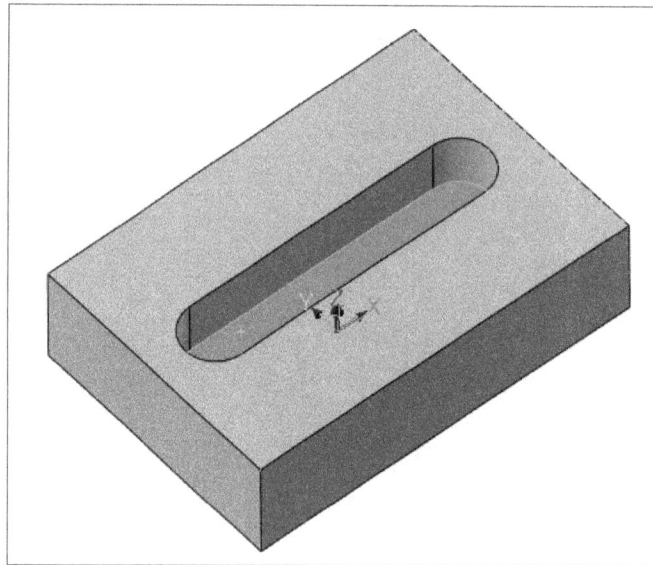

Figure-48. Flat face selected for slot

- Click on the **OK** button from the **Chain Manager**. The **2D Toolpaths - Slot Mill** dialog box will be displayed as shown in Figure-49.

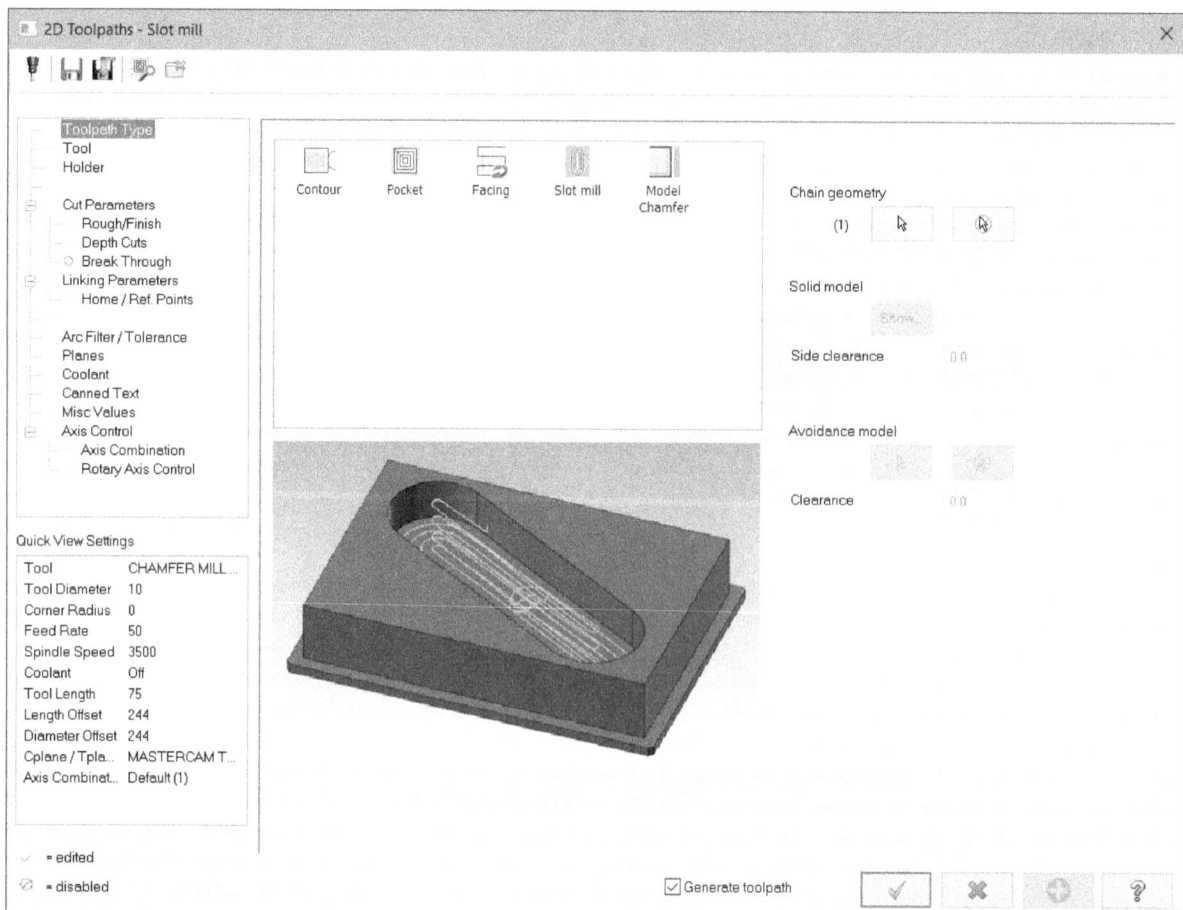

Figure-49. 2D Toolpaths-Slot Mill dialog box

- Specify the tool and the tool holder as done earlier.
- Click on the **Depth Cuts** option and specify desired depth of cuts as discussed in previous tools.
- Click on the **Linking Parameters** option. The dialog box will be displayed as shown in Figure-50.

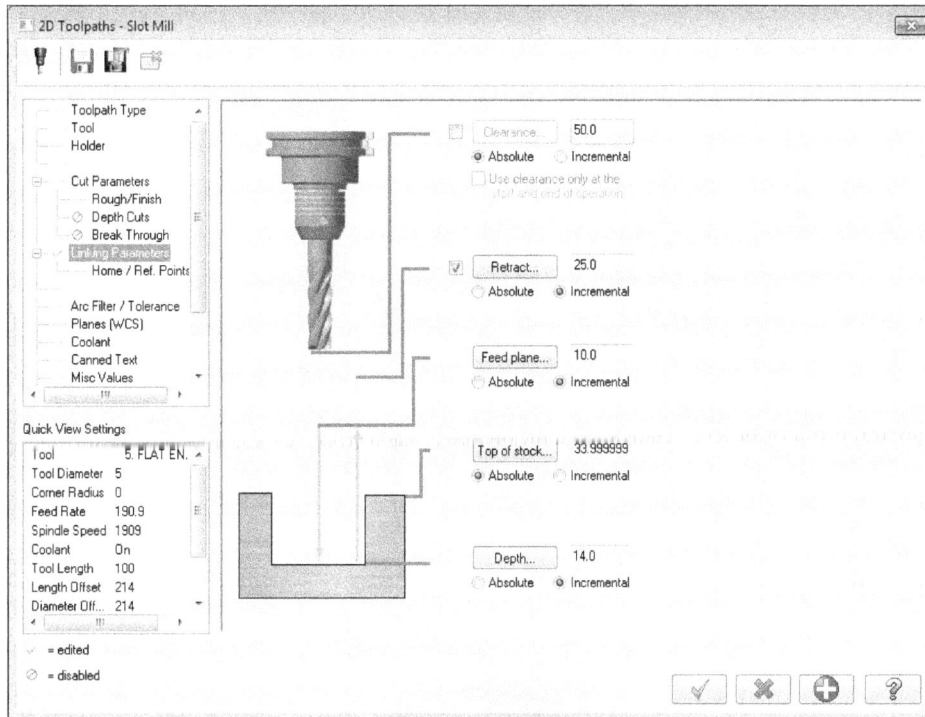

Figure-50. Depth of cut linking in dialog box

- Select the **Absolute** radio button below the **Depth** button.
- Click in the edit box next to **Depth** button and specify desired value for the depth of slot. Note that the specified value is the distance along the Z axis from the Machine Coordinate System.
- Note that you can perform arc filtering and tolerance setting for most of the toolpaths by using the options in the **Arc Filter/Tolerance** page of the dialog box; refer to Figure-51. Although, in 2D toolpaths there is no effect of specifying plane for arcs explicitly because Mastercam will automatically create arcs single plane parallel to tool-plane.

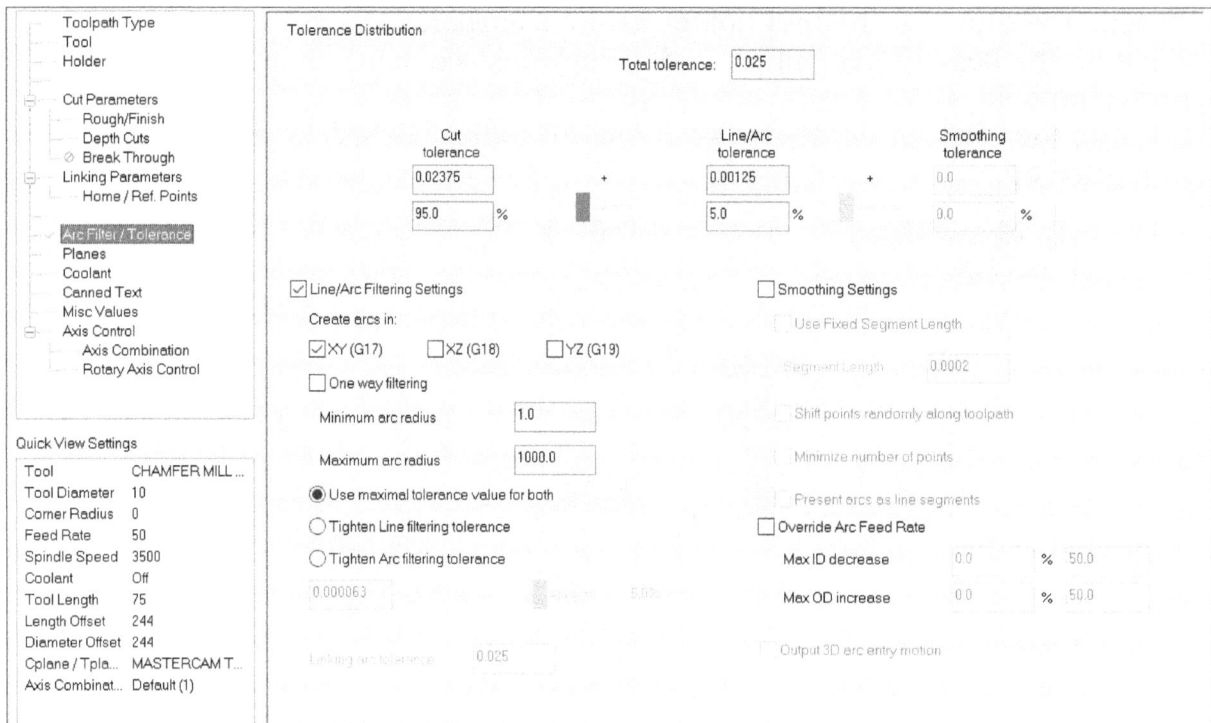
Figure-51. Arc Filter Tolerance page

- Specify the other desired parameters and click on the **OK** button from the dialog box. The toolpath will be generated; refer to Figure-52.

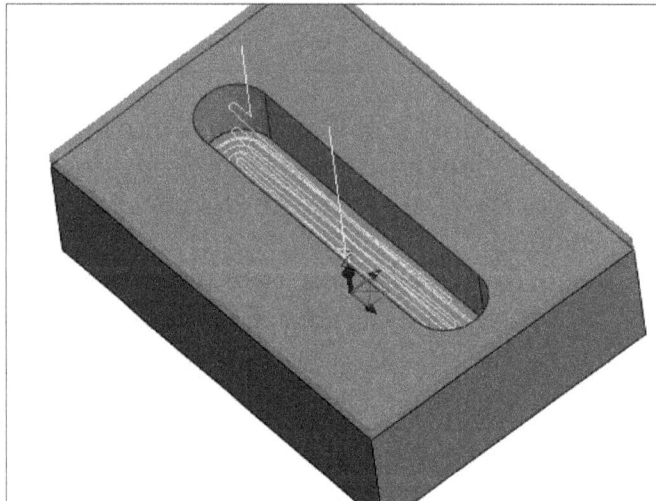
Figure-52. Toolpath for slot

WRITING LETTERS FOR ENGRAVING

If you have doubt on, how did we write the text for engraving please follow the steps below:

- Click on the **Create Letters** button from CAD drop-down in the **Mastercam2022 CommandManager** in the **Ribbon**. The **Create Letters PropertyManager** will be displayed; refer to Figure-53.

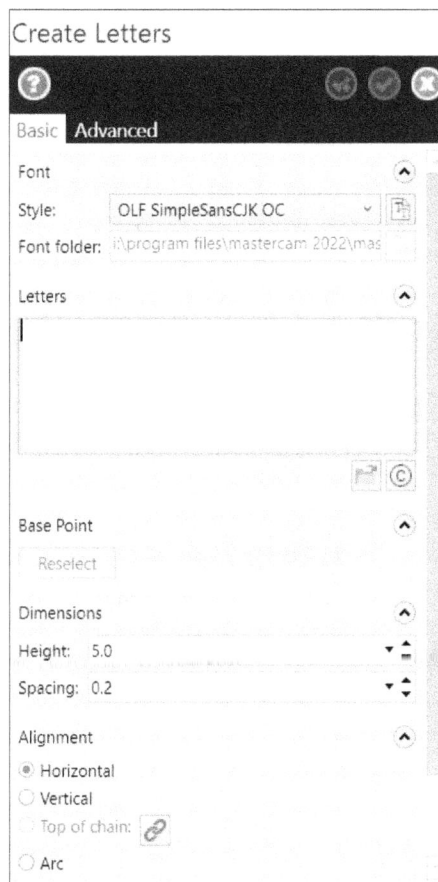

Figure-53. Create Letters PropertyManager

- Click in the **Letters** edit box in the **Parameters** rollout of the **PropertyManager** and specify desired text.
- Click on **Reselect** button in **Base Point** rollout of **PropertyManager** to specify starting point of text and its base face. The **Selection PropertyManager** will be displayed.
- Click in the **Style** drop-down and select the **Mastercam (Block) Font** option from it; refer to Figure-55. (No! we are not forcing you to select this font only, go ahead and show your creativity.)
- Change the Height and Spacing by using the corresponding spinners.

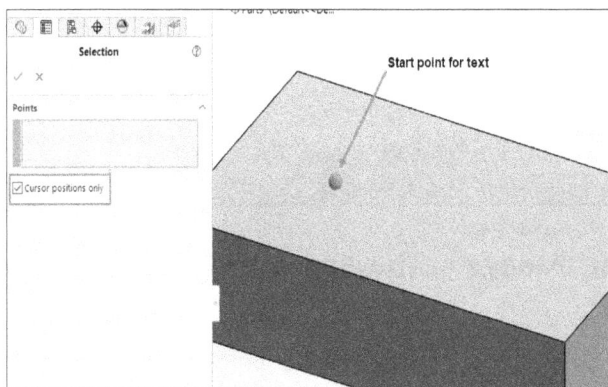

Figure-54. Plane for letter creation

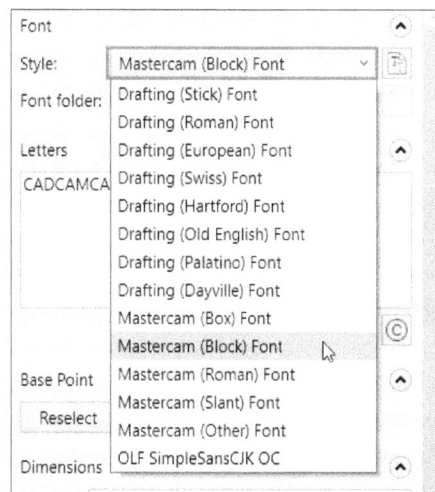

Figure-55. Font drop-down

- Scroll down in the **Create Letters PropertyManager** and specify orientation for text using options in **Alignment** rollout.
- Click on the **OK** button from the **PropertyManager** to create the letters. Note that you can also create letters using SolidWorks Text tool in Sketch CommandManager; refer to .

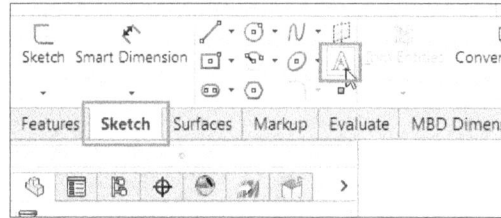

Figure-56. Text tool

ENGRAVE

The Engrave toolpaths are used to create artistic machining over the workpiece. These toolpaths are also used to print name on the press dies. The procedure to create these toolpaths is given next.

- Click on the **Engrave** tool from the **2D** drop-down. The **Chain Manager** will be displayed as discussed earlier.
- Select the profile to be engraved by selecting the entities one by one or select the **Sketch** filter button from **Chain Manager** and select the sketch of text; refer to Figure-57.

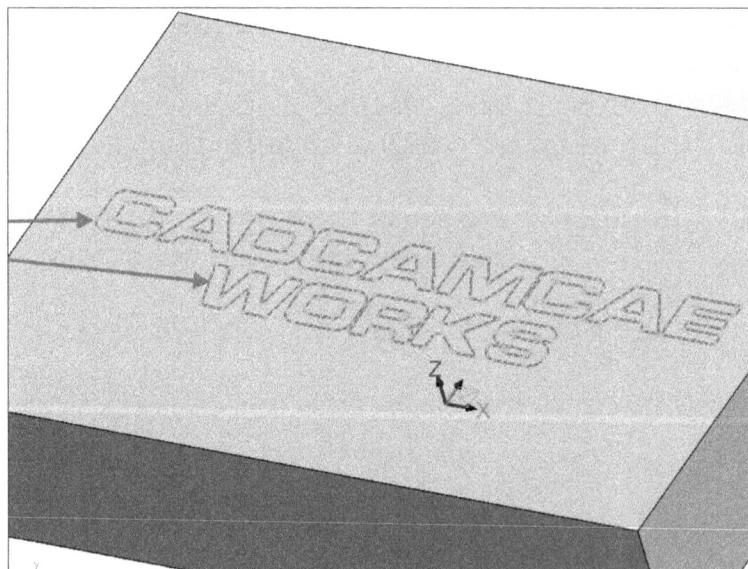

Figure-57. Text entities selected for engraving

- Click on the **OK** button from **Chain Manager**. The **Engraving** dialog box will be displayed; refer to Figure-58.

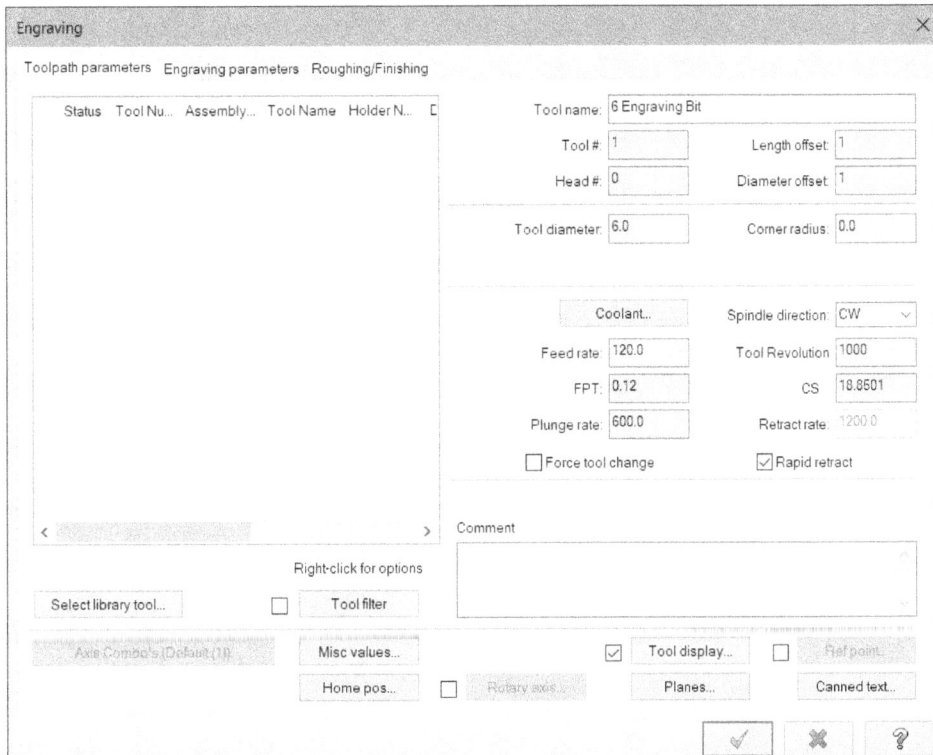

Figure-58. Engraving dialog box

- Click on the **Select library tool** button from the dialog box. The **Tool Selection** dialog box will be displayed; refer to Figure-59.

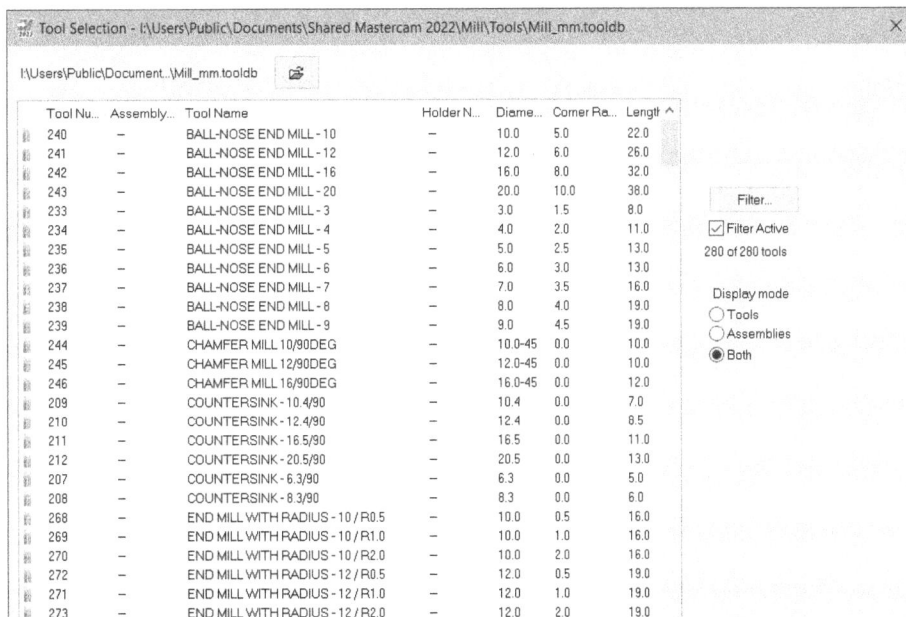

Figure-59. Tool Selection dialog box

- Select the **5 MM 90 DEGREE ENGRAVE TOOL 1 TIP** tool or any other desired engrave tool from the list of tools and click on the **OK** button from the dialog box.
- Click on the **Engraving parameters** tab of the dialog box. The dialog box will be displayed; refer to Figure-60.

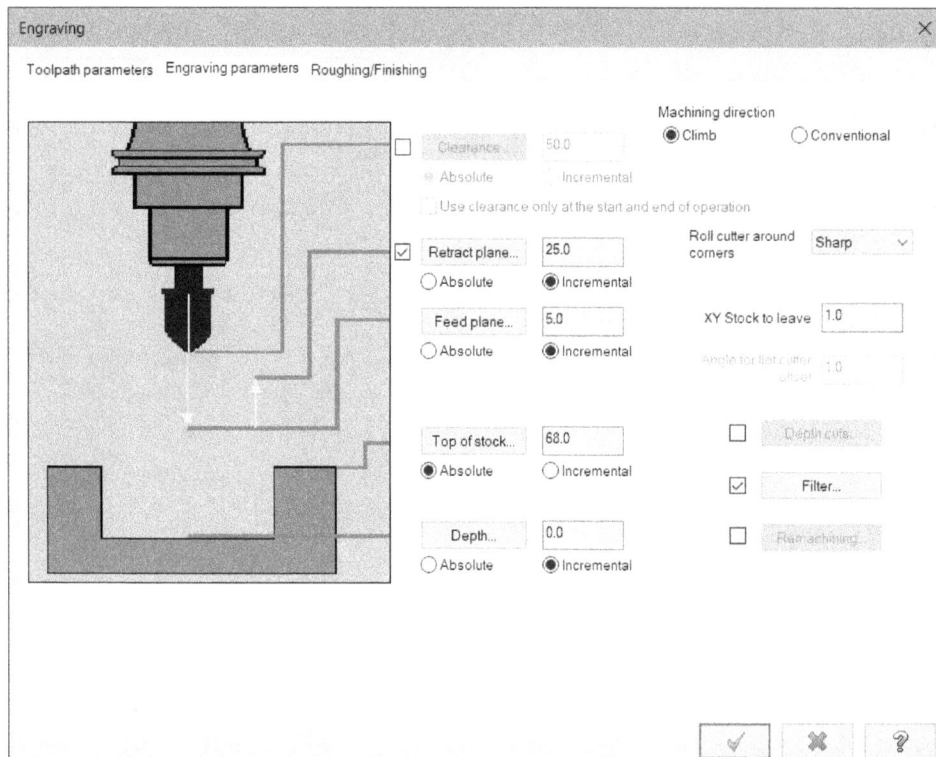

Figure-60. Engraving parameters tab

- Specify desired parameters edit boxes. Make sure you specify correct **Top of Stock** and **Depth** values in respective edit boxes.
- Select desired radio button to specify machining direction from the **Machining direction** area.
- If you want to finish engraving in current toolpath then specify the **XY Stock to leave** edit box as **0**.
- Select the **Depth cuts** check box and then specify the depth of cuts after selecting the **Depth cuts** button.
- If you want to specify a limit on arc creation and Cut tolerance then select the **Filter** check box and specify the respective parameters by selecting the **Filter** button.
- Click on the **Roughing/Finishing** tab in the dialog box. The dialog box will be displayed as shown in Figure-61.

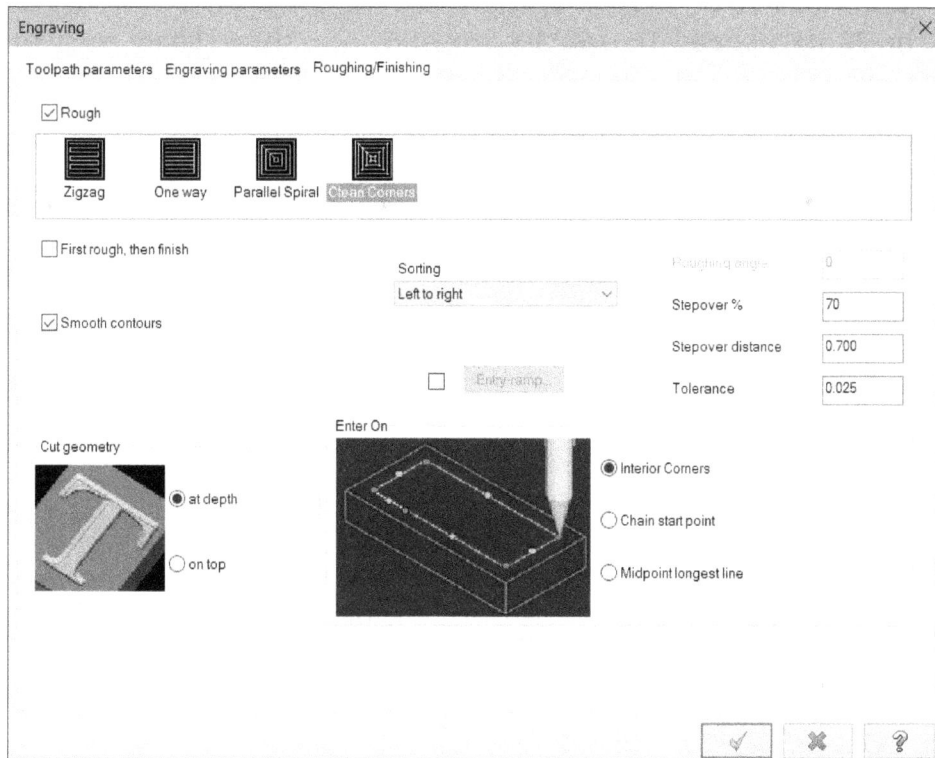

Figure-61. Roughing/Finishing tab

- Make sure that the **at depth** radio button is selected in the **Cut geometry** area of the dialog box if you want to cut the material following the inner loops of text sketch.
- Specify the other parameters as required. Note that selection of cutting method can highly affect the toolpath generation and engraving quality.
- Click on the **OK** button from the dialog box. The toolpath will be generated; refer to Figure-62.

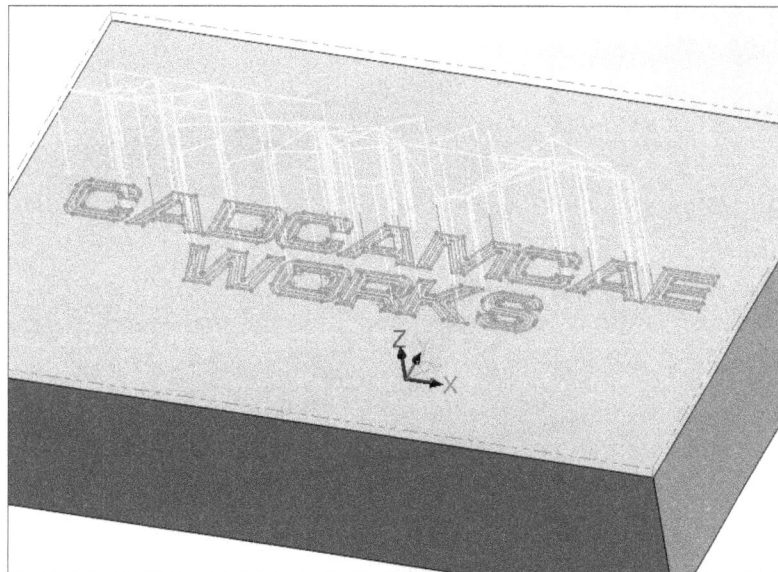

Figure-62. Toolpath for engraving

2D HIGHSPEED

2D High Speed toolpaths remove the material in the same way as the 2D Toolpaths but these toolpaths help to make your current processes more efficient and automated, minimizing programming and cycle times. These toolpaths are mainly created for finishing of material left by 2D roughing toolpaths. The tools in this category are available in the **2D Highspeed Toolpaths** drop-down in the **Ribbon**; refer to Figure-63.

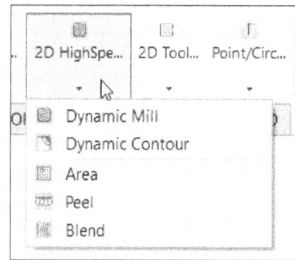

Figure-63. 2D Highspeed Toolpaths drop-down

The tools in this drop-down are discussed next.

Dynamic Mill Toolpath

The Dynamic Mill toolpath is used to perform various dynamic milling tasks like removing material inside the boundaries while avoiding islands, remove material left by other tools, and removing material near the outer boundaries. The procedure to use this toolpath is given next.

Machining Inside Boundary

*   Click on the **Dynamic Mill** tool from the **2D HighSpeed** drop-down in the **Ribbon**. The **Enter new NC name** dialog box will be displayed if this is your first toolpath. Specify desired name and click on the **OK** button. The **Chain Manager PropertyManager** will be displayed.
*   Select the face that you want to machine; refer to Figure-64.

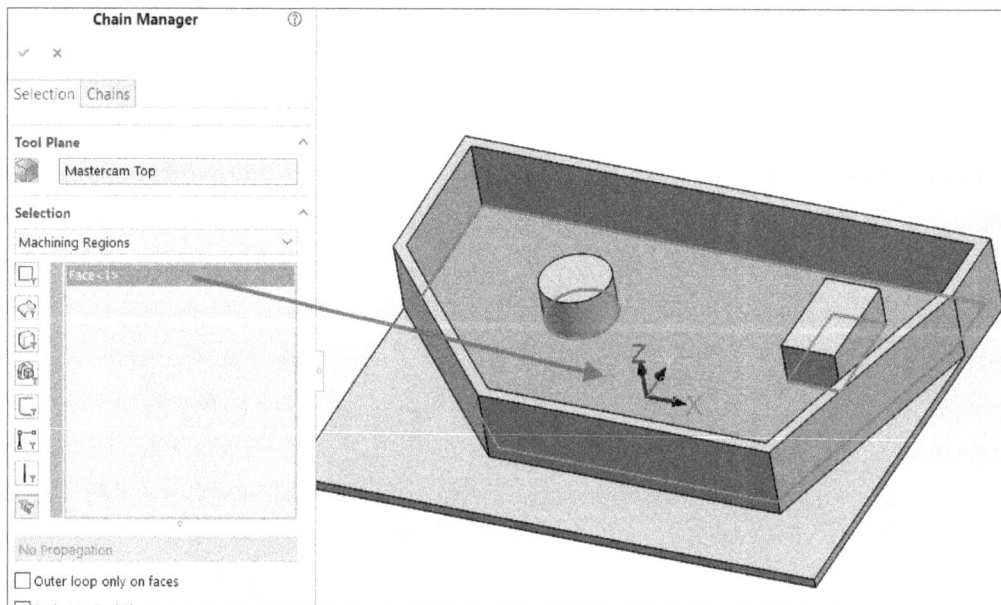

Figure-64. Face selected for dynamic mill machining

*   Note that there are islands in the part which should be counted as avoidance region to dynamic mill this part. Select the **Avoidance Regions** option from the **Selection** drop-down in the **PropertyManager** and select the faces of islands; refer to Figure-65.

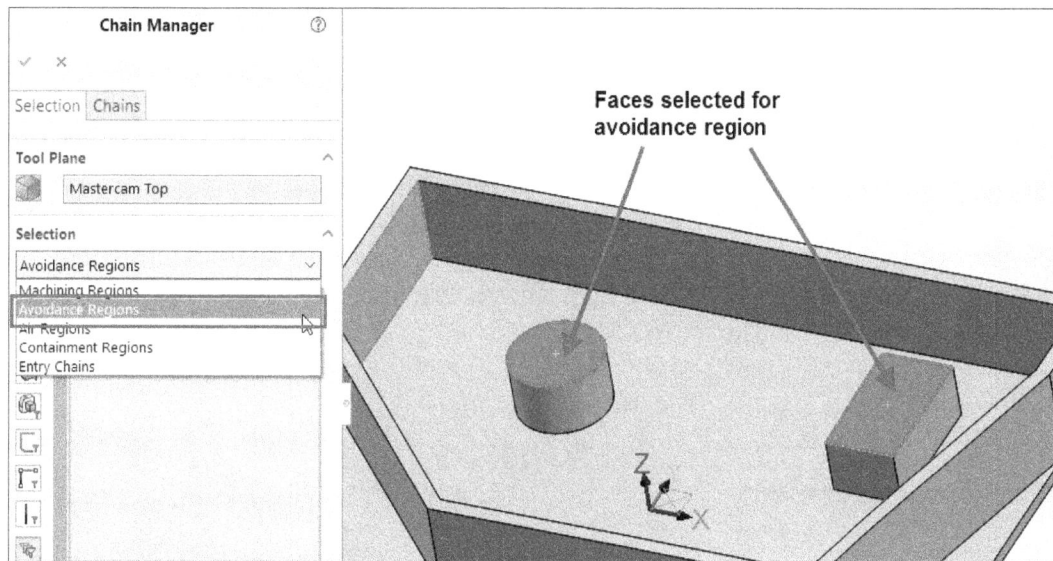

Figure-65. Avoidance region selection

- Click on the **OK** button from the **Chain Manager**. The **2D High Speed Toolpath -
Dynamic Mill** dialog box will be displayed; refer to Figure-66.

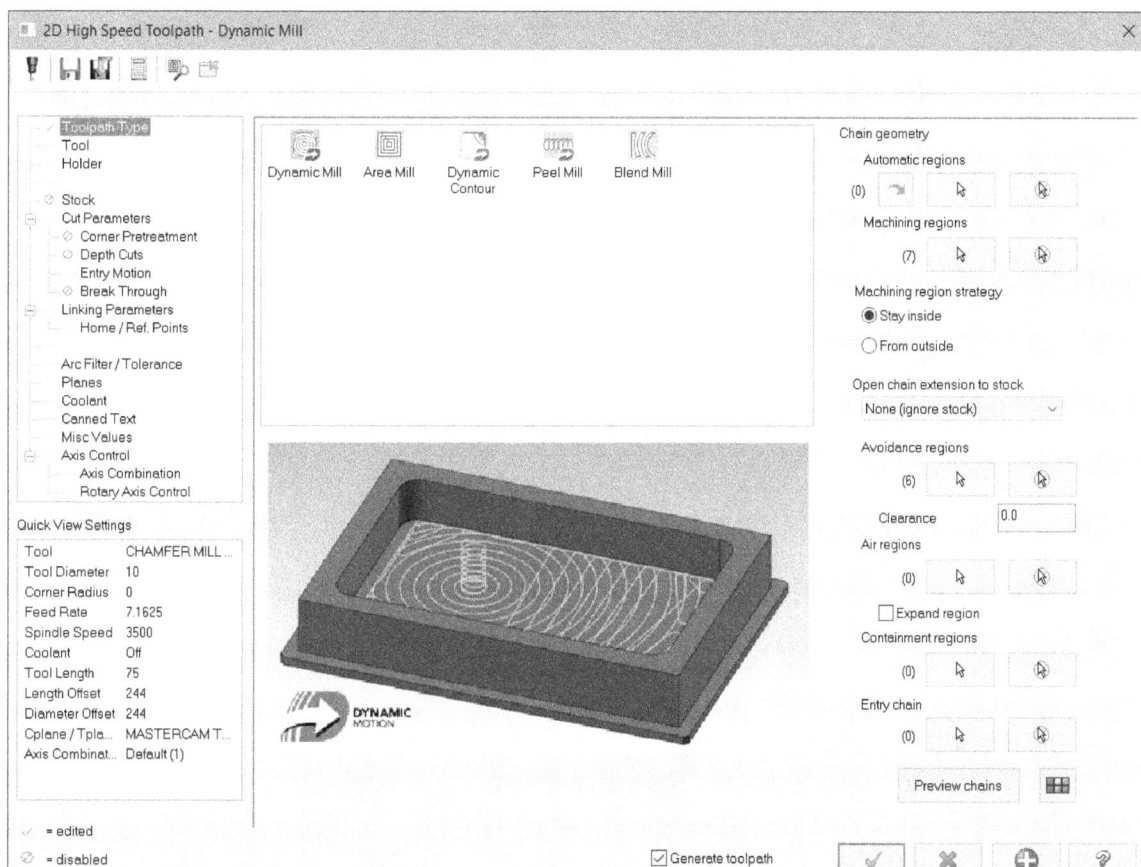

Figure-66. 2D High Speed Toolpaths-Dynamic Mill dialog box

- In current example, we are creating toolpath for milling inside selected boundary
so make sure the **Stay inside** radio button is selected in the **Machining region
strategy** area of this page. If we need to machine a part which has open boundaries
then we can select the **From outside** radio button.
- If you have selected any open chain for machining then you can select desired
radio button from the **Open chain extension to stock** area to allow Mastercam

to extend the chain if there is stock left nearby. By default **None** radio button is selected so Mastercam does not extend the open chain. If you select the **Tangent** radio button then software will extend the chain tangentially. If you select the **Shortest distance** radio button then software will extend chain to nearby shortest distance.

- If you have not selected machining regions and avoidance regions earlier then you can click on the respective selection buttons in this page and select them.
- Select the tool and tool holder from respective pages in this dialog box.
- Click on the **Cut Parameters** option from the left area. The dialog box will be displayed as shown in Figure-67.

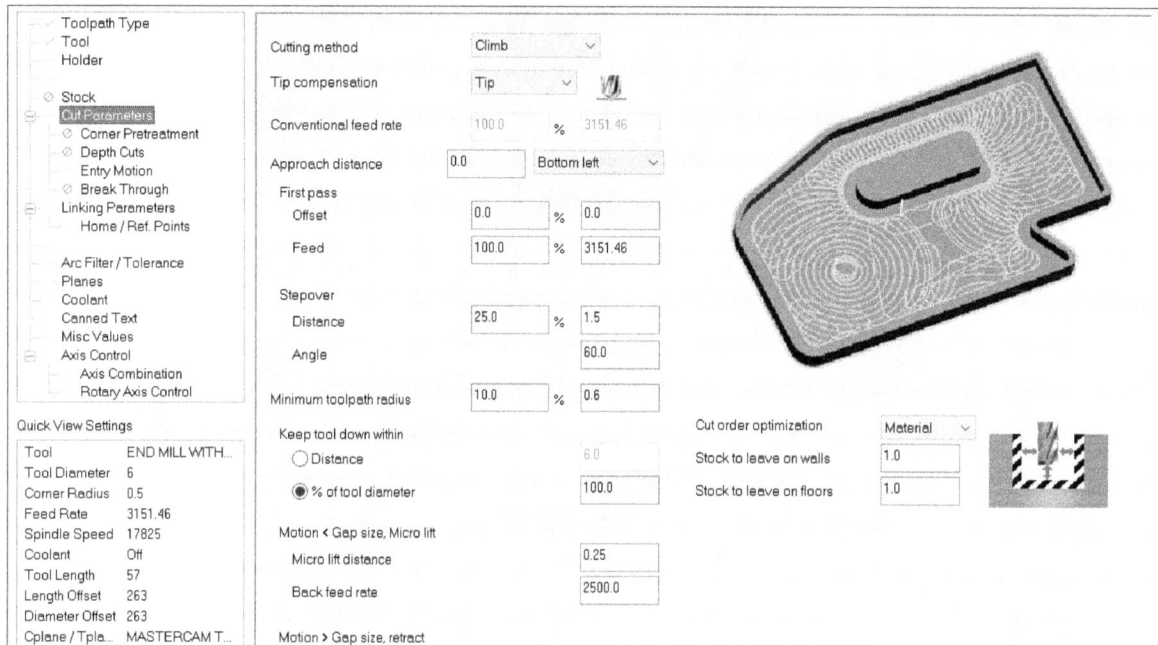

Figure-67. Cut Parameters for Dynamic Mill

- Select the cutting method and tool compensation from the respective drop-downs in the dialog box.
- In the **Approach distance** edit box, specify the distance value from which the tool will approach to the start location of cutting toolpath.
- In the **First pass** offset edit box, specify the distance to offset toolpath from original path to reduce load on the tool during first pass.
- In the **First pass feed reduction** edit box, specify the reduction percentage to reduce feed rate during first pass to reduce load on tool.
- Specify stepover, minimum toolpath radius, gap size, and micro lift data as required in respective fields.
- Click in the **Cut order optimization** drop-down and select desired option to define how Mastercam selects the start point of cutting passes.
- Click on the **Depth Cuts** option from the left area in the dialog box and specify the depth of cut parameters as discussed in previous tools. Note that in case of Dynamic mill toolpaths, you can also perform face milling on the islands coming in toolpaths. To do so, select the **Island facing** check box in this page and specify the related parameters below it.

- Select the **Subprogram** check box in this page if you want to use sub program in NC codes. Using sub-program we can reduce the repetition of same code at different depth levels.
- Specify the other parameters in this page as required.
- Click on the **Entry Motion** option from the left area and carefully define the motion of tool while entering in the workpiece for cutting material. If you want to change the feed rate for entry motion then you can do so by using options in **Entry feeds/speeds** area of this page.
- Set the **Linking Parameters** and other parameters as discussed earlier and then click on the **OK** button to generate toolpath.

Rest Material Options

- If you want to create toolpath for rest mill then click on the **Stock** option from the left area of the **2D High Speed Toolpath - Dynamic Mill** dialog box. The options in the dialog box will be displayed as shown in Figure-68.

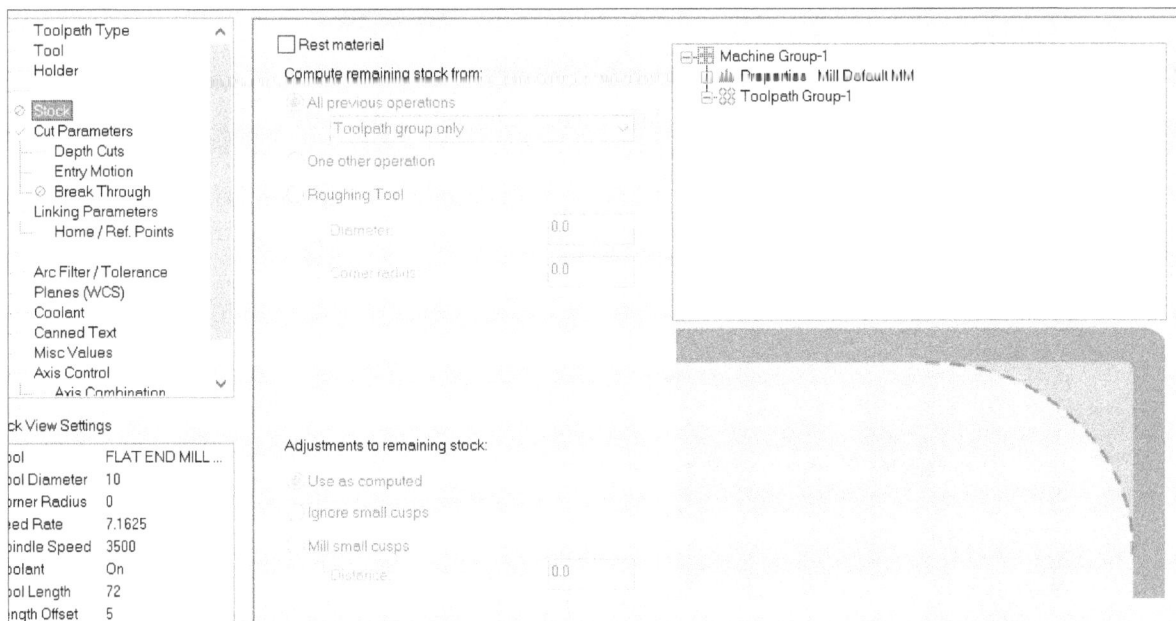

Figure-68. Rest mill options in Dynamic Mill dialog box

- Select the **Rest material** check box and then select desired option from **Compute remaining stock from** area of the dialog box. Specify the other parameters as required.
- Now, set the other parameters as discussed earlier and click on the **OK** button to generate toolpath.

Machining Outside Boundary

- After clicking on the tool, select the face outside the boundary as machining region **Chain Manager PropertyManager**; refer to Figure-69.

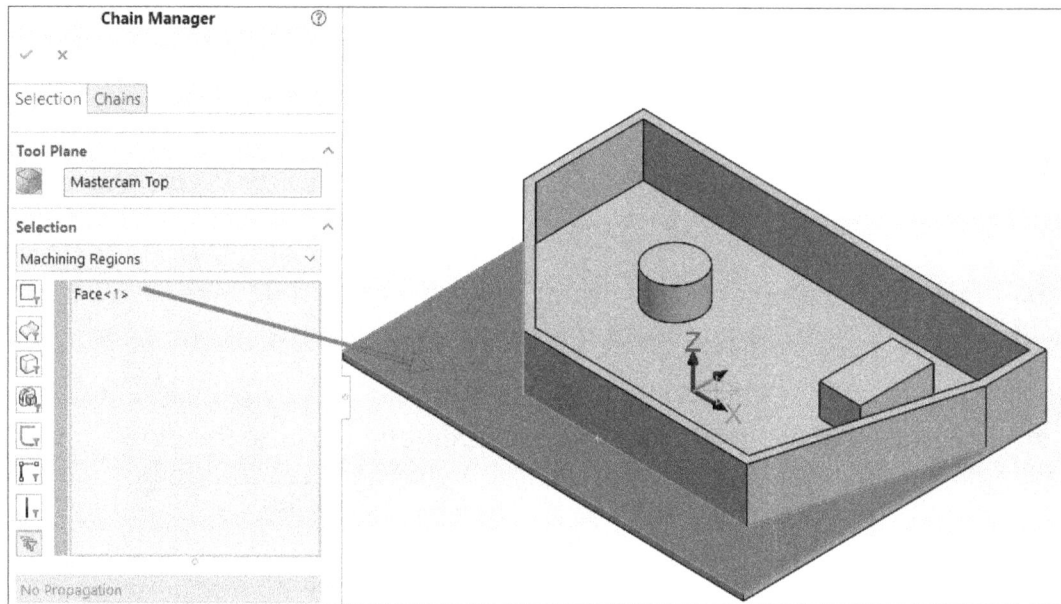

Figure-69. Face selected for machining outside with Dynamic Mill

- Click on the **OK** button from **Chain Manager PropertyManager**. The **2D High Speed Toolpath - Dynamic Mill** dialog box will be displayed. Note that if there are islands on the face then you need to select them as avoidance region as discussed earlier.
- Select the **From outside** radio button from the **Machining region strategy** area of the dialog box; refer to Figure-70.

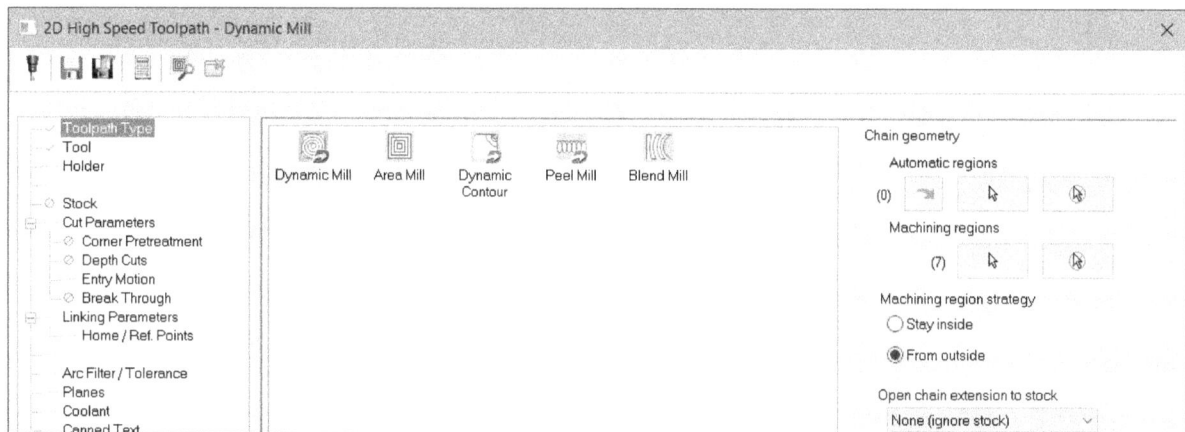

Figure-70. From outside machining region strategy

- Specify rest of the parameters as discussed earlier for this tool and click on the **OK** button to create toolpath.

Dynamic Contour Toolpath

The Dynamic Contour mill toolpaths are used to remove material by following the profile specified. The procedure to create this toolpath is given next.

- Click on the **Dynamic Contour** tool from the **2D Highspeed Toolpaths** drop-down. The **Chain Manager** will be displayed as discussed earlier.
- Select the wall faces of the model to set the profile of cut; refer to Figure-71.

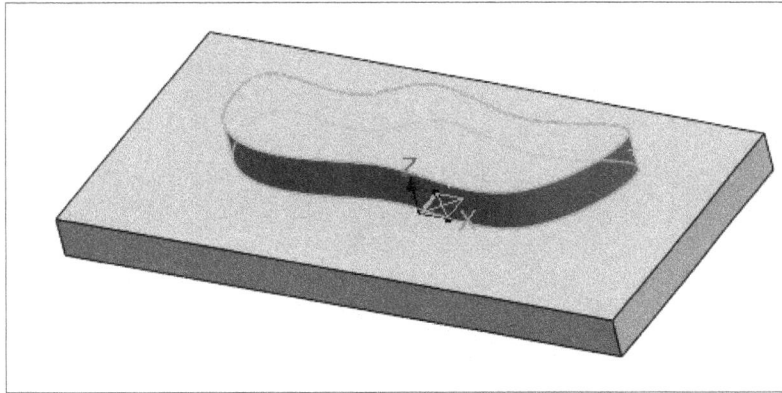

Figure-71. Face selected for dynamic contour machining

- Rest of the procedure is same as discussed for contour milling earlier in the book.

Area Toolpath

The Area toolpath is created to remove material from the area bounded by selected boundaries. The area toolpath is generally used to machine close loops from inside or outside. The **Area** tool in the **2D HighSpeed** drop-down is used to create area toolpath. The procedure to use this tool is same as for Dynamic Mill tool. This toolpath is generally used to finish pockets with high surface finish due to cleaner movement of tool; refer to Figure-72.

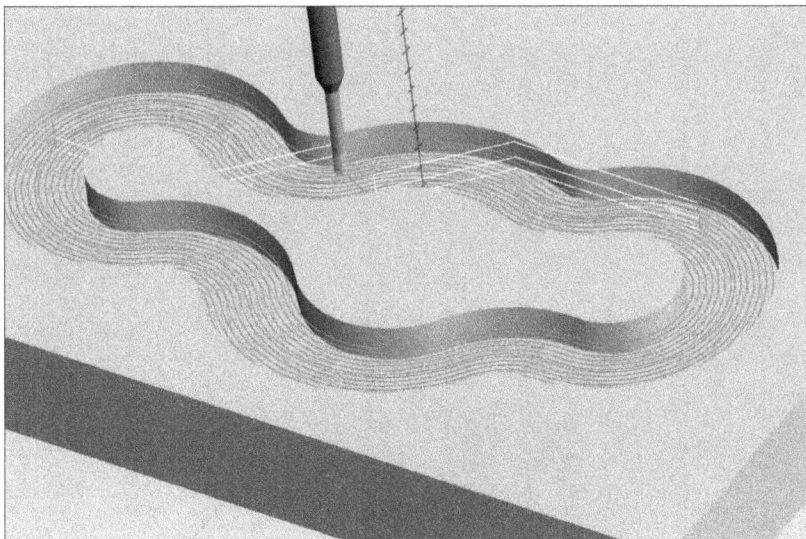

Figure-72. Area mill toolpath

Peel Toolpath

The Peel toolpath works as the name suggests. The peel toolpaths peels away the material from the workpiece along the given open boundaries. The procedure to use this toolpath is given next.

- Click on the **Peel** tool from the **2D HighSpeed** drop-down. (If the tool is not displaying in drop-down then activate any of the 2D HighSpeed tool and later switch to Peel Mill toolpath type from the 2D High Speed Toolpath dialog box; refer to Figure-74.) The **Chain Manager** will be displayed as discussed earlier.

• Select the edges of the walls between which you want to perform the peel milling operation; refer to Figure-73.

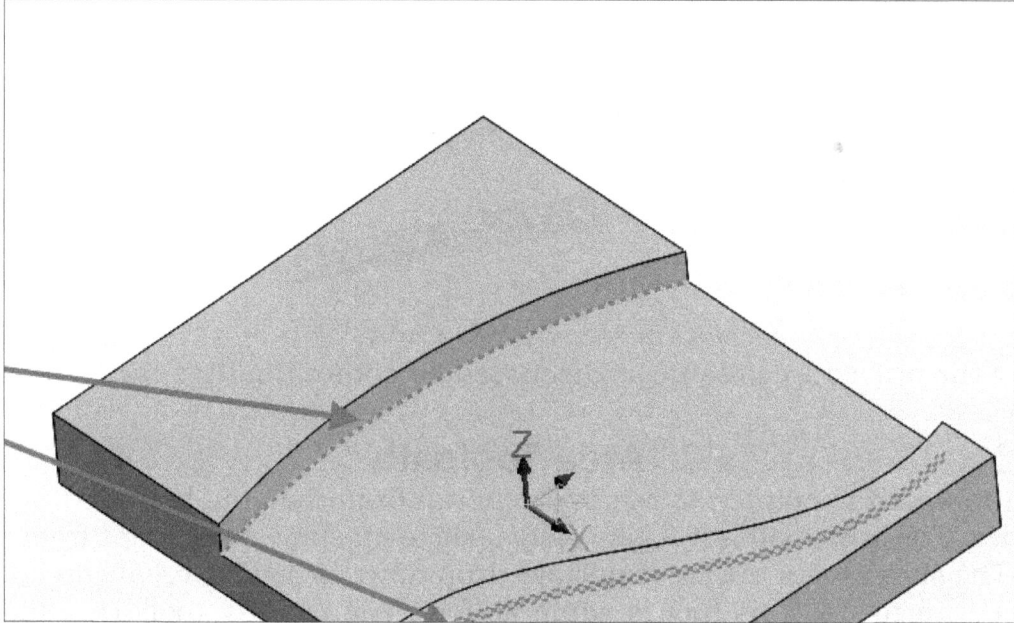

Figure-73. Edges to be selected

• Click on the **OK** button from the **Chain Manager**. The **2D High Speed Toolpath - Peel Mill** dialog box will be displayed; refer to Figure-74.

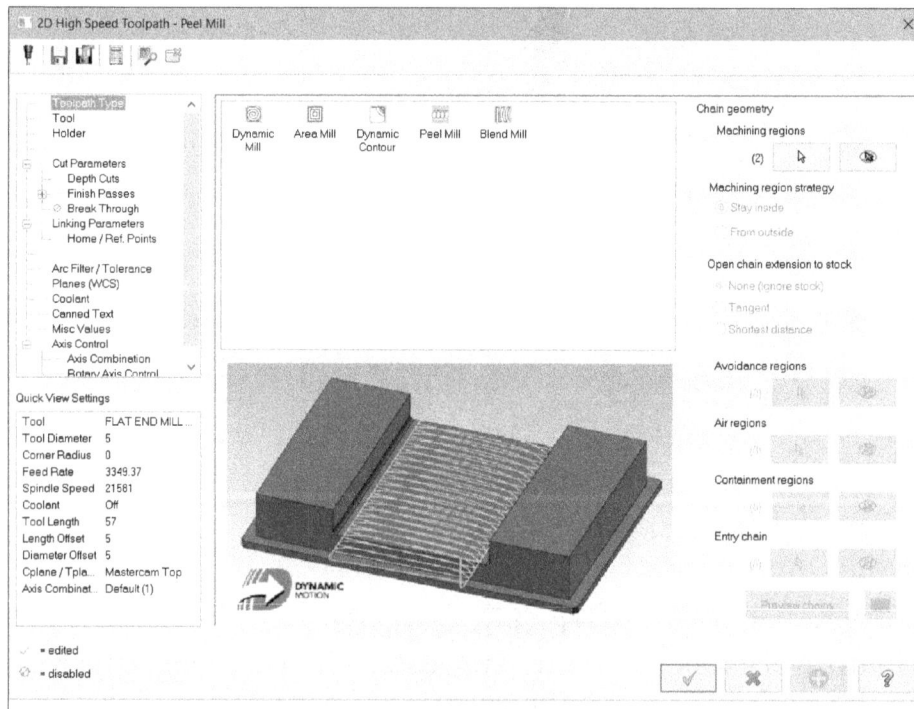

Figure-74. 2D High Speed Toolpaths-Peel Mill

• Specify the parameters in the dialog box as discussed earlier. Note that you can extend the entry and exit path of tool by using the options in **Cut Parameters** page of this dialog box.

- After specifying parameters, click on the **OK** button from the dialog box. The toolpaths will be generated; refer to Figure-75.

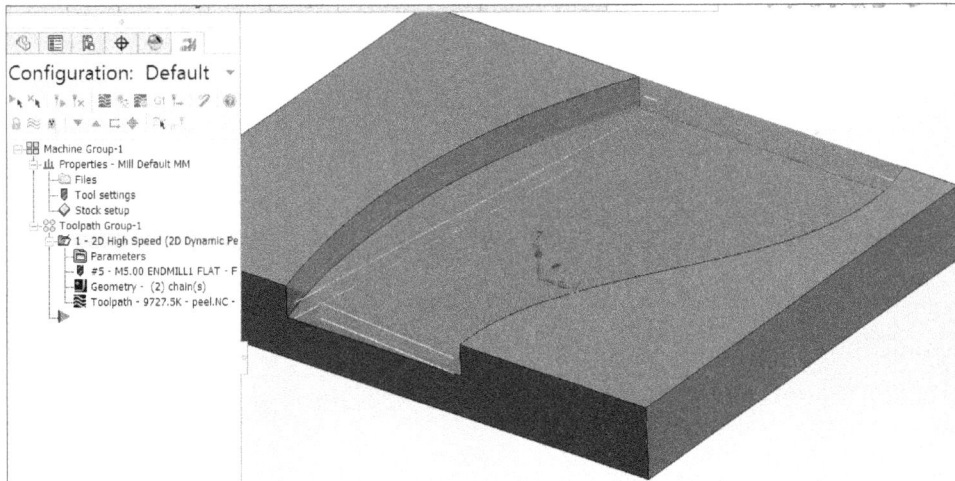

Figure-75. Toolpath for peel

Blend Mill

The **Blend mill** toolpath is used to mill smoothly between two open chains of the workpiece. This toolpath is more effect when you have wall on one side and other sides are open. The procedure to use this tool is given next.

- Click on the **Blend Mill** tool from the **2D HighSpeed** drop-down. (If the tool is not displaying in drop-down then activate any of the 2D HighSpeed tool and later switch to Blend Mill toolpath type from the 2D High Speed Toolpath dialog box; refer to Figure-77.) The **Chain Manager** will be displayed as discussed earlier.
- Select the edges of the open chains; refer to Figure-76.

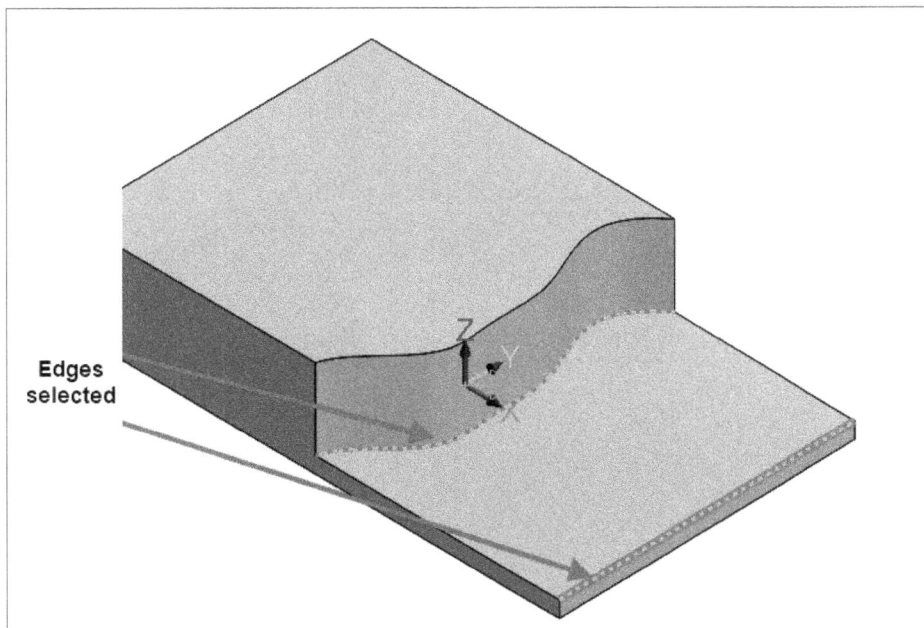

Figure-76. Edges of open chains to be selected

- Click on the **OK** button from the **Chain Manager**. The **2D High Speed Toolpath - Blend Mill** dialog box will be displayed; refer to Figure-77.

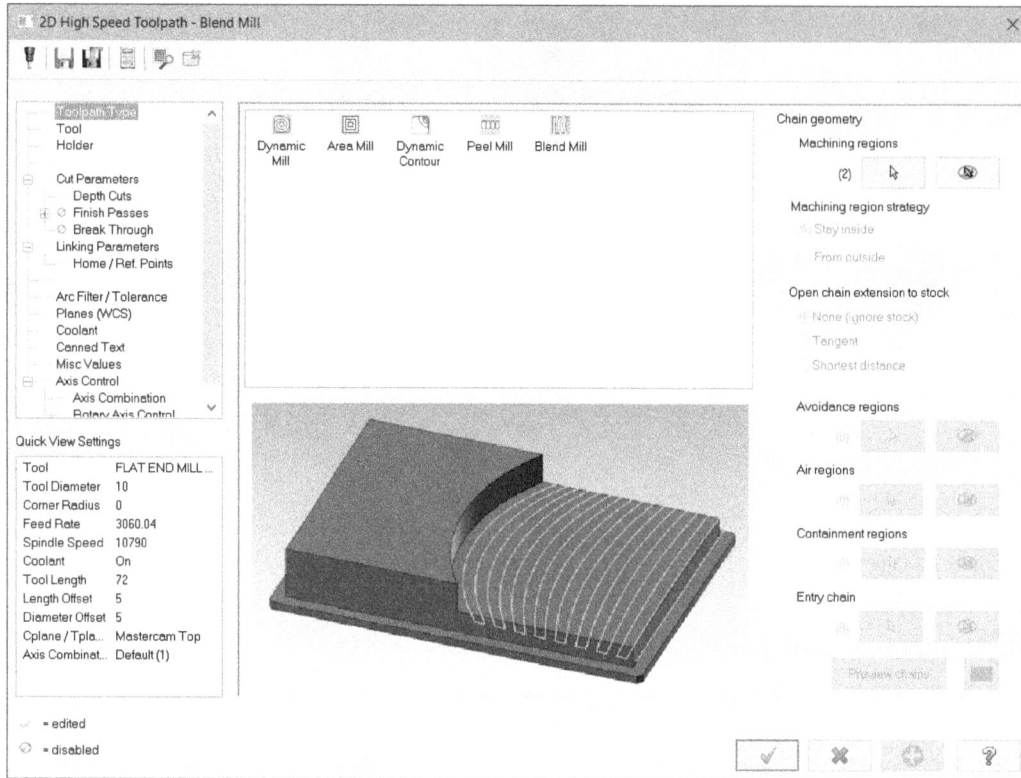

Figure-77. 2D Highspeed Toolpaths–Blend Mill

- Specify the parameters as discussed earlier. Note that you need to specify the **Cut Parameters** carefully to get good surface finish; refer to Figure-78.
- After specifying the parameters, click on the **OK** button from the dialog box. The toolpaths will be generated; refer to Figure-79.

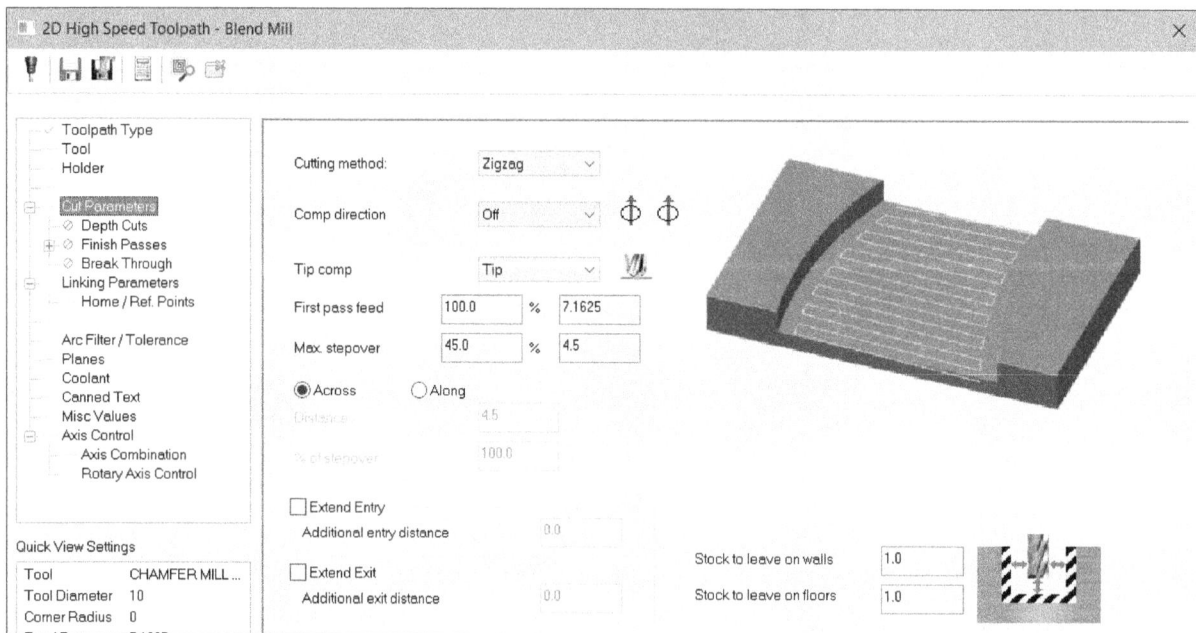

Figure-78. Cutting parameters

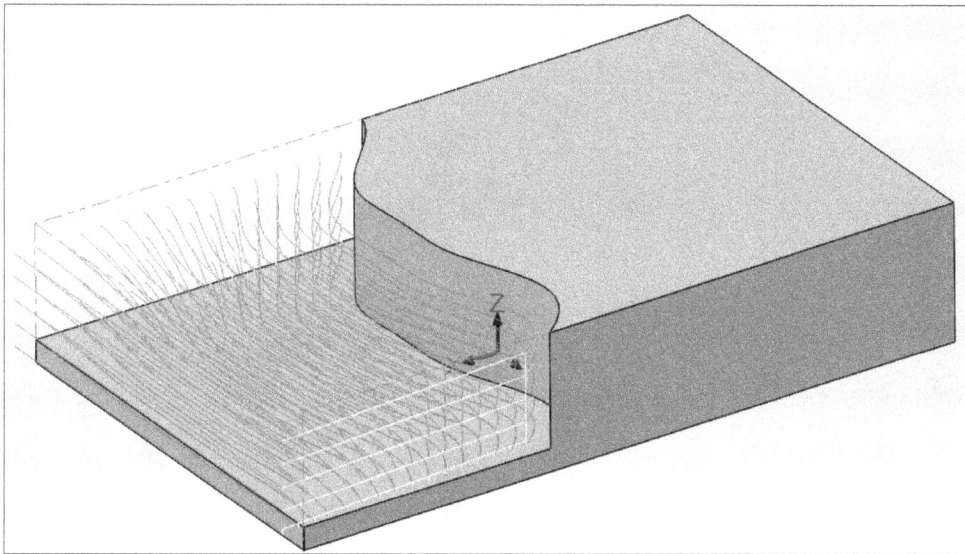

Figure-79. Toolpath for blend

PRACTICAL 1

Machine the part as given in Figure-80. The stock for the part is a rectangular boundary block.

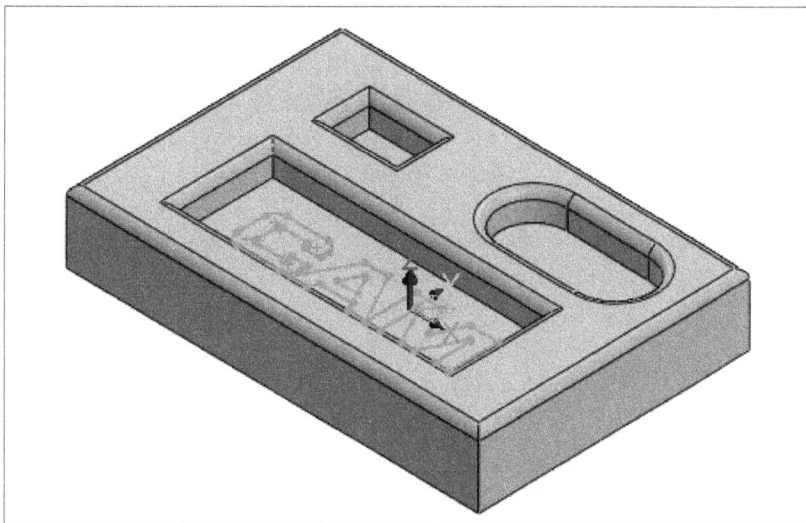

Figure-80. Practical 1

Opening the Model

- Download the files of resource kit from website and double-click on the model file for this practical.
- Enable the Mastercam2022 plug-in if not enabled by default.

Machining Plane setup

- Click on the **Planes Manager** tab from **Design Tree**. The **Planes Manager** will be displayed; refer to Figure-81.

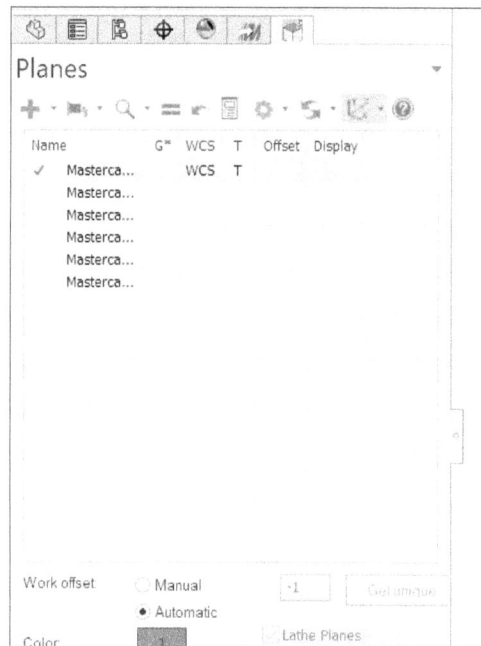

Figure-81. Plane Manager dialog box

- Make sure that the Mastercam Top plane is selected as work plane and tool plane in the dialog box.
- Note that the origin is set at the center of the bottom face of the model by default.

Creating the Stock

- Click on the **Mastercam Toolpath Manager** tab from the **PropertyManager**. The **Mastercam Toolpath Manager** will be displayed; refer to Figure-82.

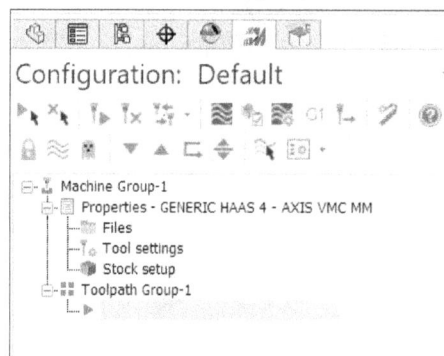

Figure-82. Mastercam Toolpath Manager

- Click on the **+** sign next to **Properties** option and click on the **Stock setup** option. The **Machine Group Properties** dialog box will be displayed with the **Stock Setup** tab selected; refer to Figure-83.

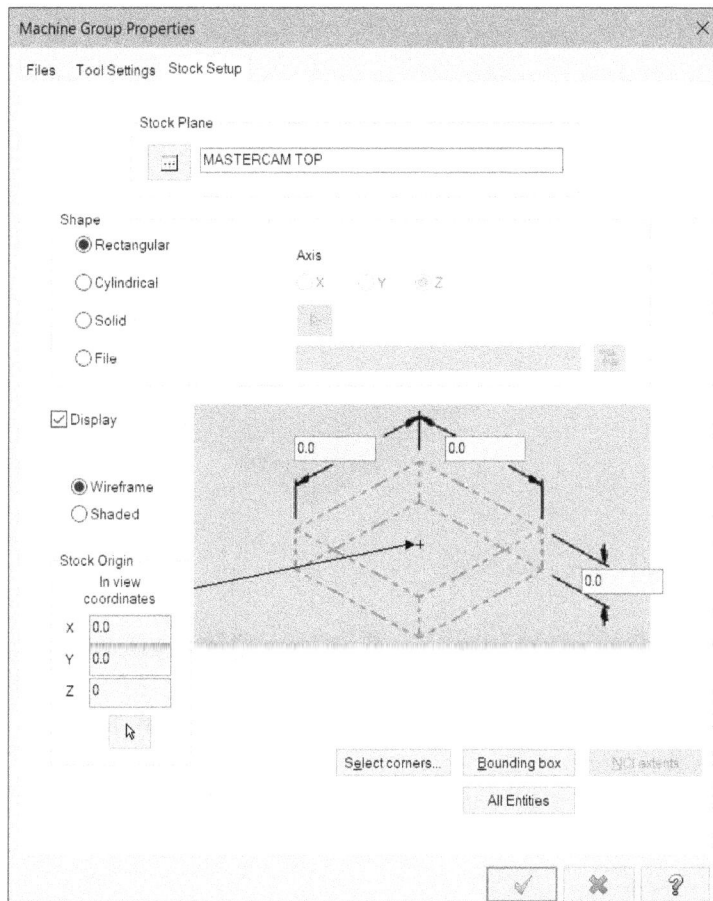

Figure-83. Machine Group Properties dialog box with Stock setup page

- Select the **Solid** radio button and then click on the **Bounding box** button. The Selection PropertyManager will be displayed.
- Select the model to create its bounding box and click on the **OK** button. The **Bounding Box PropertyManager** will be displayed; refer to Figure-84.

Figure-84. Bounding Box or Cylinder PropertyManager

- Select **Rectangular** radio button from **Shape** rollout and increase value in the **Z** spinner of **Size** section by **1**. Click on the **OK** button from the **PropertyManager**. The stock will be created; refer to Figure-85.

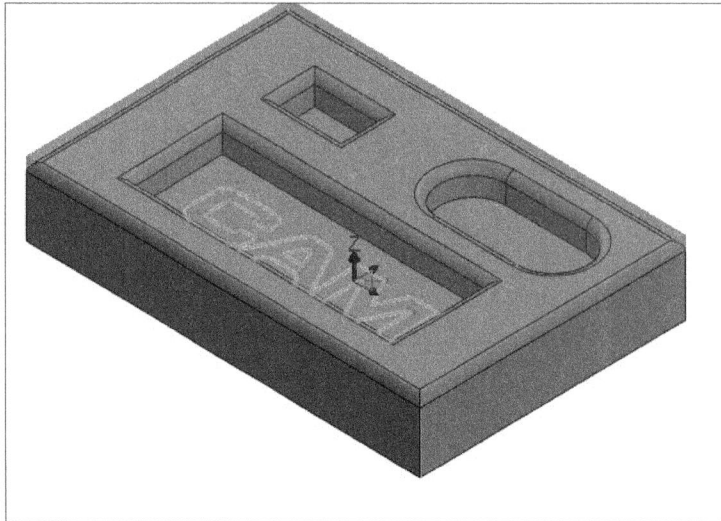

Figure-85. Stock created

Performing the Facing Operation

- Click on the **2D** drop-down from the **Ribbon**. The list of tools will be displayed.
- Click on the **Face** tool from the drop-down. The **Chain Manager** will be displayed and you are asked to select the entities for facing operation.
- Select the boundary edges of the model as shown in Figure-86.

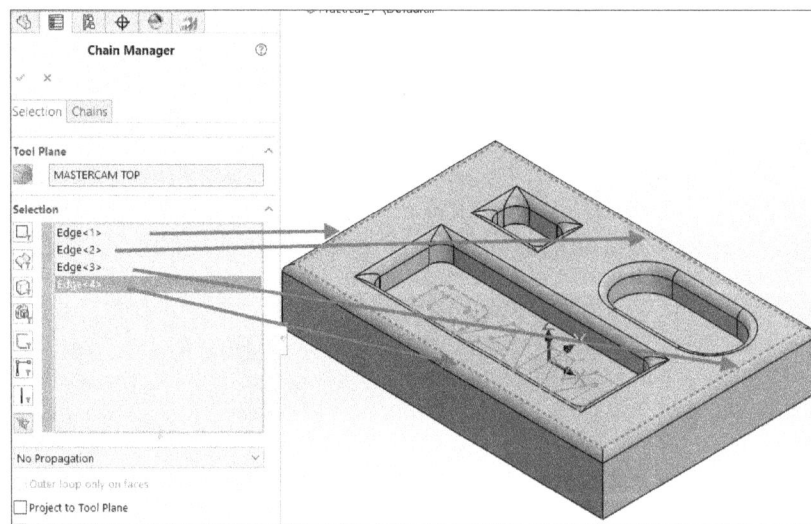

Figure-86. Edges selected for facing

- Click on the **OK** button from the **Chain Manager**. The **2D Toolpaths-Facing** dialog box will be displayed; refer to Figure-87.
- Click on the **Tool** option from the left list box in the dialog box. The options related to tool will be displayed; refer to Figure-88.

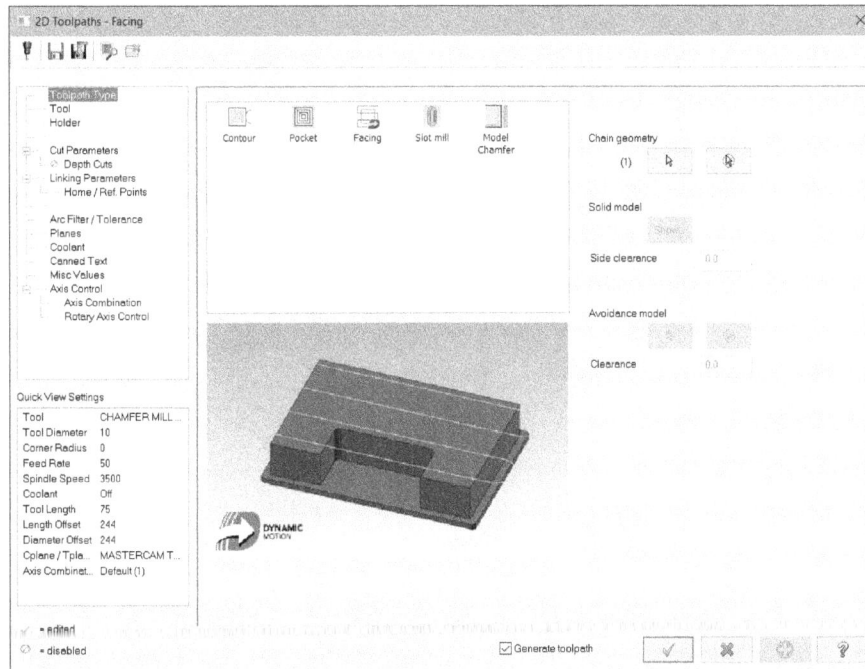

Figure-87. 2D Toolpaths–Facing dialog box

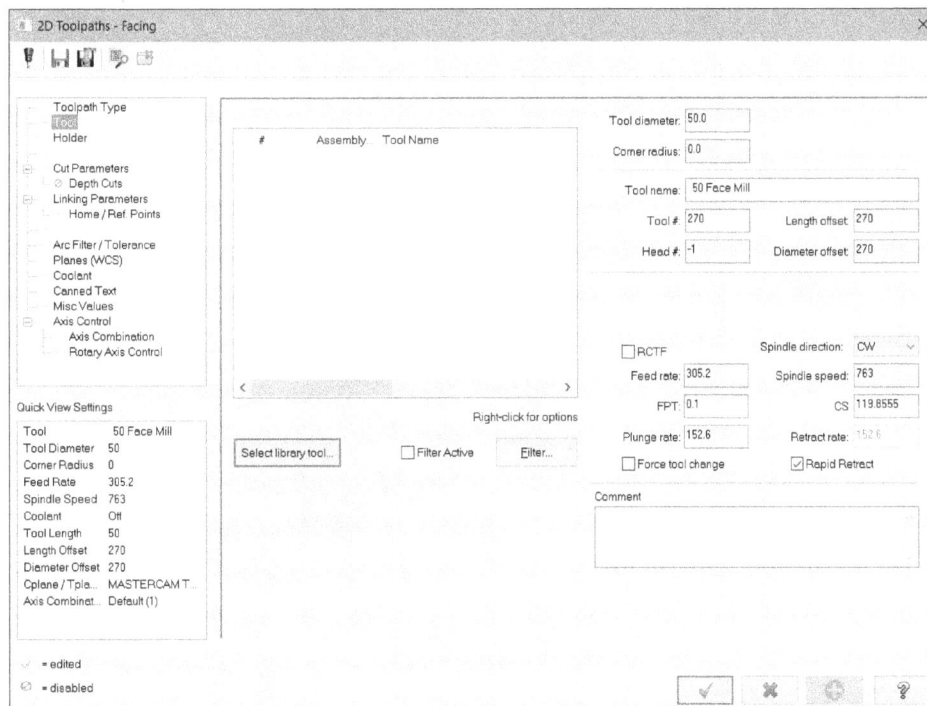

Figure-88. Tool page for Face

- Click on the **Select library tool** button from the dialog box. The **Tool Selection** dialog box will be displayed.
- Select the **50 diameter face mill** tool from the dialog box and click on the **OK** button from the dialog box.
- Click on the **Holder** option from the left list box and select desired tool holder.
- Click on the **Linking Parameters** option from the left list box in the dialog box.
- Select the **Incremental** radio button below the **Depth** button and specify the value as **-1** in the **Depth** edit box.
- Click on the **OK** button from the dialog box. The toolpath will be generated; refer to Figure-89.

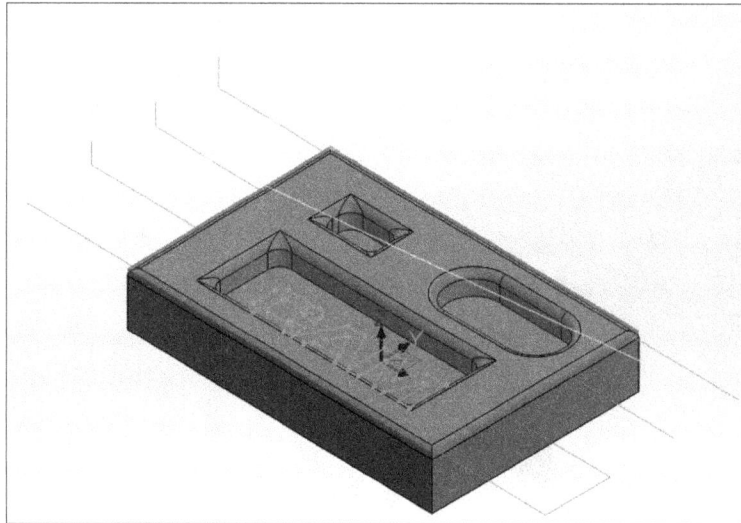

Figure-89. Facing toolpaths generated

Performing the pocket milling operation

- Select the **Pocket** tool from the **2D** drop-down. The **Chain Manager** will be displayed and you are prompted to select the faces for pocket milling.
- Select the faces as shown in Figure-90.
- Click on the **OK** button from the **Chain Manager**. The **2D Toolpaths-Pocket** dialog box will be displayed; refer to Figure-91.

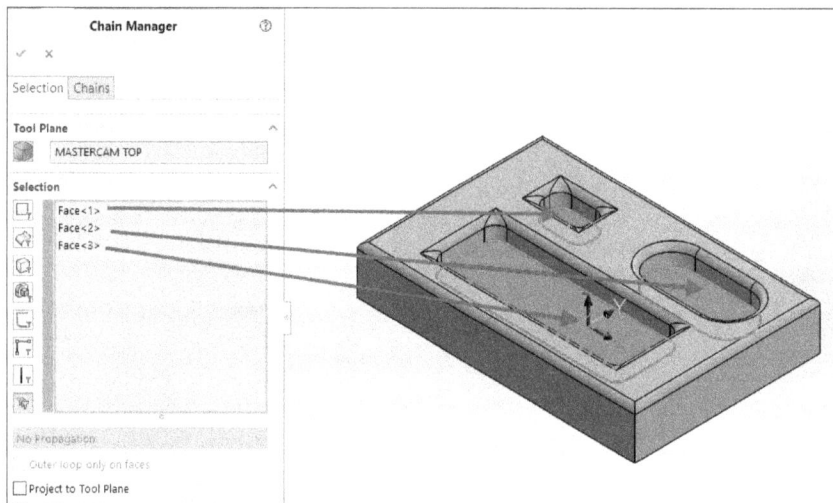

Figure-90. Faces selected for pocket milling

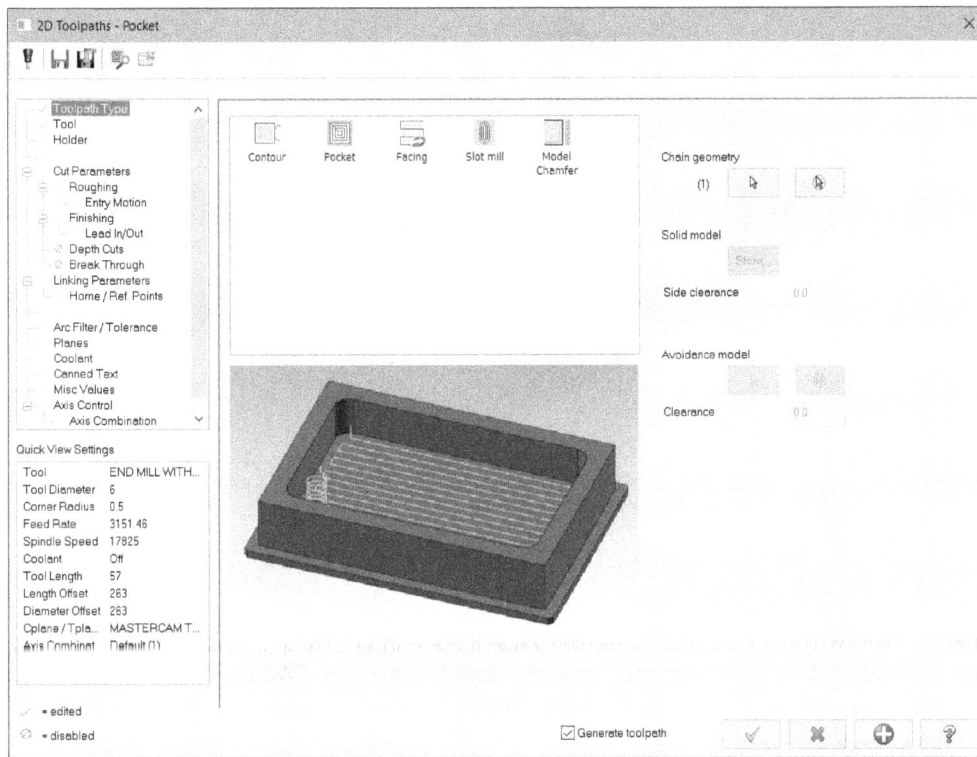

Figure-91. 2D Toolpaths-Pocket dialog box

- Click on the **Tool** option from the left box and click on the **Select library tool** button from the dialog box. The **Tool Selection** dialog box will be displayed.
- Select the **6 mm Flat Endmill** tool from the tool list and click on the **OK** button from the **Tool Selection** dialog box.
- Select desired tool holder by using the **Holder** option from the left box.
- Click on the **Linking Parameters** option from the left box.
- Select the **Absolute** radio button below the **Depth** button and then select the **Depth** button. The **Selection PropertyManager** will be displayed and you are asked to select a point to specify the depth.
- Select the point as shown in Figure-92.

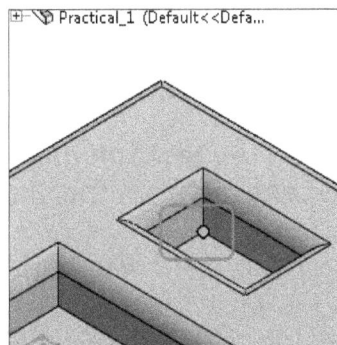

Figure-92. Point to be selected

- Click on the **OK** button from the **Selection PropertyManager**. The value in the **Depth** edit box will be changed automatically.
- Click on the **OK** button from the **2D Toolpaths-Pocket** dialog box. The toolpath will be generated; refer to Figure-93.

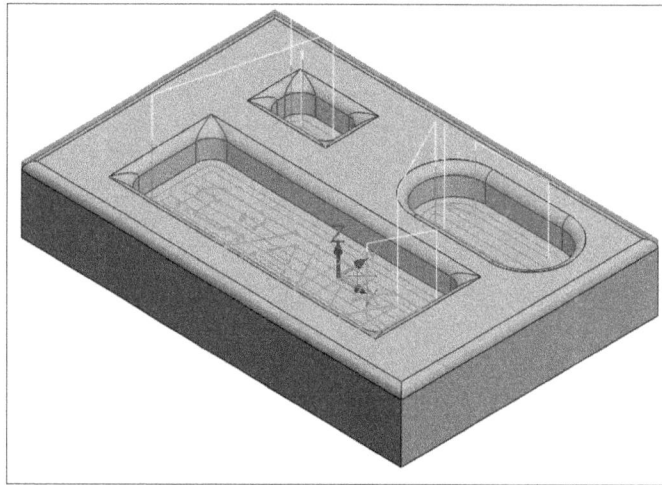
Figure-93. Pocket toolpaths generated

Performing the Engraving operation

- Click on the **Engrave** tool from the **2D** drop-down. The **Chain Manager** will be displayed.

- Click on the **Sketch entity filter** button from the **Chain Manager** and select the text by box selection; refer to Figure-94.

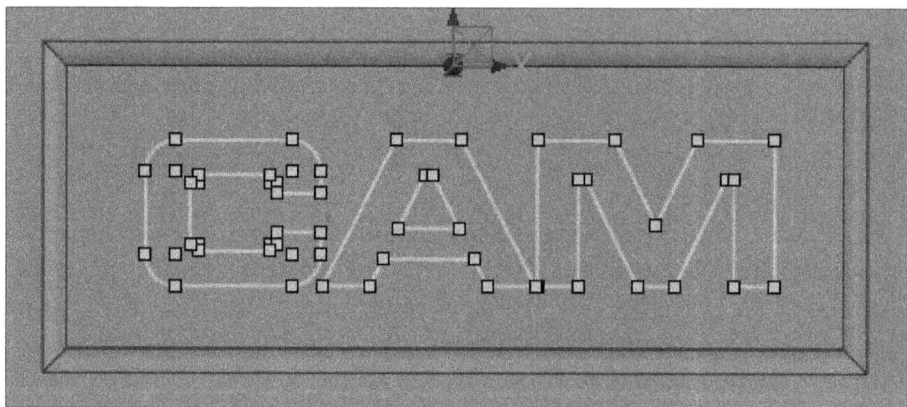
Figure-94. Sketch selected

- Click on the **OK** button from the **Chain Manager**. The **Engraving** dialog box will be displayed; refer to Figure-95.

- Click on the **Select library tool** button from the dialog box and select **5 mm Flat Endmill** tool from the list.

- Click on the **OK** button from **Tool Selection** dialog box displayed.

- Click on the **Engraving parameters** tab from the dialog box. The parameters related to engraving will be displayed.

- Select the **Incremental** radio button for the depth and specify **-1** in the **Depth** edit box.

- Click on the **Roughing/Finishing** tab and select the **at depth** radio button from the **Cut geometry** area of the dialog box.

- Click on the **OK** button from the **Engraving** dialog box. The toolpath will be generated; refer to Figure-96. Note that the toolpath will not be visible because it is at **1** mm depth in the part (not the stock). To view toolpath, you need to hide the model by right-clicking on it and selecting the **Hide** button from the shortcut menu; refer to Figure-97 then the toolpath will be displayed. To display model again, select it from **Model Tree** and click on the **Show** button from Mini-toolbar.

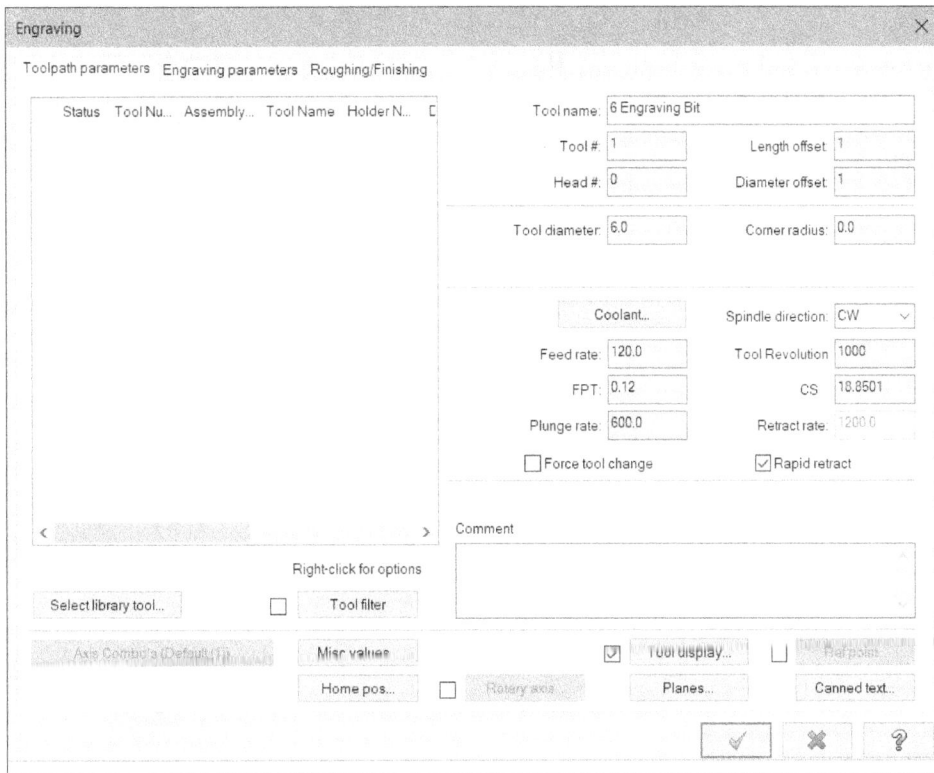

Figure-95. Engraving dialog box

Figure-96. Engraving toolpath generated

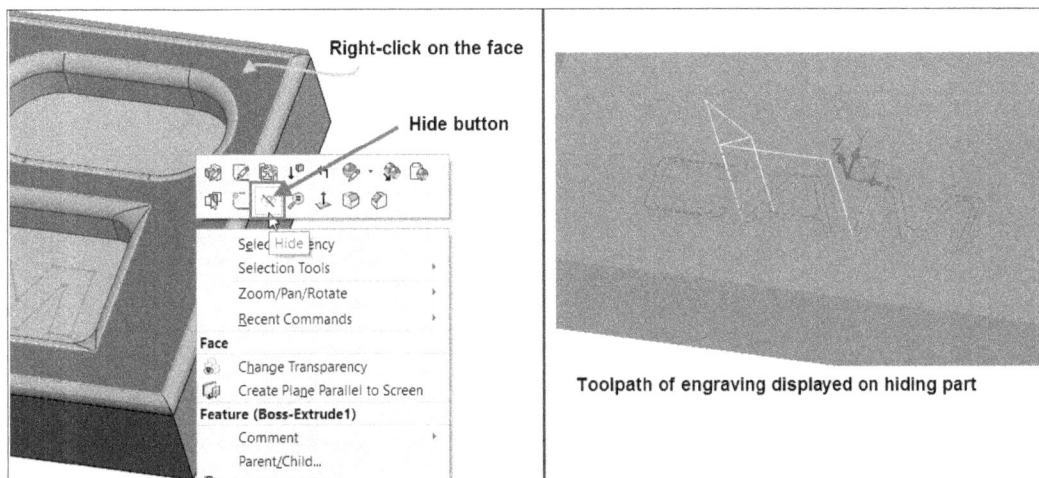

Figure-97. Shortcut menu to hide part

Creating rounds on the edges

- Click on the **Contour** tool from the **2D** drop-down. The **Chain Manager** will be displayed.
- Select the edges of the model that are to be rounded; refer to Figure-98.

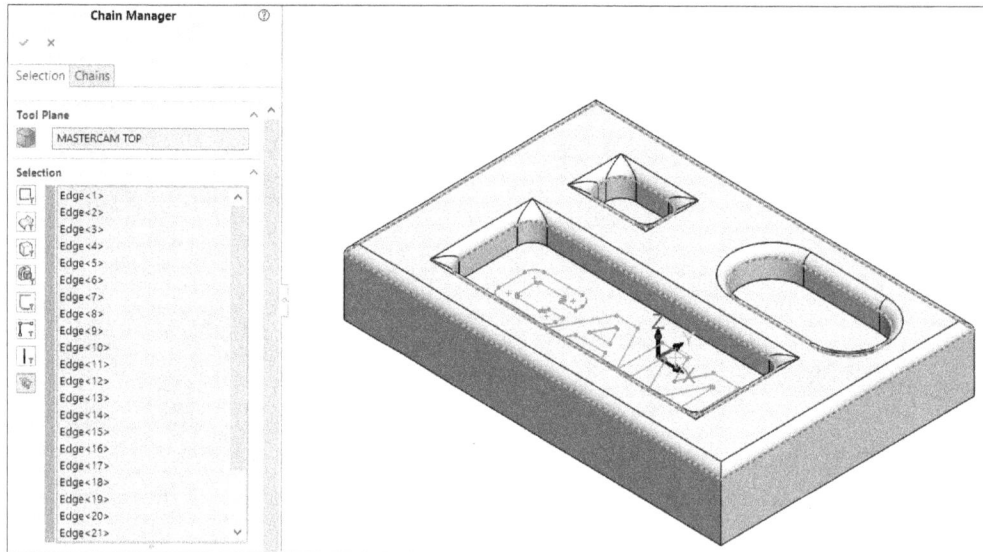

Figure-98. Edges selected for round

- Click on the **OK** button from the **Chain Manager**. The **2D Toolpaths-Contour** dialog box will be displayed.
- Select a tool for rounding having round radius of 3 mm. Note that you can create your own round tool by using the procedure explained earlier or by modifying the existing tool; refer to Figure-99.

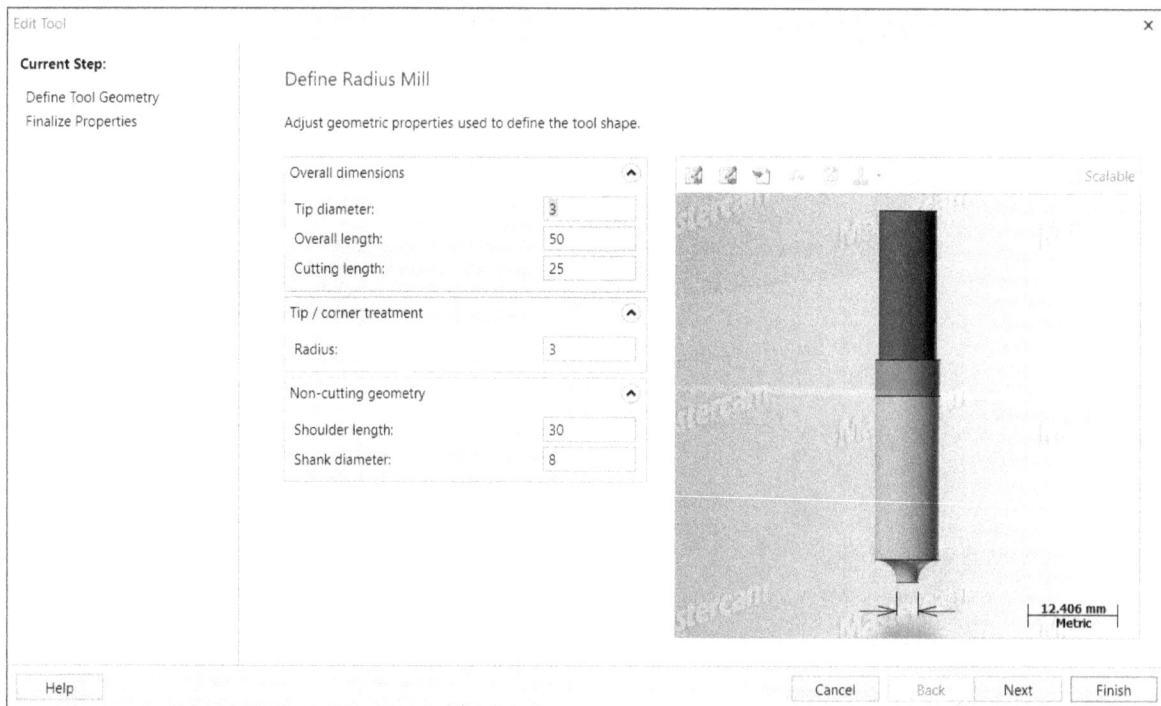

Figure-99. Edit Tool dialog box

- Click on the **Cut Parameters** option from the left box in the **2D Toolpaths-Contour** dialog box and select **Right** in the **Compensation direction** drop-down.
- Click on the **OK** button from the dialog box. The toolpath will be generated; refer to Figure-100.

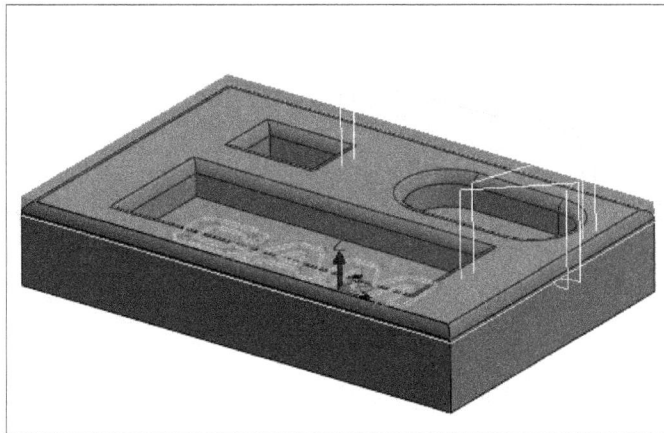
Figure-100. Toolpath for radius creation

Note that in this tutorial we have not changed the **Depth Cuts** parameter and we have machined each operation in single pass but in real-machining you need to set depth of each pass as per your tool catalog.

PRACTICE 1

Machine the part given in Figure-101. Stock for the part is a bounding cylinder. Use a Mill-Turn machine to do turning as well as milling.

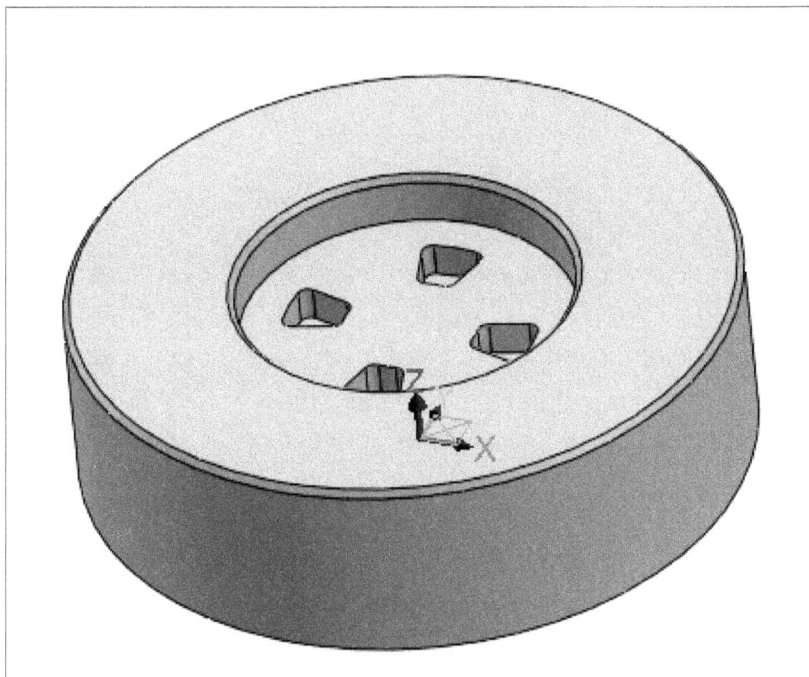
Figure-101. Practice 1

FOR STUDENT NOTES

Chapter 7

Milling Toolpaths-II

Topics Covered

The major topics covered in this chapter are:

- *FBM Mill, FBM Drill and Wizard Holes (Feature Based Toolpaths)*
- *3D Roughing*
- *3D Finishing*

INTRODUCTION

The toolpaths discussed in previous chapter were for 2D milling. But milling is not confined to 2D. The milling machine can cut material in 3D space along 5 different axes. In this chapter, we will discuss tools to create 3D toolpaths. Also, there are some special tools like FBM Mill and FBM Drill in Mastercam that can recognise the models features and automate the procedure of toolpath creation. Note that these tools can generate only 2D so the other toolpaths need to be generated by you. The tools are discussed next.

FEATURE-BASED TOOLPATHS

Feature-based toolpaths are used to automate the process of 2D toolpath generation based on the features in the part. These tools are available in the **Feature-Based Toolpaths** drop-down in the **Mastercam2022** tab of the **Ribbon**; refer to Figure-1. These tools are discussed next.

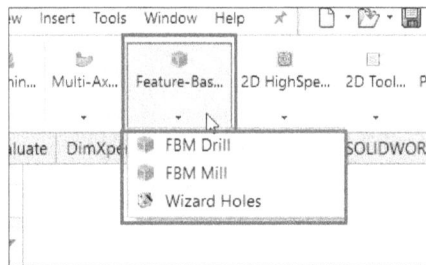

Figure-1. Feature-Based Toolpaths drop-down

FBM Mill

The **FBM Mill** tool is used generate 2D milling toolpaths based on recognized features of the model. The procedure to use this tool is given next.

*   Open the part on which you want to perform 2D milling and create stock; refer to Figure-2.

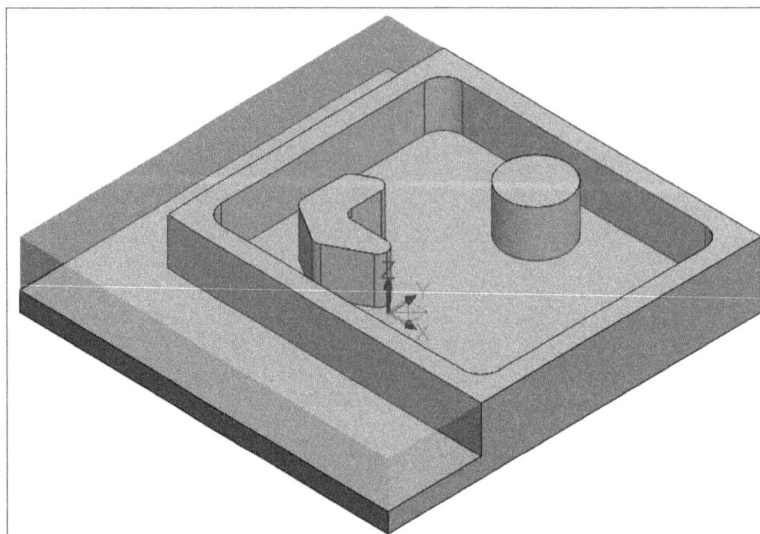

Figure-2. Model and stock for FBM

- Click on the **FBM Mill** tool from the **Feature-Based Toolpaths** drop-down in the **Ribbon**. The **FBM Toolpaths-Mill** dialog box will be displayed; refer to Figure-3.

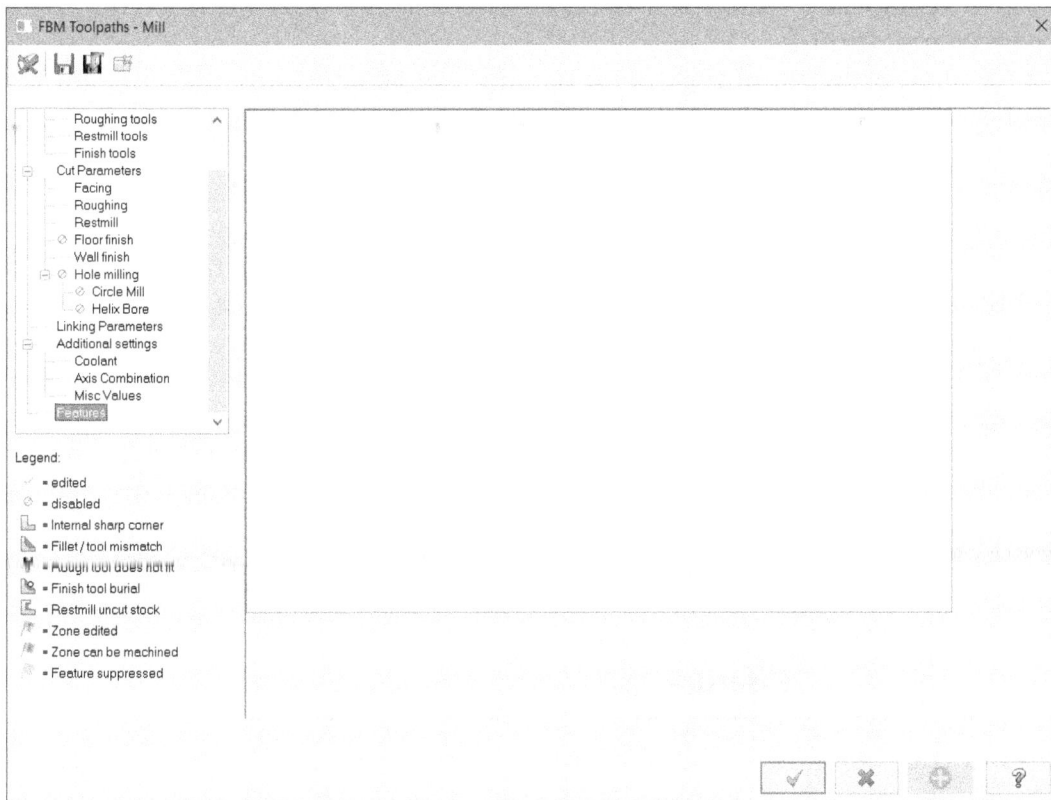

Figure-3. FBM Toolpaths-Mill dialog box

Setup Parameters

- Click on the **Setup** option from the left box in the dialog box. The options in dialog box will display as shown in Figure-4.

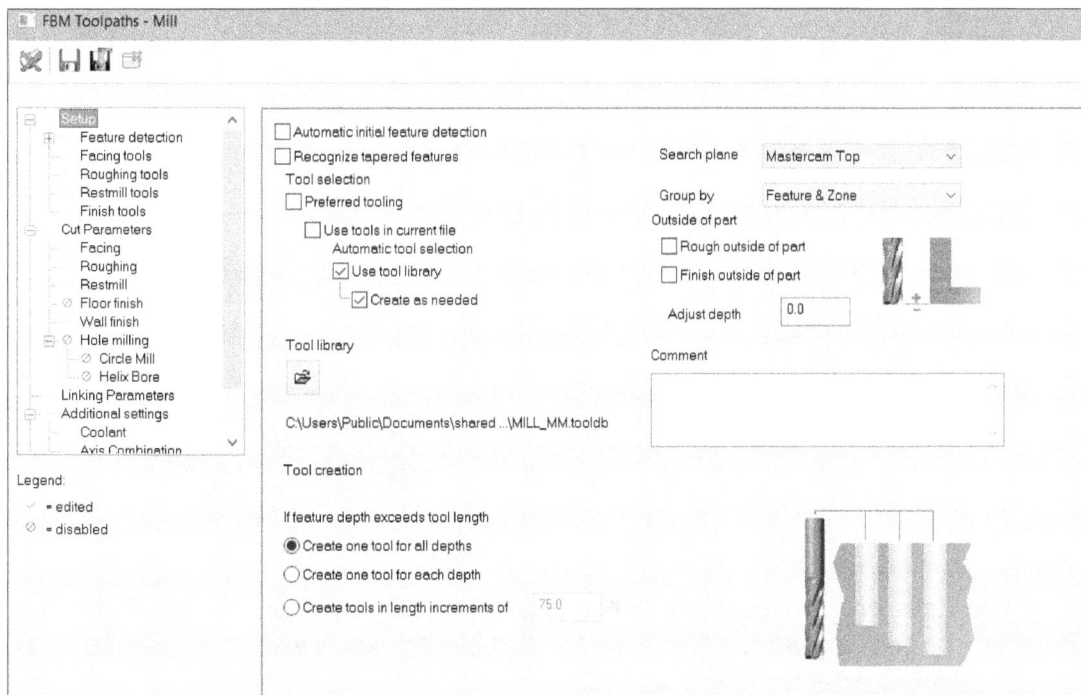

Figure-4. Setup options for FBM-Mill

- Select the **Automatic initial feature detection** check box to allow Mastercam automatically detect initial features for the toolpaths.
- Select the **Recognize tapered features** check box if you want to include tapered features in machining.
- Select the **Preferred tooling** check box so that Mastercam searches for preferred tools of each operation (facing, roughing etc.).
- Select the **Use tools in current file** check box if you want to restrict Mastercam to use only those tools which have been used in current file.
- Select the **Use tool library** check box so that Mastercam searches selected tool library for tools. After selecting this check box, click on the **Open** button from the **Tool library** area of the this page to select desired tool library.
- Select the **Create as needed** check box if you want Mastercam to automatically create tools as per the requirement if tools are not available in the tool library or you have not opted to use tool library. After selecting this check box, select the desired option from the **Tool creation** area of the this page to define how tool will be created. Generally this check box is kept clear so that you can use only standard tools which are easily available in market.
- Select the check boxes in the **Outside of part** area to rough and finish the part outside the boundaries.

Feature Detection Parameters

- Click on the **Feature detection** option from the left area of the dialog box. The options in the dialog box will be displayed as shown in Figure-5.

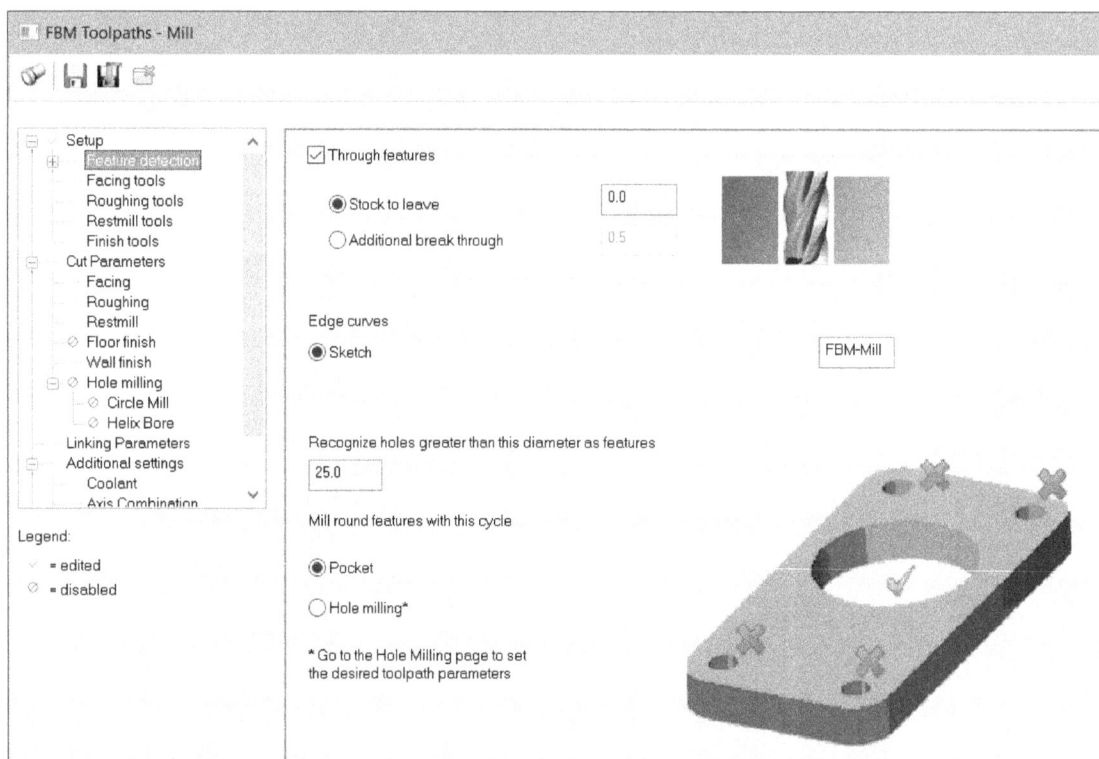

Figure-5. Feature detection options

- Specify the desired parameters in the page. Specify the minimum diameter in the **Recognize holes greater than this diameter as features** edit box to make Mastercam recognise features as holes.
- Select the desired milling strategy from the **Mill round features with this cycle** area of this page. If you select the **Hole milling** radio button then **Hole milling**

page will also become available in this dialog box. On selecting Hole milling strategy, Circle mill and Helix bore toolpaths will be used. If you select the **Pocket** radio button then the pocket toolpaths will be used for machining.

Slug Cutting Options

The slug cutting is used when we have large stock of plastic or wood to be removed. In such cases, Mastercam uses Contour toolpaths in place of pocket/area toolpaths to remove material. This might be desirable when working with wood and composite materials on large vacuum table machines. Instead of pocketing the entire area, FBM Mill uses the parameters you define to generate a contour toolpath that cuts the outermost passes of the profile, leaving behind a slug that is held in place by the vacuum table.

- To enable slug cutting, click on the **Slug cutting** option from the expanded **Feature detection** node in the left of the dialog box. The options will be displayed as shown in Figure-6.

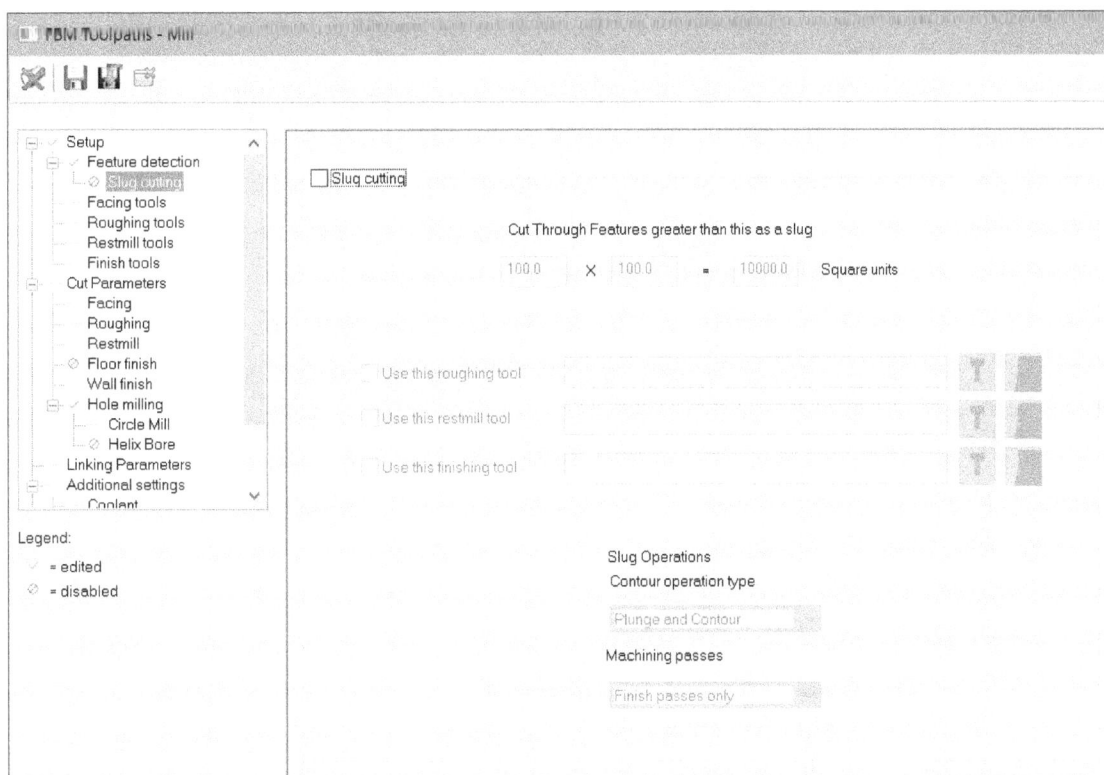

Figure-6. Slug cutting options

- Select the **Slug cutting** check box and specify related parameters to define slug cutting.
- Note that you should keep these option inactive if you are working with metals.

Facing/Roughing/Restmill/Finish Tools parameters

The options in **Facing tools**, **Roughing tools**, **Restmill tools**, and **Finish tools** pages of the dialog box are almost same. Here, we will discuss the options in the **Finish tools** page, you can apply the knowledge on other similar pages.

- Click on the **Finish tools** option from the left area of the dialog box. The options in the dialog box will be displayed as shown in Figure-7.

- Right-click in the **Preferred tool** list box of this dialog box. A shortcut menu will be displayed; refer to Figure-8.

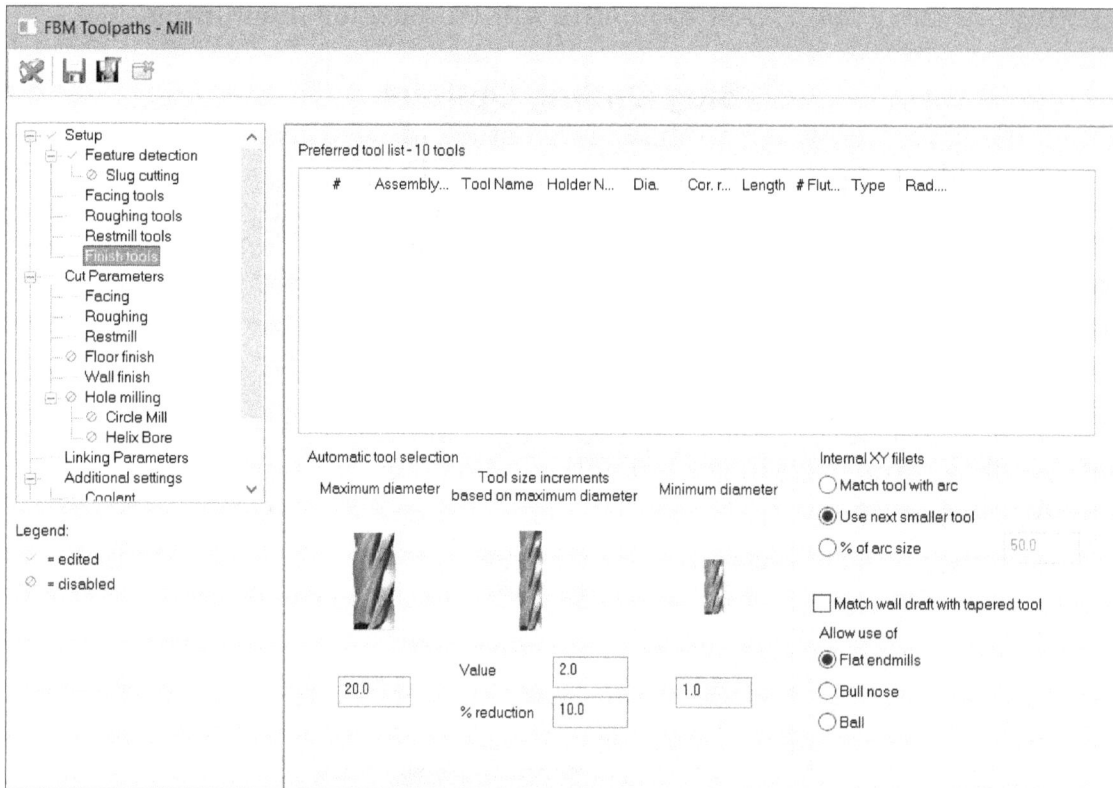

Figure-7. Finish tools page in FBM Toolpaths–Mill dialog box

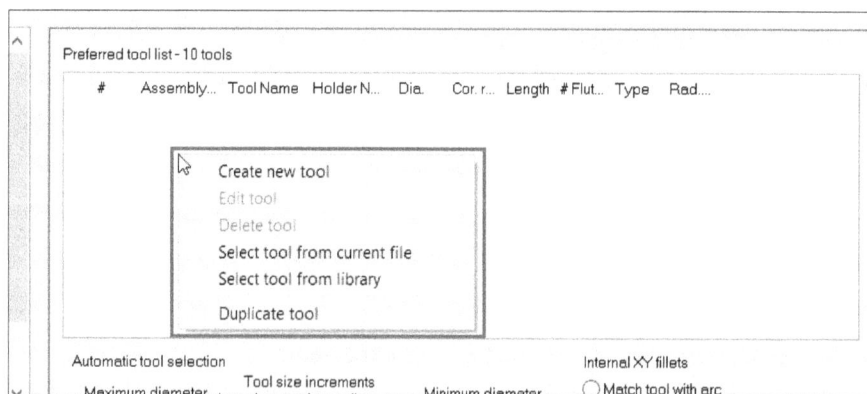

Figure-8. Shortcut menu for tool

- Create tool or select tool from the library as discussed earlier. The selected/created tools will be added in the list.
- Specify the parameters related to tool selection in the **Automatic tool selection** area of the dialog box.
- Select the desired radio button from the **Allow use of** area to filter the tool types

to be selected.
- Select the **Match wall draft with taper** tool check box if walls of your part are tapered.
- Select the desired radio button from the **Internal XY fillets** area to set parameters for internal fillets.

Similarly, specify the parameters in other pages of this dialog box.

Features Detection

- After setting parameters, click on the **Detect** button 🔲 from the toolbar in the dialog box. Mastercam will divide different regions of part into zones based on machining and the **Features** page will be displayed; refer to Figure-9.

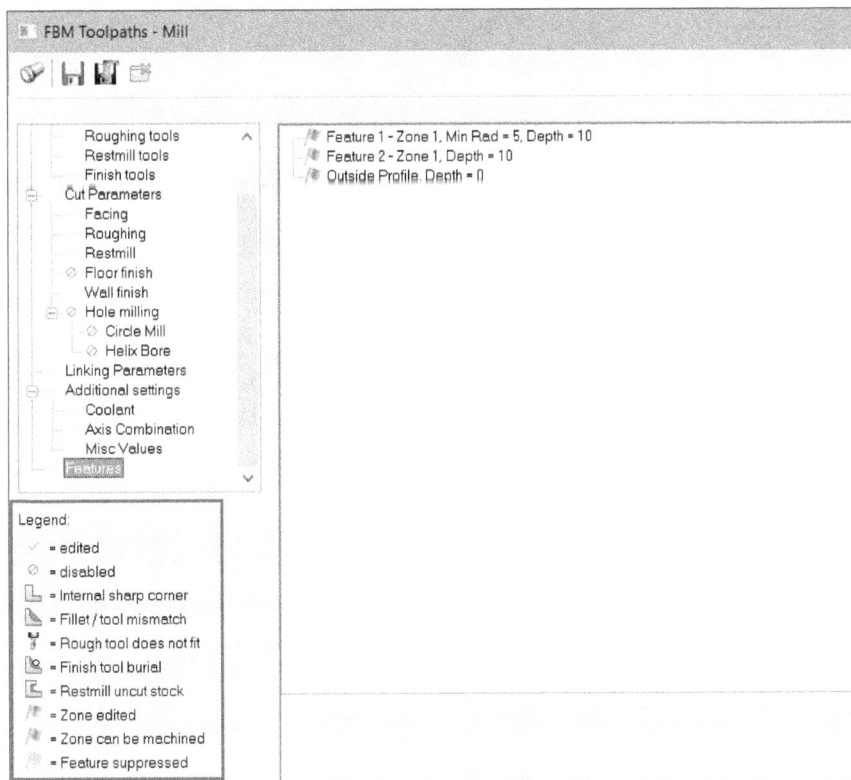

Figure-9. Options in Features page

- Check the legends on this page to know the status of various zones to be machined. For example, in Figure-9 all the zones are green flagged which means they can be machined by FBM Mill. If there is a problem in any zone and you think that it can be machined by 2D then modify the parameters of the dialog box and click on the Detect button again to verify.

- Click on the **OK** button from the dialog box to generate toolpaths. The toolpaths will be generated; refer to Figure-10.

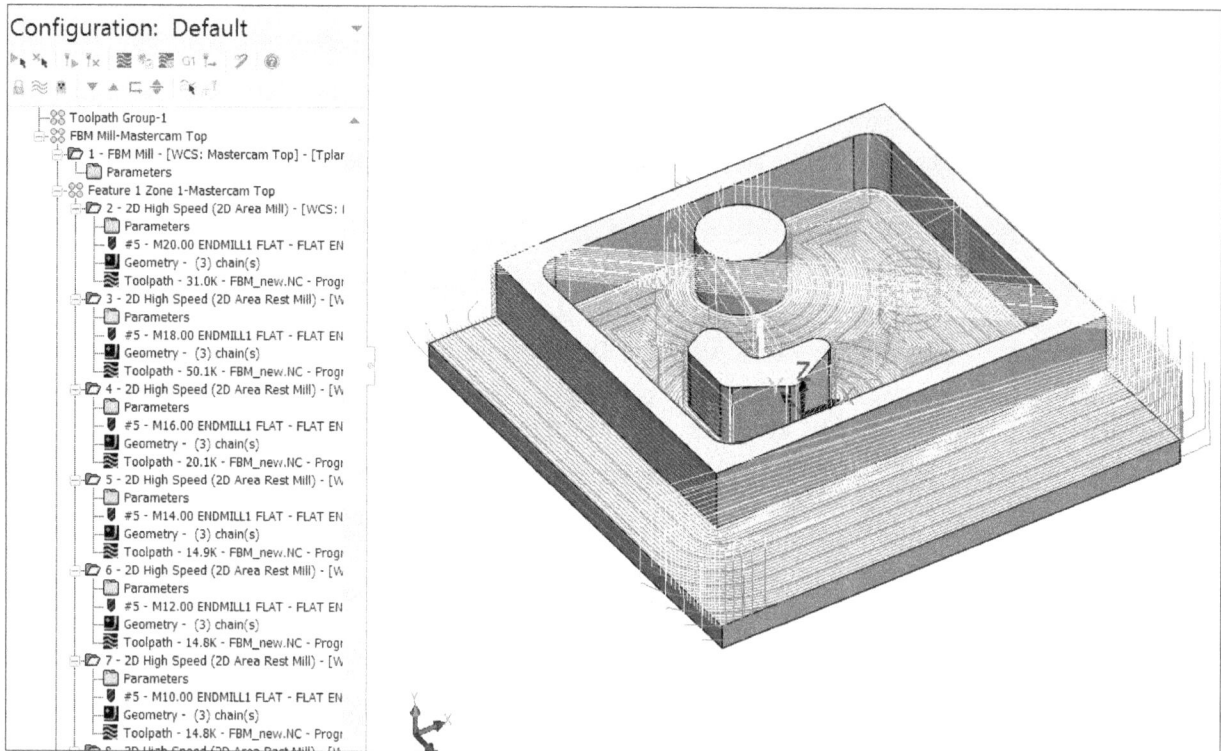

Figure-10. FBM-Mill toolpaths generated

FBM Drill

The **FBM Drill** tool is used to automatically recognize holes by depth and diameters, and automate the process of toolpath generation. The procedure to use this tool is given next.

- Open the part on which you want to perform FBM drilling and create stock.
- Click on the **FBM Drill** tool from the **Feature-Based Toolpaths** drop-down in the **Ribbon**. The **FBM Toolpaths-Drill** dialog box will be displayed; refer to Figure-11.

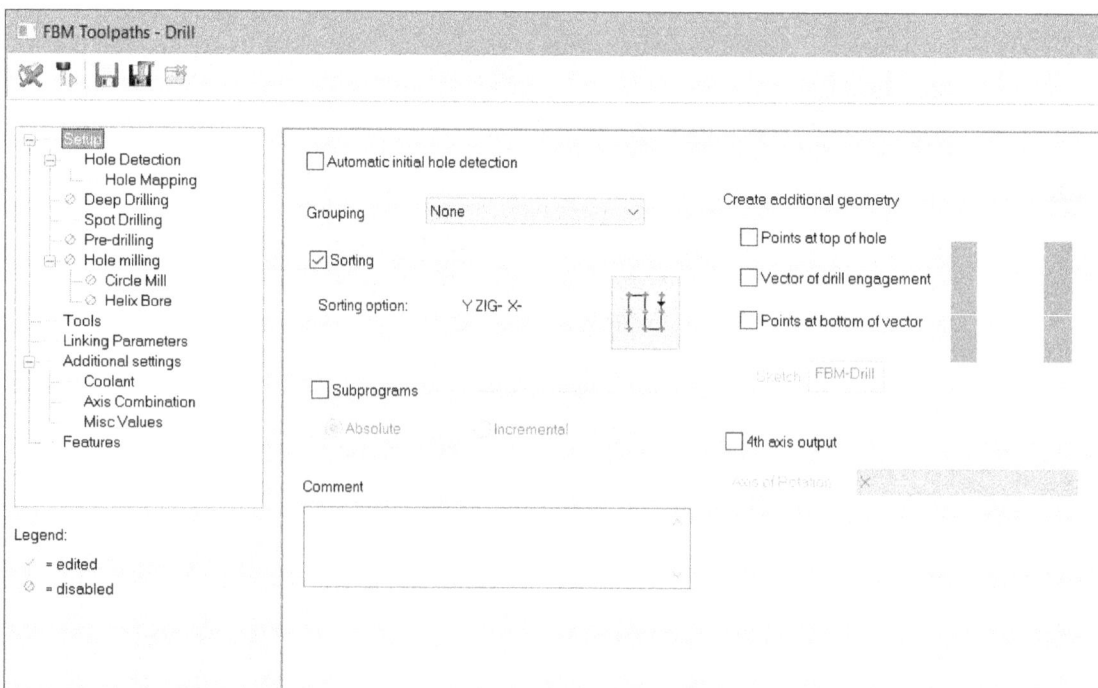

Figure-11. FBM Toolpaths-Drill dialog box

Setup Parameters

- Click on the **Setup** button from the left area of the dialog box. The options will be displayed as shown in Figure-11.
- Select the **Automatic initial hole detection** check box to let Mastercam automatically detect holes for machining.
- Select desired grouping option from the **Grouping** drop-down. Select **Tool** option if you want to group the operations based tool used for them. Select the **Plane** option to group operations based on their planes. Select **None** to not group operations.
- Click on the **2D Sort Points** button ⊞ from the Sorting area of the dialog box to sort drilling operation.
- Select the **Subprogram** check box to reduce main program length.
- If you want to create additional geometries in SolidWorks like drill point, vector, or bottom point then select the respective check boxes from the **Create additional geometry** area of the dialog box.
- Select the **4th axis output** check box and specify desired rotation axis if your machine can rotate the part while milling.

- Set the parameters on other pages in the same way discussed earlier and click on the **Detect** button ▨ at the top in the dialog box. The holes will be detected and will be displayed in the **Features** page; refer to Figure-12.

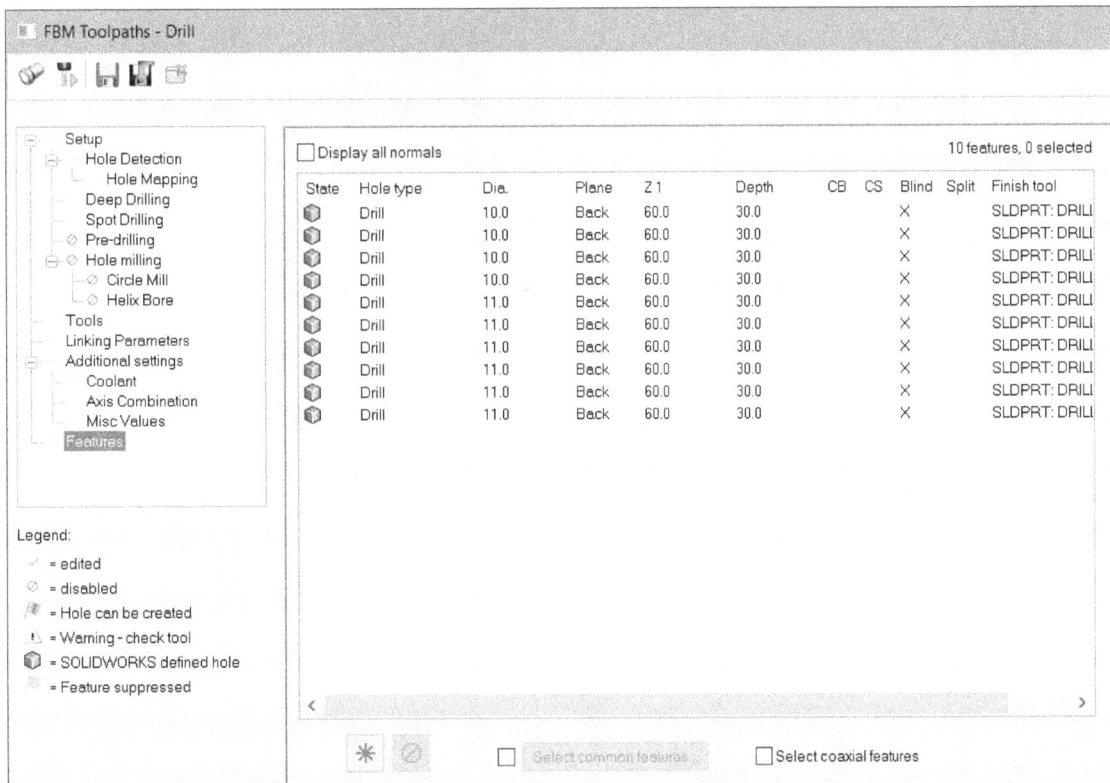

Figure-12. Drill holes detected

- Click on the **OK** button from the dialog box to create drill toolpaths.

Wizard Holes

The **Wizard Holes** tool works in the same way as **FBM Drill** tool but in case of **Wizard Holes** tool, you need to select all the holes which we want to be machined. Mastercam will automatically define tools and toolpaths. Note that this tool can only be used for holes created by SolidWorks **Hole Wizard** tool. The procedure to use this tool is given next.

- Click on the **Wizard Holes** tool from the **Feature-Based Toolpaths** drop-down in the **Mastercam2022 CommandManager** of the **Ribbon**. The **Selection PropertyManager** will be displayed asking you to select holes.
- Select the holes created by **Hole Wizard** tool which you want to be machined; refer to Figure-13 and click on the **OK** button from the **PropertyManager**. The **Wizard Holes** dialog box will be displayed; refer to Figure-14.

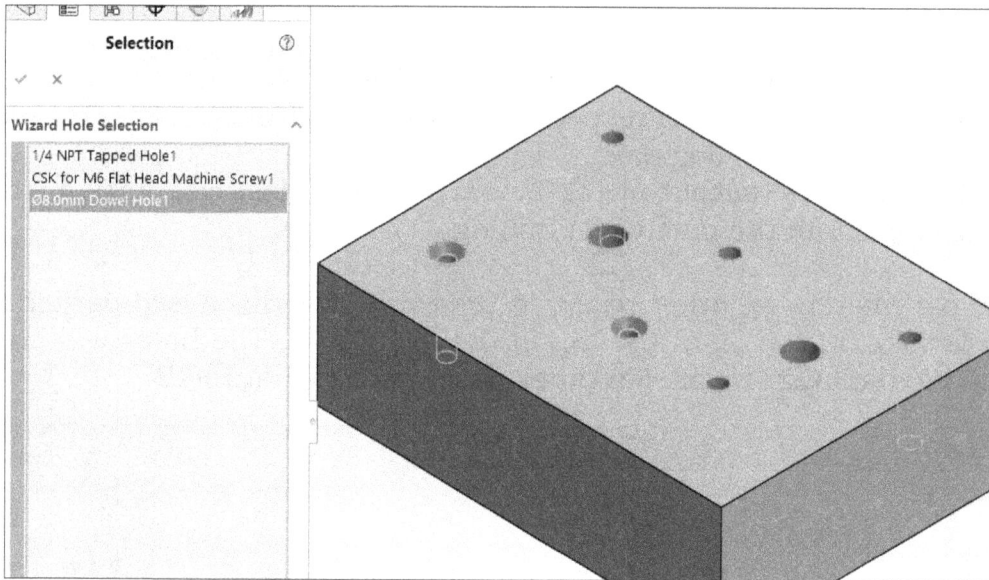

Figure-13. Selecting holes for Wizard Holes tool

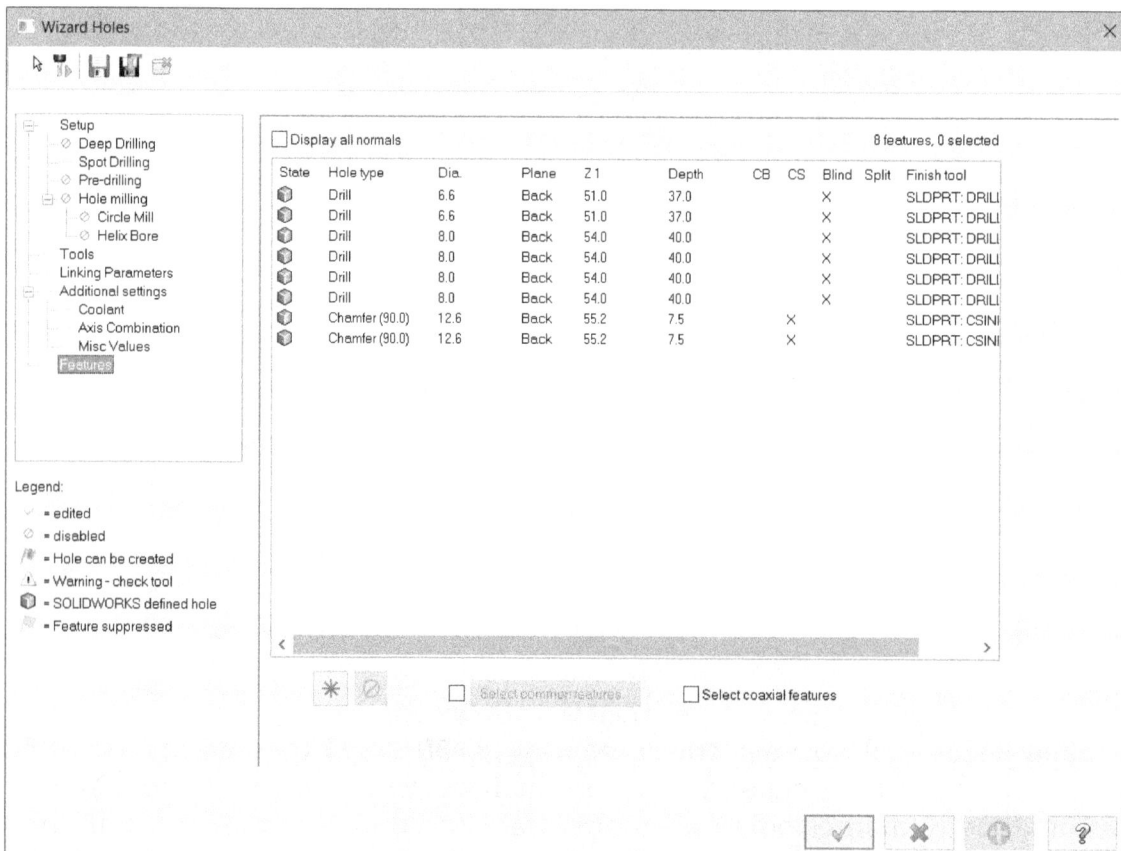

Figure-14. Wizard Holes dialog box

- Set the parameters as done for **FBM Drill** tool and click on the **OK** button. The toolpaths for drilling will be generated.

SURFACE ROUGHING

Surface toolpaths are used to mill the components which have non planar faces. Like every machining strategy, these 3D toolpaths are also available for roughing and finishing. We will start with surface Roughing and later we will work on surface Finishing. The tools for surface roughing are available in the **Roughing** drop-down in the **Ribbon**; refer to Figure-15.

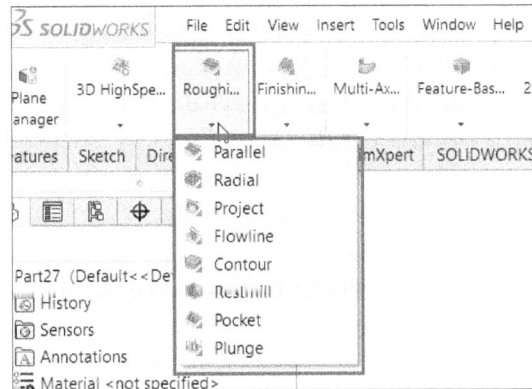

Figure-15. Roughing drop-down

Parallel Rough Surface Toolpath

The Parallel Rough Surface toolpath is used when you need to remove material by parallel line motion of tool; refer to Figure-16. The procedure to use this tool is given next.

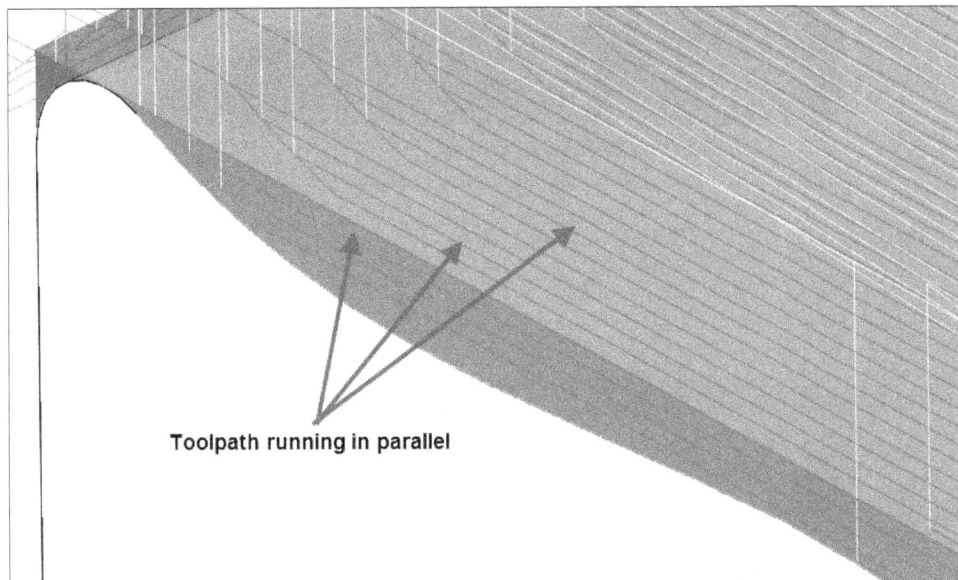

Figure-16. Parallel surface roughing

- Click on the **Parallel** tool from the **Roughing** drop-down in the **Mastercam2022 CommandManager** of the **Ribbon**. The **Select Boss/Cavity** dialog box will be displayed; refer to Figure-17.

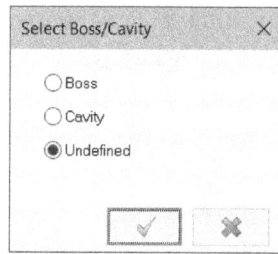

Figure-17. Select Boss/Cavity dialog box

- Select the **Boss** radio button if the surface is coming upward from the base plane. Select the **Cavity** radio button if the surface is going downward from the base plane. Select the **Undefined** radio button if you are not able to decide then Mastercam will automatically use suitable strategy. Note that there can be high difference between tool movements based on **Boss** or **Cavity** radio button selection; refer to Figure-18.

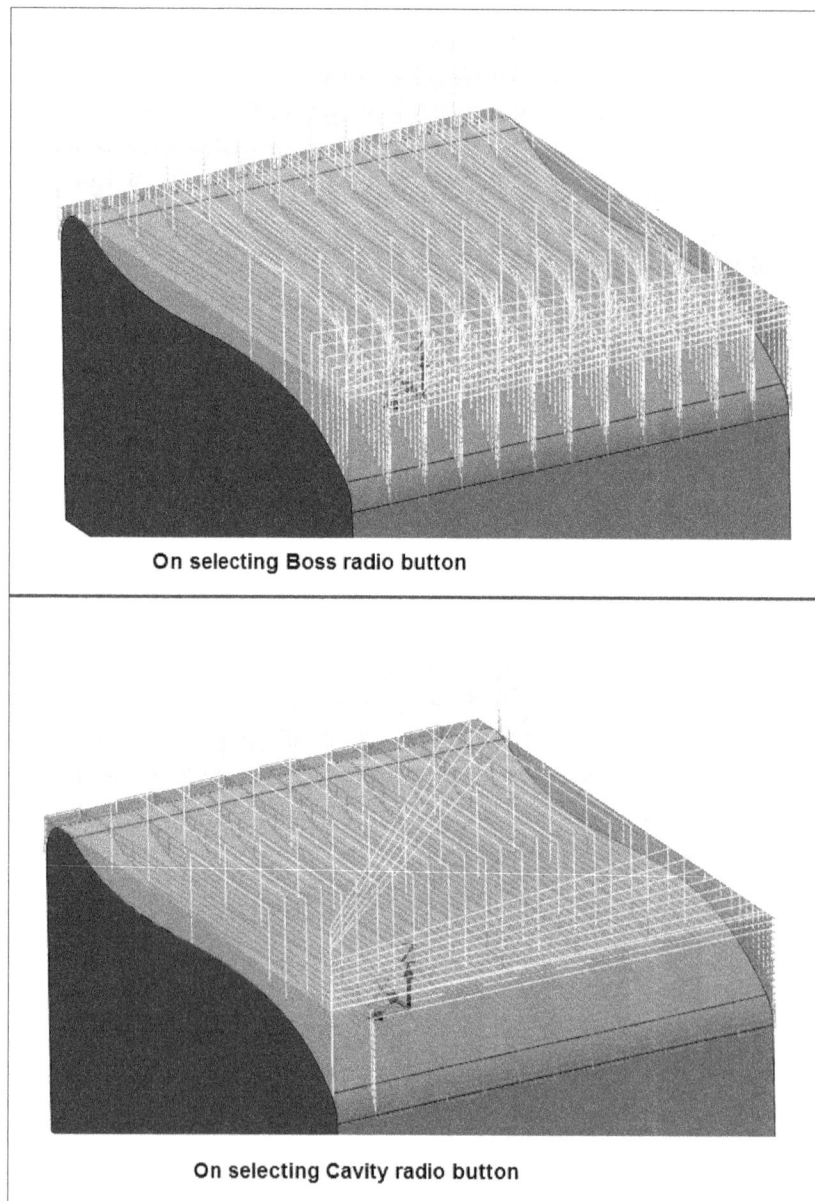

On selecting Boss radio button

On selecting Cavity radio button

Figure-18. Difference between toolpaths on selecting Boss and Cavity radio button

- After selecting the radio button, click on the **OK** button. The **Selection PropertyManager** will be displayed; refer to Figure-19.
- Select the faces that you want to machine.

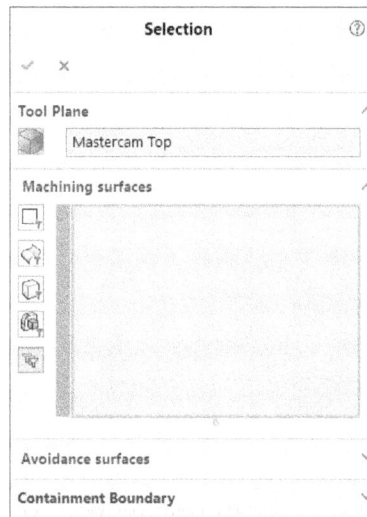

Figure-19. Selection PropertyManager

- If you want to avoid any feature/face/boundary then expand the **Check surfaces** rollout and click in the selection box. You will be asked to select the surfaces that should be avoided while machining. Select the faces of the feature that you want to avoid.
- If you want to define a boundary within which tool can perform machining then expand the **Containment Boundary** rollout and click in the selection box. You will be asked to select the faces/edge/curves to define boundary. Select the geometry as required.
- After selecting desired geometries; refer to Figure-20, click on the **OK** button from the **PropertyManager**. The **Surface Rough Parallel** dialog box will be displayed; refer to Figure-21.

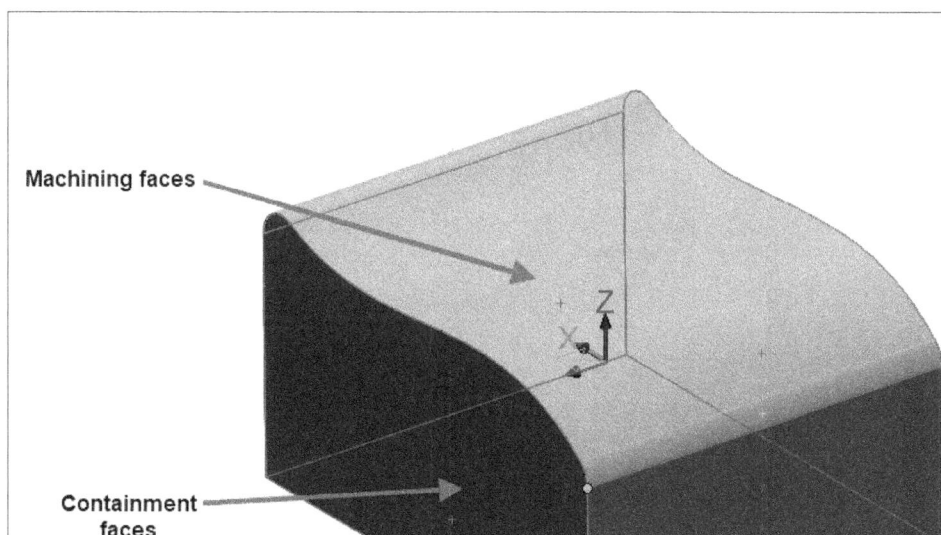

Figure-20. Faces selected for parallel rough toolpath

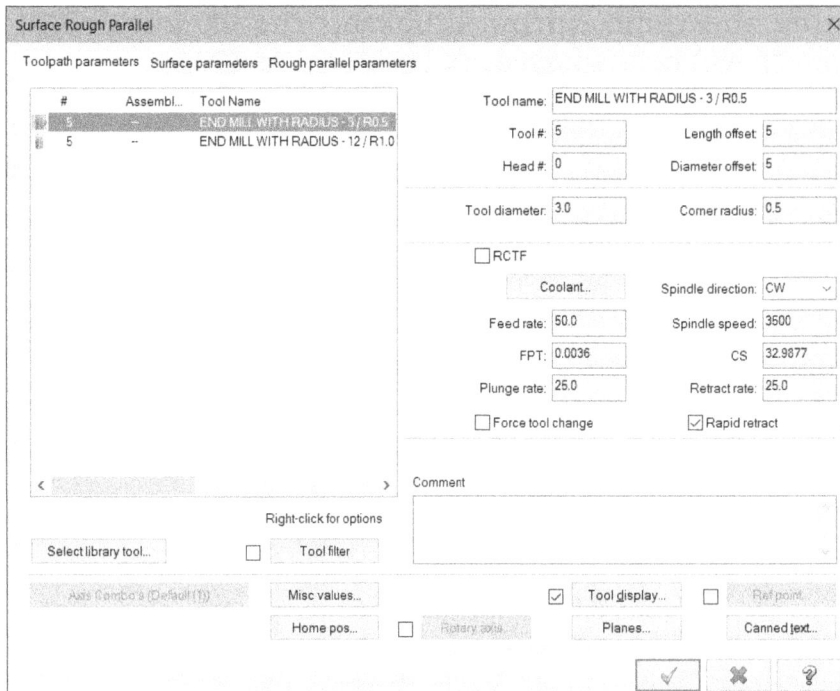

Figure-21. Surface Rough Parallel dialog box

Toolpath Parameters

- Click on the **Select library tool** button and select the desired tool by double-clicking from the **Tool Selection** dialog box displayed. Generally, bull endmill or ball endmill cutter is used for surface milling.
- Select the **RCTF** check box if you want to use radial chip thinning calculations to set feed rate and spindle speed.
- Specify the other parameters in this tab as discussed earlier.

Surface Parameters

- Click on the **Surface parameters** tab in the dialog box. The options will be displayed as shown in Figure-22.

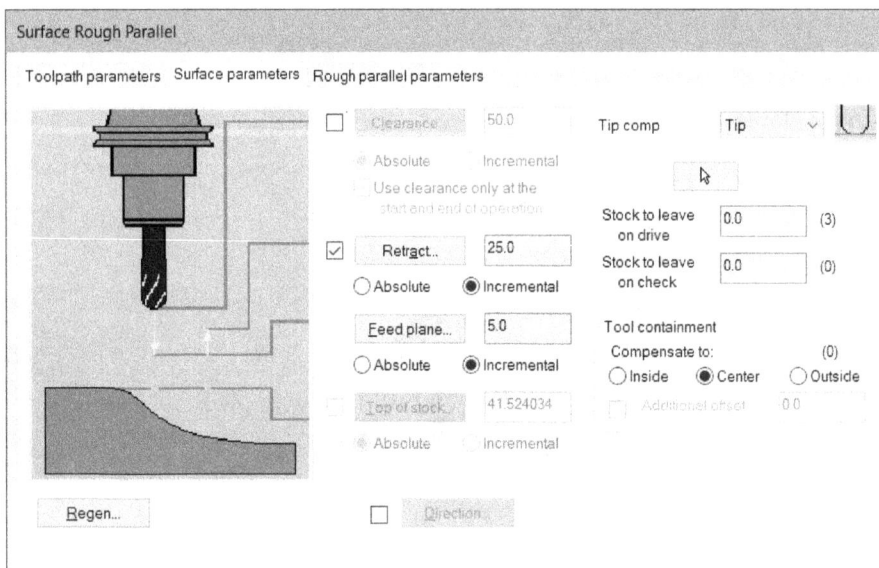

Figure-22. Surface Parameters tab

- Specify the retraction distance and feed plane distance in **Retract** and **Feed plane** edit boxes respectively.
- Select the check box before **Clearance** button and specify desired clearance if required. Clearance is used to retract the tool at higher level to avoid collision and allow free movement of workpiece.
- Specify the amount of stock to leave on faces to be machined and check surfaces in the **Stock to leave on drive** and **Stock to leave on check** edit boxes.
- Select the desired options for tool compensation and tool tip compensation from respective areas.

Rough Parallel Parameters

- Click on the **Rough parallel parameters** tab in the dialog box. The options will be displayed as shown in Figure-23.

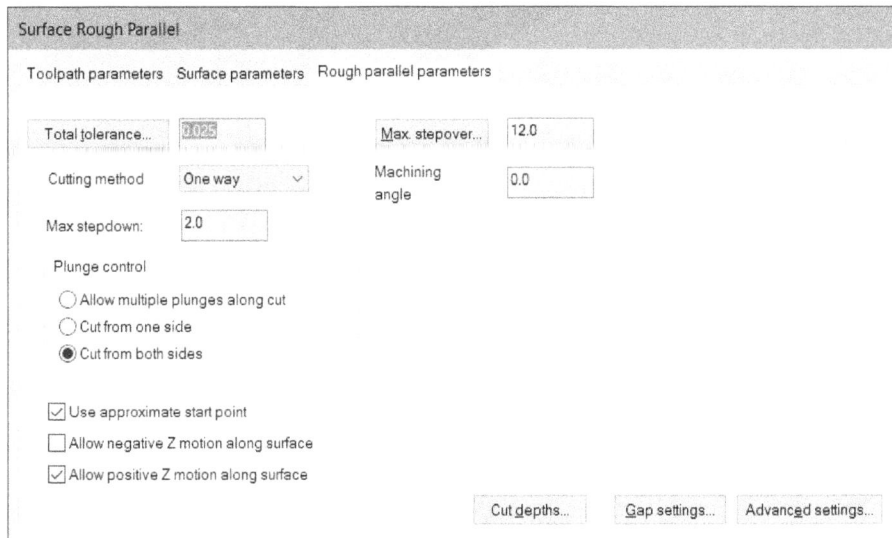

Figure-23. Rough parallel parameters tab

- Specify the cutting parameters and tolerances in this tab.
- Click on the **Cut depths** button and specify the depth of cut parameters in the **Cut Depths** dialog box displayed.
- Note that you should carefully define tolerances to get good surface finish on the part. To define tolerances, click on the **Total tolerance** button and specify parameters in the **Arc Filter/Tolerance** dialog box displayed on choosing the tool.
- After specifying desired parameters, click on the **OK** button from the dialog box. The toolpath will be generated; refer to Figure-24.

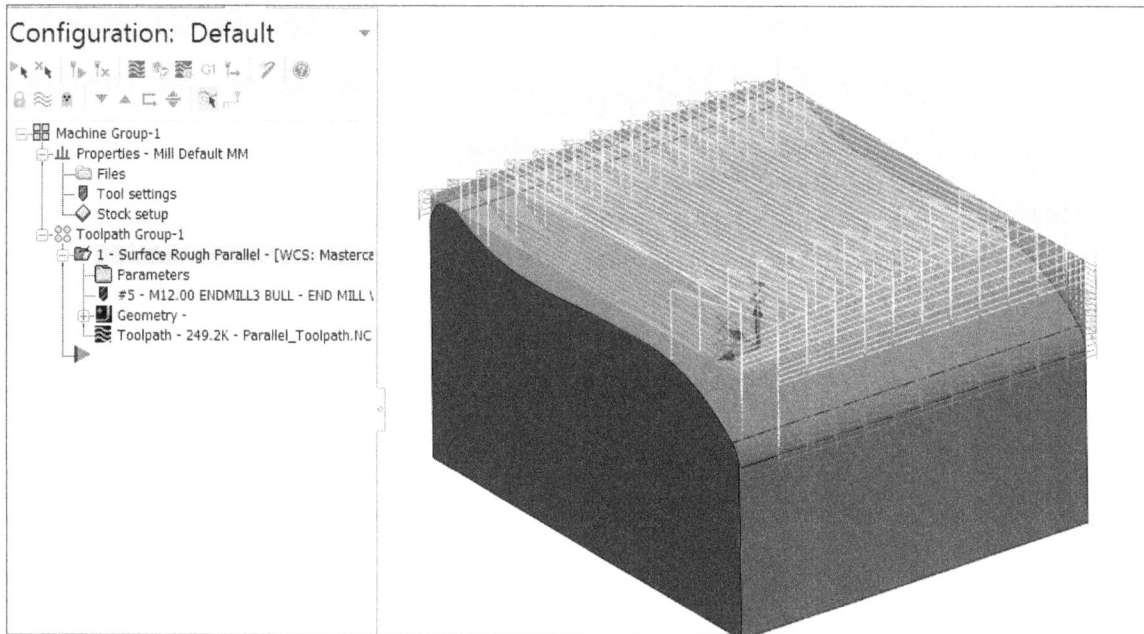

Figure-24. Toolpath generated for parallel rough

Radial Rough Surface Toolpath

The Radial Rough Surface toolpath is used to create radial surface toolpaths like in spokes of alloy wheel rims; refer to Figure-25. The procedure to use this tool is given next.

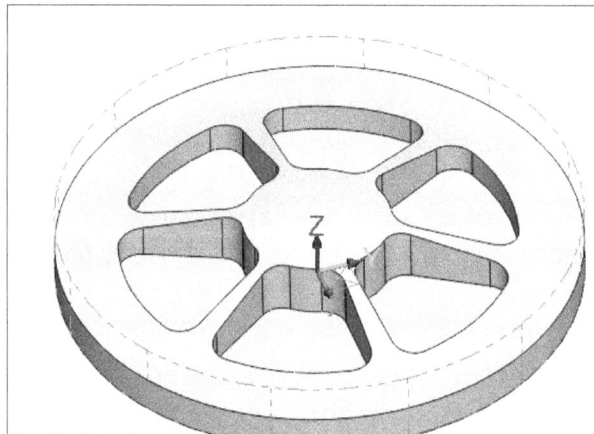

Figure-25. Model for radial rough toolpath

- Click on the **Radial** tool from the **Roughing** drop-down in the **Ribbon**. The **Select Boss/Cavity** dialog box will be displayed.
- Select the desired radio button as discussed earlier and click on the **OK** button. The **Selection PropertyManager** will be displayed.
- Select the faces/surfaces of the model to be machined; refer to and click on the **OK** button. The **Surface Rough Radial** dialog box will be displayed; refer to Figure-26.

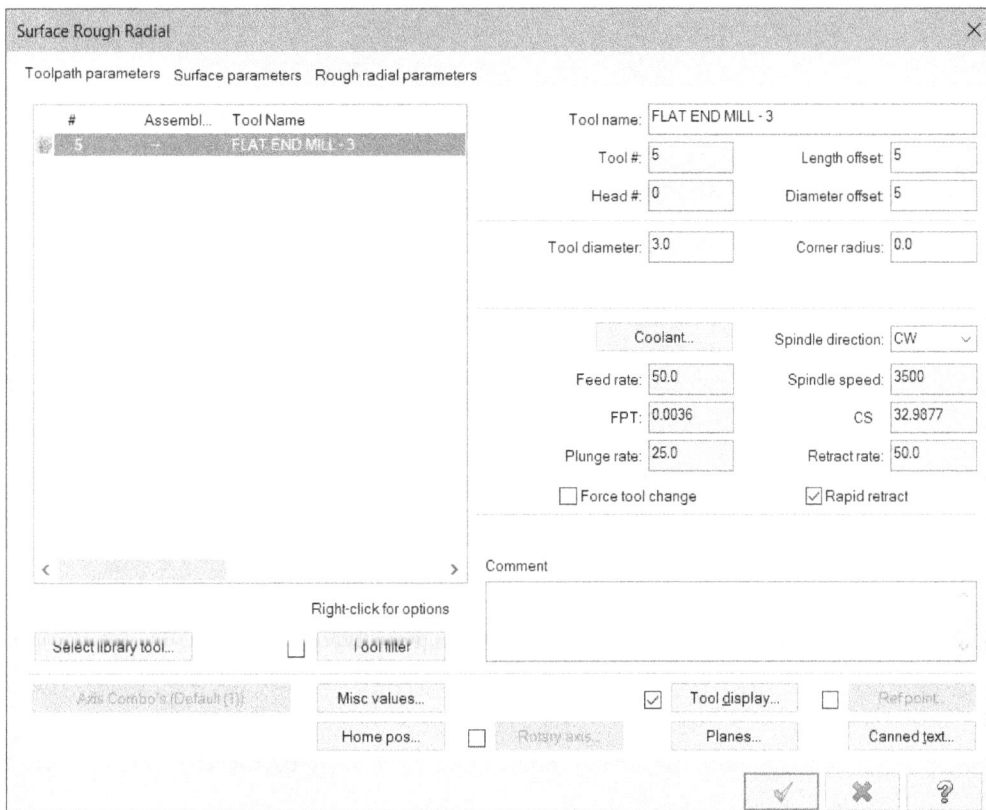

Figure-26. Surface Rough Radial dialog box

- Specify the parameters in different tabs of this dialog box as discussed earlier and click on the **OK** button. The toolpath will be generated; refer to Figure-27.

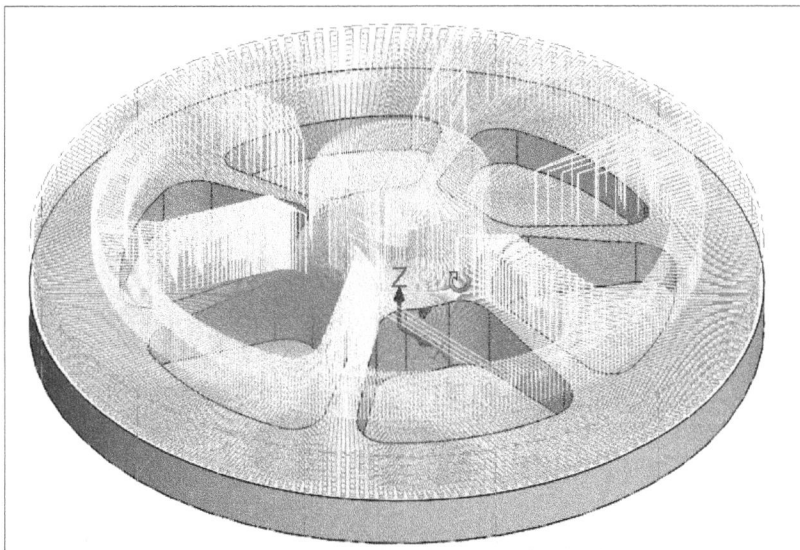

Figure-27. Surface rough radial toolpath generated

Project Rough Surface Toolpath

The Project Rough Surface toolpath is used to machine a toolpath projected on the surface of workpiece. You can also engrave text or curves on the face of part by using this tool; refer to Figure-28. The procedure to use this toolpath is given next.

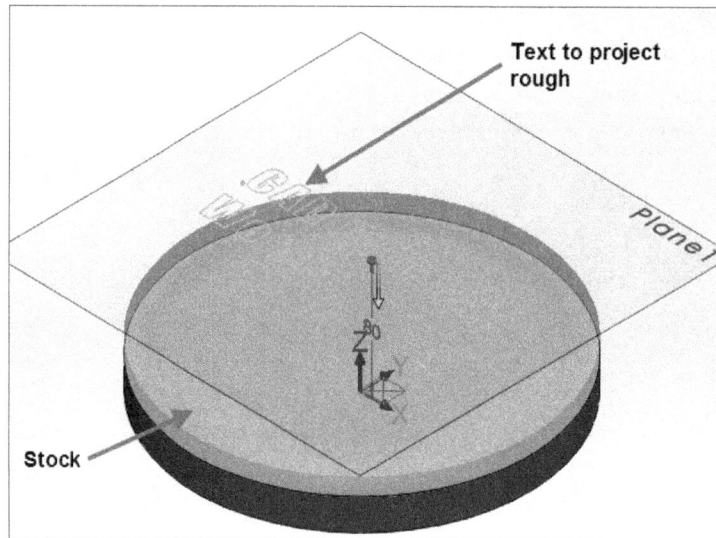

Figure-28. General arrangement for Project rough tool

- Open the model with stock and make sure that you have text or curve to be machined on the face of part. Click on the **Project** tool from the **Roughing** drop-down in the **Ribbon**. The **Select Boss/Cavity** dialog box will be displayed.
- Select the desired option as discussed earlier and click on the **OK** button. The **Selection PropertyManager** will be displayed.
- Select the face of model on which you want to create project rough surface toolpath; refer to Figure-29 and click on the **OK** button. The **Surface Rough Project** dialog box will be displayed; refer to Figure-30.

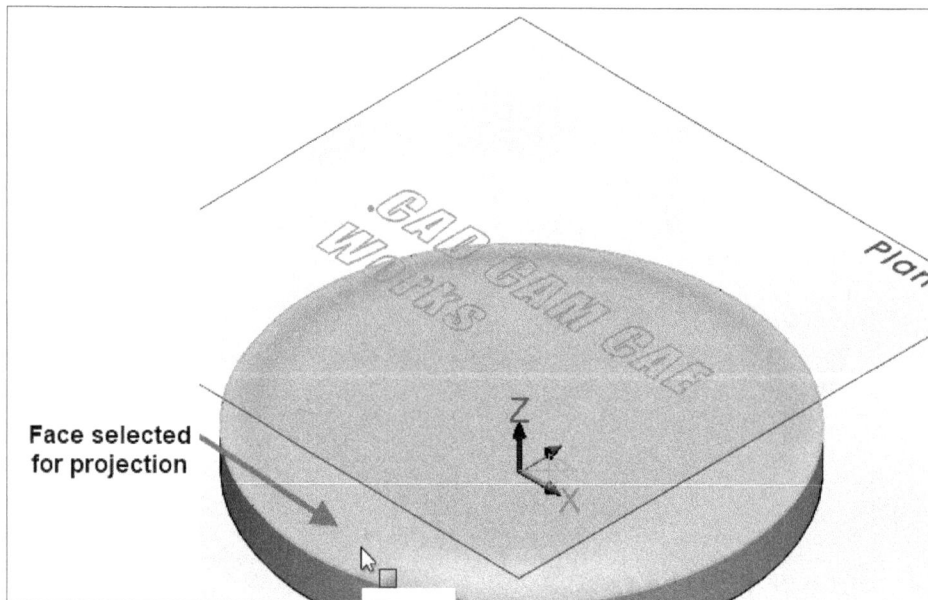

Figure-29. Face selected for project toolpath

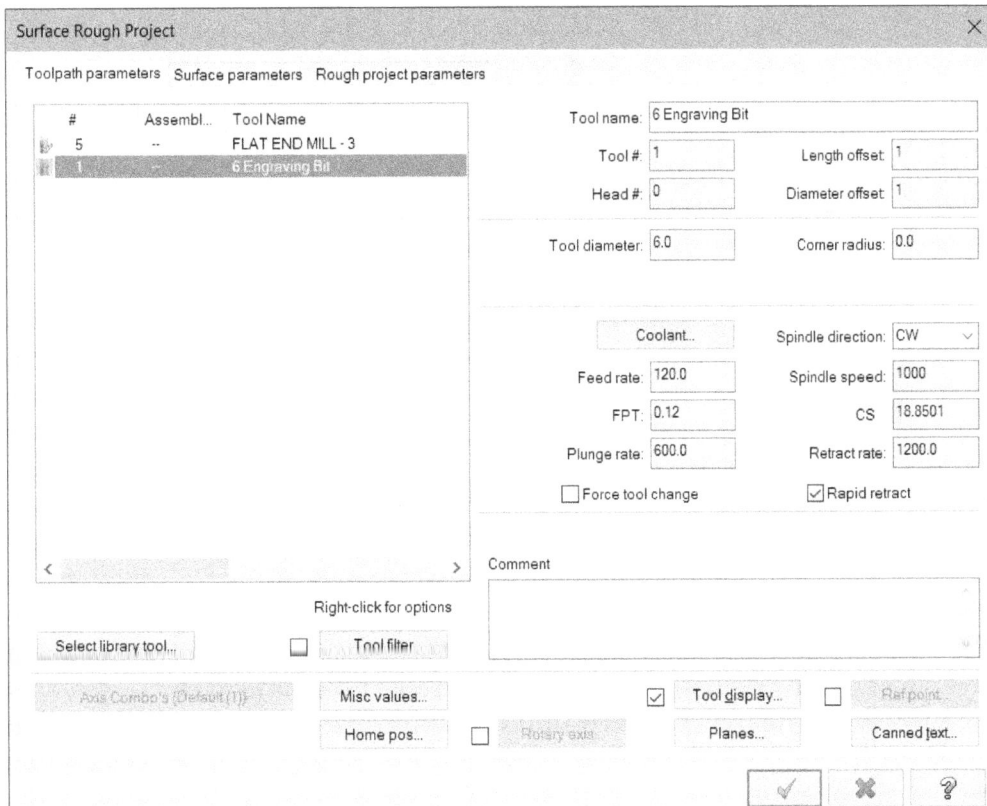

Figure-30. Surface Rough Project dialog box

- Select the desired tool and specify the desired parameters in various tabs of this dialog box. Click on the **OK** button. The **Chain Manager** will be displayed asking you to select the curves/text to be projected.
- Select the geometry or text to be projected; refer to Figure-31 and click on the **OK** button from the **Chain Manager**. The toolpath will be generated; refer to Figure-32.

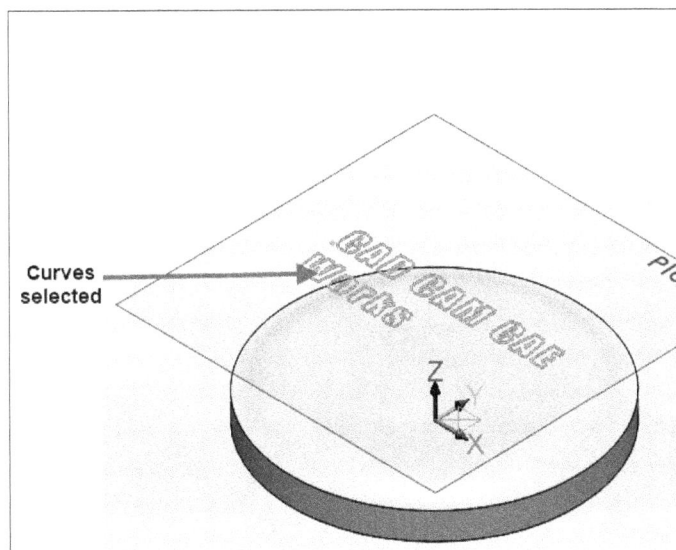

Figure-31. Sketch selected for projection

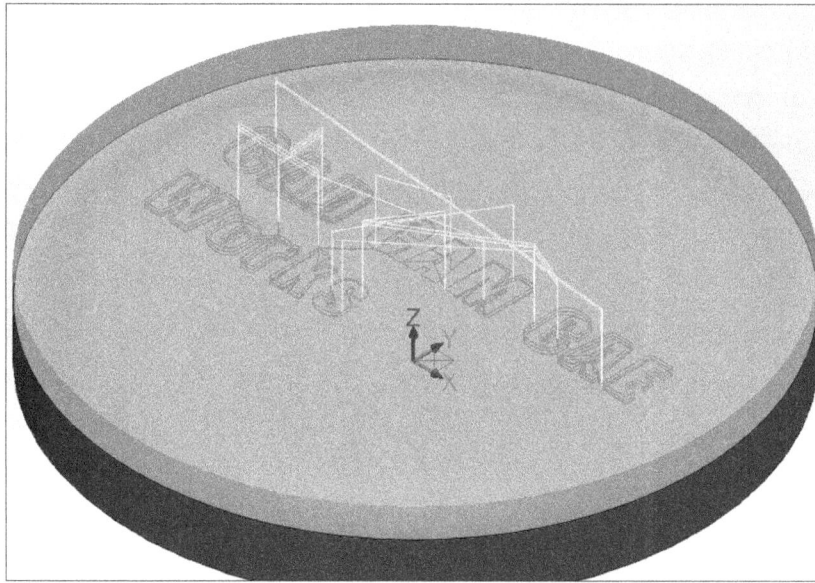

Figure-32. Toolpath generated for project rough

Flowline Rough Surface Toolpath

The Flowline Rough Surface toolpath is used when we need to remove material from channels and runner like structures of mold dies. The procedure to use this tool is given next.

- Click on the **Flowline** tool from the **Roughing** drop-down in the **Ribbon**. The **Select Boss/Cavity** dialog box will be displayed.
- Select the desired radio button and click on the **OK** button from the dialog box. The **Selection PropertyManager** will be displayed.
- Select the faces of the runners/channels in the part; refer to Figure-33 and click on the **OK** button. The **Surface Rough Flowline** dialog box will be displayed; refer to Figure-34.

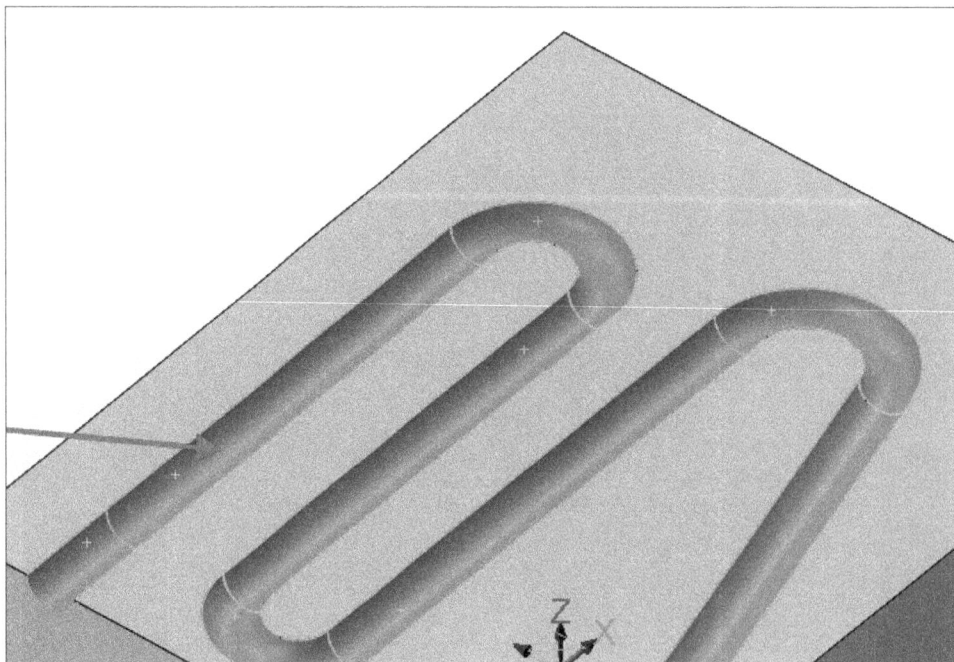

Figure-33. Faces selected for flowline toolpath

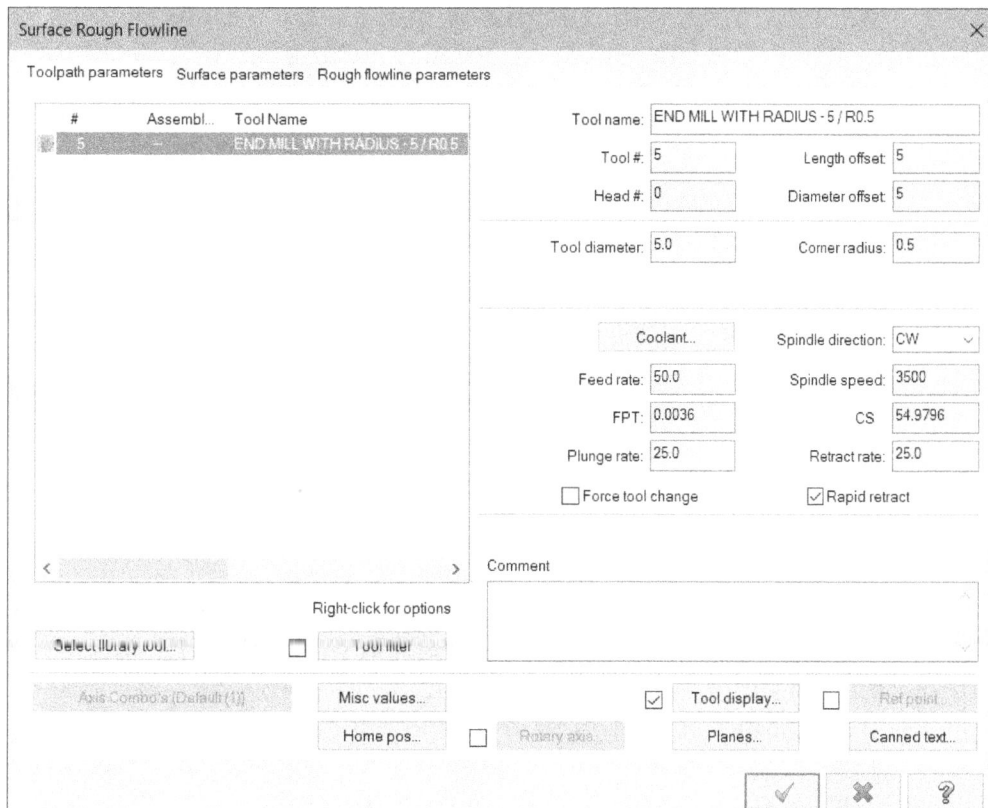

Figure-34. Surface Rough Flowline dialog box

- Select desired tool and specify related parameters in the **Toolpath parameters** and **Surface parameters** tab of the dialog box. Note that generally ball end mill or bull end mill cutter is used for this type of toolpath.

- Click on the **Rough flowline parameters** tab. The options in the dialog box will be displayed as shown in Figure-35.

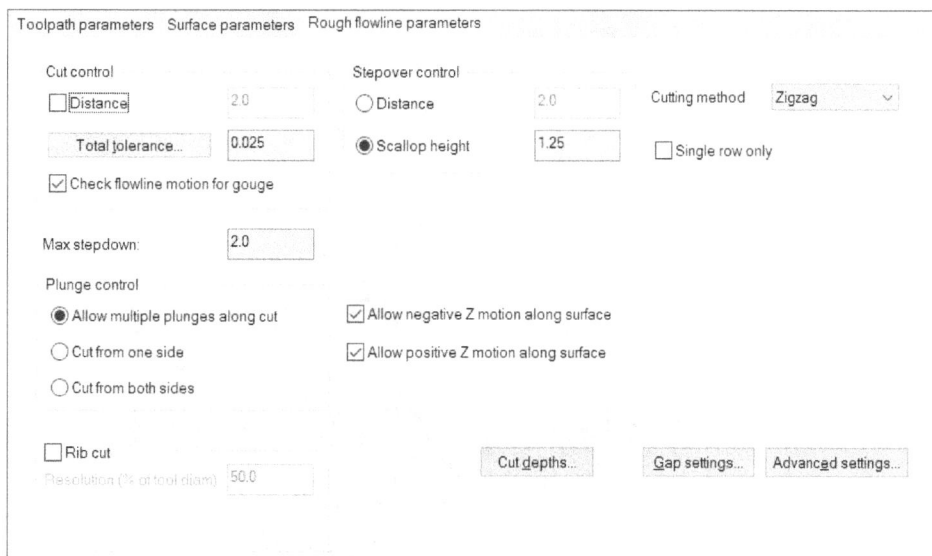

Figure-35. Rough flowline parameters tab

- Specify the desired tolerance or absolute cut length in the **Cut control** area. This tolerance is range for each toolpath step within which the tool can deviate while cutting.

- Specify the other parameters in this dialog box as discussed earlier and click on the **OK** button. The **Flowline options PropertyManager** will be displayed; refer to Figure-36.

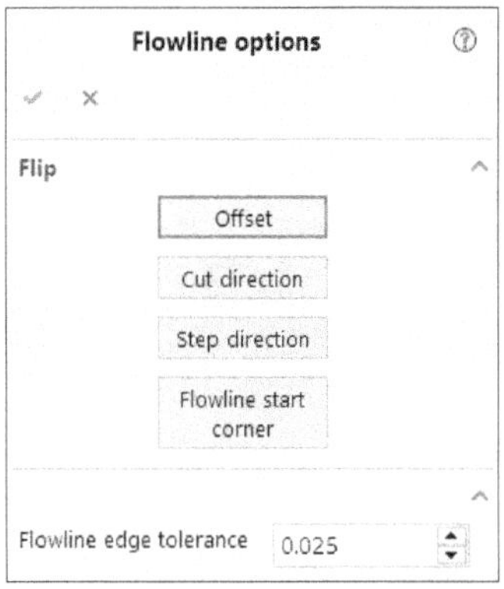

Figure-36. Flowline options PropertyManager

- Click on the **Offset** button from the **Flip** rollout of the **PropertyManager** to offset in/offset out the toolpath with respect to part boundaries. In this way, you can specify tool compensation left or right.
- Click on the **Cut direction** button from the **PropertyManager** to change the tool's cutting direction.
- Similarly, set the step direction and flowline start corner by using the respective buttons in the **PropertyManager**.
- After setting desired parameters, click on the **OK** button from the **PropertyManager**. The toolpath will be generated; refer to Figure-37.

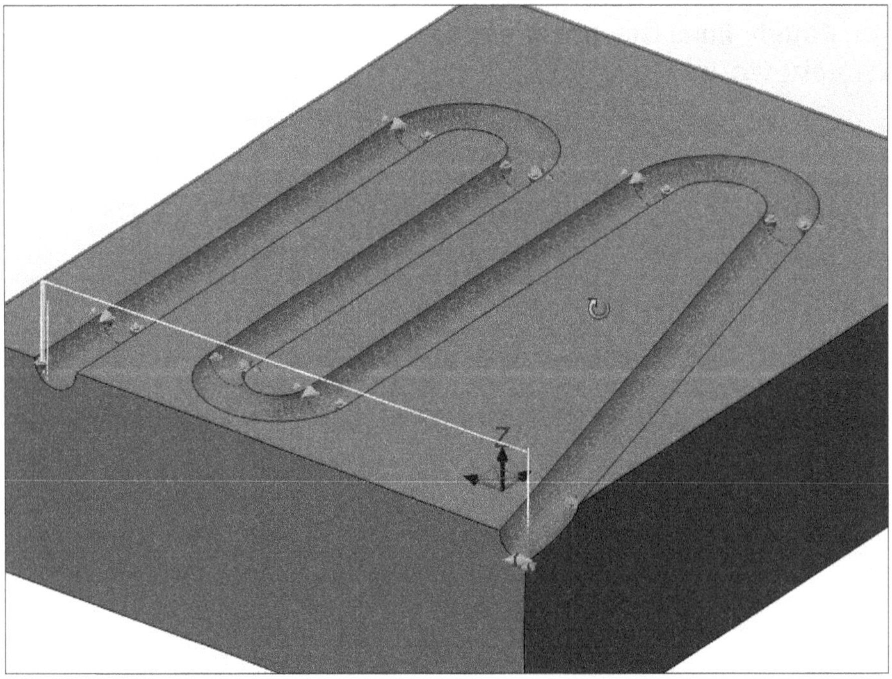

Figure-37. Surface rough flowline toolpath

Contour Rough Surface Toolpath

The Contour Rough Surface toolpath is used to machine steep walls of the part. This toolpath allows the tool to move downward gradually along Z axis. The procedure to use this tool is given next.

- Click on the **Contour** tool from the **Rough Toolpaths** drop-down in the **Ribbon**. The **Selection PropertyManager** will be displayed.
- Select the faces of the contour to be machined; refer to Figure-38 and click on the **OK** button. The **Surface Rough Contour** dialog box will be displayed; refer to Figure-39.

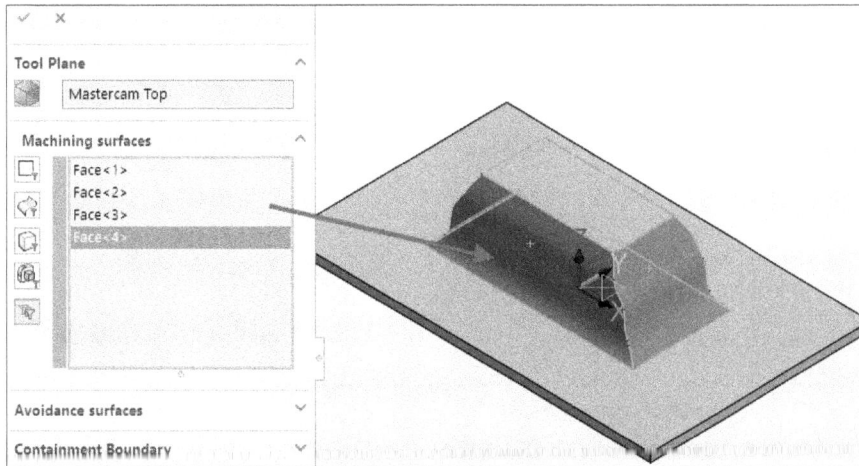

Figure-38. Faces selected for contour surface toolpath

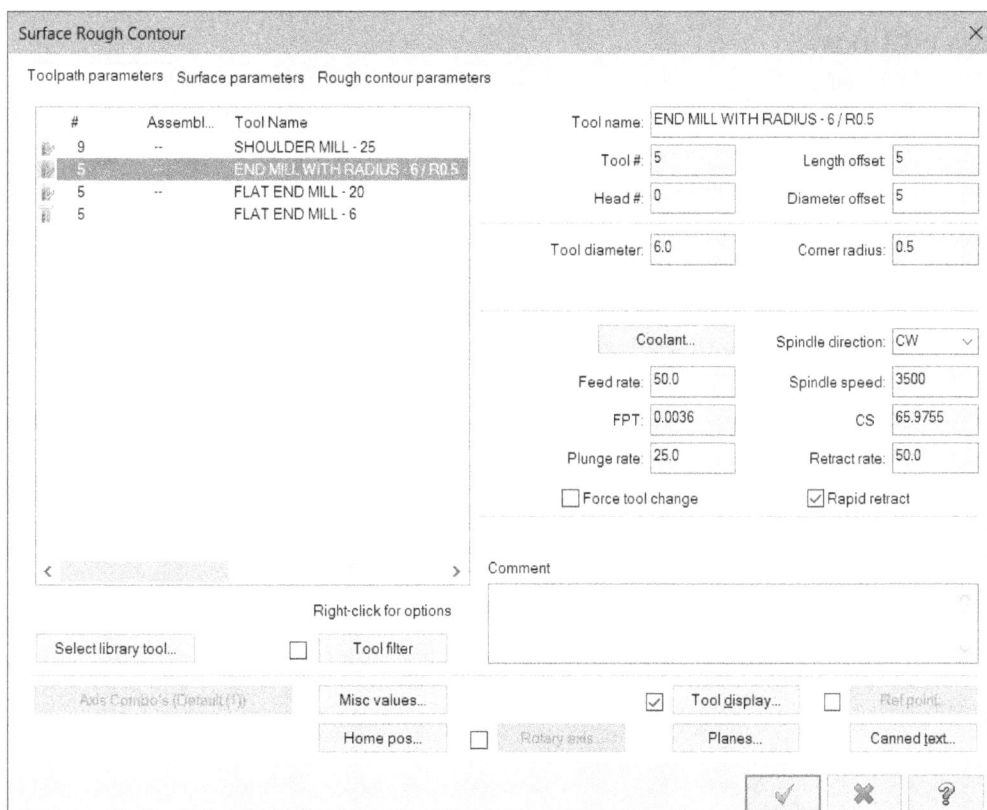

Figure-39. Surface Rough Contour dialog box

- Select the desired tool (generally ball mill cutter) and specify desired parameters in the **Tool parameters** and **Surface parameters** tabs of the dialog box.
- Click on the **Rough contour parameters** tab in the dialog box. The options will be displayed as shown in Figure-40.

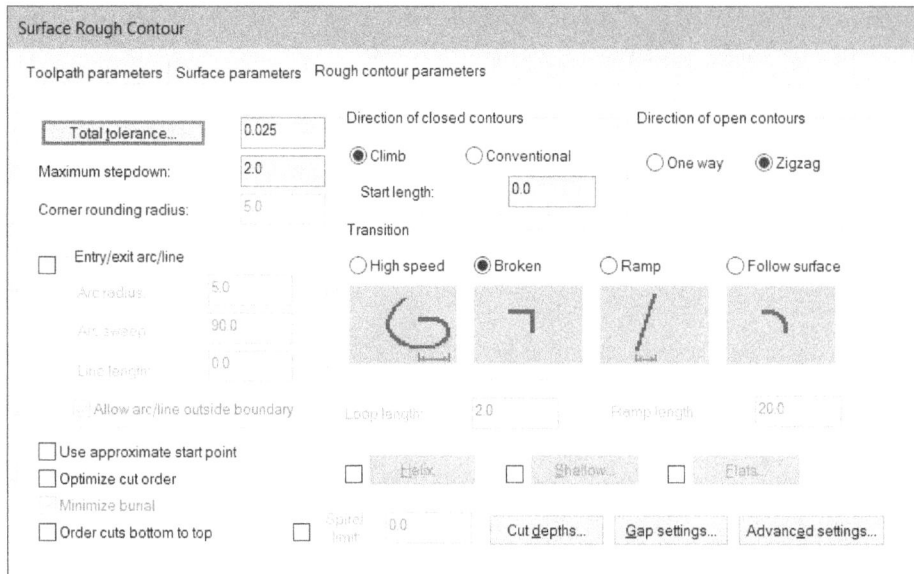

Figure-40. Rough contour parameters tab

- Specify total tolerance to define maximum allowed deviation from the toolpath in the **Total tolerances** edit box.
- Specify maximum allowed step-down value for toolpath levels in the **Maximum stepdown** edit box.
- Select the **Entry/exit arc/line** check box if you want to specify the arc radius, arc sweep and line length parameters for entry and exit locations of the toolpath. After selecting check box specify the desired parameters in the **Arc radius**, **Arc sweep**, and **Line length** edit boxes. If you want to the toolpath to be created outside the boundaries then select the **Allow arc/line outside boundary** check box.
- If you want to use start point selected earlier by using the **Selection PropertyManager** then select the **Use approximate start point** check box.
- Select the **Optimize cut order** and **Minimize burial** check boxes to increase tool life if required.
- Select the desired radio button from the **Transition** area and similarly, specify the other parameters.
- Click on the **OK** button. The toolpath will be generated; refer to Figure-41.

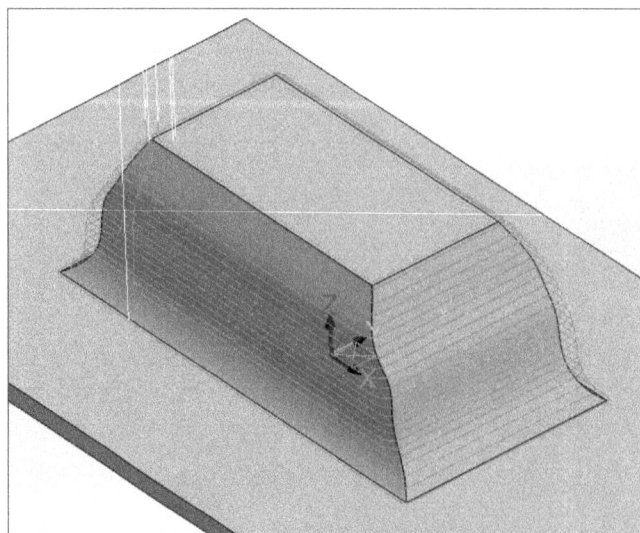

Figure-41. Contour rough surface toolpath generated

Restmill Surface Rough Toolpath

The Restmill Surface Rough toolpath is used to remove material left by other toolpaths. The procedure to create this toolpath is given next.

- Click on the **Restmill** tool from the **Roughing** drop-down in the **Ribbon**. The **Selection PropertyManager** will be displayed.
- Select all the faces on which you want to perform restmill operation. Note that you can select faces that were earlier selected for contour toolpath, flowline toolpath, and other toolpaths of surface roughing.
- After selecting faces, click on the **OK** button from the **PropertyManager**. The **Surface Restmill** dialog box will be displayed; refer to Figure-42.

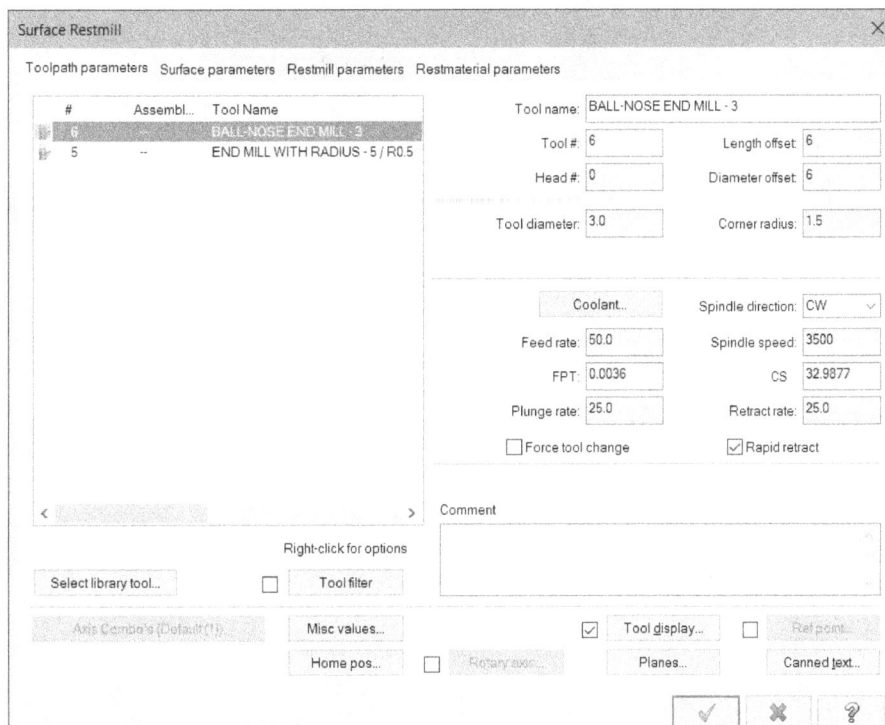

Figure-42. Surface Restmill dialog box

- Specify the desired parameters in various tabs of the dialog box and click on the **OK** button. The toolpath will be generated.

Pocket Surface Rough Toolpath

You have worked on pocket toolpaths earlier also but in those cases, there was no surface deviation. Now, we will work with pockets that have irregular surfaces. The procedure to create pocket surface rough toolpath is given next.

- Click on the **Pocket** tool from the **Roughing** drop-down in the **Ribbon**. The **Selection PropertyManager** will be displayed.
- Select the pocket surface and containment boundary; refer to Figure-43. Click on the **OK** button from the **PropertyManager**. The **Surface Rough Pocket** dialog box will be displayed; refer to Figure-44.

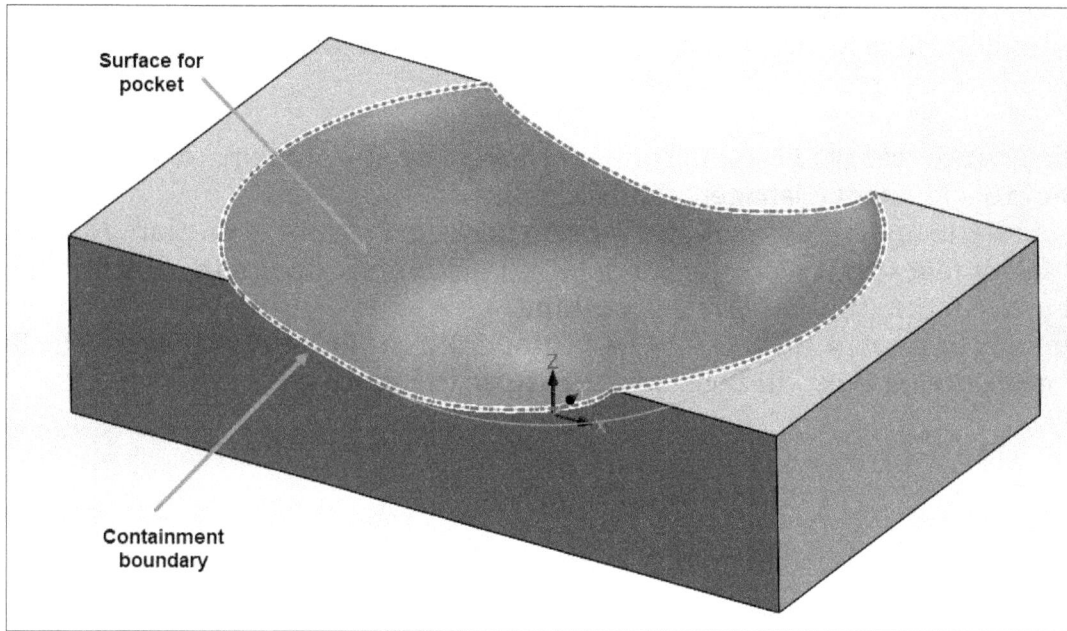

Figure–43. Selection for pocket surface rough toolpath

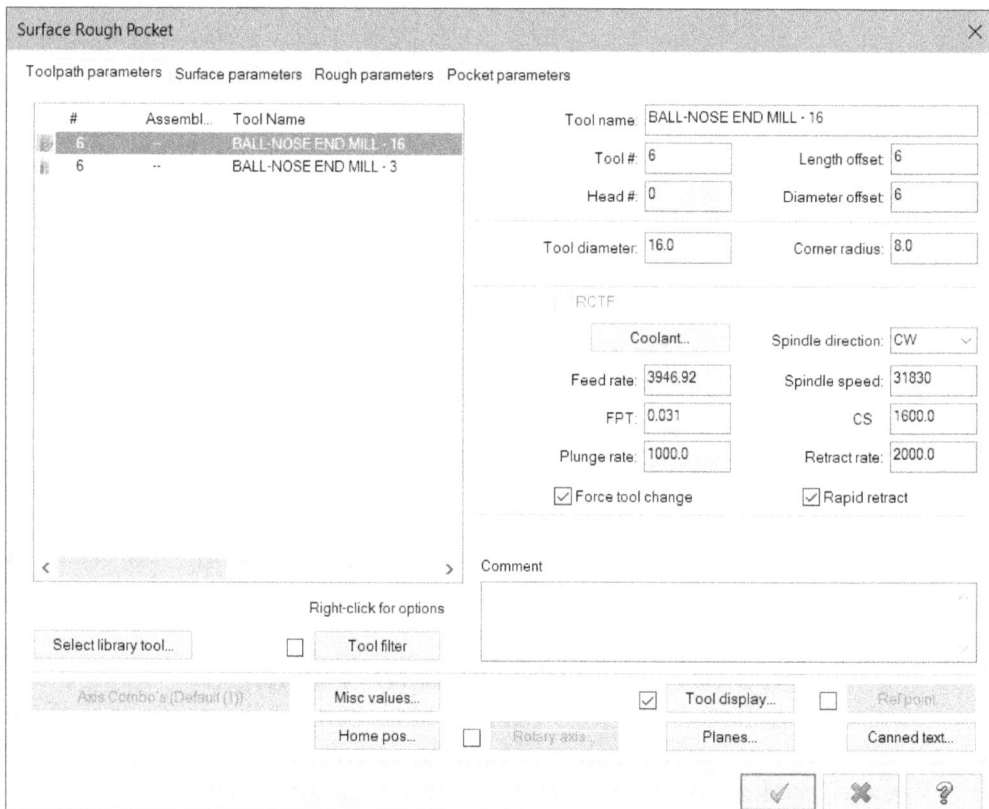

Figure–44. Surface Rough Pocket dialog box

- Specify the desired parameters in **Toolpath parameters** and **Surface parameters** tabs as discussed earlier.
- Click on the **Rough parameters** tab. The options in the dialog box will be displayed as shown in Figure-45.

- From the **Entry options** area, select the desired entry method for the tool. If there is a requirement of facing before pocket toolpath then select the check box before **Facing** button and click on the **Facing** button to specify parameters.
- Click on the **Pocket parameters** tab in the dialog box and select the desired cutting method; refer to Figure-46.

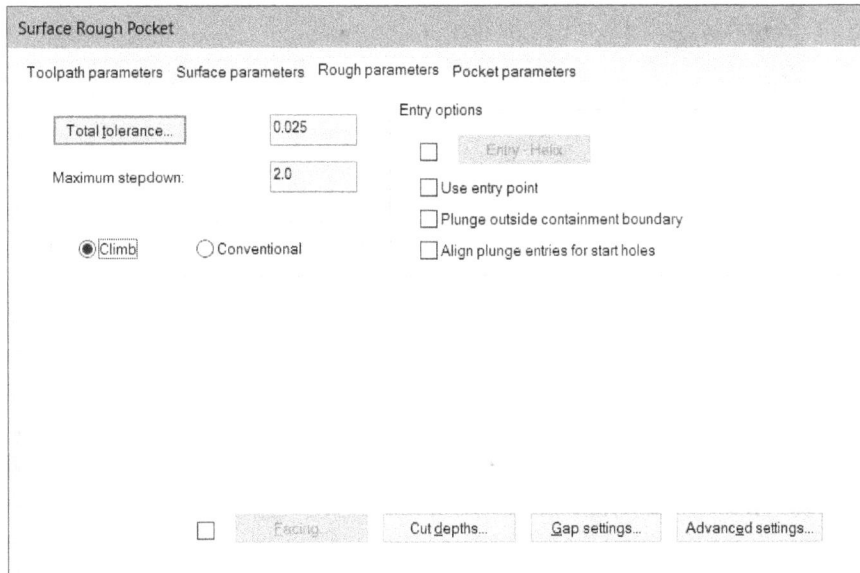

Figure-45. Options in Rough parameters tab

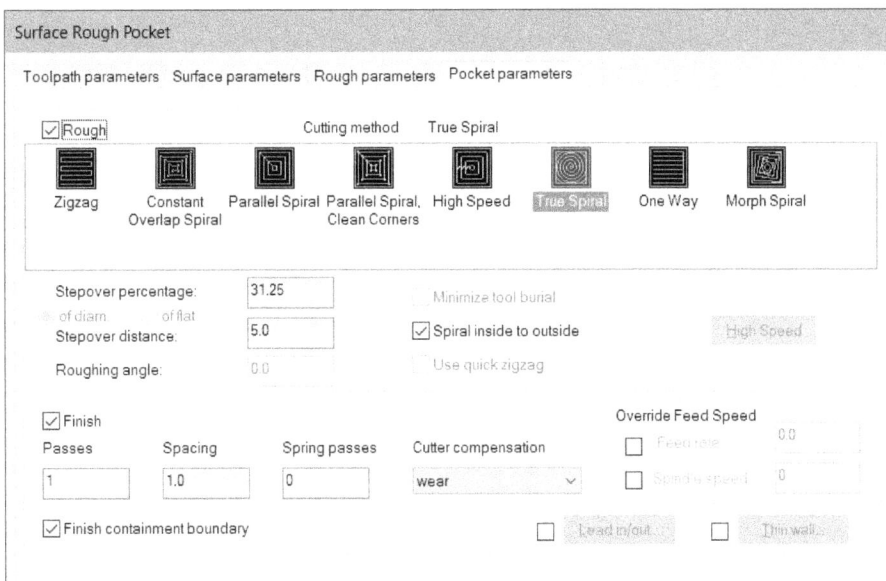

Figure-46. Pocket parameters tab

- Specify the other parameters as required and click on the **OK** button. The toolpath will be generated; refer to Figure-47.

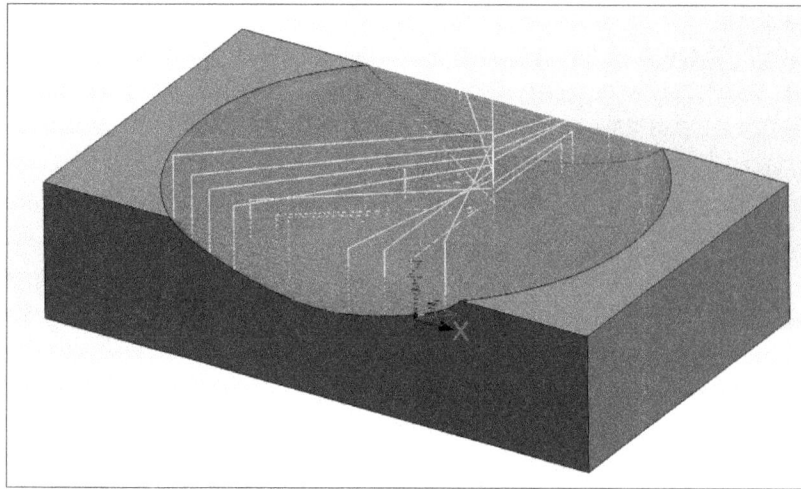

Figure-47. Pocket Surface Rough Toolpath

Plunge Rough Surface Toolpath

The Plunge Rough Surface toolpath works in the same way as 2D Plunge toolpath but this toolpath also considers the curvature of walls and base while cutting. Note that plunge toolpath can be beneficial if you have unstable machine, you need to machine deep slots, or you need to machine corners with round. In general, plunge milling is an alternate method when side milling is not possible due to vibrations. The procedure to use this tool is given next.

• Click on the **Plunge** tool from the **Roughing** drop-down in the **Ribbon**. The **Selection PropertyManager** will be displayed.
• Select the faces/bodies/features to be machined and points to define grid plane for plunging; refer to Figure-48. Click on the **OK** button from the **PropertyManager**. The **Surface Rough Plunge** dialog box will be displayed; refer to Figure-49.

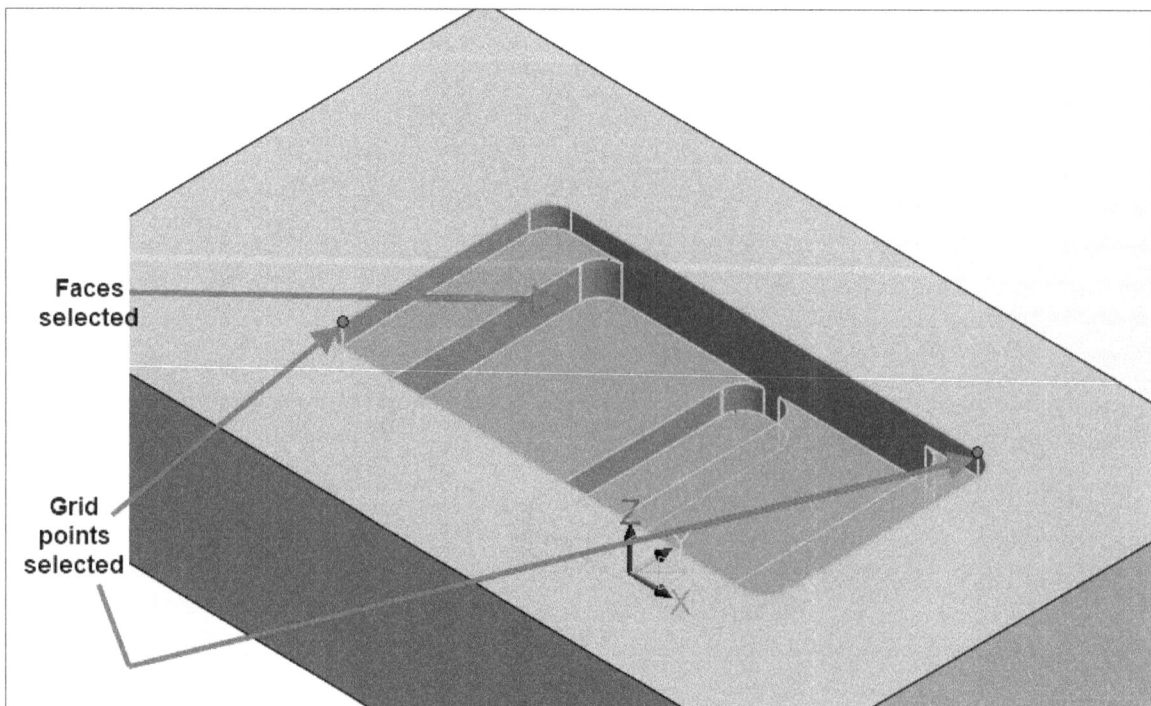

Figure-48. Features and points selected for plungemill rough

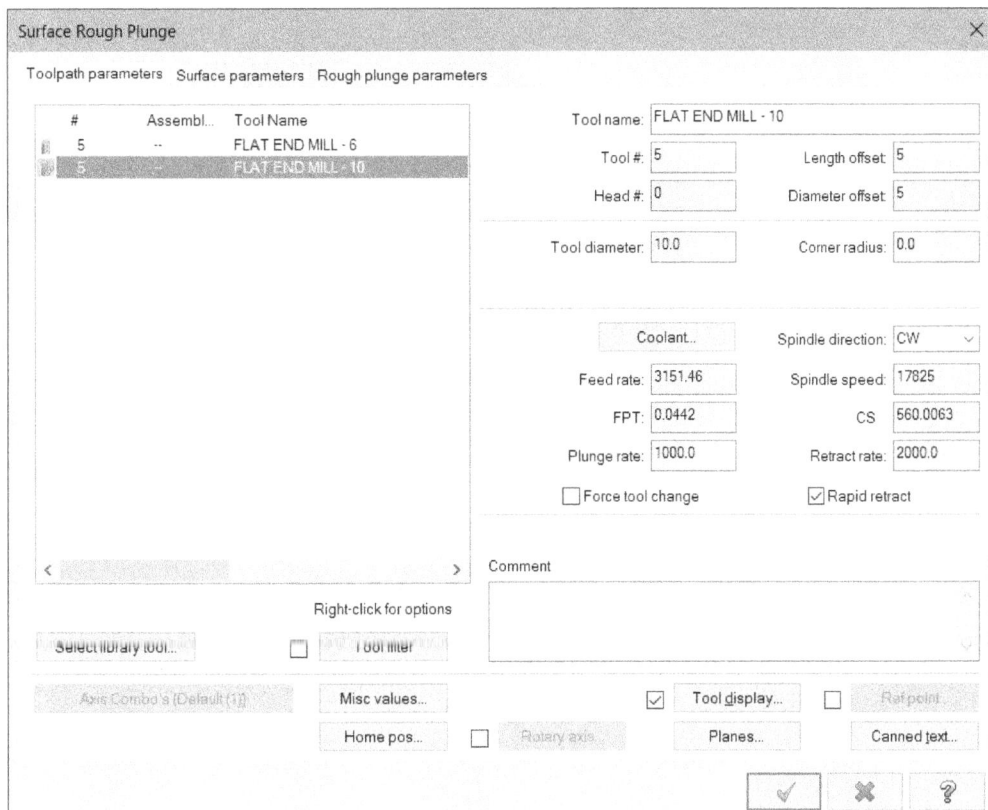

Figure-49. Surface Rough Plunge dialog box

- Specify the parameters as discussed earlier. Make sure you side a higher diameter tool selected.
- Specify the **Max stepdown** and **Maximum stepover** values in **Rough plunge parameters** tab carefully as per the tool selected; refer to Figure-50. A higher Max stepdown can increase load on tool and a higher step-over can leave the material unmachined giving you many holes as output.

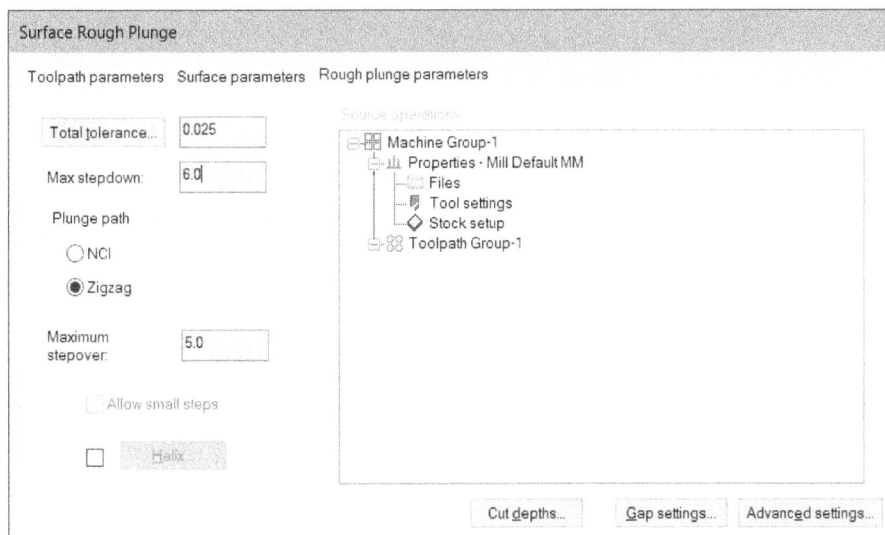

Figure-50. Rough plunge parameters tab

- After specifying parameters, click on the **OK** button. The toolpath will be generated; refer to Figure-51.

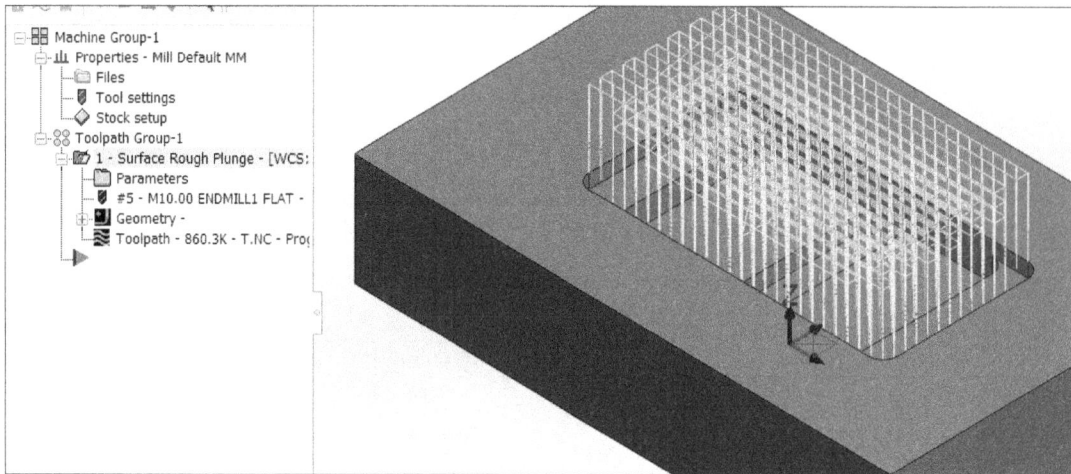

Figure-51. Plunge Surface Rough toolpath generated

SURFACE FINISHING

After performing surface roughing operations, we need toolpaths to perform finishing to remove material left over by Roughing. These tools are available in the **Finishing** drop-down of the **Ribbon**; refer to Figure-52.

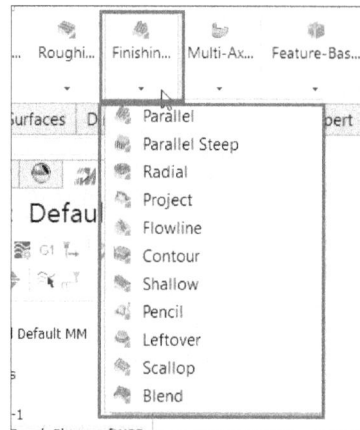

Figure-52. Finishing drop-down

Most of the tools in this drop-down are same as discussed for Roughing. Note that after performing roughing, you need a finer tool running at relatively higher speed to remove material in Finishing. Now, we will discuss the toolpaths that have not been discussed in case of surface Roughing.

Parallel Steep Surface Finish Toolpath

The Parallel Steep Surface Finish toolpath is used to remove material left by roughing and parallel surface Finishing over surfaces falling between two slope angles. The procedure to use this toolpath is given next.

- Click on the **Parallel Steep** tool from the **Finishing** drop-down in the **Ribbon**. The **Selection PropertyManager** will be displayed.
- Select the steep faces of the model which you want to finish and boundary sketch; refer to Figure-53 and click on the **OK** button. The **Surface Finish Parallel Steep** dialog box will be displayed; refer to Figure-54.

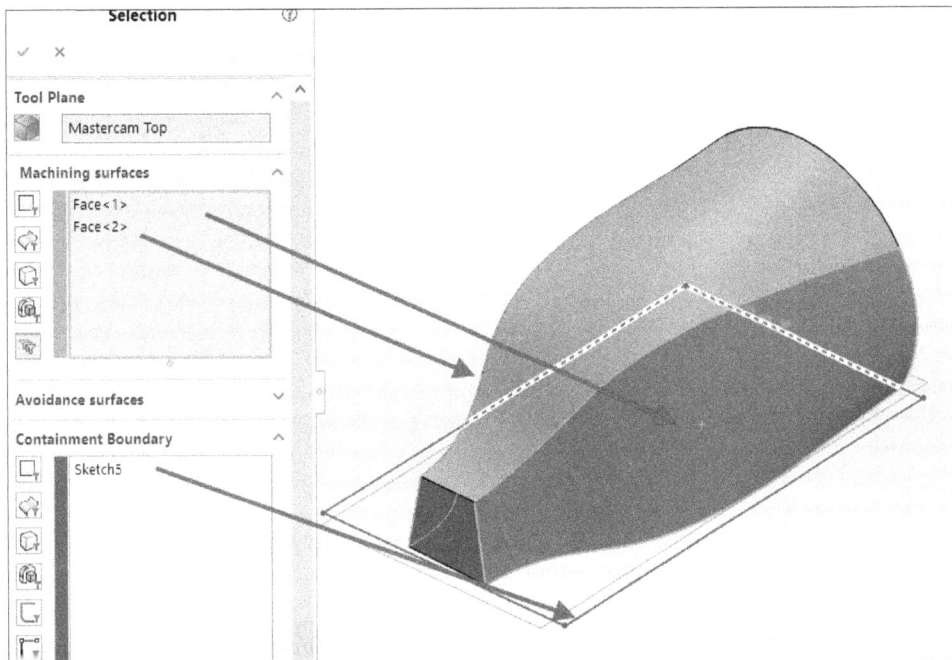

Figure-53. Faces selected for parallel steep finish toolpath

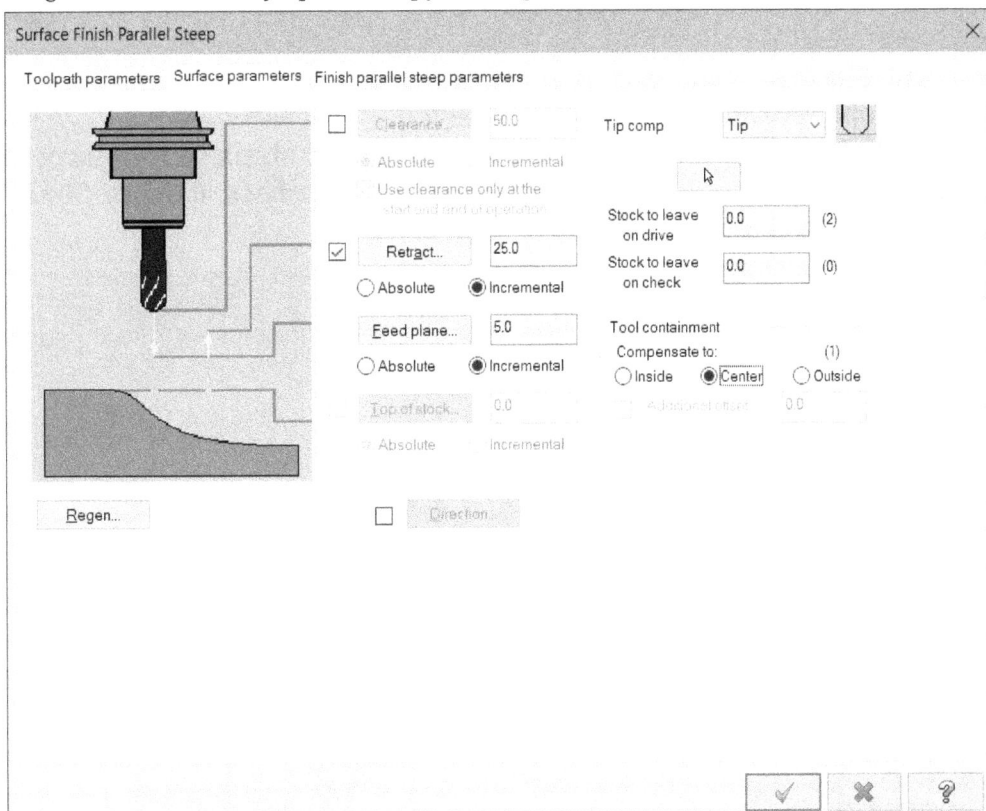

Figure-54. Surface Finish Parallel Steep dialog box

- Specify the parameters as discussed earlier and click on the **OK** button. The toolpath will be generated; refer to Figure-55. Note you need to roughing operations to remove large stock of material before using this toolpath.

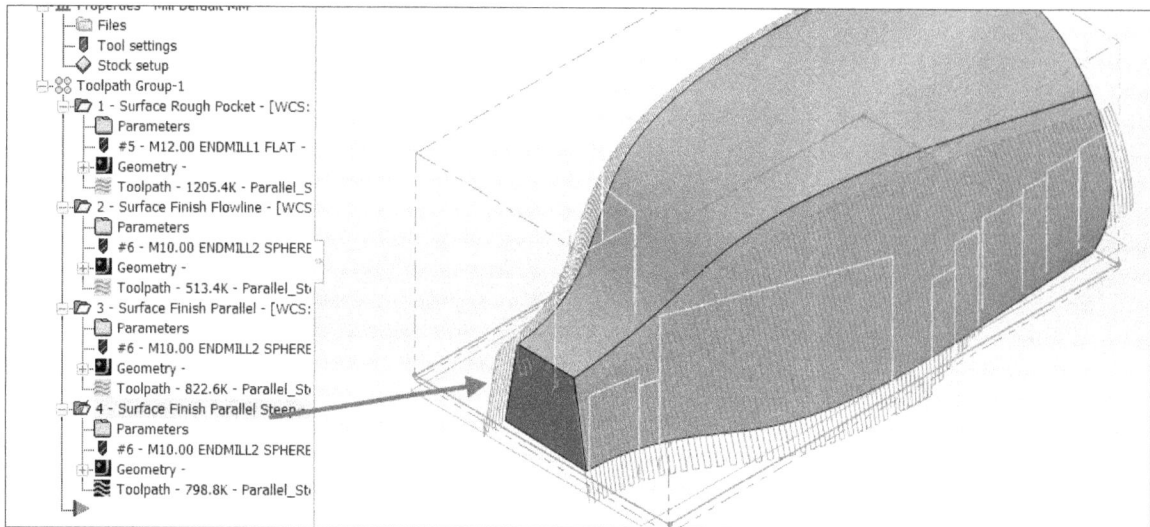

Figure-55. Parallel Steep Surface finish toolpath generated

Shallow Surface Finish Toolpath

The Shallow Surface Finish toolpath is used to finish material from the shallow areas based on specified angle range. The procedure to create this toolpath is given next.

- Click on the **Shallow** tool from the **Finishing** drop-down in the **Ribbon**. The **Selection PropertyManager** will be displayed.
- Select the geometry on which you want to perform shallow finishing operation; refer to Figure-56 and click on the **OK** button. The **Surface Finish Shallow** dialog box will be displayed; refer to Figure-57.

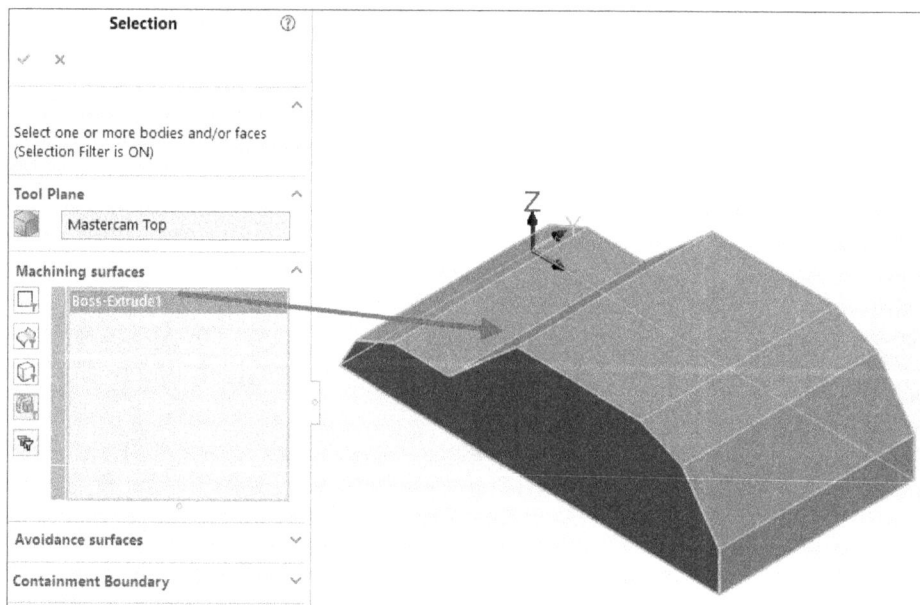

Figure-56. Feature selected for shallow finishing

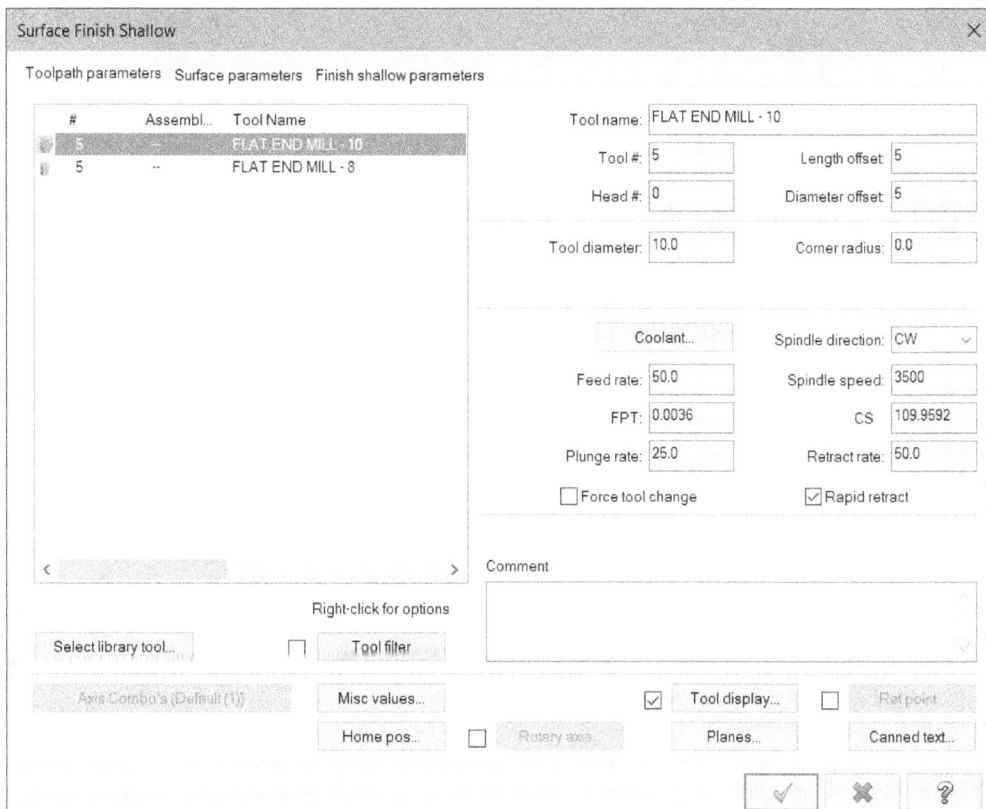

Figure-57. Surface Finish Shallow dialog box

- Specify the desired parameters in the **Toolpath parameters** and **Surface parameters** tabs as discussed earlier.
- Click on the **Finish shallow parameters** tab in the dialog box. The options in the dialog box will be displayed as shown in Figure-58.

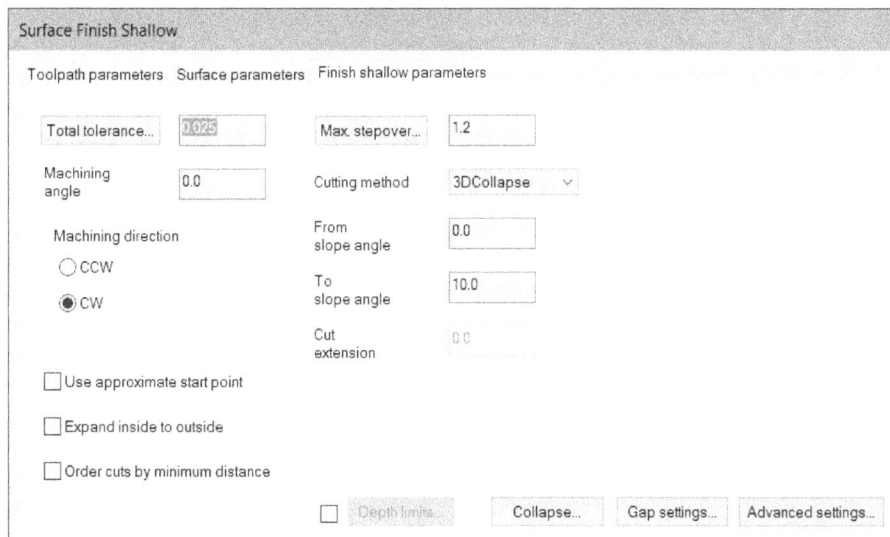

Figure-58. Finish shallow parameters tab

- Select the desired cutting method from the **Cutting method** drop-down in this tab. Select the **3DCollapse** option if you need to machine spiral cuts or there is gradual transition between the faces of parts. Select the **Zigzag** or **One way** options for all the other cases. Note that **Zigzag** is the fastest cutting method if your tooling and machine allows and **One way** is the slowest cutting method in most of the cases.
- Specify the desired angle values in the **From slope angle** and **To slope angle** edit boxes to define range of slope to be machined.

- Set the other parameters as required and click on the **OK** button. The toolpath will be generated; refer to Figure-59.

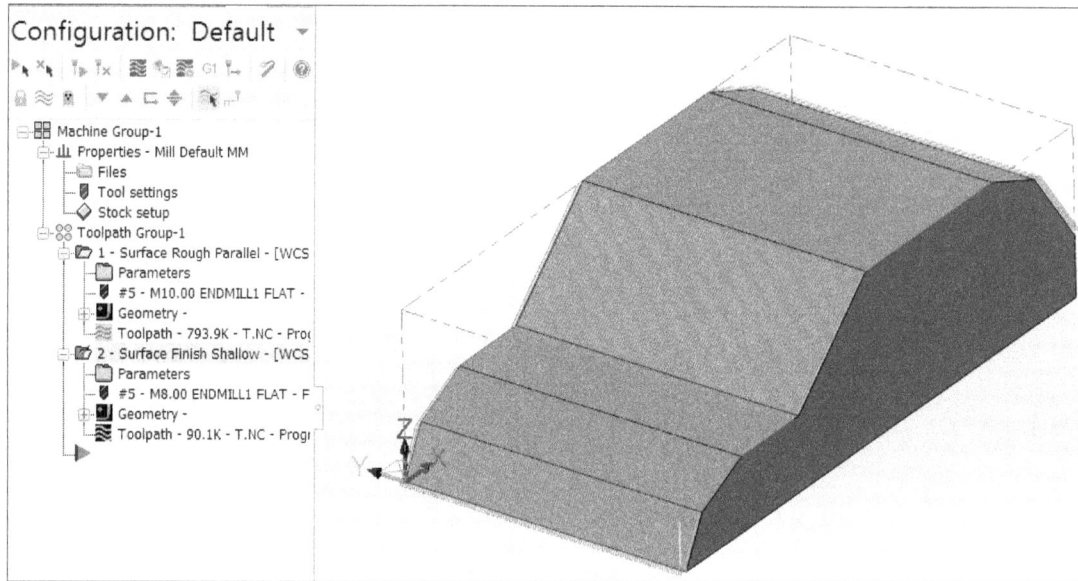

Figure-59. Shallow finish surface toolpath generated

Pencil Surface Finish Toolpath

The Pencil Surface Finish toolpath is used to remove material at the intersection of two surfaces in the part. In this toolpath, the tool moves tangent to the intersection curve of surfaces.

Leftover Surface Finish Toolpath

The Leftover Surface Finish toolpath works in the same way as Restmill toolpath in surface roughing. Use this tool to finish any material left over by other toolpaths.

Scallop Surface Finish Toolpath

The Scallop Surface Finish toolpath is used to round sharp corners in the part. The procedure to create scallop surface finish toolpath is given next.

- Click on the **Scallop** tool from the **Finishing** drop-down in the **Ribbon**. The **Selection PropertyManager** will be displayed and you will be asked to select the faces to be machined.
- Select the desired faces; refer to Figure-60 and click on the **OK** button. The **Surface Finish Constant Scallop** dialog box will be displayed; refer to Figure-61.

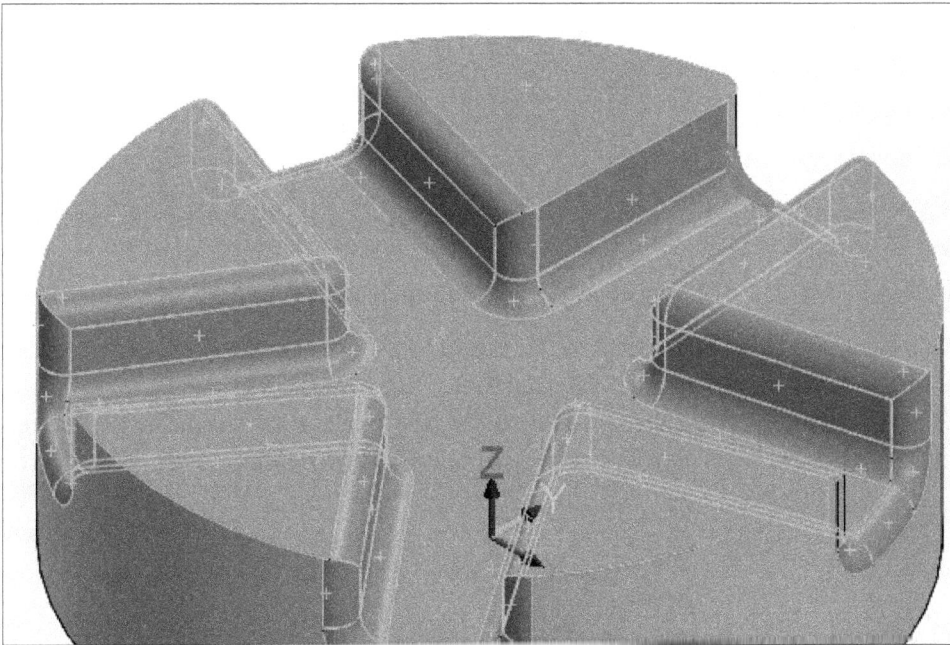

Figure-60. Faces selected for scallop toolpath

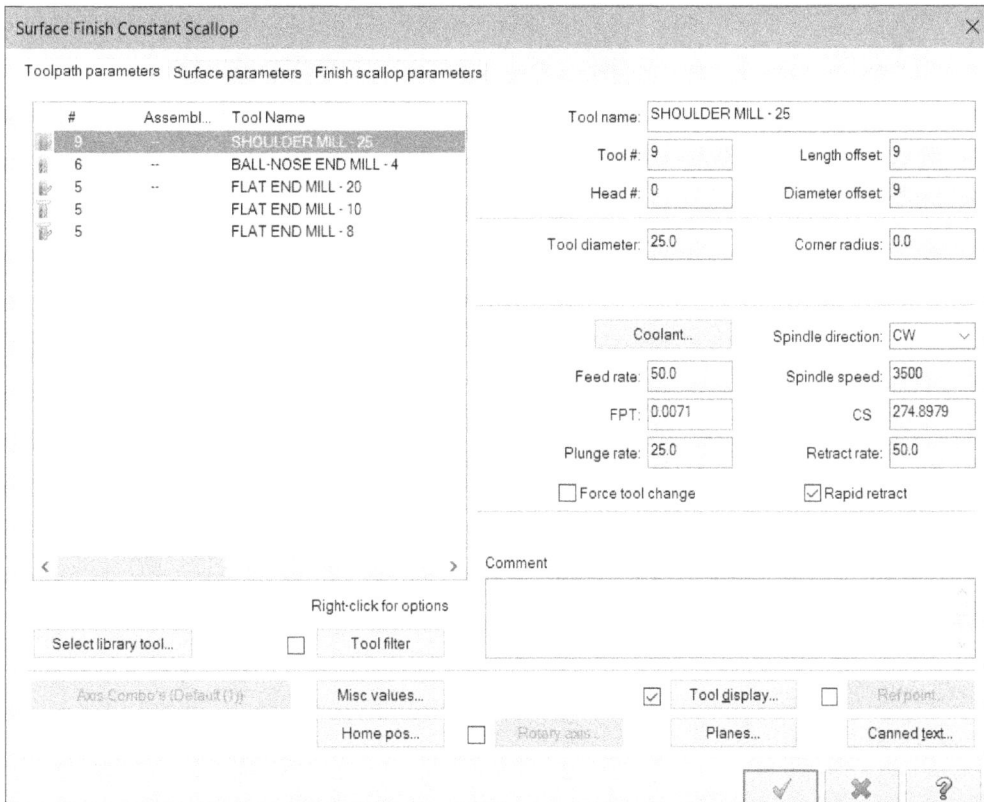

Figure-61. Surface Finish Constant Scallop dialog box

- Select the desired tool (generally a ball nose cutter is used) and specify desired parameters in the **Toolpath parameters** and **Surface parameters** tab as discussed earlier.
- Click on the **Finish scallop parameters** tab and click on the **Depth limits** button to specify the limit up to which tool can go inside the stock in single pass. The **Depth limits** dialog box will be displayed; refer to Figure-62.

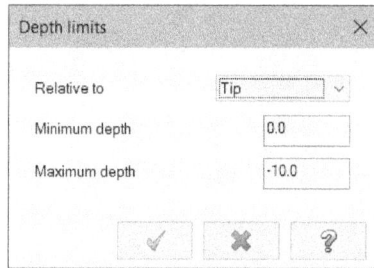

Figure-62. Depth limits dialog box

- Note that the depth is specified with negative sign in this dialog box. Specify the desired value and click on the **OK** button.
- Similarly, specify the maximum stepover and other parameters in the dialog box and click on the **OK** button. The toolpath will be created; refer to Figure-63.

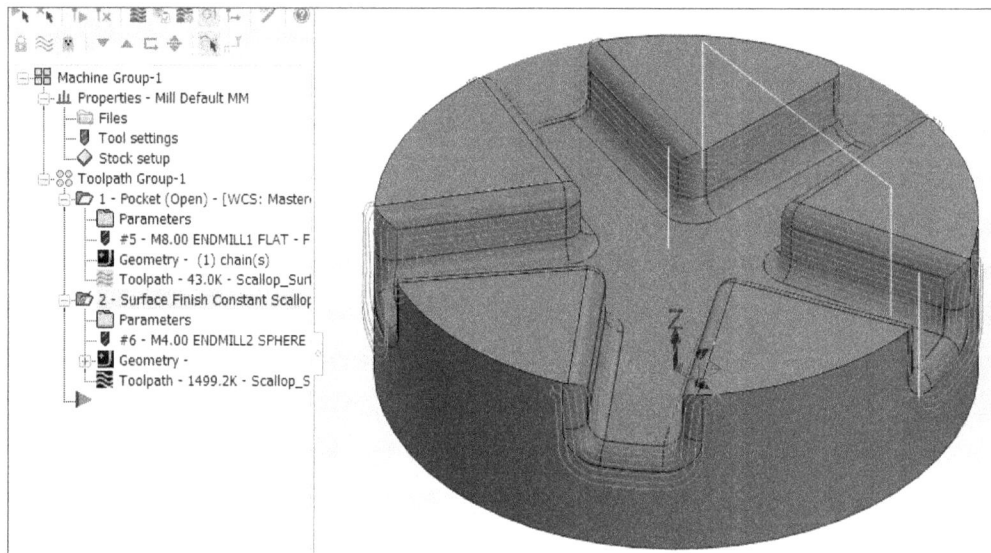

Figure-63. Surface finish scallop toolpath

Blend Surface Finish Toolpath

The Blend Surface Finish toolpath is used to remove material from the surface of part by forming a blend feature. Note that Blend is a solid/surface feature created by using boundaries of selected curves; refer to Figure-64. The procedure to create this toolpath is given next.

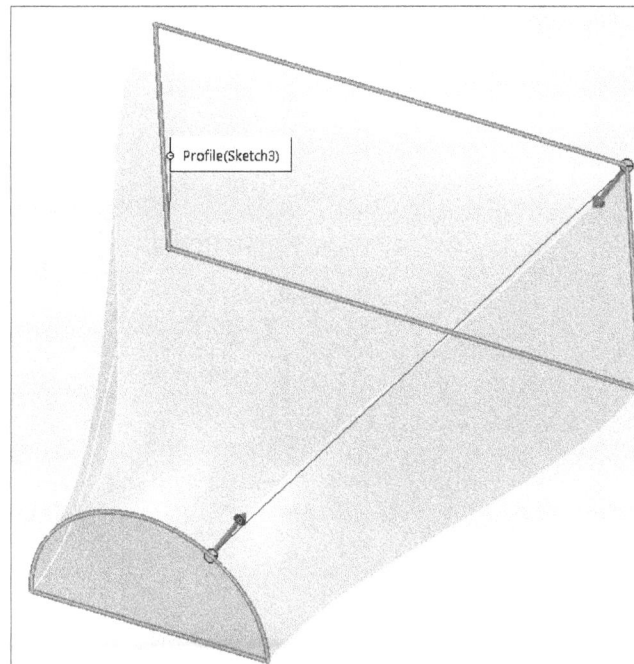

Figure-64. Blend feature example

- Click on the **Blend** tool from the **Finishing** drop-down in the **Ribbon**. The **Selection PropertyManager** will be displayed.
- Select the face of model to be machined; refer to Figure-65 and click on the **OK** button. The **Surface Finish Blend** dialog box will be displayed; refer to Figure-66.

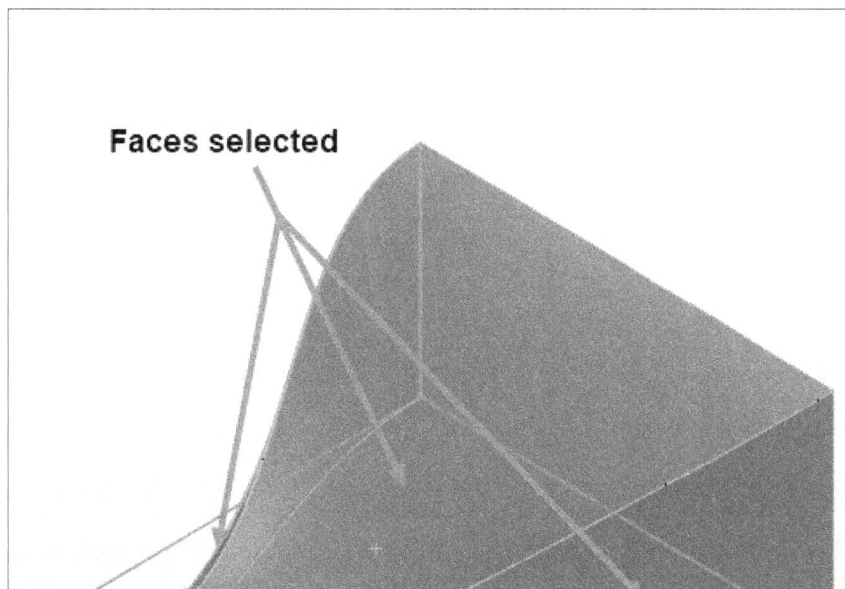

Figure-65. Faces selected for blend surface finish toolpath

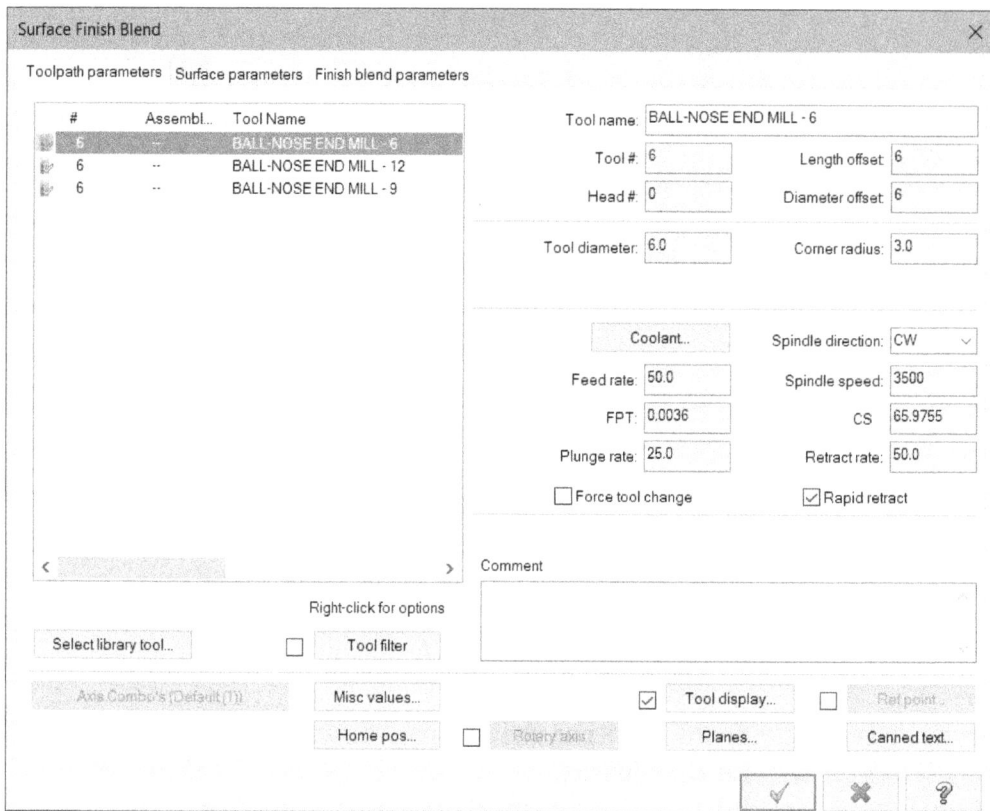

Figure-66. Surface Finish Blend dialog box

- Select the desired ball nose end mill and specify desired parameters in the **Toolpath parameters** and **Surface parameters** tabs in the dialog box.
- Click on the **Finish blend parameters** tab in the dialog box. The options in the dialog box will be displayed as shown in Figure-67.

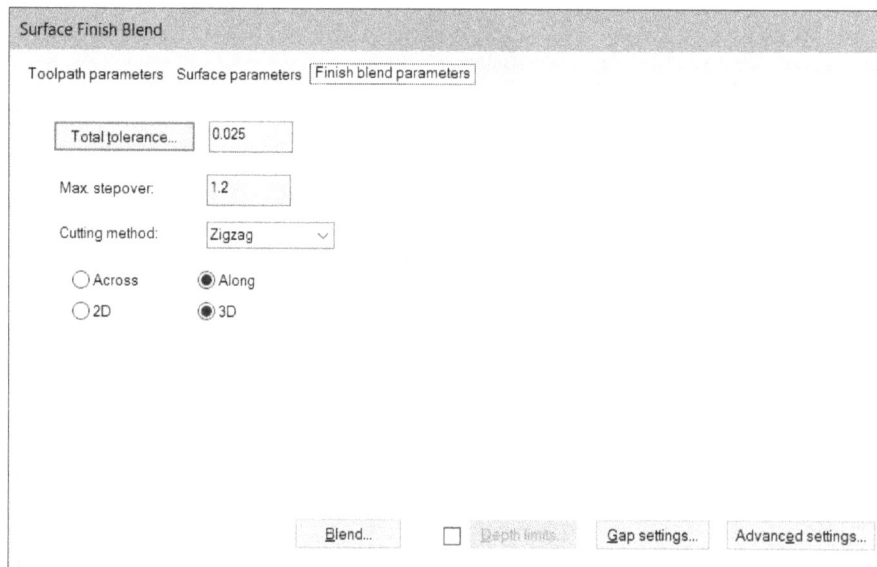

Figure-67. Finish blend parameters tab

- Select the desired radio button from the tab to define shape of blend toolpath.

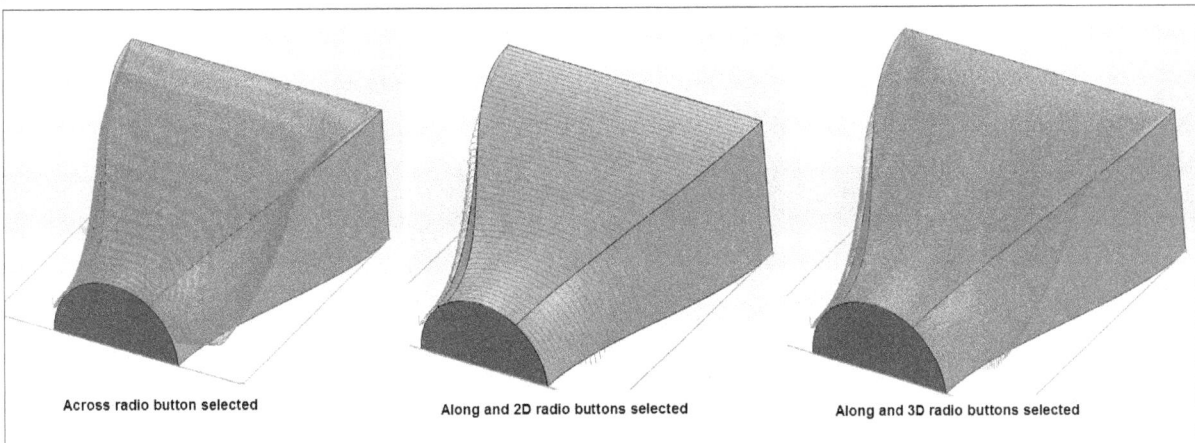

Figure-68. Selection of 2D 3D Across and Along radio button

- Specify the other parameters in this tab as required and then click on the **OK** button from the dialog box. The **Chain Manager** will be displayed.
- Select the two boundaries of the part that form blend feature; refer to Figure-69. Click on the **Chains** tab and verify that there are two chains with same directions. Click on the **OK** button from the **Chain Manager**. Refer to Figure-68 for different outputs of blend surface finish toolpaths.

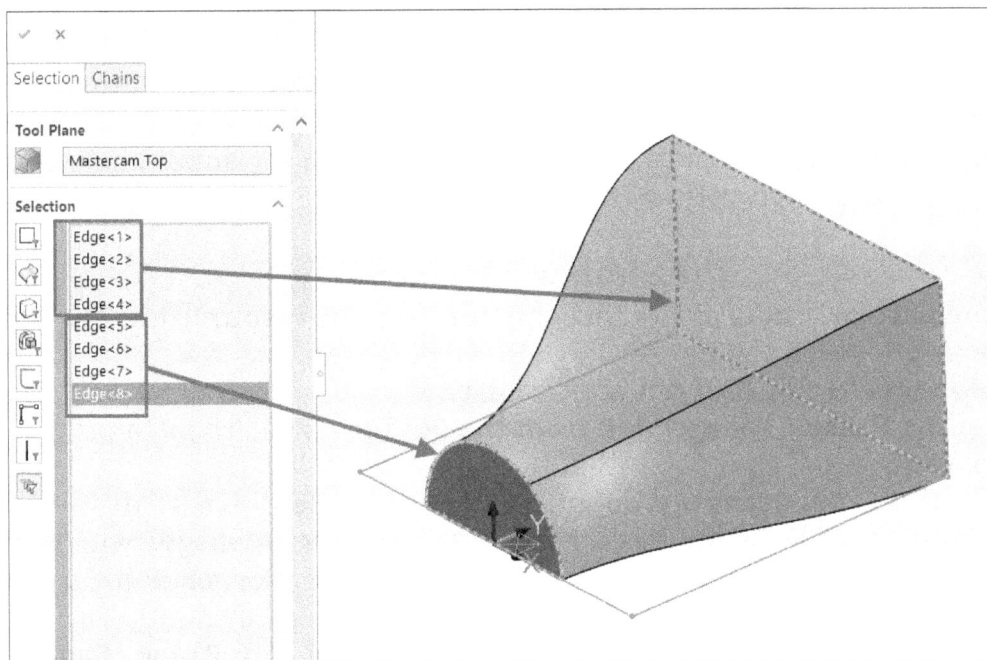

Figure-69. Edges selected to form blend feature

PRACTICAL

Create a machining program for part given in Figure-70. The stock of part is a block of size 170x120x42. The controller for machine is Haas. You can use default controller if Haas is not available.

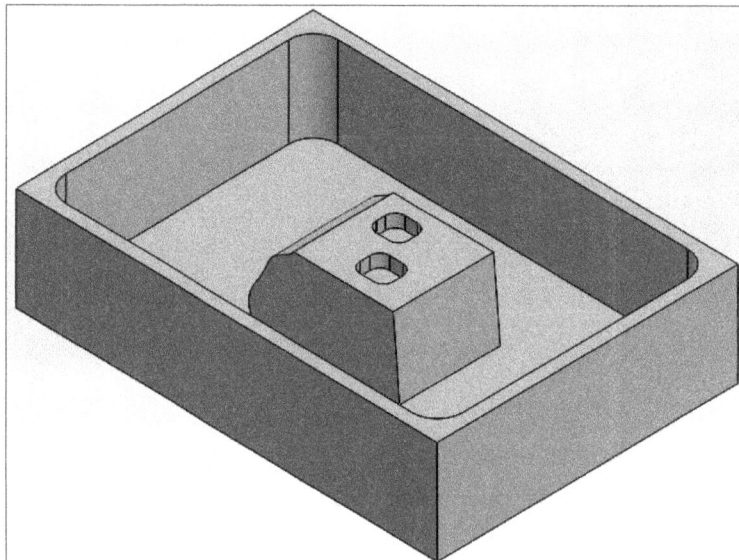

Figure-70. Part for Practical

Steps:

Identifying toolpath strategies

We need to perform facing operation to make surface of part even. Next, we need to perform pocket toolpath with bigger tool and then smaller tool for finishing. We also need to perform surface rough and surface finish parallel to machine boss feature in the middle of part. In the end, perform pocket finish for small pockets at the top of boss feature. Let begin with setting up the job.

Preparation for Machining

*   Open the part for practical 1 of this chapter from the resource kit. The part should display as shown in Figure-70.
*   Start Mastercam add-in if not set to start yet.
*   Click on the **Planes Manager** tab from the **Design Tree**. The **Planes Manager** will be displayed.
*   Click on the **From geometry** option in **Create a new plane** drop-down of the **Planes Manager**. The **Define Mastercam Plane PropertyManager** will be displayed.
*   Select the top face of the model as reference plane, corner point as origin and Top-Face as name of the plane; refer to Figure-71.
*   Click on the **OK** button from the **PropertyManager**. The **Plane Manager** will be displayed again.
*   Select the recently created Top-Face plane and click on the **Set all** button ⊟ from the toolbar in **Planes Manager**. The selected plane will be set as WCS and Tool plane.
*   Click on the **OK** button from the dialog box to apply settings.

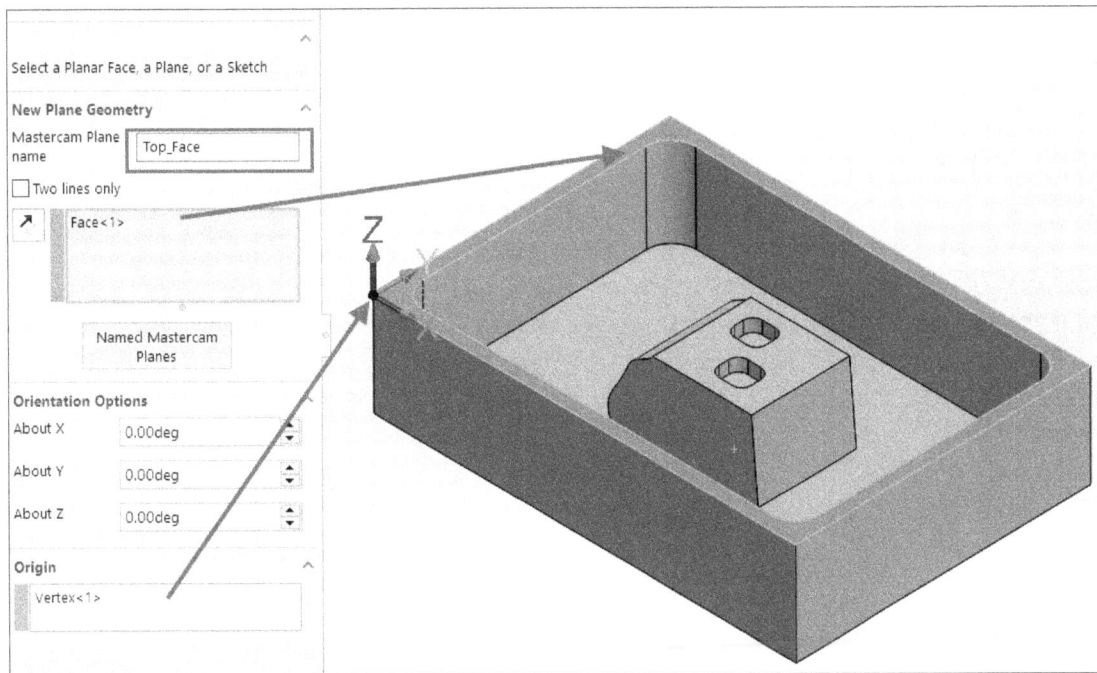

Figure-71. Defining mastercam plane

Setting Up Haas 3X Mill Machine

- Click on the **Files** option in the expanded **Properties** node of Machine group in the **Mastercam Toolpath Manager**; refer to Figure-72. The **Machine Group Properties** dialog box will be displayed; refer to Figure-73.

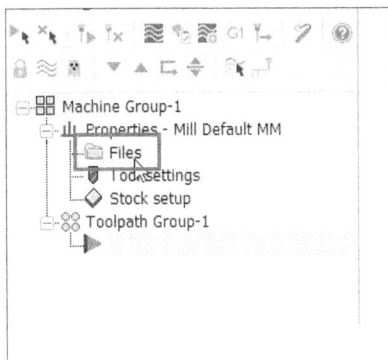

Figure-72. Files option in Mastercam Toolpath Manager

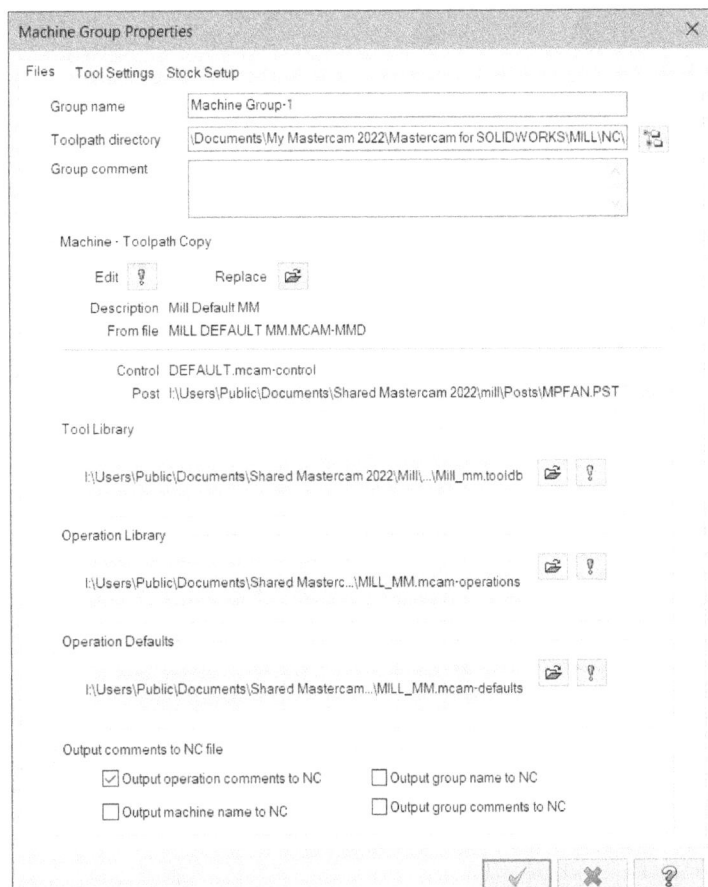

Figure-73. Files tab in Machine Group Properties dialog box

- Click on the **Replace** button ☞ in the dialog box. The **Open Machine Definition File** dialog box will be displayed.
- Select the **GENERIC HAAS 3X MILL MM.MCAD-MMD** file from the dialog box and click on the **Open** button to load Haas machine definition. The Haas 3X mill will be loaded with post processor and controller definitions.

Setting Tool Parameters and Stock

- Click on the **Tool Settings** tab in the dialog box and select the **Assign tool numbers sequentially** check box. Also, select the **Warn of duplicate tool numbers** check box so that a single tool number is not assigned to two different tools.
- Click on the **Stock Setup** tab in the dialog box. The stock setup options will be displayed.
- Click on the button in the **Stock Plane** area of the dialog box. The **Plane Selection** dialog box will be displayed; refer to Figure-74.

Figure-74. Plane Selection dialog box

- Select the newly created plane and click on the **OK** button from the dialog box. The **Machine Group Properties** dialog box will be displayed again.
- Select the **Rectangular** radio button from the **Shape** area of the dialog box and click on the **Bounding box** button. The **Bounding Box/Cylinder PropertyManager** will be displayed.
- Click on the **OK** button from the **PropertyManager** and select the **Solid** radio button.
- Edit the dimension of the rectangular box as shown in Figure-75 and click on the **OK** button from the dialog box to create the stock. The model with stock will be displayed as shown in Figure-76.

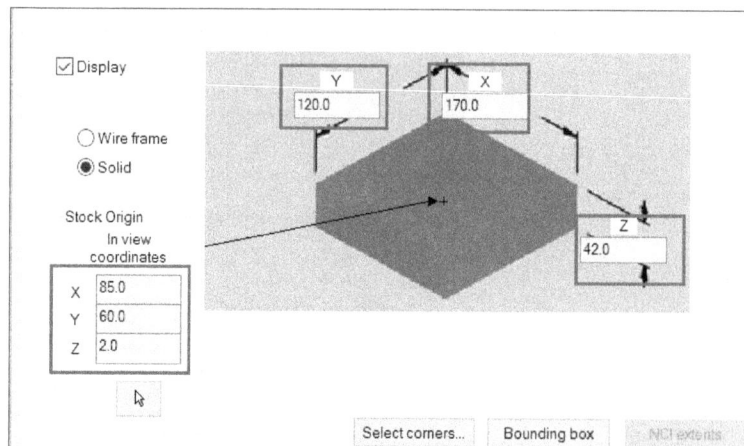

Figure-75. Values to set for stock

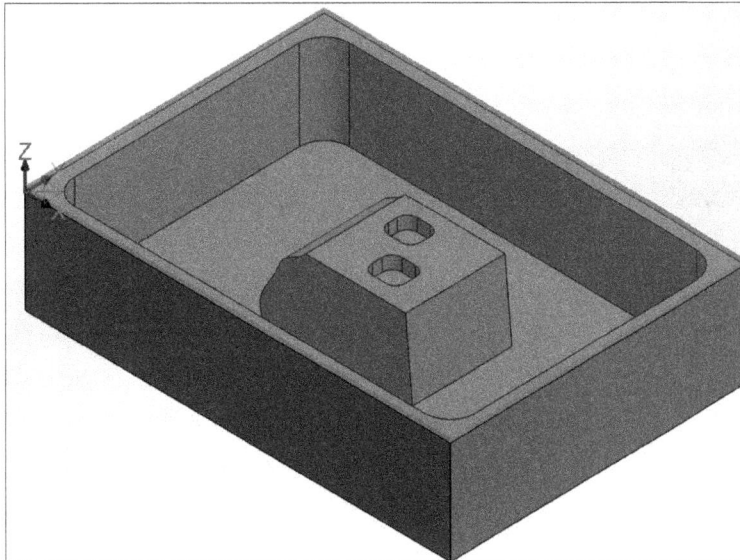

Figure-76. Stock created for practical 1

Facing

- Click on the **Facing** tool from the **2D** drop-down in the **Ribbon**. The **Chain Manager PropertyManager** will be displayed.
- Select the outer edges of the top face of model so that only one chain is formed; refer to Figure-77. Click on the **OK** button from the **Chain Manager**. The **2D-Facing** dialog box will be displayed.
- Click on the **Tool** option from the left area and then click on the **Select library** tool button from the **Tool** page in the dialog box. The **Tool Selection** dialog box will be displayed.
- Select the **FACE MILL-50/58** tool from the list; refer to Figure-78 and click on the **OK** button. The tool will get selected for operation.
- Click on the **Cut Parameters** option from the left area and set the cutting style as **Zigzag** in the **Style** drop-down of the page. Also, set the **Stock to leave on floors** as **0** in this page.
- Click on the **Linking Parameters** option from the left area and set the total cutting depth as absolute **0** in the **Depth** edit box of the page.
- Click on the **OK** button from the dialog box to create toolpath.

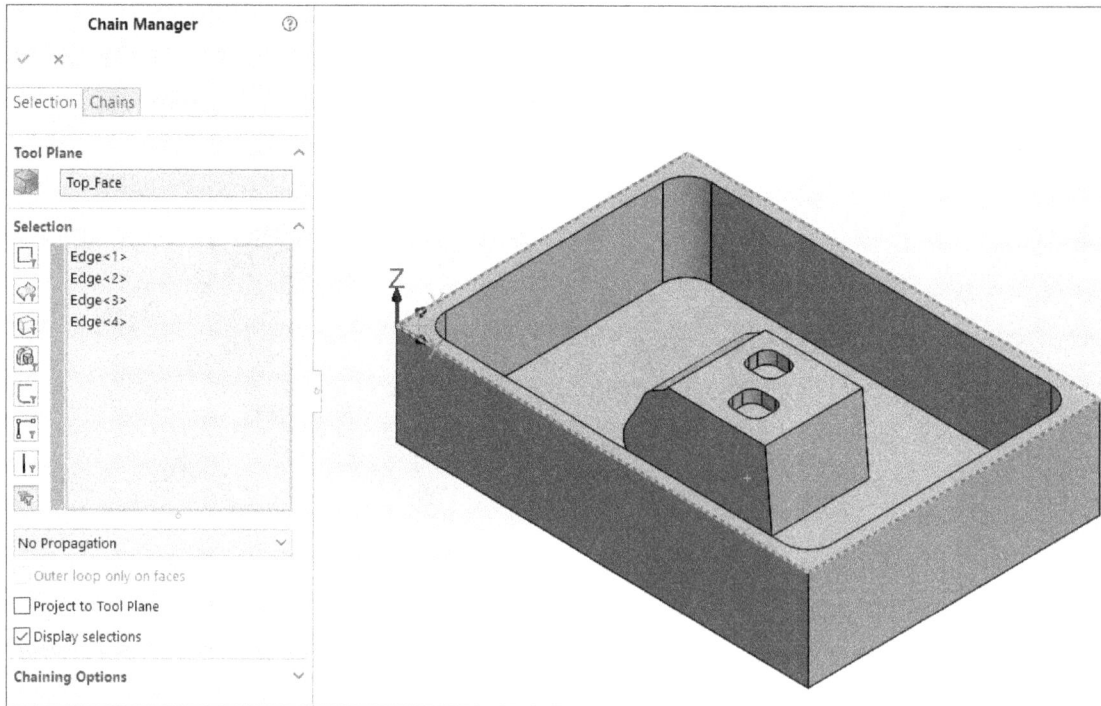

Figure-77. Edges selected for facing boundary

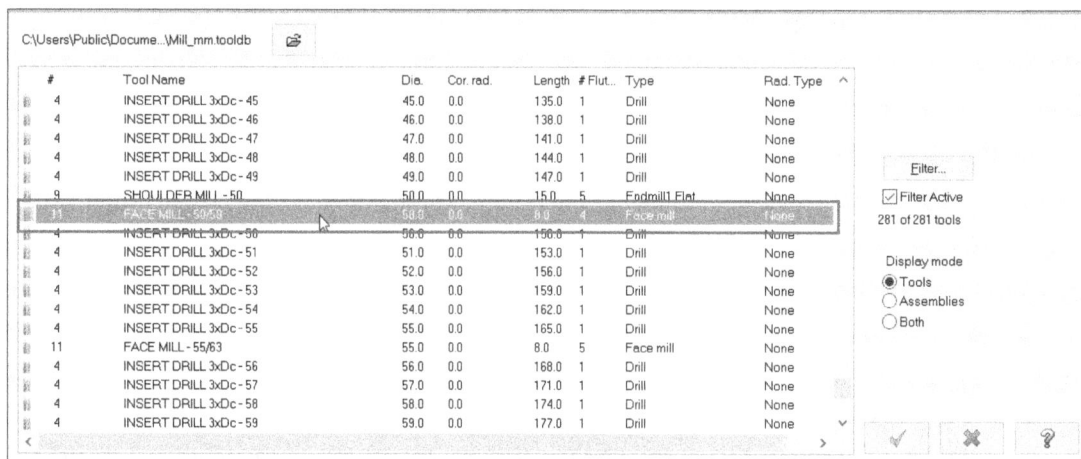

Figure-78. Face mill tool to be selected

2D Pocket Toolpath

2D Toolpaths take relatively less time when we need to remove large stock. So, we will now remove extra material of pocket from the stock.

- Click on the **Pocket** tool from the **2D** drop-down in the **Ribbon**. The **Chain Manager** will be displayed.
- Select the bottom face of the part to be machined; refer to Figure-79 and click on the **OK** button from the **Chain Manager**. The **2D-Pocket** dialog box will be displayed.
- Click on the **Tool** option from the left and select the Flat End Mill of **10** mm diameter as tool.
- In the **Cut Parameters** page, specify the stock to leave values as **0** in both **Stock to leave on walls** and **Stock to leave on floors** edit boxes.
- Click on the **Roughing** option from the left and select **Zigzag** as the cutting method from the respective page to smoothen cutting based on our model.

- Click on the **Depth Cuts** option from the left area to specify maximum depth limit for each cut and select the **Depth cuts** check box. The options in the page will become active.
- Set the **Max Rough Step** as **6** and select the **Keep tool down** check box to reduce machining time.

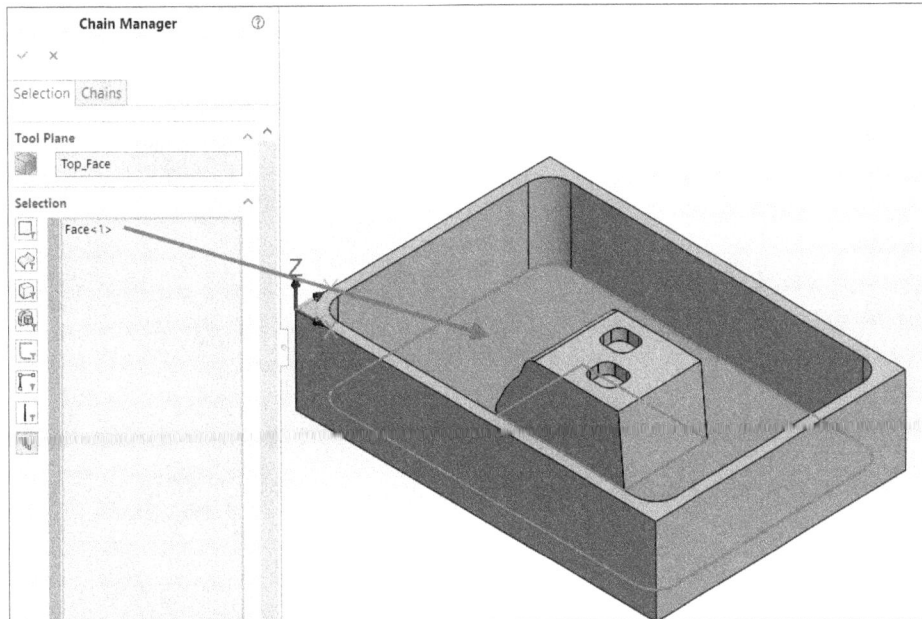

Figure-79. Face selected for 2D pocket

- Click on the **Linking Parameters** option from the left area to specify total cut depth, retraction and other parameters. Select the **Retract** value as incremental **35** and click on the **OK** button. The toolpath will be generated; refer to Figure-80. Note that specifying this retract value will make sure that the tool do not collide with stock during retraction.

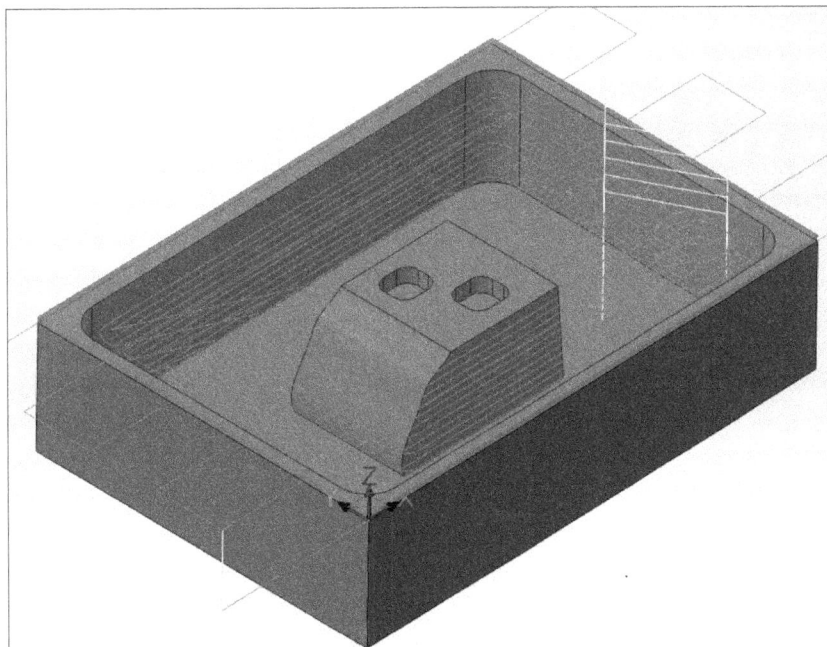

Figure-80. 2D Pocket toolpath created

Surface Rough Parallel Toolpath

- Click on the **Parallel** tool from the **Roughing** drop-down in the **Ribbon**. The **Selection PropertyManager** will be displayed.
- Select the faces/features of the model to be machined; refer to Figure-81 and click on the **OK** button from the **PropertyManager**. The **Surface Rough Parallel** dialog box will be displayed; refer to Figure-82.

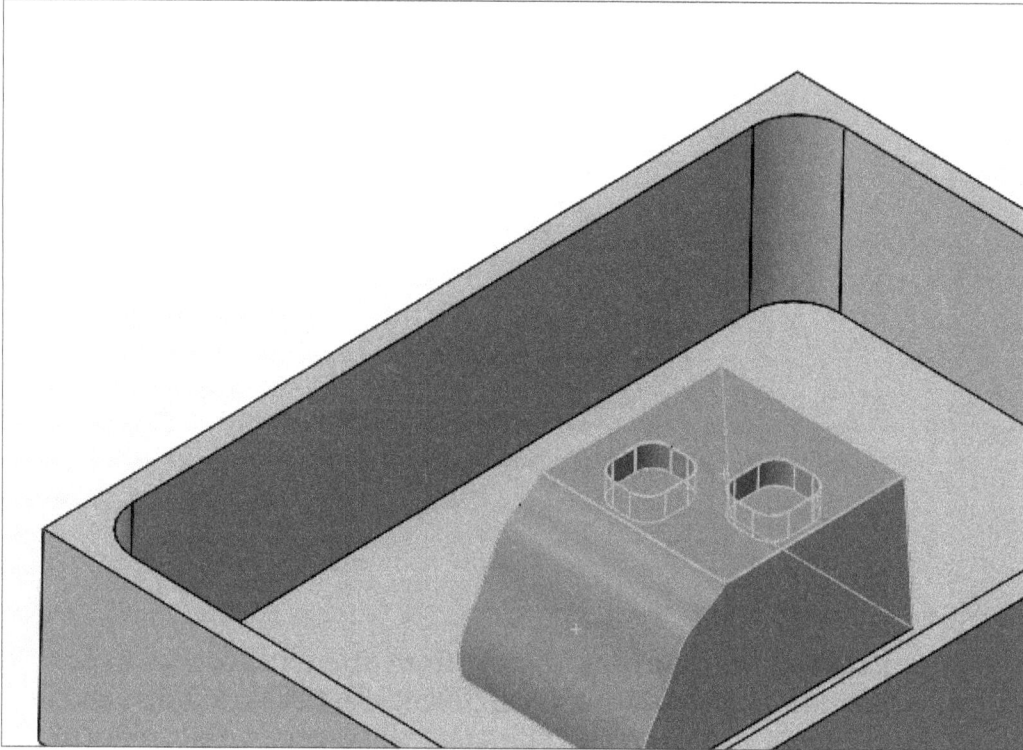

Figure-81. Faces and features selected for parallel rough

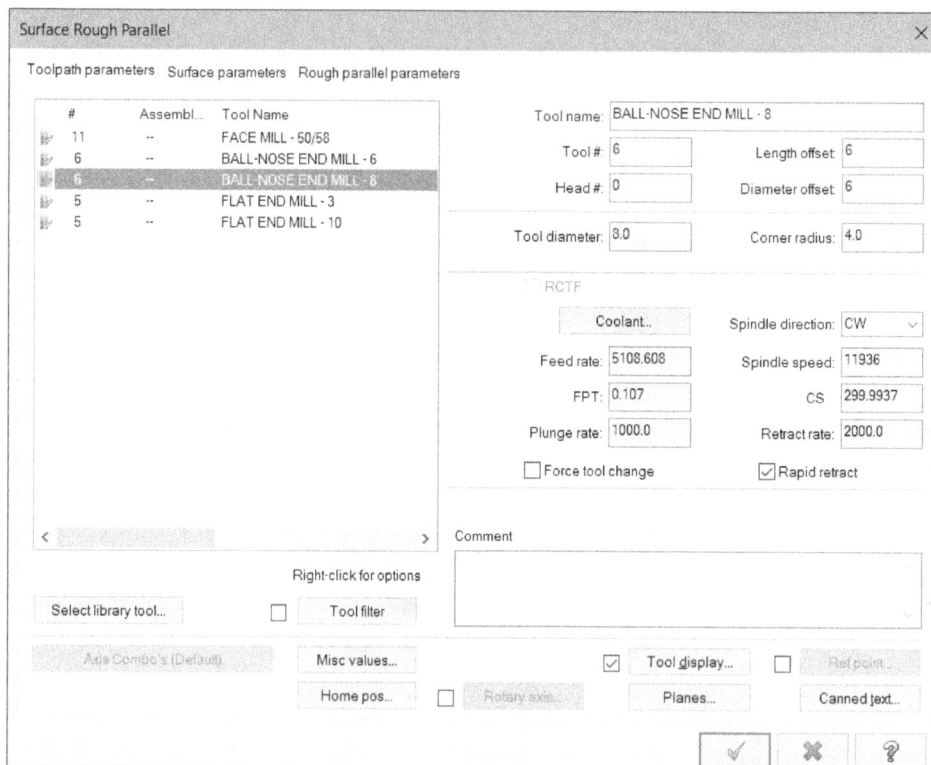

Figure-82. Surface Rough Parallel dialog box

- Select the Ball Nose end mill of **8** diameter as tool.
- Click on the **Rough parallel parameters** tab and, specify **Max stepdown** as **2** and **Max stepover** as **4** in respective edit boxes in this tab.
- Click on the **OK** button from the dialog box. The toolpath will be generated; refer to Figure-83.

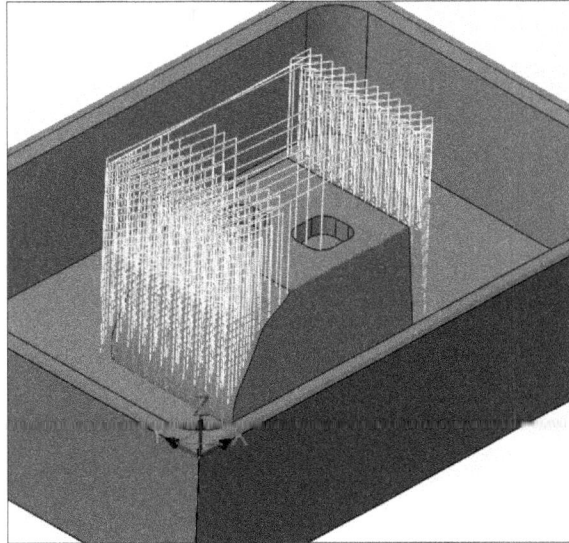

Figure-83. Surface parallel rough toolpath generated

Surface Finish Parallel Toolpath

- Click on the **Parallel** tool from the **Finishing** drop-down in the **Ribbon**. The **Selection PropertyManager** will be displayed.
- Select all the faces and features earlier used for Surface rough parallel toolpath and click on the **OK** button. The **Surface Finish Parallel** dialog box will be displayed.
- Select the Ball Nose End Mill of diameter **6** as cutting tool.
- Click on the **Finish parallel parameters** tab and specify the **Max. stepover** value as **1**.
- Click on the **OK** button. The **Surface finish parallel** toolpath will be generated; refer to Figure-84.

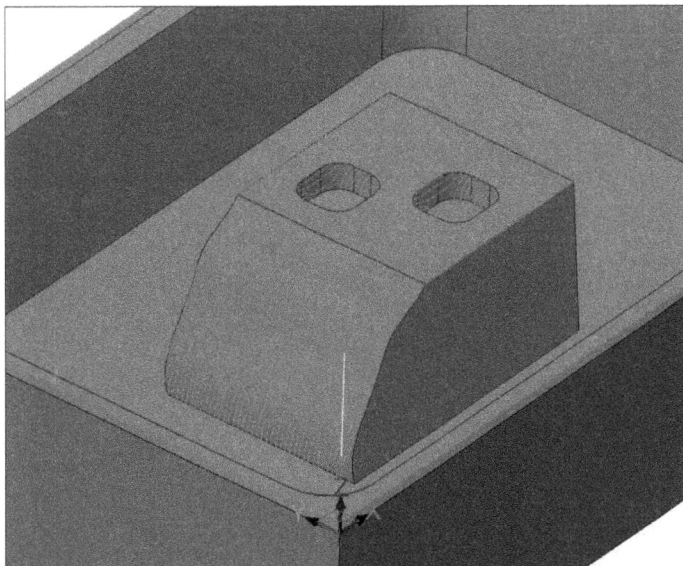

Figure-84. Surface finish parallel toolpath

Pocket Toolpath

Create a 2D pocket toolpath for small pockets at the top face of boss feature using **3** mm diameter flat end mill; refer to Figure-85.

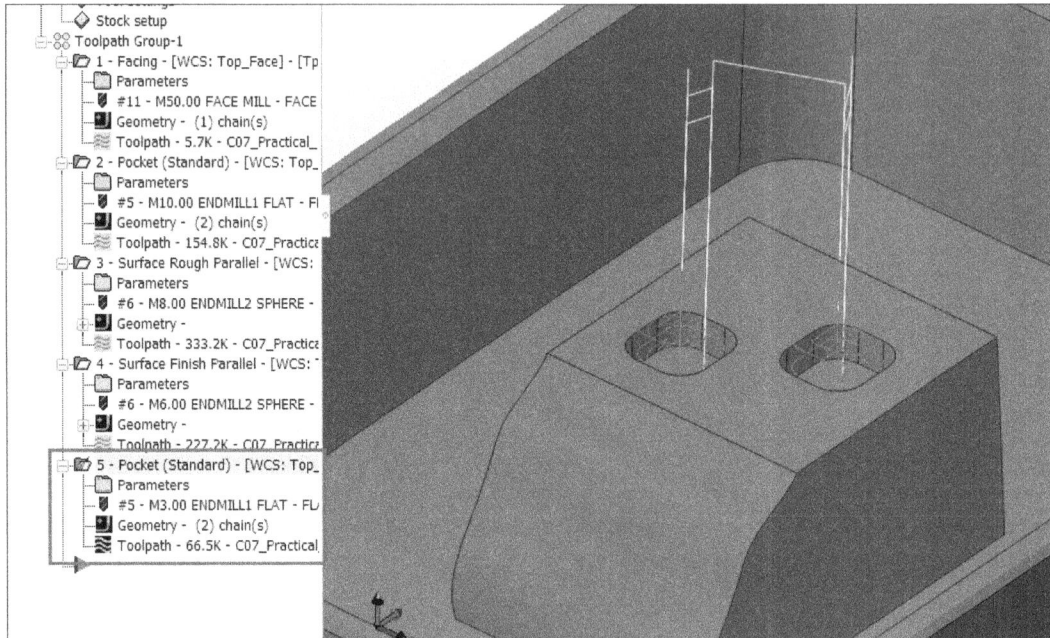

Figure-85. Toolpath for small pockets

Generate the numeric codes by selecting the toolpath group and then clicking on the **Post selected operations** button from the **Mastercam Toolpath Manager**. You verify the toolpaths by using the **Verify selected operations** button from the **Mastercam Toolpath Manager**.

PRACTICE

Create NC program for both sides of the model shown in Figure-86. The dimensions of stock for model are: **Diameter = 700 mm and Length of cylinder = 270** mm; refer to Figure-87.

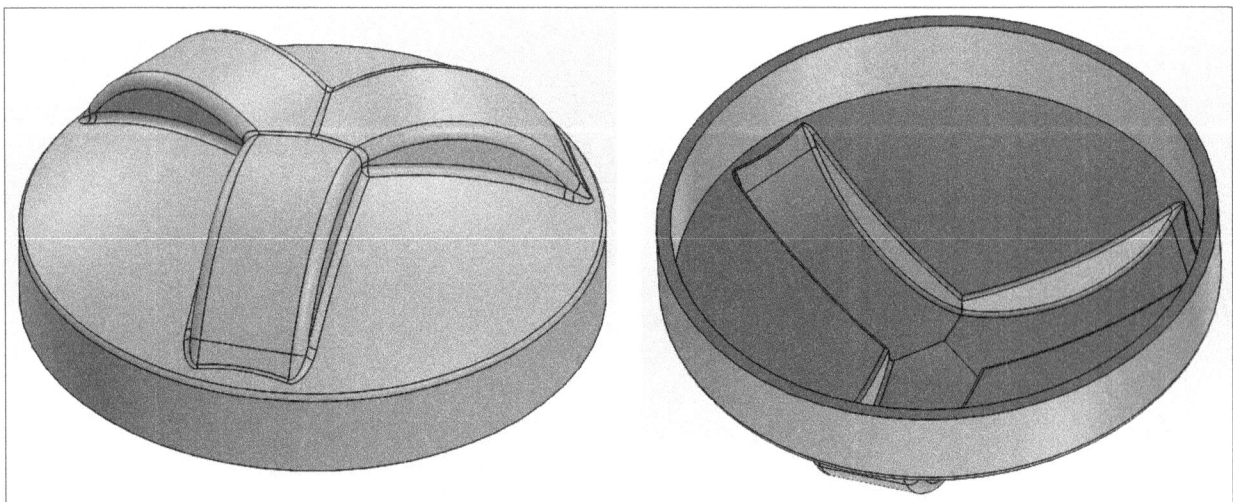

Figure-86. Practice Model

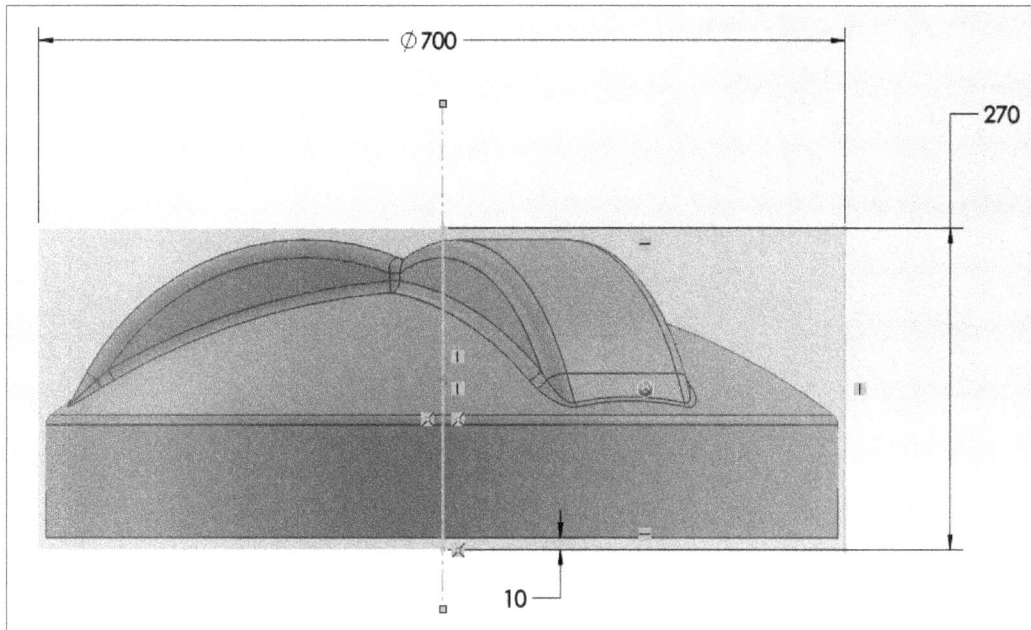

Figure-87. Stock dimensions for Practice

FOR STUDENT NOTES

Chapter 8

Milling Toolpaths-III

Topics Covered

The major topics covered in this chapter are:

- *3D HighSpeed*
- *Advanced Toolpath Operations*

INTRODUCTION

In previous chapter, we worked with 3D surface roughing and finishing toolpaths. We will now discuss the 3D High Speed Toolpaths and Multi-Axis Toolpaths in this chapter. Lets start with 3D High Speed Toolpaths.

3D HIGH SPEED TOOLPATHS

The 3D High Speed toolpaths are used to machine parts using the tool movement in 3D space. You can think of these toolpaths as 3D versions of 2D High Speed toolpaths. The 3D High Speed toolpaths are available in the **3D HighSpeed** drop-down; refer to Figure-1. The options in this drop-down are divided into two categories: Roughing toolpaths and finishing toolpaths. Under roughing toolpaths come **Dynamic OptiRough** and **Area Rough**. Under finishing toolpaths come **Waterline**, **Scallop**, **Horizontal Area**, **Raster**, **Pencil**, **Spiral**, **Radial**, **Hybrid**, and **Project**. These tools are discussed next.

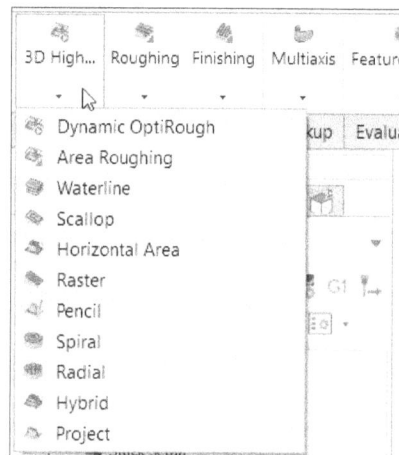

Figure-1. 3D HighSpeed drop-down

Dynamic OptiRough Toolpath

The Dynamic OptiRough toolpath is used to rough the part using Mastercam's optimum cutting strategy which removes large stock in single operation. The procedure to use this toolpath is given next.

* Click on the **Dynamic OptiRough** tool from the **3D HighSpeed** drop-down in the **Ribbon**. The **3D High Speed Toolpaths-Dynamic OptiRough** dialog box will be displayed; refer to Figure-2.

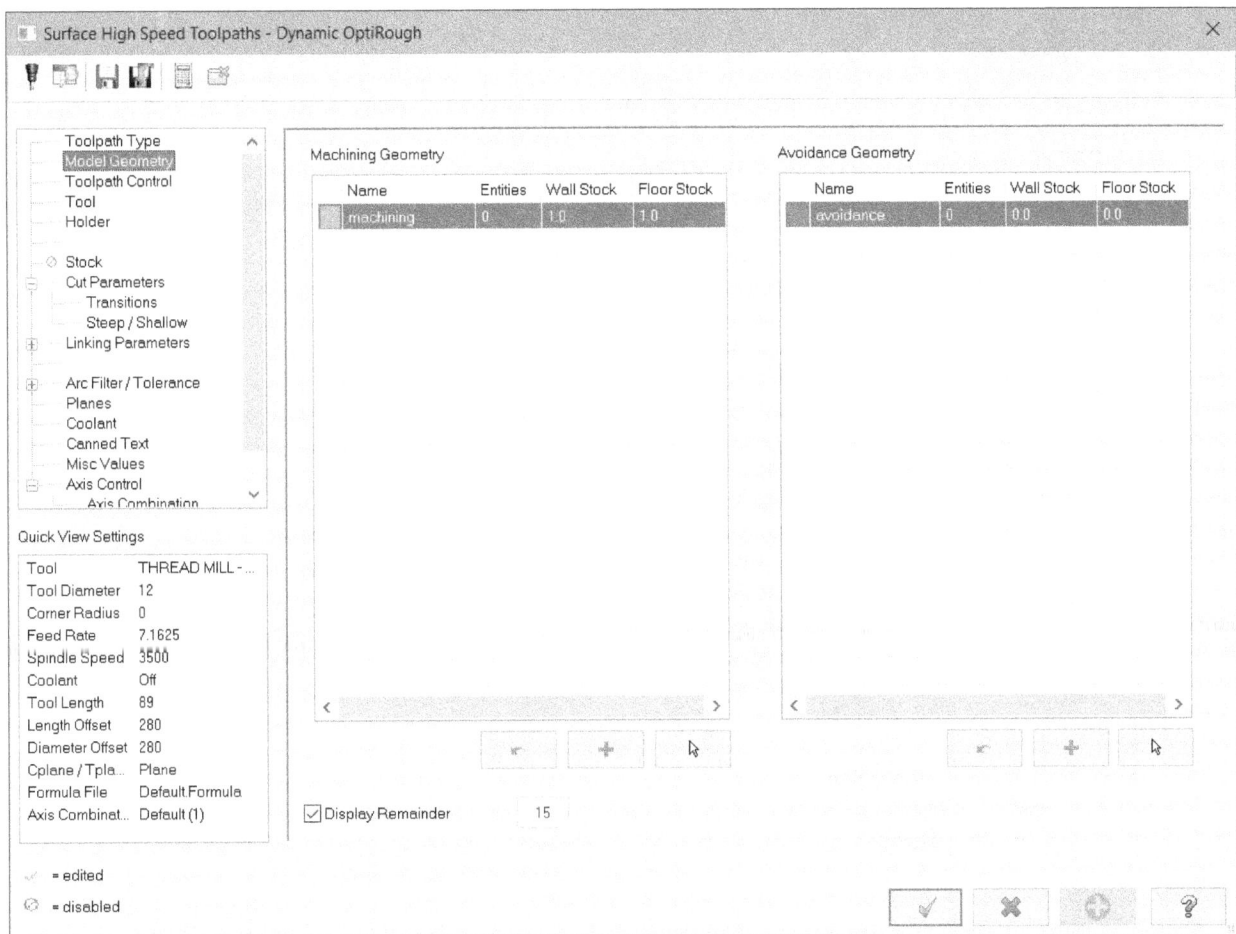

Figure-2. Surface High Speed Toolpath-Dynamic OptiRough dialog box

- Click on the **Select entities** button from the **Machining Geometry** area to select the objects for 3D machining and click on the **End Selection** button. Similarly, you can use the **Select entities** button from the **Avoidance Geometry** area of the dialog box to select objects to be avoided by 3D dynamic rough machining.

- Select the **Toolpath Type** option from the left area of the dialog box and make sure the **Roughing** radio button and the **Dynamic OptiRough** option are selected in this page. If you want to perform finishing then select the **Finishing** radio button from this page then respective toolpaths will be displayed.

- Click on the **Toolpath Control** option from the left area of the dialog box to set parameters related to containment boundary.

- In the **Containment boundary** area, there are two radio buttons: **Stay inside** and **From Outside**. Select the **Stay inside** radio button if you want to machine a pocket feature and select the **From outside** radio button if you want to machine a boss feature with open boundaries. Note that if you have a pocket with islands then you should select the **Stay inside** radio button from this area.

- Select desired tool and tool holder from the **Tool** and **Holder** pages in the dialog box.

- If you want to use the dynamic optirough toolpath to remove material left by other toolpaths then select the **Stock** option from the left area. The options will be displayed as shown in Figure-3.

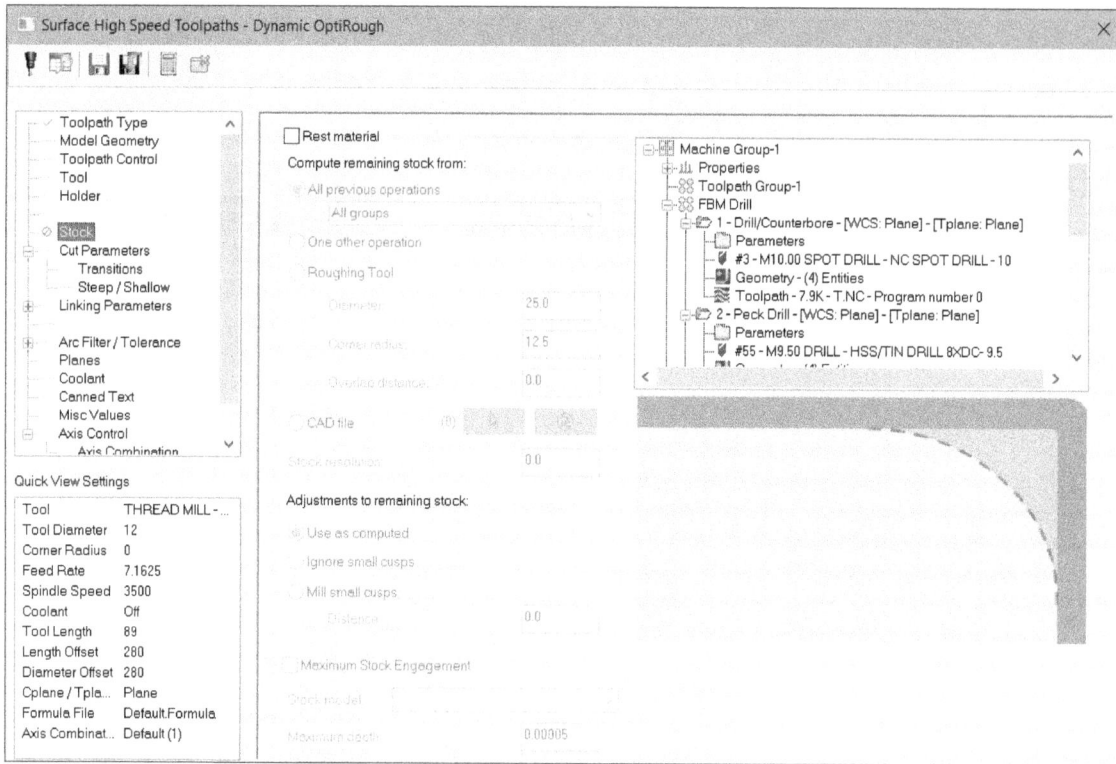

Figure-3. Stock page in the dialog box

- Select the **Rest material** check box from the page and specify desired parameters as discussed earlier. If you do not want to perform rest mill then keep this check box cleared.
- Select desired option from the **Adjustments to remaining stock** area to specify the value of stock to be left after the roughing operation.

Cut Parameters

- Click on the **Cut Parameters** option from the left area. The options will be displayed as shown in Figure-4.

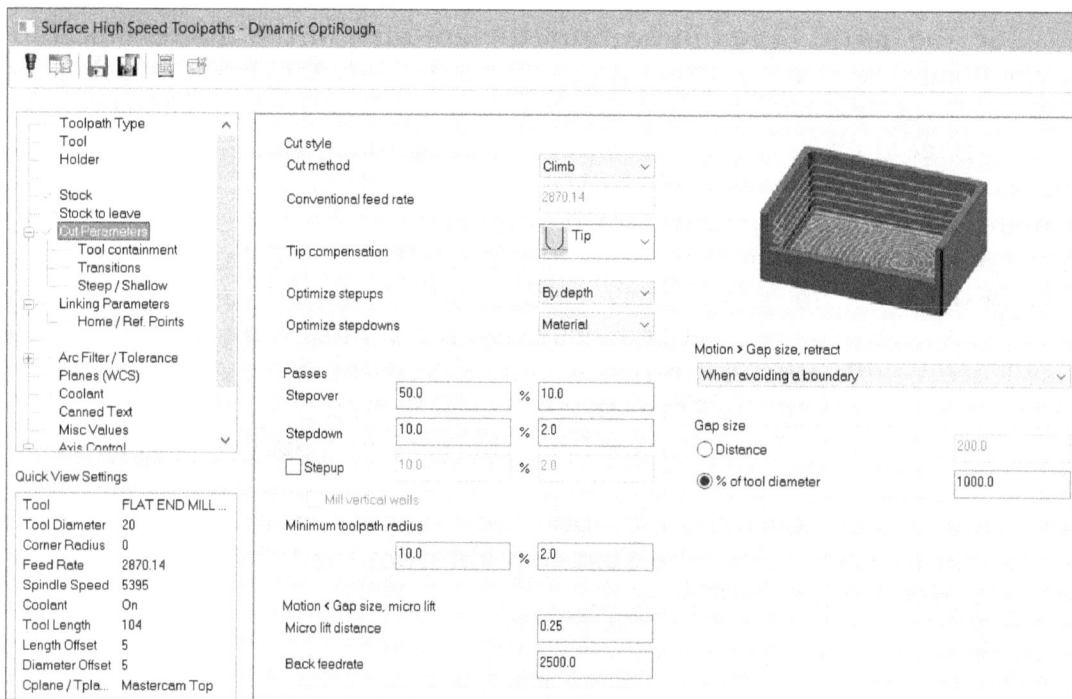

Figure-4. Cut Parameters page in Dynamic OptiRough dialog box

- Most of the parameters in this dialog box have been discussed in previous chapter, so no repetition here! Click in the **Optimize stepups** drop-down and select desired option. If you select the **By depth** option then tool will cut at equal z level in all pockets selected in the part. If you select the **Next closest** option then the tool will first perform step down operations in all the pockets at equal Z level and then it will perform step up operation in all the pockets at equal Z level. If you select the **By pocket** option then the tool will one by one rough each pocket in step down and then it will use the Next closest strategy to step up; refer to Figure-5.

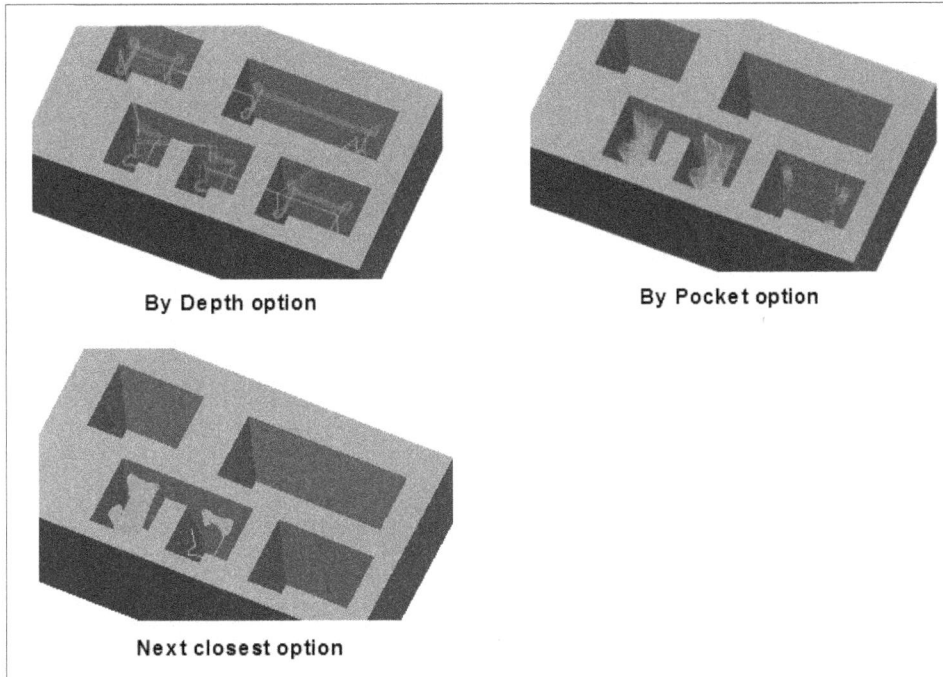

By Depth option   By Pocket option

Next closest option

Figure-5. Optimize stepup options

- Similarly, select desired option from the **Optimize stepdowns** drop-down. Select the **Material** option to start cutting from the material near to tool. Select the **Air** option to start cutting from the lesser material area. Select the **None** option to start cutting from the location where tool of previous toolpath left cutting.
- Specify the **Stepup**, **Stepdown**, and **Stepover** values in the **Passes** area of the dialog box.
- If the free motion of tool required in cutting toolpath is less than Gap size specified then set desired microlift in the **Motion < Gap size, micro lift** area of the page.
- If the free motion required is more than the Gap size specified then set desired retraction parameters in the **Motion > Gap size, retract** area of the page. Specify desired gap size in the **Gap size** area of this page.

Toolpath Control

- The options in this page are used to specify how tool compensation will be applied with respect to containment boundary selected; refer to Figure-6.

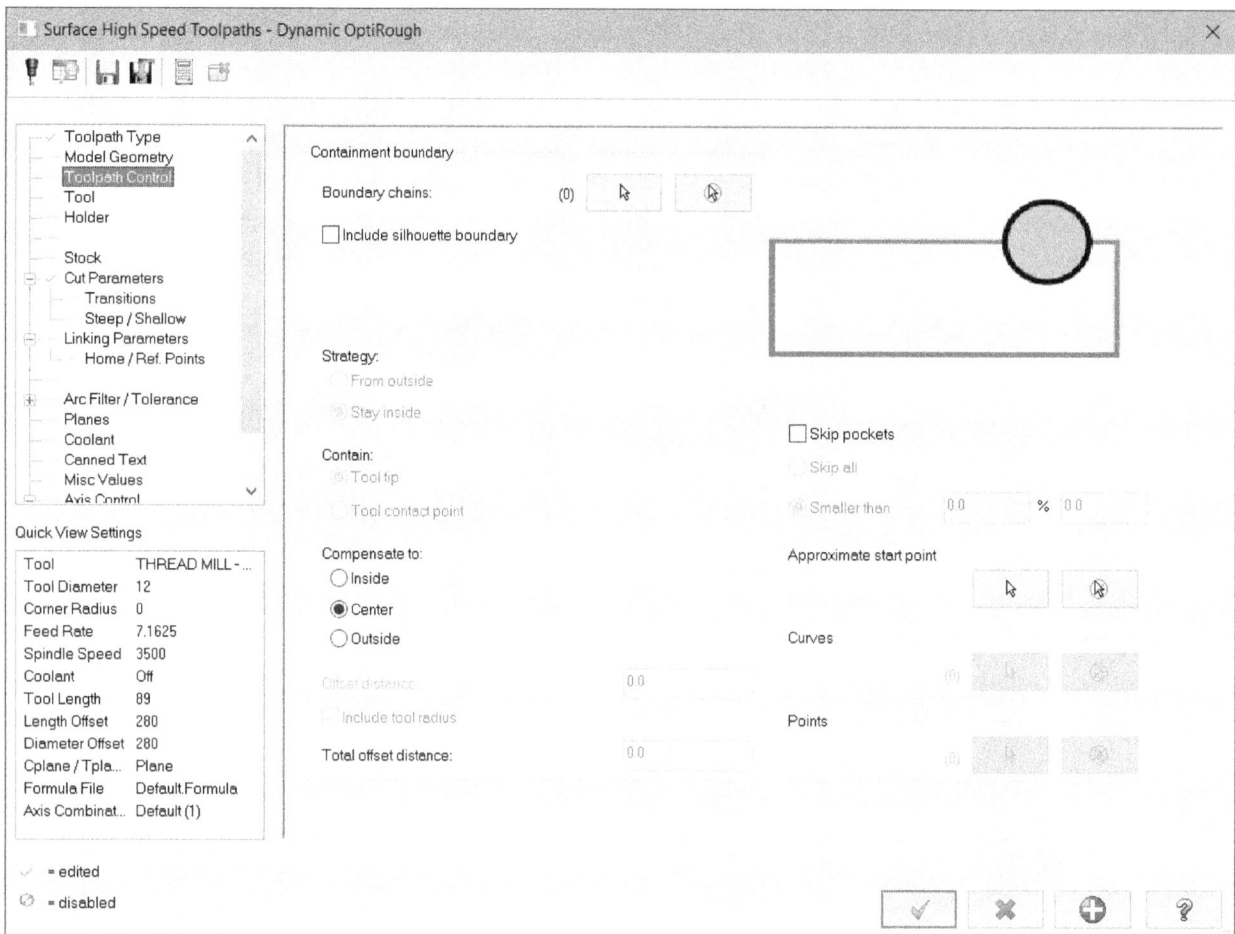

Figure-6. Tool containment page

- Select the **Inside** radio button from the **Compensate to** area to keep tool completely inside the containment boundary (generally required in pocket cuts). Select the **Outside** radio button to keep tool completely outside the containment boundary. Similarly, select the **Center** radio button to keep tool tip over the containment boundary. Note that you can also specify offset values if **Inside** or **Outside** radio button is selected.

- Specify the tool entry method in the **Transition** page of this dialog box as discussed in previous chapters.
- If your part has steep or shallow faces then click on the **Steep/Shallow** option from the left area and specify the depth in Z upto which your tool should machine such faces.

Linking Parameters

- Click on the **Linking Parameters** option from the left area of the dialog box. The options in the page will be displayed as shown in Figure-7.

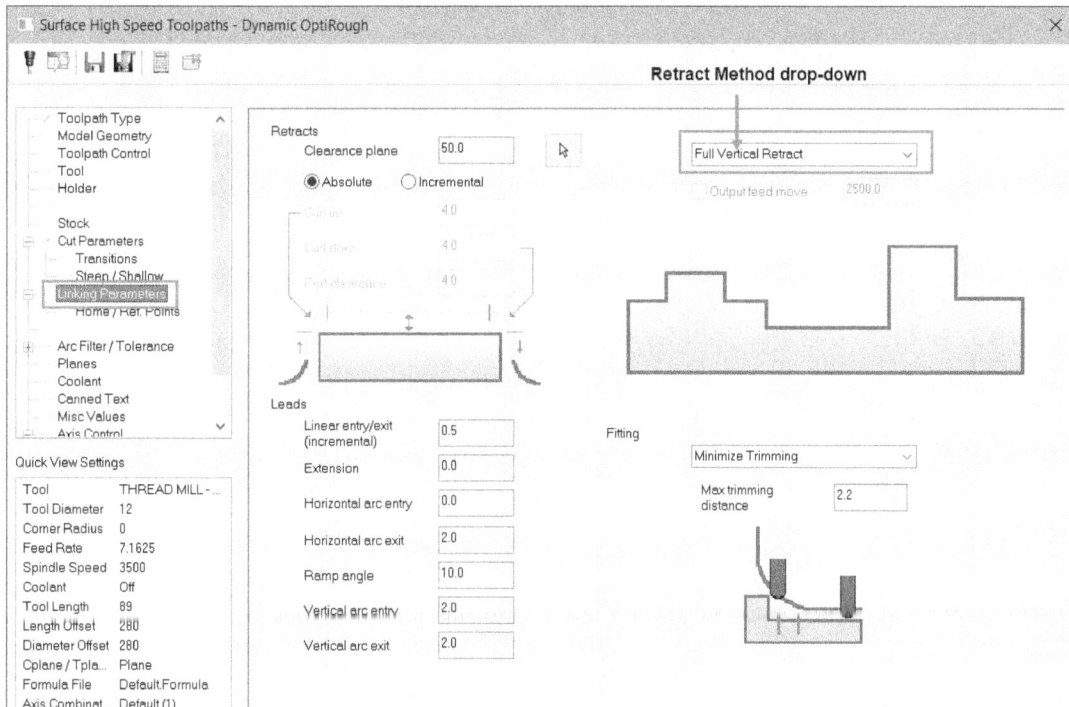

Figure-7. Linking Parameters page

- Specify the clearance plane distance in the **Clearance plane** edit box.
- Select desired option from the **Retract Method** drop-down to specify how tool will retract in non-cutting moves. On selecting the **Minimum Vertical Retract** option, the tool will retract at distance specified in **Part clearance** edit box and follow the trajectory of part edges. On selecting the **Minimum Distance** option, the tool will retract directly for larger distance but follow the trajectory of part for smaller distances.
- Select desired option from the **Fitting Method** drop-down in the **Fitting** area. The options of this area are used to specify the movement of tool along the corners.
- Specify the other parameters as required and click on the **OK** button. The toolpath will be created; refer to Figure-8.

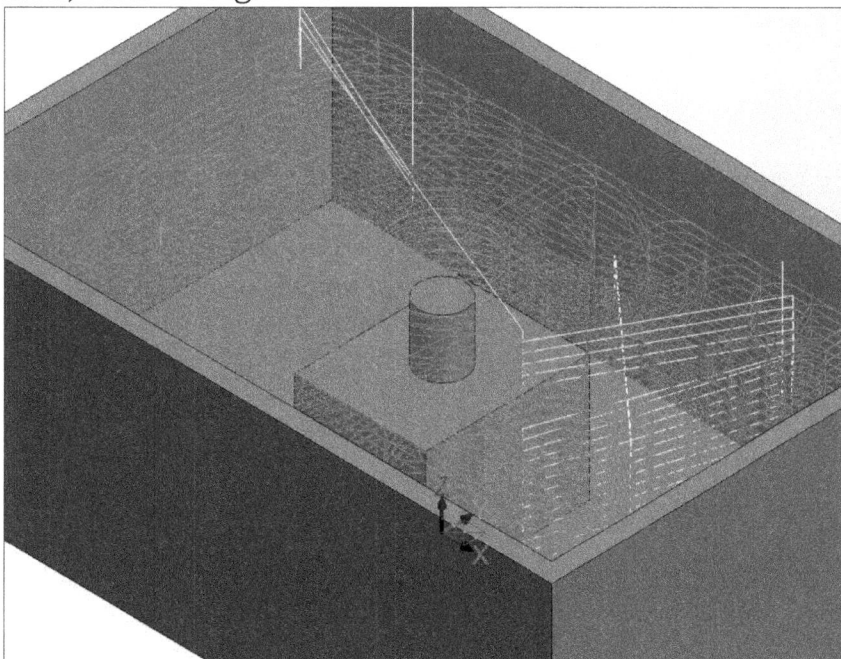

Figure-8. Dynamic optirough toolpath

Area Rough Toolpath

The Area Rough toolpath is used to mill the selected face area using 3D cutting paths. The most common question is asked here why area rough then dynamic rough toolpaths are available for same work. Answer is Dynamic rough toolpaths are useful when stock is similar in shape to the part but Area Rough is useful when you have a block and you want to rough machine it to get the desired shape. The procedure to use this tool is given next.

*   Click on the **Area Rough** tool from the **3D HighSpeed** drop-down. The **3D High Speed Toolpaths - Area Roughing** dialog box will be displayed; refer to Figure-9.

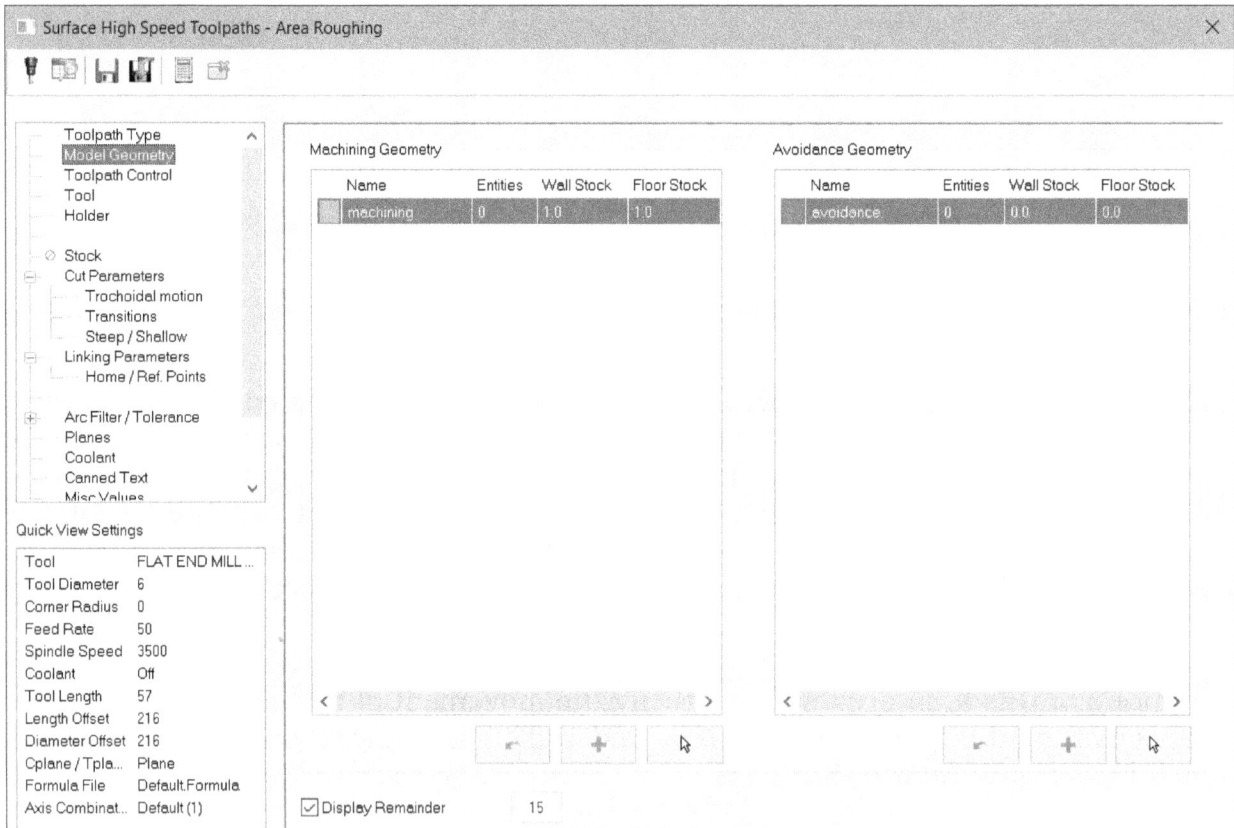

Figure-9. 3D High Speed Toolpaths–Area Roughing dialog box

*   Select the faces that you want to be machined and the faces you want to avoid by using the options in the **Machining Geometry** and **Avoidance Geometry** areas respectively; refer to Figure-10 and click on the **OK** button.
*   Click on the **Toolpath Control** option from the left area of the dialog box. The page in **3D High Speed Toolpaths - Area Roughing** dialog box will be displayed as shown in Figure-11.

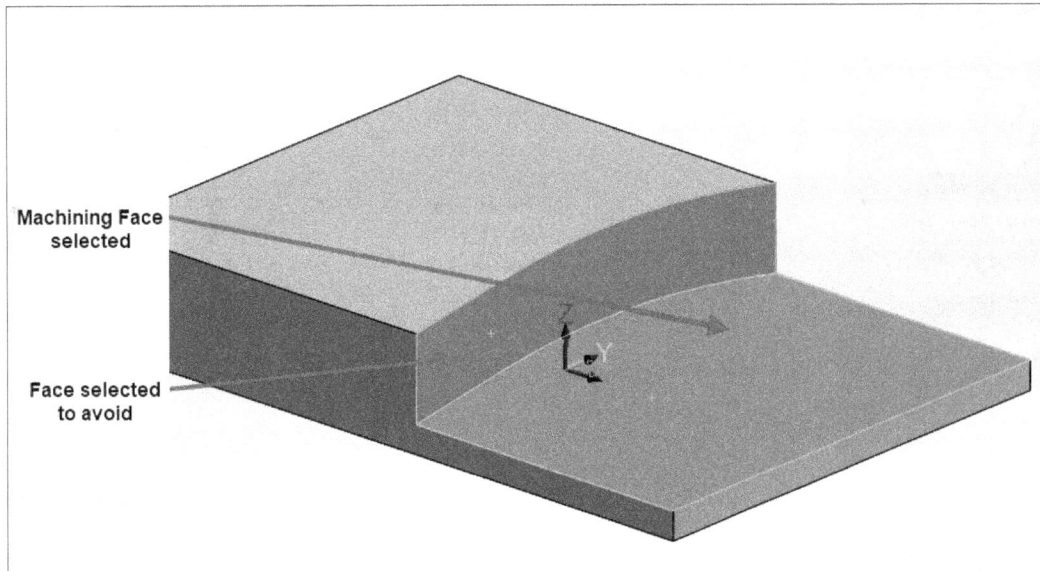

Figure-10. Faces selected for area rough

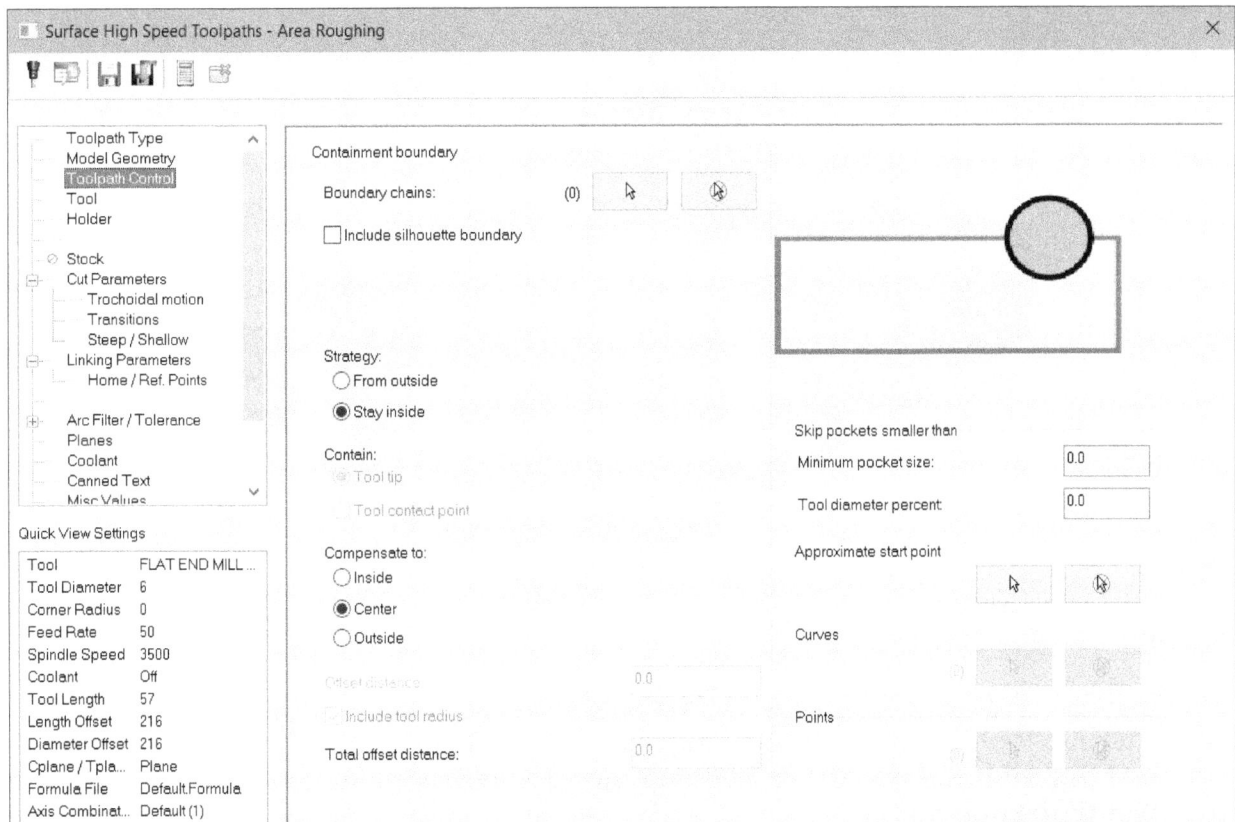

Figure-11. Surface High Speed Toolpath-Area Rough dialog box

- From the **Containment boundary** area, select the **Stay inside** radio button if you want to machine a pocket. If you want to machine a boss feature or the part has open boundary then select the **From Outside** radio button.
- Select the **Trochoidal motion** option from the left area to define the options related to motion of tool while cutting. The options in the dialog box will be displayed as shown in Figure-12. To minimize burial of tool, select the **Minimize burial** radio button from the page and specify desired parameters to create trochoidal motion of tool. Preview of the motion is displayed in right of the page.

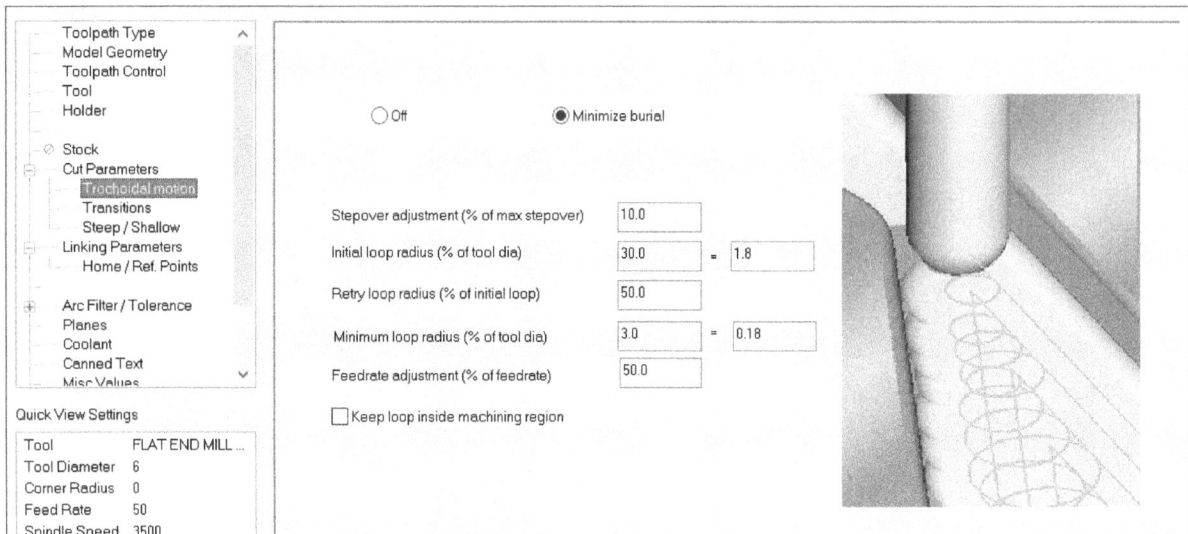

Figure-12. Trochoidal motion options

- Similarly, specify the other parameters and click on the **OK** button to create the toolpath; refer to Figure-13.

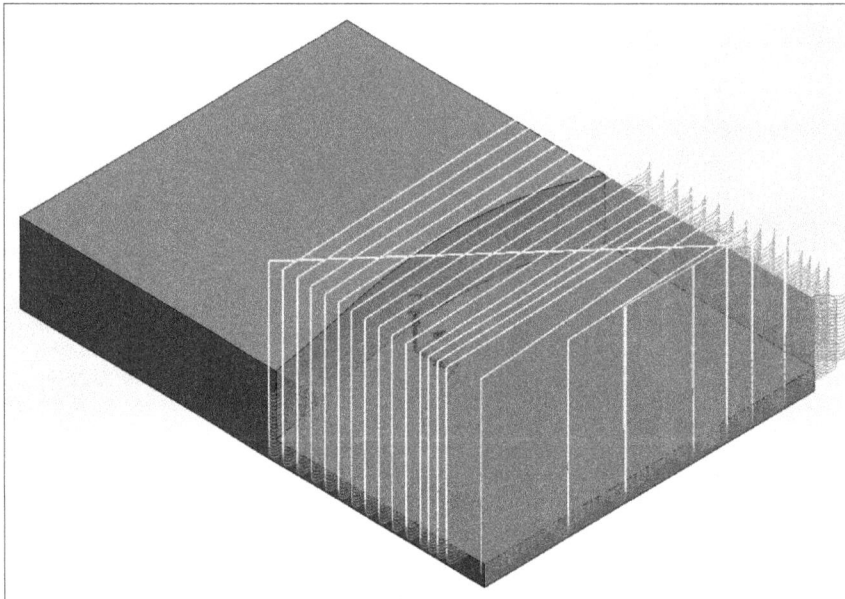

Figure-13. Area rough toolpath

Waterline

The Waterline toolpath as the name suggests is used to create toolpath as water flows. It is used to finish the part layer by layer along Z axis while following the contour of part. The Waterline toolpath is best with steeper angled surfaces. The procedure to use this tool is given next.

- After performing roughing operations, click on the **Waterline** tool from the **3D HighSpeed** drop-down in the **Ribbon**. The **3D High Speed Toolpaths-Waterline** dialog box will be displayed; refer to Figure-14.

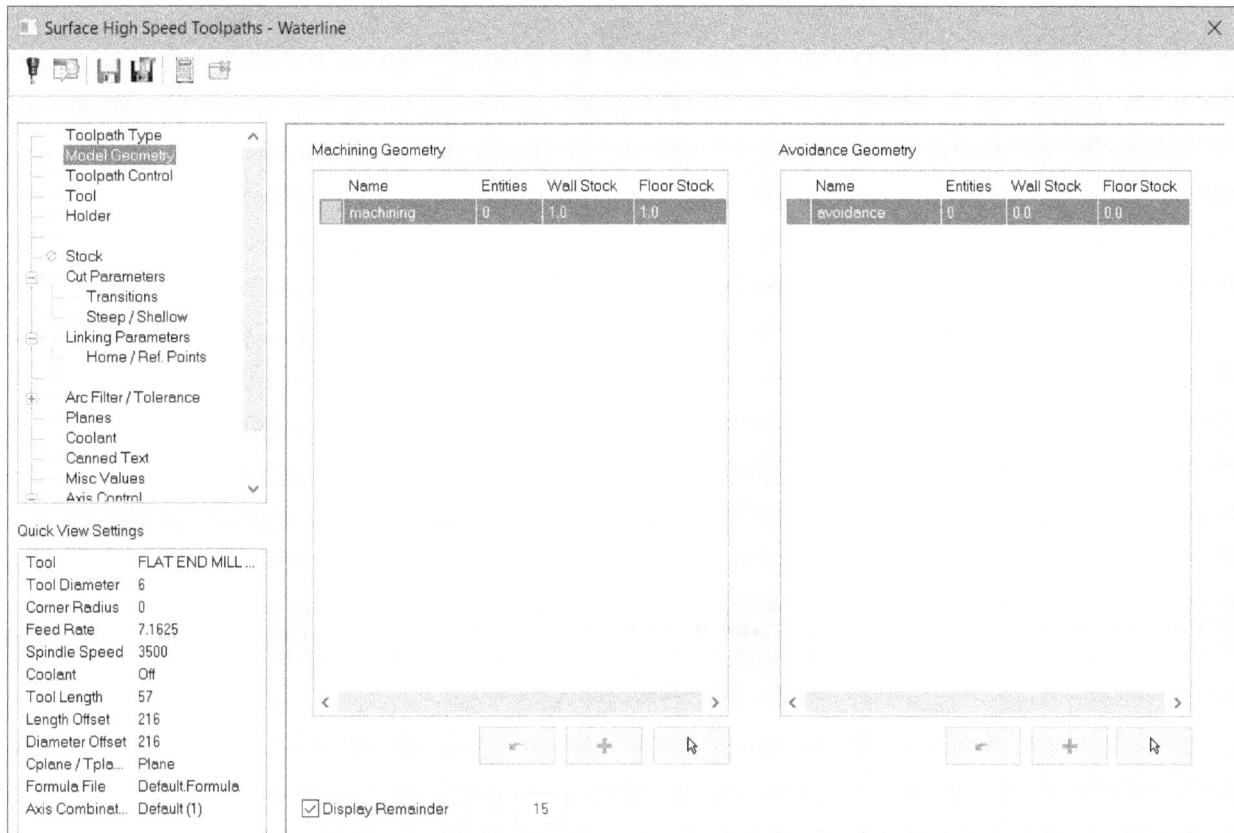

Figure-14. 3D High Speed Toolpaths-Waterline dialog box

- Click on the **Select entities** button from the **Machining Geometry** area of the dialog box. You will be asked to select the entities.
- Select the faces/features/bodies to be machined; refer to Figure-15 and click on the **OK** button. The **3D High Speed Toolpaths-Waterline** dialog box will be displayed; refer to Figure-16.

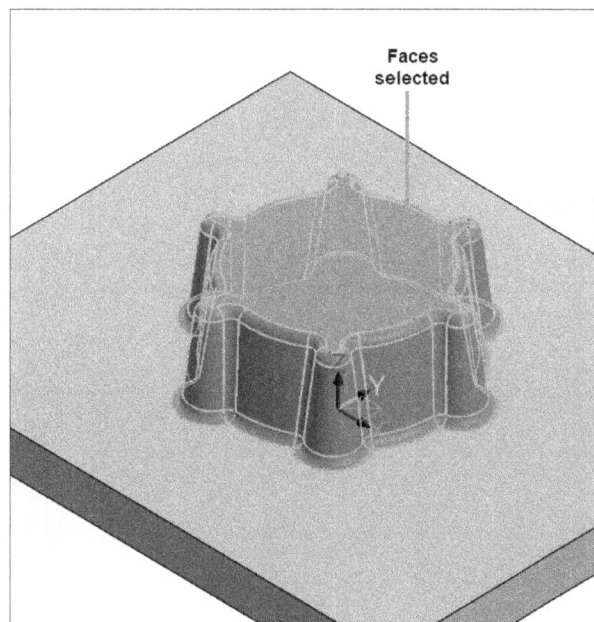

Figure-15. Features selected for machining

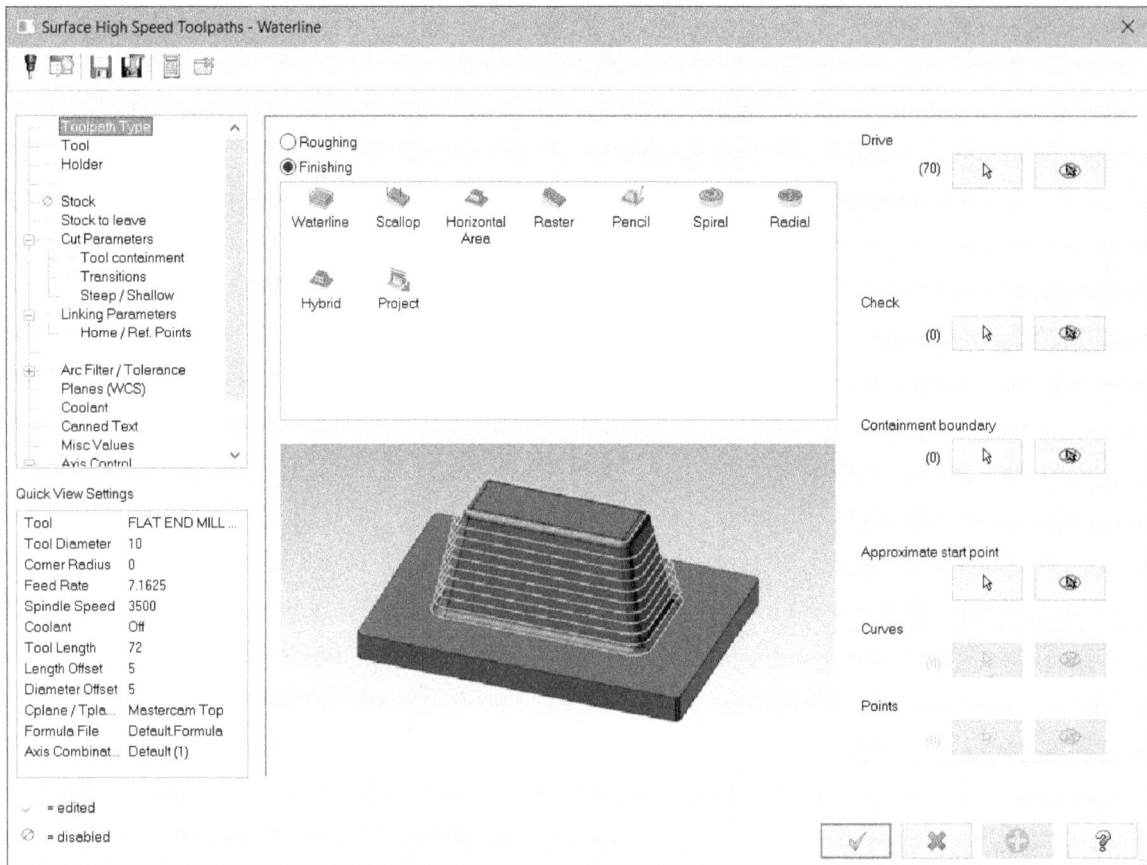

Figure-16. Surface High Speed Toolpath-Waterline dialog box

• Select the tool and tool holder as required from the respective pages.

Stock Options

• To define how Mastercam should calculate the stock amount, click on the **Stock** option from the left area in the dialog box. The Stock page will be displayed as shown in Figure-17.

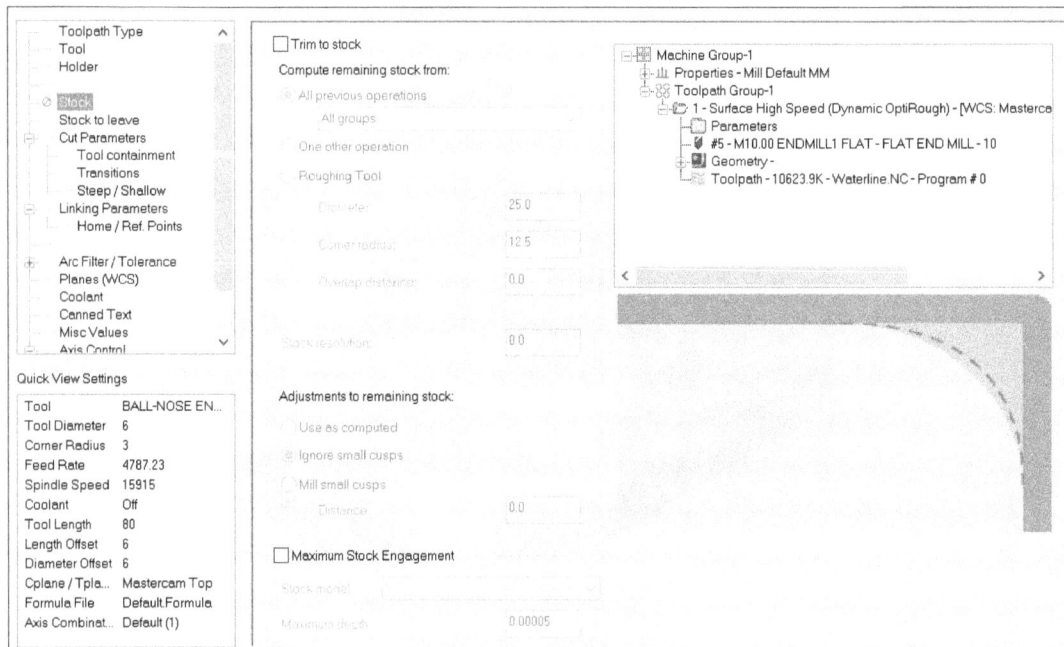

Figure-17. Stock page in 3D High Speed Toolpaths-Waterline dialog box

- Select the **Trim to stock** check box from the dialog box to define stock for current operation. On selecting this check box, the extra material left after previous operations will be assumed as stock for current operation. If you want to specifically select an operation up to which you want to trim the stock then select the **One other operation** radio button from the **Trim to stock** area and select desired operation from the operation box in the right.
- Set the other parameters like stock resolution and adjustments to remaining stock as required. Stock resolution is the accuracy value used to approximate the remaining material. Typically, stock resolution is set to 20x the total tolerance. Higher value means loose stock used for rough machining. Lower value means tight stock for finishing.
- Select the **Maximum Stock Engagement** check box if you want to keep the tool engaged in stock up to specified depth while cutting. After selecting check box, specify desired depth value in the **Maximum depth** edit box. Note that most of the time, the values specified here improve the surface quality while cutting faster.

Note that if you do not define any option in the **Stock** page, still the toolpath will be generated but Mastercam will take all the default values to perform these operations.

- Click on the **Stock to leave** option from the left area in the dialog box and specify the value of stock to be left after this operation. If this is the finishing cut then specify 0 in all the edit boxes.

Cut Parameters

- Click on the **Cut Parameters** option from the left area of dialog box to display options related to cutting strategy; refer to Figure-18.

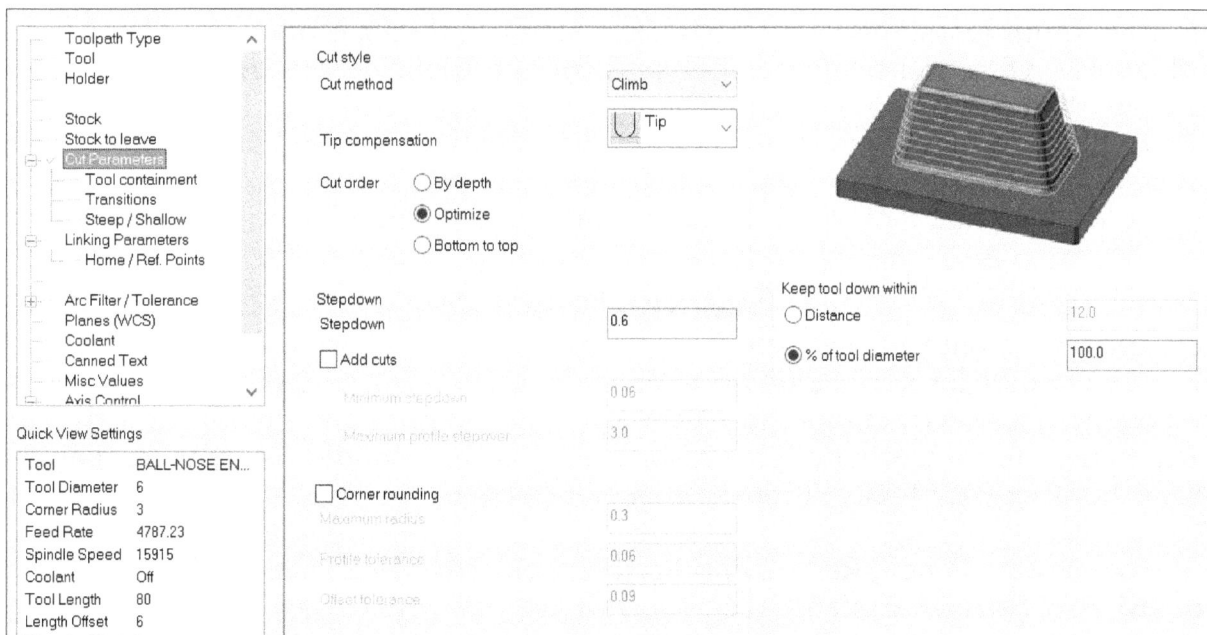

Figure-18. Cut Parameters page in 3D High Speed Toolpaths-Waterline dialog box

- Select desired cutting method from the **Cut method** drop-down in the **Cut style** area. Note that most of the time, **Zigzag** method is time saving if your tool and surface finish allows to use.

- Select desired tool tip compensation option from the **Tip compensation** drop-down in the dialog box.
- From the **Cut order** area, select desired radio button to define cutting order for the operation. Select the **By depth** radio button to machine all cut passes along Z axis level by level. Select the **Optimize** radio button to keep the tool in an area and keeps it there until all cuts in that area are finished. Select **Bottom to top** radio button to machine from the bottom of the part to the top.
- Specify desired parameters in **Stepdown**, **Corner rounding**, and **Keep tool down within** areas to refine surface finish.
- Set the other parameters as discussed earlier and click on the **OK** button. The toolpath will be generated; refer to Figure-19.

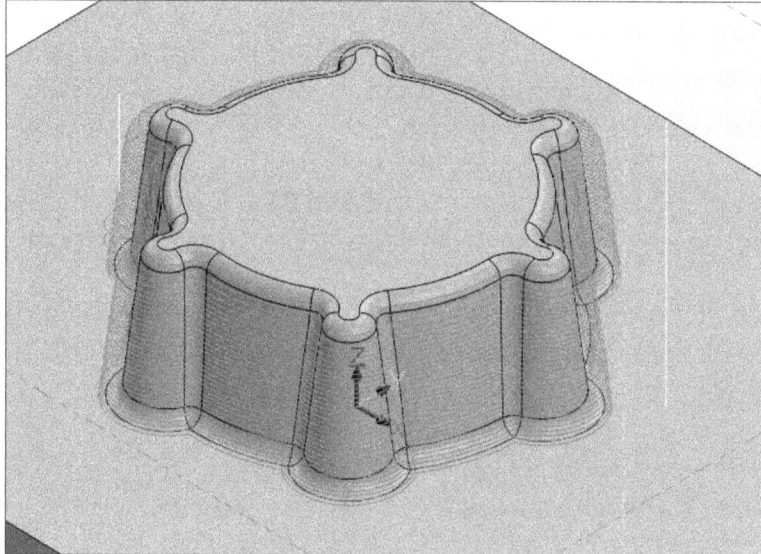

Figure-19. Waterline toolpath generated

Note that if your tool skips some area while machining then it is not fault of Mastercam. In this case, you need to change the settings and tool selected by you. Like in our example, the rounds will not be machined by 6 mm ball end mill. To finish them, we need to use a lower radius tool which needs an extra waterline toolpath. Now, you may ask why we did not use the lower radius tool at first then the answer is saving time. A lower radius tool will take more time to finish same amount of material.

Scallop Toolpath

The Scallop toolpath in high speed works similar to Scallop toolpath discussed in previous chapter for surface finish toolpaths. In High Speed toolpaths, the you get the benefit of less retractions. Note that if your geometry is very complex with many features near to each other or check surfaces then you should use smaller diameter tools otherwise your tool will hit the stock or part.

In the same way, Pencil, Radial, and Project toolpaths are similar to their respective toolpaths in Surface Finishing toolpaths category which has been discussed in previous chapters.

Horizontal Area Toolpath

The Horizontal Area toolpath is used to finish flat faces. This toolpath functions in the same way as Facing works in 2D Toolpaths. The procedure to create this toolpath is given next.

- After roughing, click on the **Horizontal Area** tool from the **3D HighSpeed** drop-down in the **Ribbon**. The **3D High Speed Toolpaths-Horizontal Area** dialog box will be displayed; refer to Figure-20.
- Select the machining geometry and avoidance geometry objects as discussed earlier; refer to Figure-21.

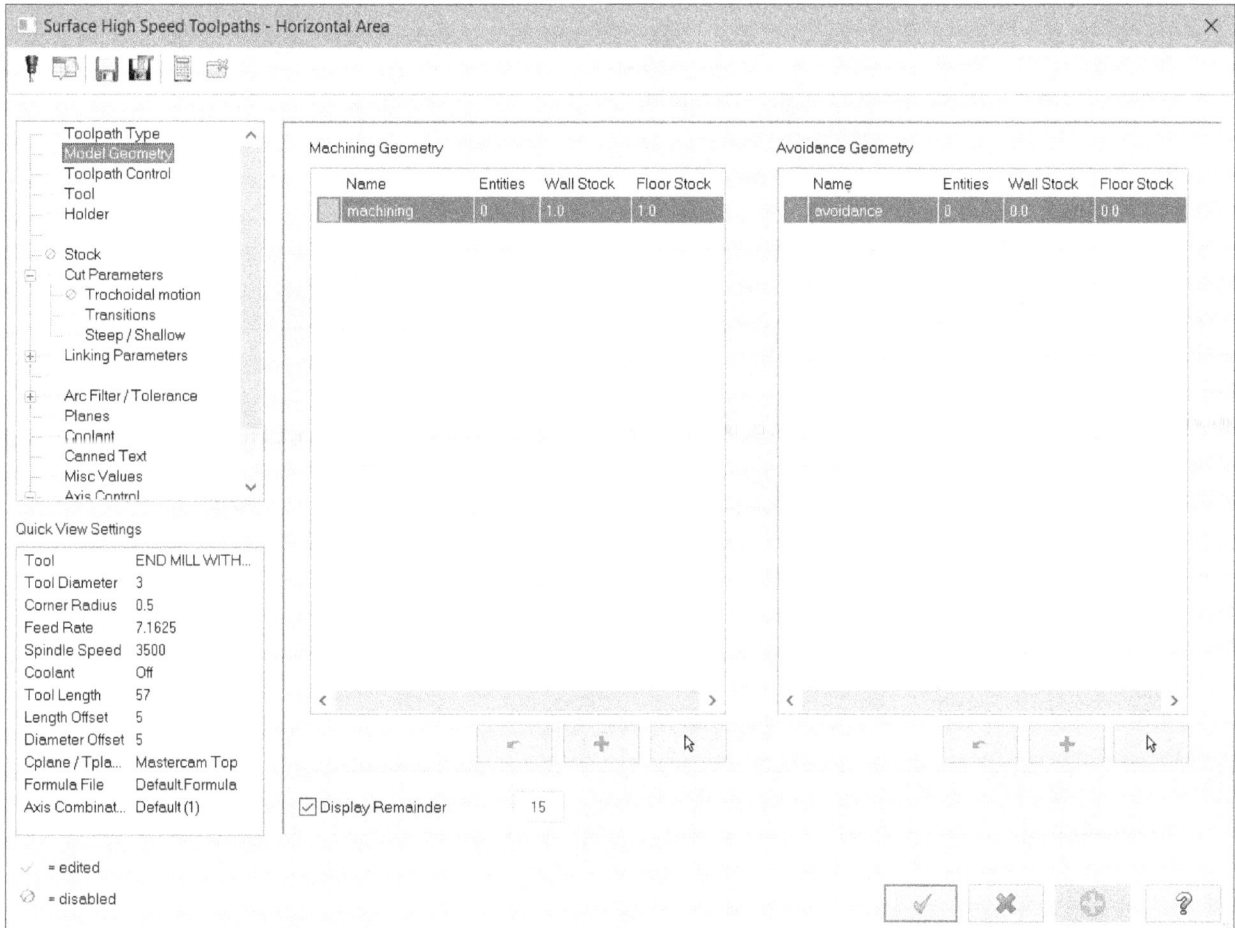

Figure-20. 3D High Speed Toolpaths-Horizontal Area dialog box

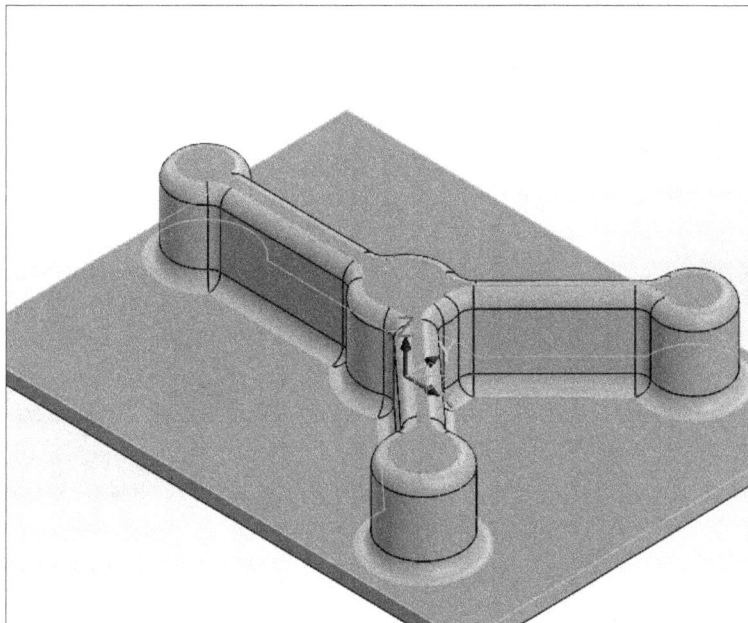

Figure-21. Faces selected for horizontal area HST

- Specify the parameters as discussed for other High Speed Toolpaths. Note that you can skip selecting the check surfaces here as toolpath will move only the selected faces taking their edges as boundaries.
- After specifying the parameters, click on the **OK** button from the dialog box. The toolpath will be generated; refer to Figure-22.

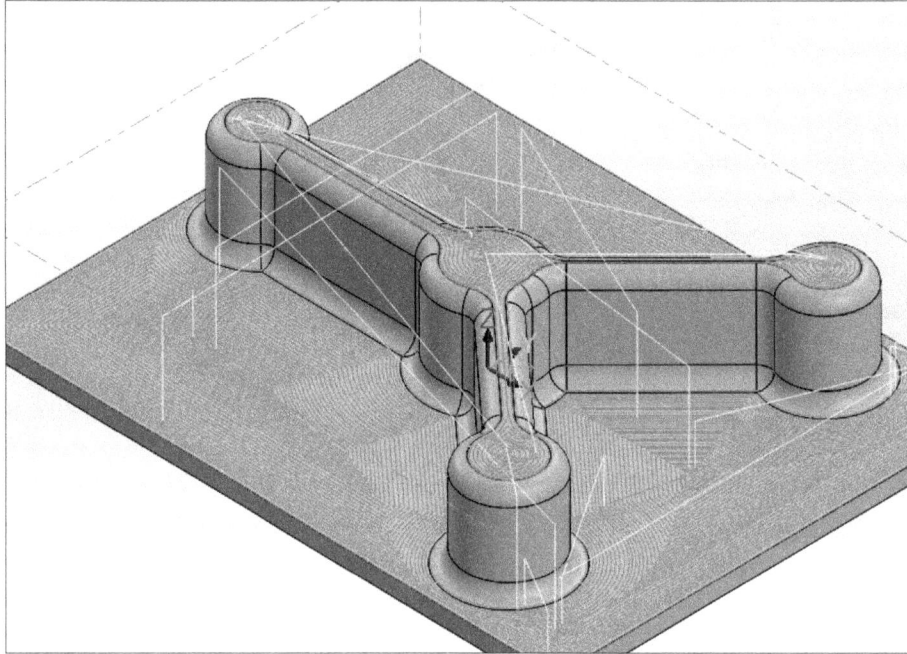

Figure-22. Toolpath for horizontal area HST

Raster Toolpath

The Raster toolpath is used to cut material in parallel directions. The working of this toolpath is similar to Parallel toolpath discussed earlier in previous chapter; refer to Figure-23. The benefit of using this toolpath is in transition and low retraction.

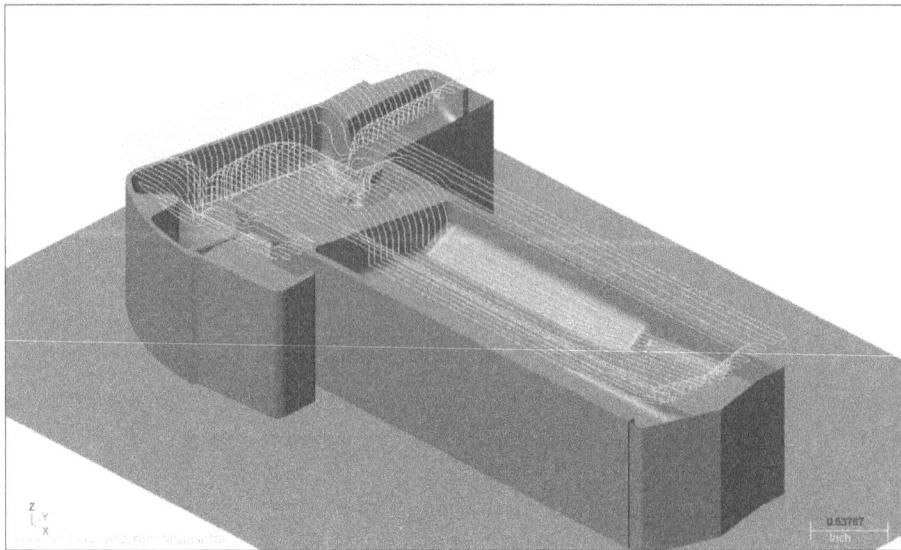

Figure-23. Raster toolpath example

Spiral Toolpath

The Spiral toolpath is used to cut material in spiral fashion. This toolpath is useful when you need to finish rough objects like shown in Figure-24. The method to use this tool is same as discussed earlier.

Figure-24. Spiral High Speed Toolpath

Hybrid Toolpath

The Hybrid toolpaths take advantage of both waterline toolpaths and scallop toolpaths; refer to Figure-25. After roughing if you need finishing passes which machine in waterline and then scallop fashion then you should use this toolpath. The parameters for hybrid toolpath are same as discussed earlier.

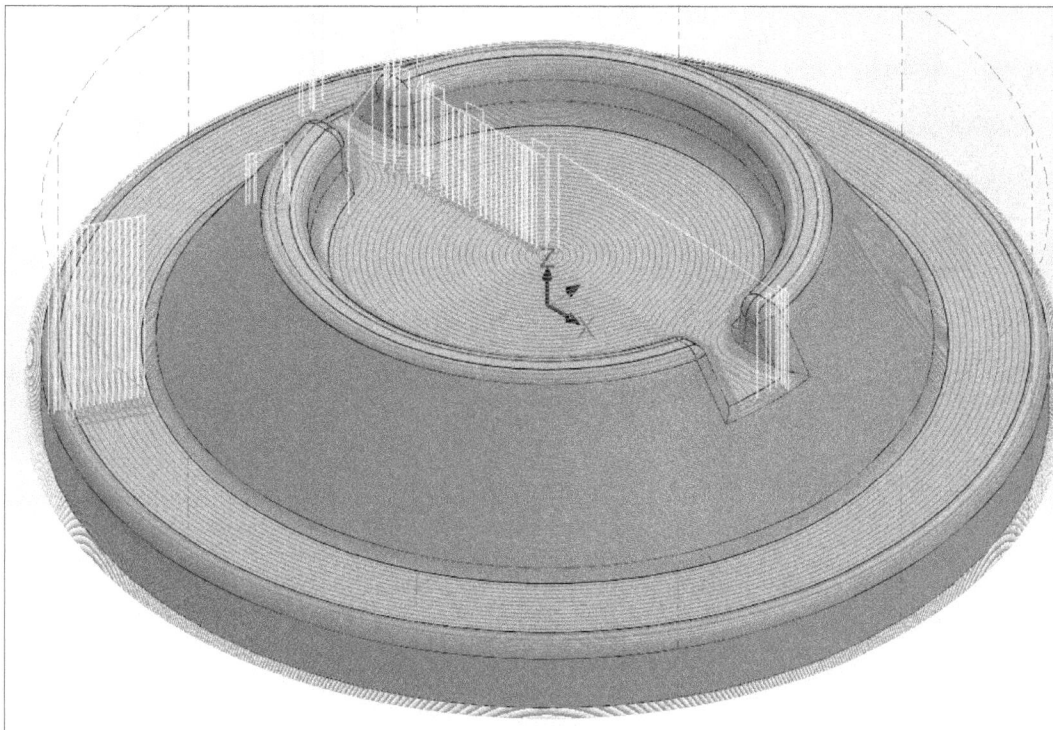

Figure-25. Hybrid High Speed Toolpath created

PRACTICAL 1

Create NC program for model shown in Figure-26. The stock for model is given in Figure-27.

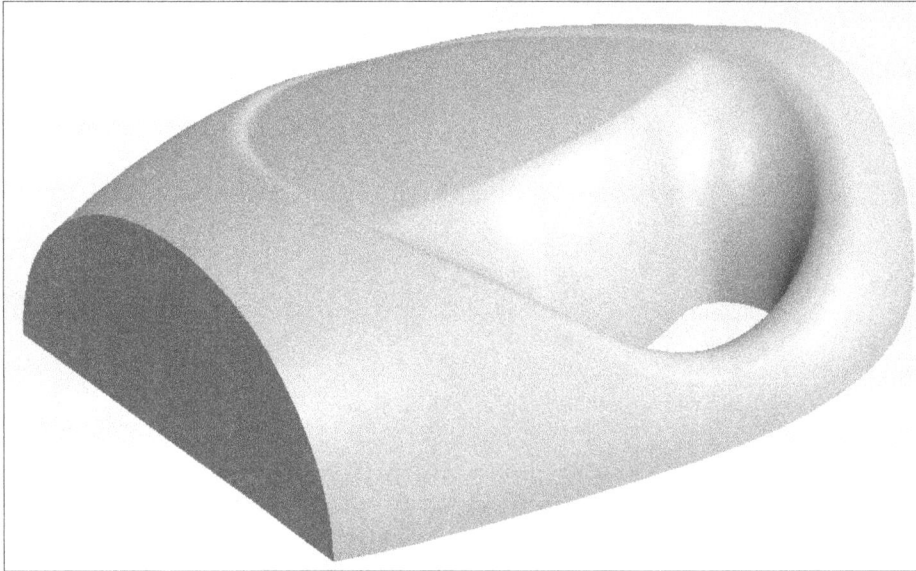

Figure-26. Model for Practical

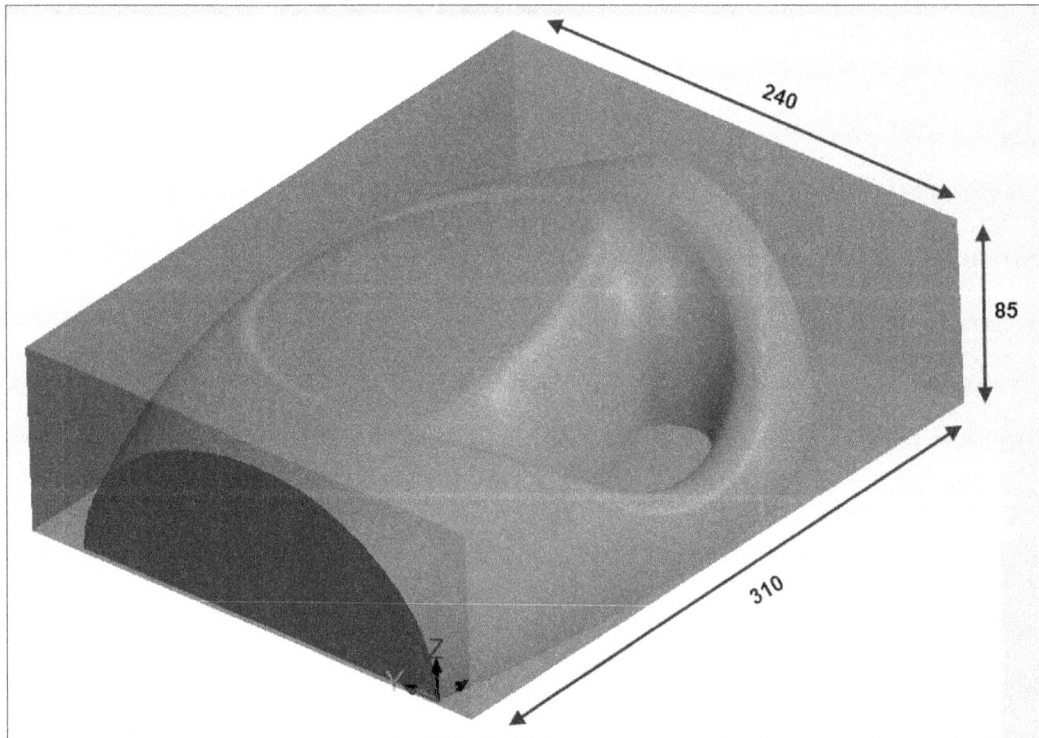

Figure-27. Stock for practical

Steps

Identifying toolpath strategies

We will perform Dynamic Optirough operation to rough complete part in one go. Then we will create stock after this operation to visualize the result of operation. To finish the part, we will use Hybrid 3D HighSpeed toolpath. In the end, we will verify the result of Hybrid toolpath only using the stock created earlier.

Preparation for Machining

*   Open the part for practical 1 of this chapter from the resource kit. The part should display as shown in Figure-26.
*   Start **Mastercam** add-in if not set to start yet.
*   Click on the **Planes Manager** tab from the **Design Tree**. The **Planes Manager** will be displayed.
*   Click on the **From geometry** option in **Create a new plane** drop-down of **Planes Manager**. The **Define Mastercam Plane PropertyManager** will be displayed.
*   Select the bottom face of the model as reference, corner point as origin and reverse the direction of plane if required; refer to Figure-28.

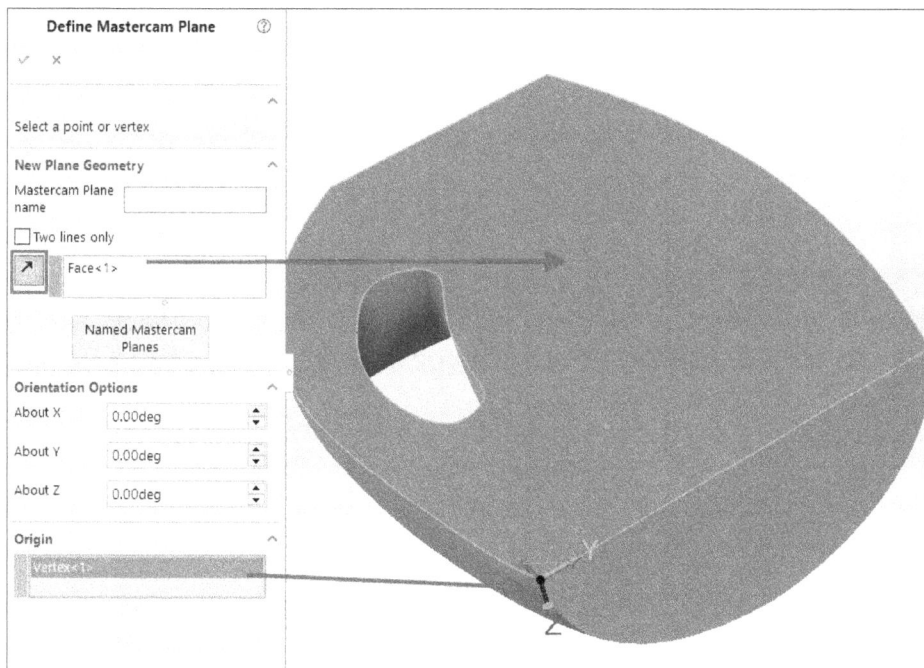

Figure-28. Selection for defining plane

*   Specify the name of plane as Bottom-Face and click on the **OK** button from **PropertyManager**.
*   The **Planes Manager** will be displayed again. Select the Bottom-Face plane and click on the **Set all** button ☐ to make it WCS and Toolplane.
*   Click on the **Stock setup** option from the expanded **Properties** node in the **Mastercam Toolpath Manager**. The **Machine Group Properties** dialog box will be displayed with **Stock Setup** tab selected.
*   Set the parameters in the dialog box as shown in Figure-29.

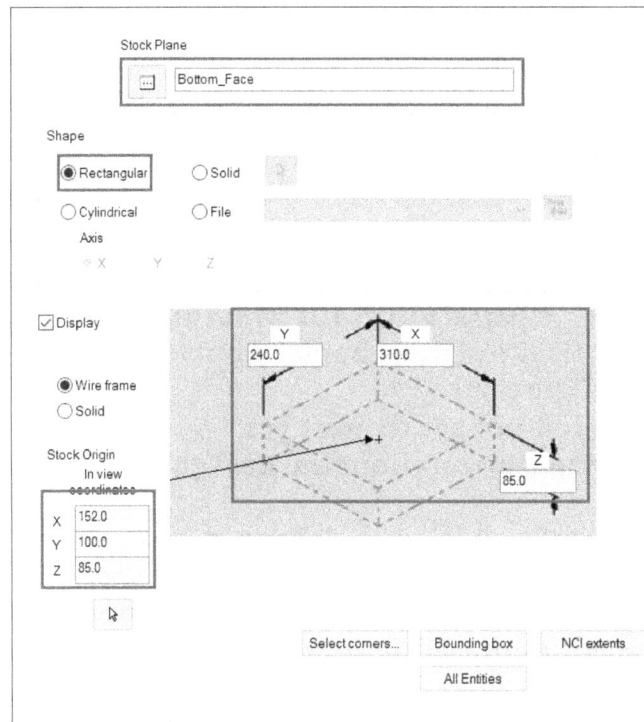

Figure-29. Parameters for stock

- Click on the **OK** button from the dialog box to create the stock.

Dynamic Optirough 3D High Speed Toolpath

- Click on the **Dynamic OptiRough** tool from the **3D HighSpeed** drop-down in the **Ribbon**. The **3D High Speed Toolpaths - Dynamic OptiRough** dialog box will be displayed with options to select geometry. Click on the **Select entities** button from the **Machining Geometry** area of dialog box. The **Selection PropertyManager** will be displayed as discussed earlier.
- Select the whole body of part using the **Body** filter; refer to Figure-30 and click on the **OK** button. The **3D High Speed Toolpaths-Dynamic OptiRough** dialog box will be displayed again.

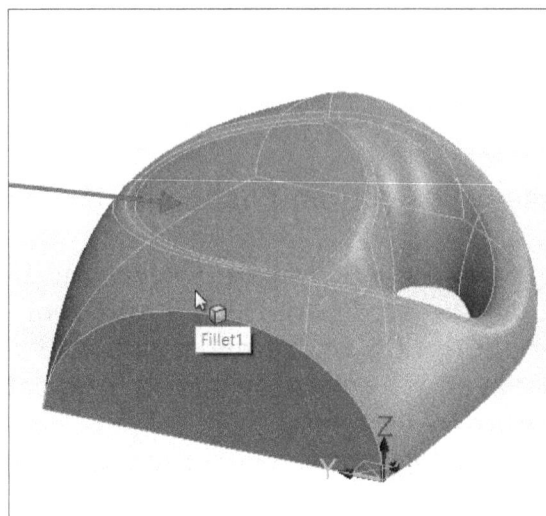

Figure-30. Part body selected for dynamic optirough toolpath

- Select Ball Nose End Mill tool of **12** mm diameter and select the desired holder from respective pages in the dialog box.
- Set the desired cutting parameters in the **Cut Parameters** page, if you want to change.
- Click on the **Linking Parameters** option and set the **Clearance plane** value as **65 Incremental** on the page.
- Click on the **OK** button from the dialog box. The toolpath will be generated; refer to Figure-31.

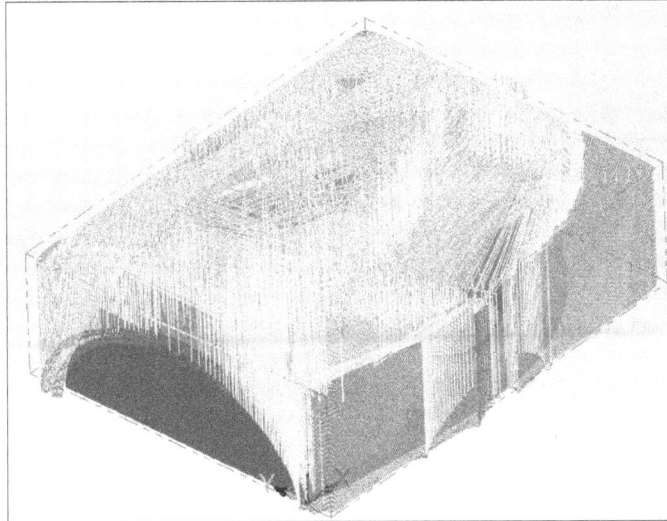

Figure-31. Dynamic optirough toolpath created

Creating Stock after Dynamic OptiRough

- Click on the **Stock Model** button from the right in the **Mastercam2022 CommandManager** of the **Ribbon**. The **Stock model** dialog box will be displayed.
- Set the same parameters for stock as done while defining stock for the part earlier. Specify the name of stock in the **Name** edit box as **After roughing**.
- Click on the **Source Operations** option from the left area in the dialog box. The options will be displayed as shown in Figure-32.

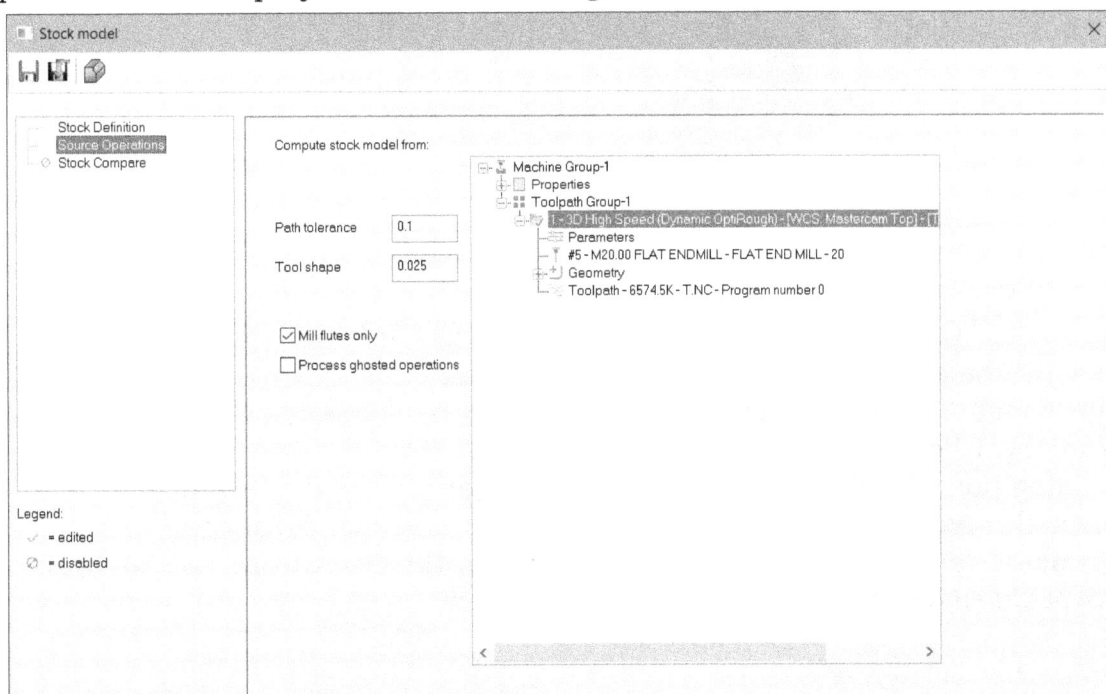

Figure-32. Source Operations page of Stock model dialog box

- Select the roughing operation created earlier from the right-box and click on the **OK** button. The system will start processing stock.
- Click on the **Multi-threading Manager** tool from the **Ribbon** to check progress of operation. The **Multi-Threading Manager** dialog box will be displayed showing the current progress in creating stock; refer to Figure-33. You may need to wait for a few minutes depending on your system configuration till the operation gets completed.

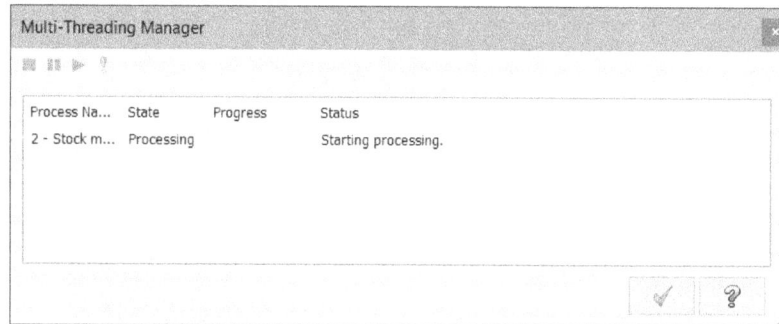

Figure-33. Multi Threading Manager dialog box

- Once the process is complete, the operation will be removed from the **Multi-Threading Manager** dialog box. Click on the **OK** button to exit. The stock will be created; refer to Figure-34.

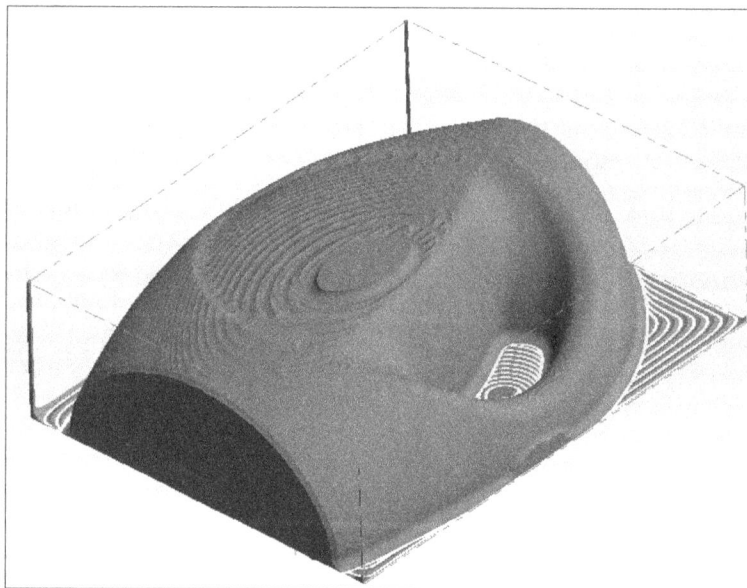

Figure-34. Stock after dynamic optirough operation

Creating Hybrid 3D High Speed Finish Toolpath

- Click on the **Hybrid** tool from the **3D HighSpeed** drop-down in the **Ribbon**. The **3D High Speed Toolpaths-Hybrid** dialog box will be displayed with options to select geometry. Click on the **Select entities** button from the **Machining Geometry** area of dialog box. The **Selection PropertyManager** will be displayed.
- Select the **Body** filter from the left area in the **PropertyManager** and select the part as done earlier in this practical. Click on the **OK** button. The **3D High Speed Toolpaths-Hybrid** dialog box will be displayed.

- Select the Ball Nose End Mill tool of diameter **8** mm and select the desired holder from respective pages.
- In the **Stock to leave** page specify the stock to leave values as **0** in edit boxes.
- Click on the **Cut Parameters** option from the left area and specify the values as: **Z stepdown = 0.5**, **Limiting angle = 85**, and **3D stepover = 1.5**.
- Click on the **Linking Parameters** option from the left in the dialog box and specify the **Clearance plane** distance as **50** Incremental.
- Click on the **OK** button from the dialog box. The toolpath will be created; refer to Figure-35. Note that you can see the progress of toolpath creation by using the **Multi-threading Manager** tool as discussed earlier.

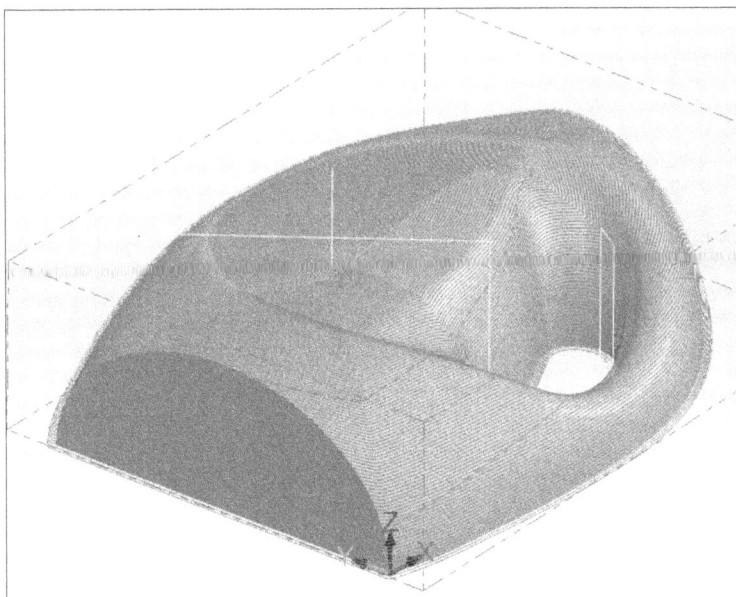

Figure-35. Hybrid toolpath created

Verifying Finish Operation only

- Click on the **Backplot/Verify Options** button ▣ from the **Mastercam Toolpath Manager**. The **Backplot/Verify Options** dialog box will be displayed.
- Select the **Stock model** radio button from the **Stock** area in the dialog box and make sure the **After roughing** option is selected in the drop-down; refer to Figure-36.
- If you want to add any fixture to model then you can do so by using the options in **Fixture** tab.
- Click on the **OK** button from the dialog box.
- Select the **Hybrid Surface High Speed toolpath** and click on the **Verify selected operation** button from the **Mastercam Toolpath Manager**. The **Mastercam Simulator** window will be displayed with roughed part as stock; refer to Figure-37.

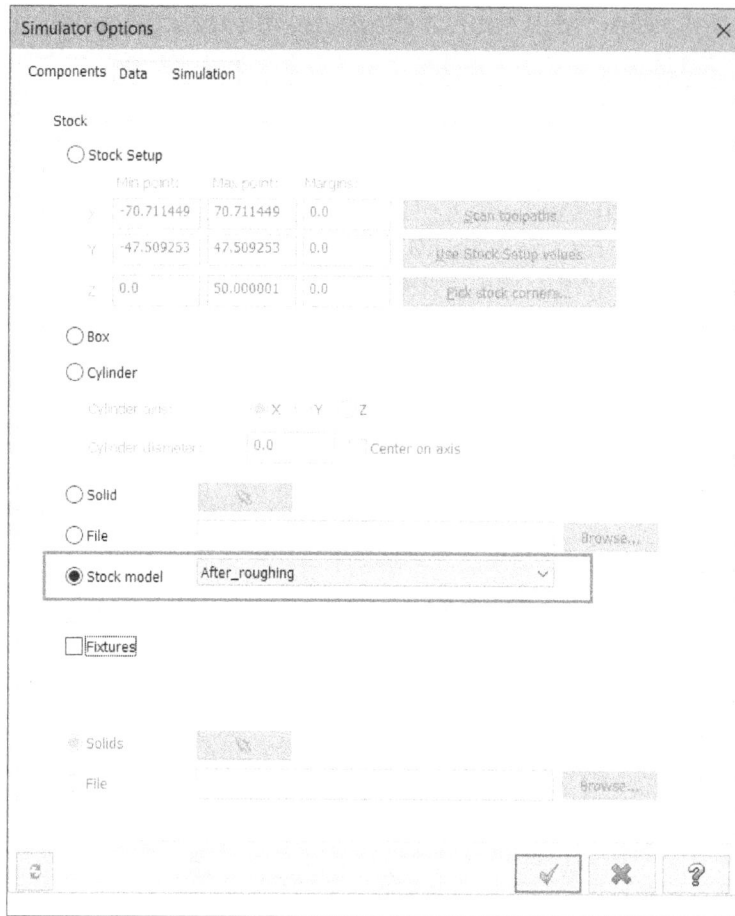

Figure-36. Stock model to be selected

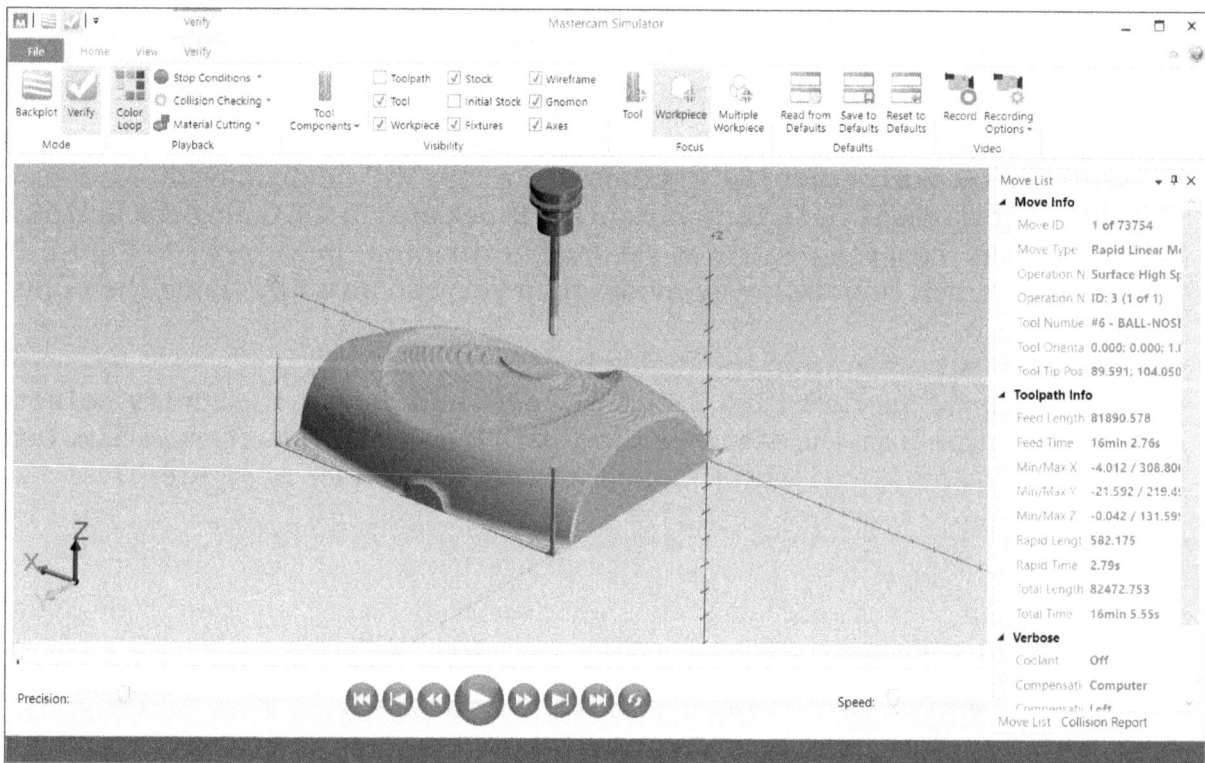

Figure-37. Mastercam Simulator for finishing

- Click on the **Play** button to check the simulation. After verifying, you will notice that some material is left at the bottom of the part; refer to Figure-38. Now, we will create last finishing toolpaths to remove material leftover.

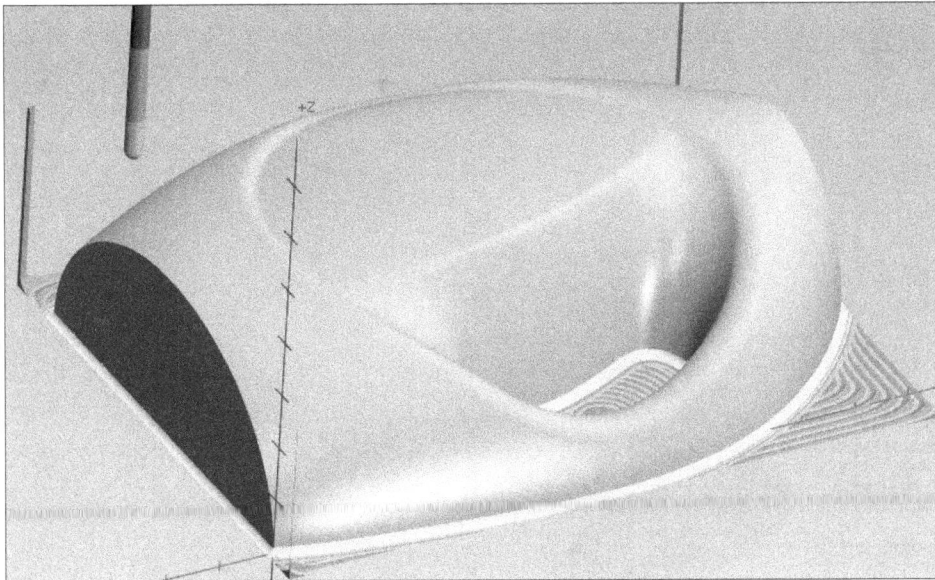

Figure-38. Model after finishing operation

2D Contour toolpaths to finish edges

- Click on the **Contour** tool from the **2D Toolpaths** drop-down in the **Ribbon**. The **Chain Manager** will be displayed.
- Select the outer boundary edges to form a loop; refer to Figure-39 and click on the **OK** button. The **2D Toolpaths - Contour** dialog box will be displayed.

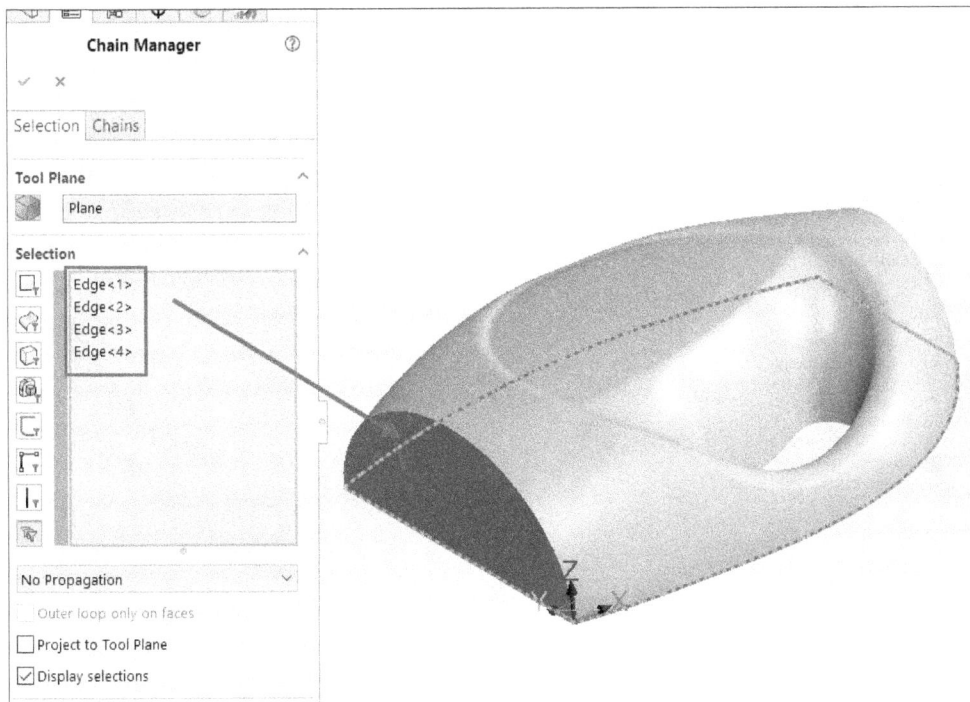

Figure-39. Edges selected for 2D contour toolpath

- Select the Flat End Mill of diameter **10** as tool and **B3C4-0011** as tool holder (to avoid collision between stock and holder) from the respective pages in the dialog box.

- Click on the **Cut Parameters** option from the left to change cutting parameters. Make sure **Stock to leave on walls** and **Stock to leave on floors** are set to **0**.
- Set the **Compensation direction** as **Right** in the drop-down and click on the **OK** button to create the toolpath. The toolpath will be generated; refer to Figure-40.

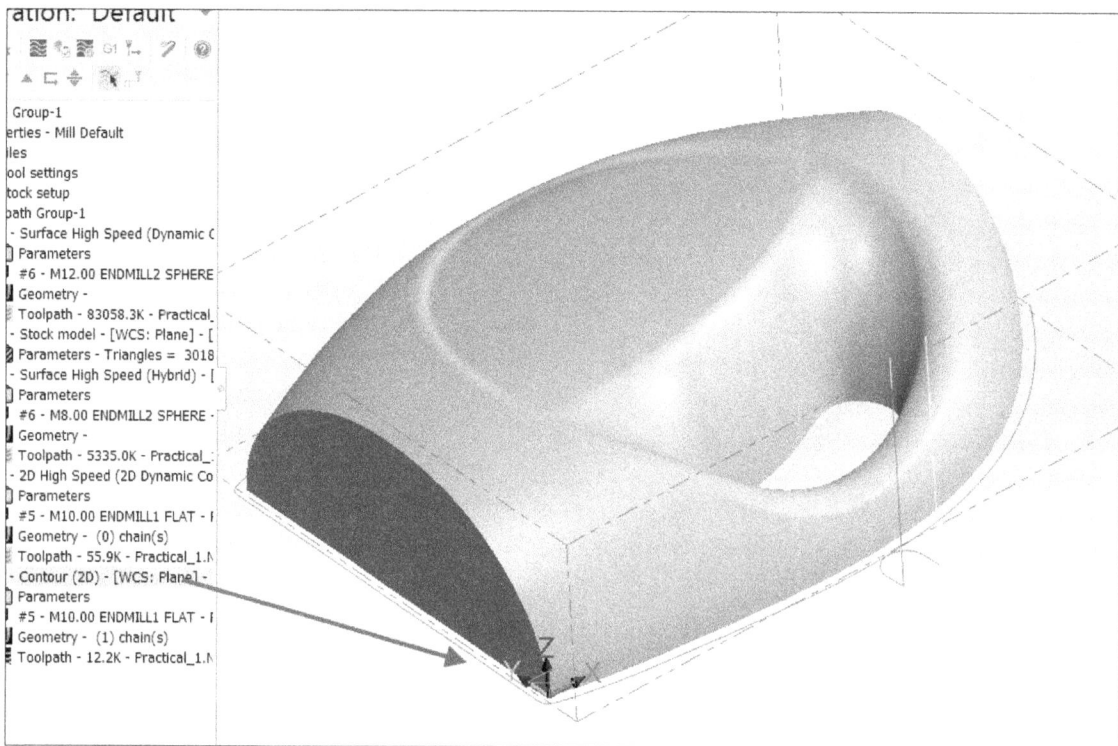

Figure-40. Contour toolpath created

Similarly, create 2D contour toolpath for inner using the same tool and settings but with **Compensation direction** set to **Left**; refer to Figure-41.

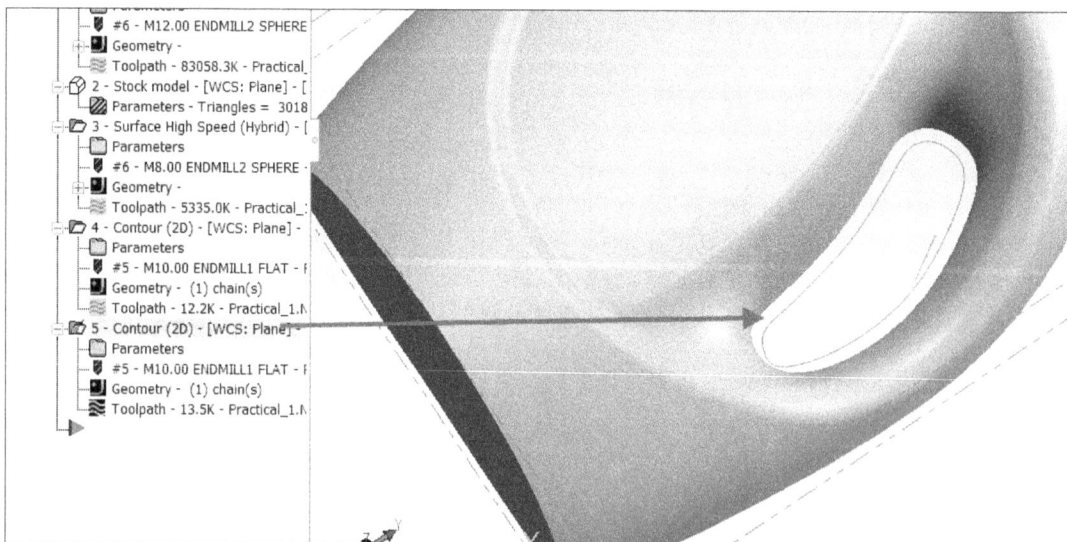

Figure-41. Second contour toolpath created

PRACTICE

Create NC program for milling the part shown in Figure-42. The stock for this part is a rectangular block of 70 mm x 110mm x 45 mm.

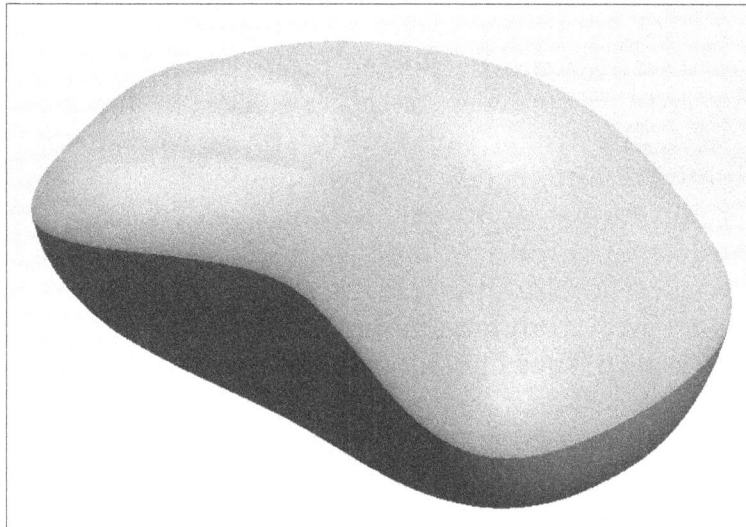

Figure-42. Model for Practice

TRANSFORM TOOLPATHS

Transforming toolpaths is a dangerous fun. If transformed in wrong way, it can cause accident so be careful while using toolpath transformation tools. Using the **Transform** tool in the **Mastercam2022 CommandManager**, you can translate, rotate, and mirror toolpath(s). Here, we will discuss the procedure of rotating the toolpath and you can apply the same procedure for all other operations.

- Click on the **Transform** tool from the **Mastercam2022 CommandManager** in the **Ribbon**. The **Transform Operation Parameters** dialog box will be displayed; refer to Figure-43.

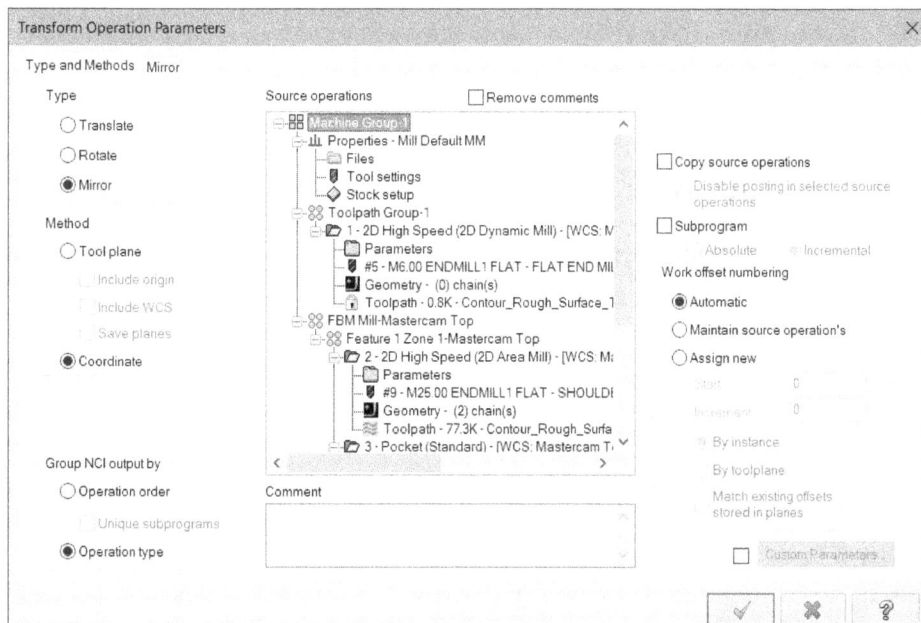

Figure-43. Transform Operation Parameters dialog box

- Select desired radio button from the **Type** area of the dialog box to define whether you want to translate, rotate, or mirror selected toolpaths. Select the **Translate** radio button if you want to move selected toolpaths to a different location. Select the **Rotate** radio button if you want to rotate selected toolpaths by specified value about an axis. Select the **Mirror** radio button if you want to create a mirror image of selected toolpaths about specified mirror plane.

- Select desired radio button from the **Method** area to define whether you want to create new tool planes and related geometries or you want to insert new coordinates in previous toolpaths. Select the **Tool plane** radio button to create a new tool plane after performing transformation operation. The **Include origin**, **Include WCS**, and **Save planes** check boxes will become available on selecting this radio button. Select the check boxes to include them with tool plane. Select the **Coordinate** radio button to copy coordinates of transformed toolpaths in original toolpath operation.

- Select desired radio button from the **Source** area of the dialog box to specify whether you want to copy only numeric codes or you want to copy geometry with toolpath. Select **NCI** radio button to copy only numeric codes generated by toolpath. Select the **Geometry** radio button to copy toolpath along with geometry while performing transformation operation.

- Select desired radio button from the **Group NCI output by** area to define how numeric codes will be grouped after transformation. Select the **Operation order** radio button to group output codes based on their operation order in original program. Select the **Operation type** radio button to group output codes based on what type of operation is being performed. For example, if there are pocket toolpaths and contour toolpaths in machining then all the pocket toolpaths will be grouped together and all the contour toolpaths will be grouped together.

- Select the **Remove comments** check box to remove comments given with toolpath after performing transformation.

- Select the **Create new operations and geometry** check box if you want to create a copy of operations and geometries after performing transformation. If this check box is not selected then all the transformation will be applied on the original operations and geometries. After selecting this check box, select the **Keep this transform operation** check box to apply all the transformations to newly created operations and geometries while keeping the originals unchanged.

- Select the **Copy source operations** check box to create a duplicate copy of selected toolpaths directly above source. After selecting this check, select the **Disable posting in selected source operations** check box to prevent source operations from being posted twice by disabling posting for the source operation.

- Select the **Subprogram** check box to create repeating sections of program as subprograms. Note that this option is not available if you have opted to create new operations and geometries. After selecting this check box, select the **Absolute** radio button if you want to use absolute coordinates or select the **Incremental** radio button to use incremental coordinate values when creating the program.

- Select desired radio button from the **Work offset numbering** area to define how work offset numbers will be created. Select the **Automatic** radio button to automatically assign next available offset number. Select the **Maintain source operation's** radio button to use same offset numbers as defined in source operations. Select the **Assign new** radio button to create a new set of work offset numbers. After selecting the **Assign new** radio button, specify related parameters in the edit boxes below it. Select the **Match existing offsets stored in planes** check box to match offset numbers with existing offsets associated with planes.

- Based on the radio button selected in the **Type** area of the dialog box, a new tab will be added in the dialog box. For example, if the **Mirror** radio button is selected in the **Type** area then **Mirror** tab will be added in the dialog box. The options of **Mirror**, **Rotate**, and **Translate** tabs are discussed next.

Mirror Transformation

If the **Mirror** radio button is selected then **Mirror** tab will be displayed in the dialog box. Click on the **Mirror** tab to modify parameters related to mirror transformation; refer to Figure-44. Options of this tab are discussed next.

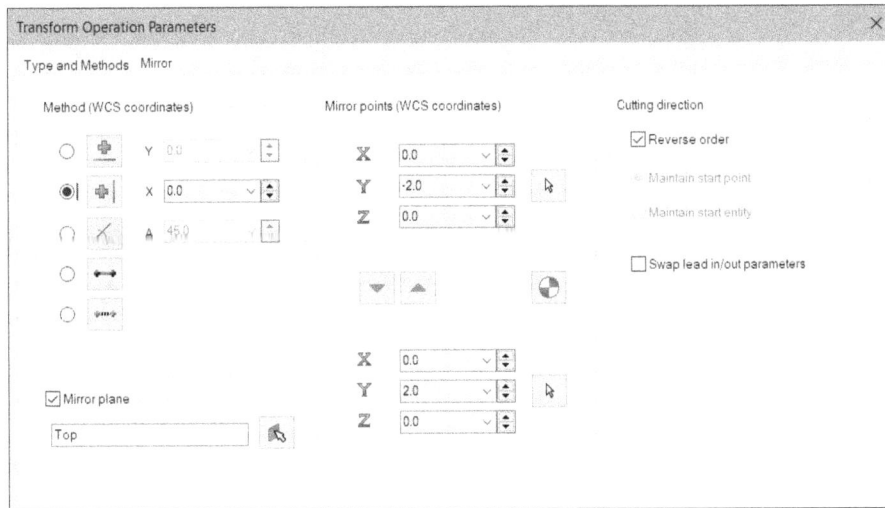

Figure-44. Mirror tab

- Select the **Mirror plane** check box from the **Mirror** tab in dialog box to use a plane as mirror reference.
- Click on the **Select Plane** button from the **Mirror plane** area of the dialog box. The **Plane Selection** dialog box will be displayed.
- Select desired plane from the dialog box and click on the **OK** button. Note that mirror plane and tool plane should be same for creating mirror copy.
- Select desired radio button from the **Method (WCS coordinates)** area of the dialog box to define location and orientation of WCS for measuring location of mirror plane/point.
- Specify desired coordinates in the **Mirror points (WCS coordinates)** area to define the start and end locations of mirror line.
- Select the **Reverse order** check box from the **Cutting direction** area to reverse the start and end points of toolpaths. Select the **Maintain start point** radio button to keep the starting point same as in original toolpath. Select the **Maintain start entity** radio button to use same geometries as selected for original toolpath.
- After setting desired parameters, click on the **OK** button to create mirror copy.

Rotation Transformation

If the **Rotate** radio button is selected in the **Type** area of the **Type and Methods** tab in the dialog box then **Rotate** tab will be displayed in the dialog box. Click on the **Rotate** tab to modify parameters related to rotate transformation; refer to Figure-45. The options of this tab are discussed next.

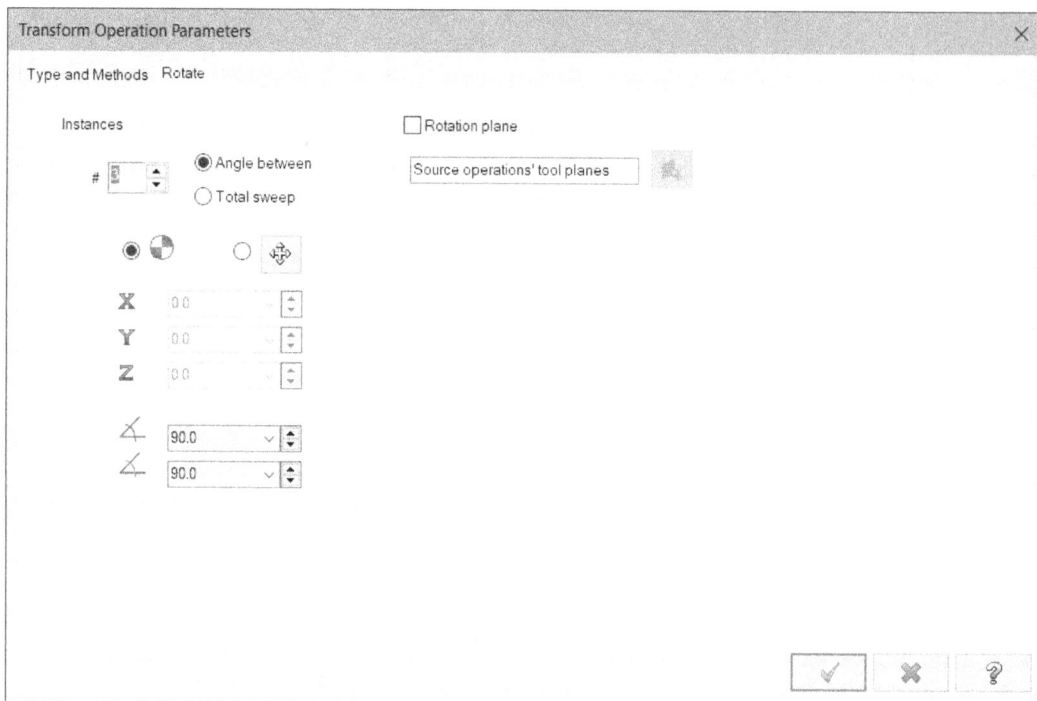

Figure-45. Rotate tab

- Specify the number of copies to be created in the **Instances** spinner.
- Select the **Angle between** radio button from the **Instances** area to define angular gap between two consecutive instances of the toolpath. Select the **Total sweep** radio button if you want to specify total angular span in which the instances will be placed equally spaced.
- Specify desired values in the first and second angle edit boxes to define angle values for both upper and lower angle ranges, respectively.
- Select the **Rotation plane** check box to use selected plane as reference for rotation. On selecting the check box, selection box below the check box will become active.
- Click on the **Select plane** button from **Rotation plane** area and select desired plane to be used for rotation.
- After setting desired parameters, click on the **OK** button to perform transformation; refer to Figure-46.

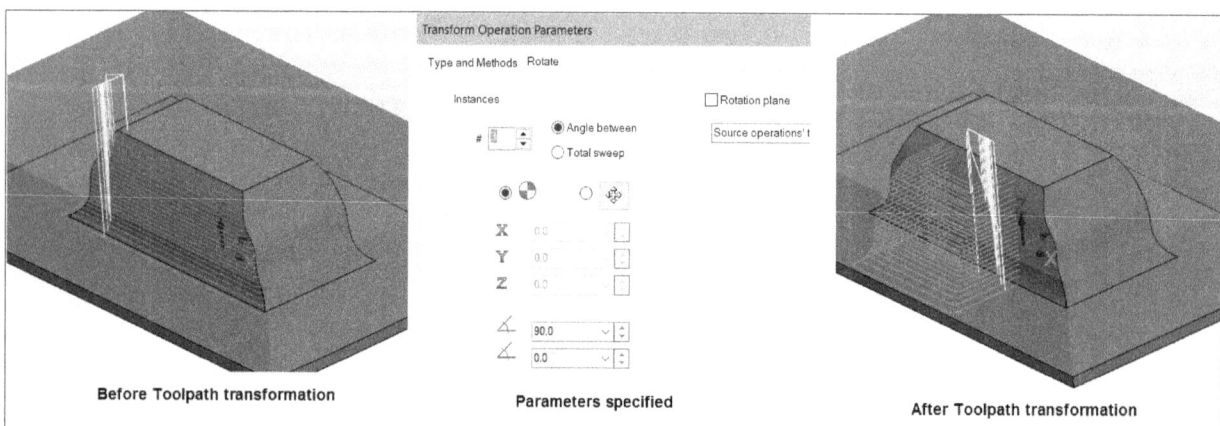

Figure-46. Rotating Toolpath

You may ask what are the benefits of such operations. Answer is: When the part placed on the machine is slightly rotated or tilted on the machine due to some hardware conditions then you can rotate or tilt your toolpaths to compensate hardware faults. (Yes I know, repairing some machine problems are quite expensive!!)

Translate Transformation

If the **Translate** radio button is selected then **Translate** tab will be displayed in the dialog box. Click on the **Translate** tab to modify parameters related to translate transformation; refer to Figure-47.

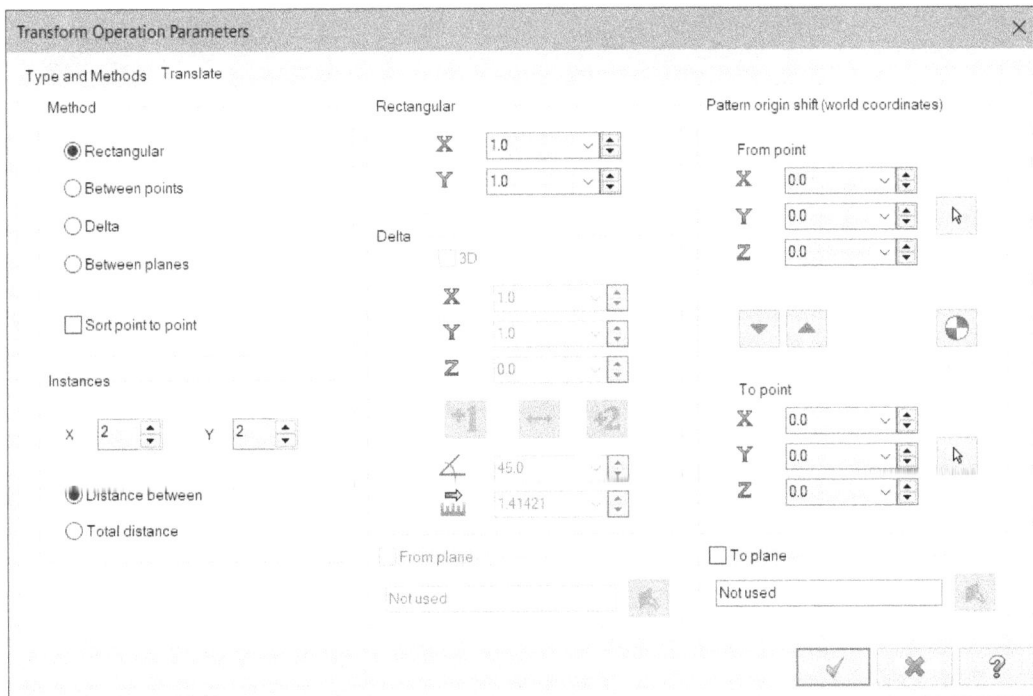

Figure-47. Translate tab

- Select desired radio button from the **Method** area to specify method to be used for moving selected toolpaths to different location. Select the **Rectangular** radio button from the **Method** area to specify distance to be moved in **X** and **Y** edit boxes of **Rectangular** area. Select the **Between points** radio button from the **Method** area to use **X**, **Y**, and **Z** edit boxes for **From point** and **To point** in the **Pattern origin shift (world coordinates)** area of the dialog box to move selected toolpaths. Select the **Delta** radio button to specify offset distance from values specified in the **Pattern origin shift (world coordinates)** area of the dialog box. Select the **Between planes** radio button from the **Method** area to use planes selected in the **From plane** and **To plane** areas of the dialog box.

- If the **Rectangular** radio button is selected in the **Method** area then specify the parameters in the **Instances** area of the dialog box to define number of instances to be created after translation.

- After setting desired parameters, click on the **OK** button to perform transformation.

TOOLPATH TRIM

The **Toolpath Trim** tool is used to cut the toolpaths at specified boundaries. In Master for SolidWorks, there are limited functions available for this tool so you may not be able to trim all the toolpaths. We have seen the **Toolpath Trim** tool working on all transformed toolpaths. The procedure to use this tool is given next.

- Select any transformed toolpath (even with zero rotation applied and source operation posting off; refer to Figure-48) and click on the **Toolpath Trim** tool from the **Mastercam2022 CommandManager**. The **Chain Manager** will be displayed.

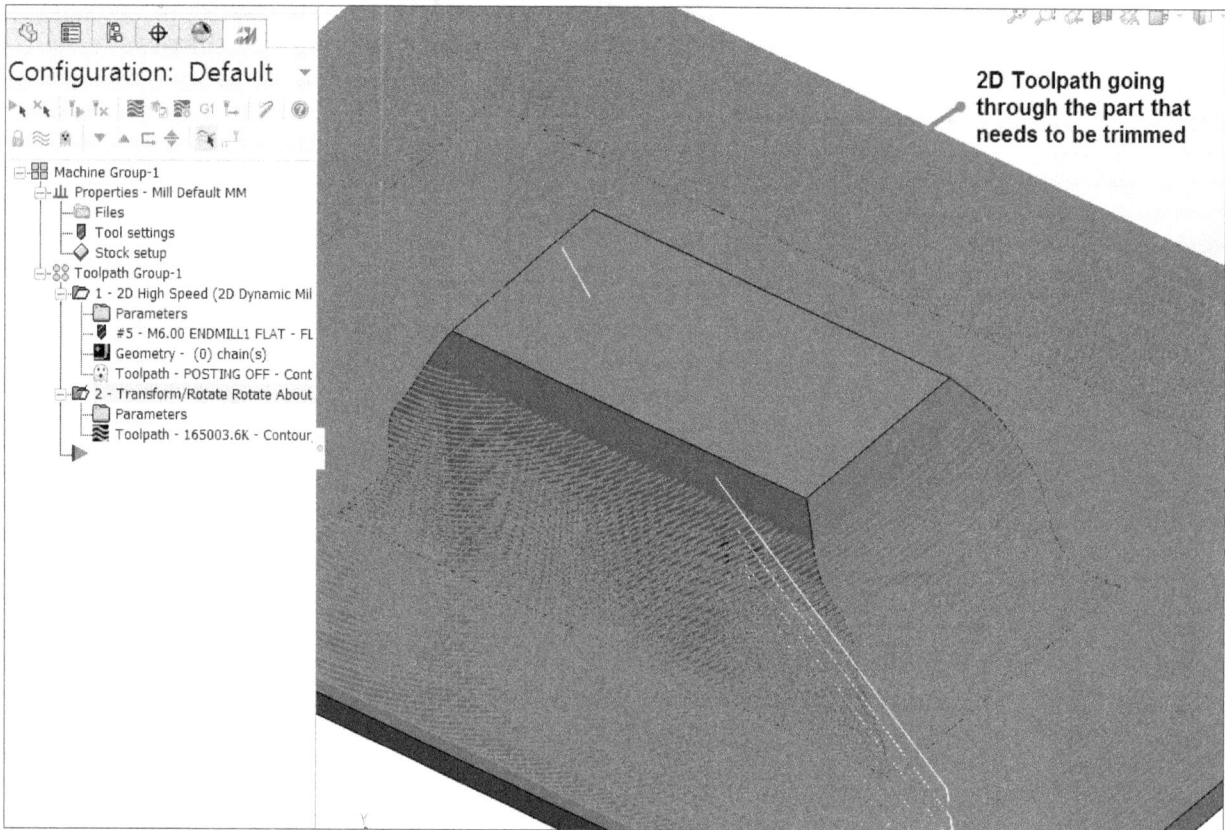

Figure-48. Toolpath to be trimmed

- Select the face/edge/boundaries by which you want to trim the toolpath; refer to Figure-49 and click on the **OK** button from the **Chain Manager**. The **Selection PropertyManager** will be displayed and you will be asked to select a bias point.
- Select a point on the side which you want to be saved after trimming; refer to Figure-50. Click on the **OK** button. The **Trimmed** dialog box will be displayed; refer to Figure-51.

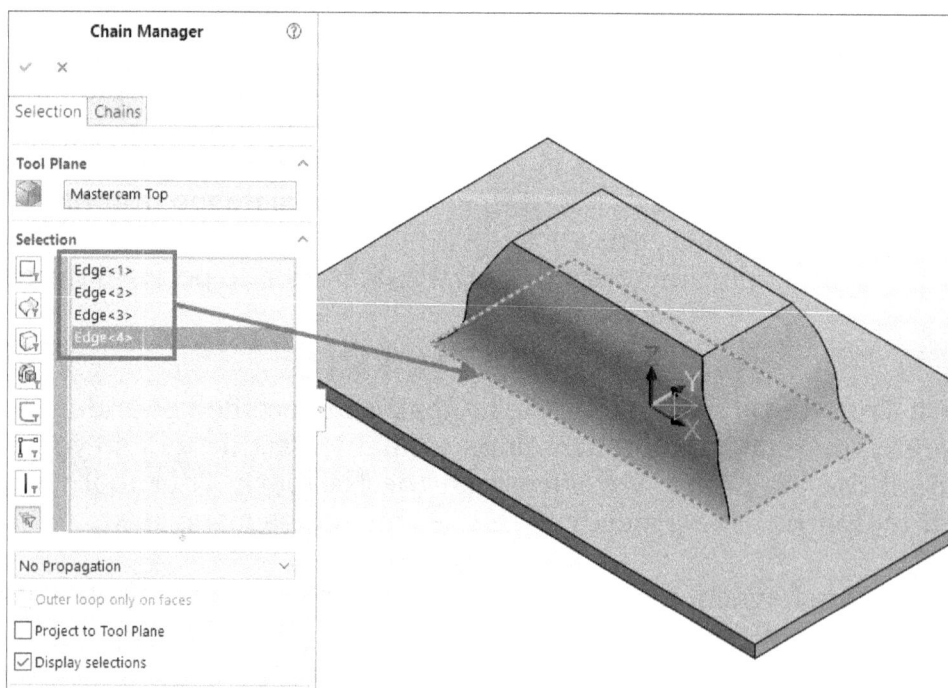

Figure-49. Boundary edges selected for trimming

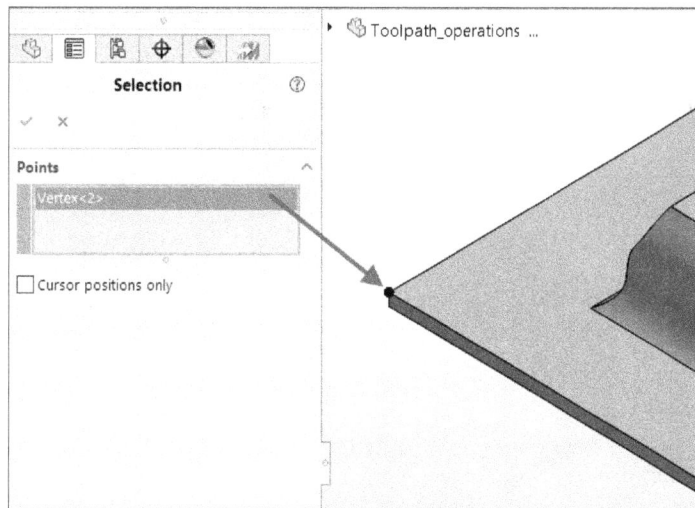

Figure-50. Point selected as bias point

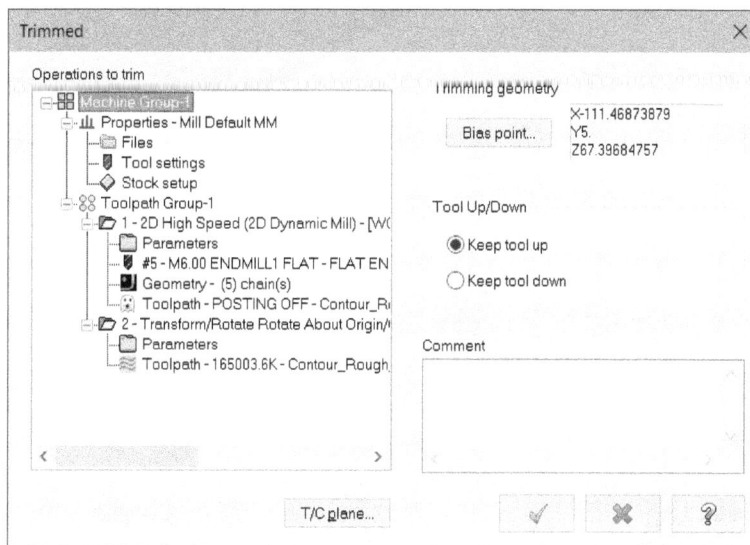

Figure-51. Trimmed dialog box

- Select all the operations from the **Operations to trim** box in the dialog box that you want to be trimmed. Note that the tool/construction plane for trimming should be same as of operations to be trimmed.
- Select the **Keep tool up** radio button if you want the tool to retract at boundaries. Select the **Keep tool down** radio button if you do not want the tool to retract at boundaries. Note that selecting the **Keep tool down** radio button can cause the tool to violate trimming boundaries.
- Click on the **OK** button. The selected toolpaths will get trimmed; refer to Figure-52.

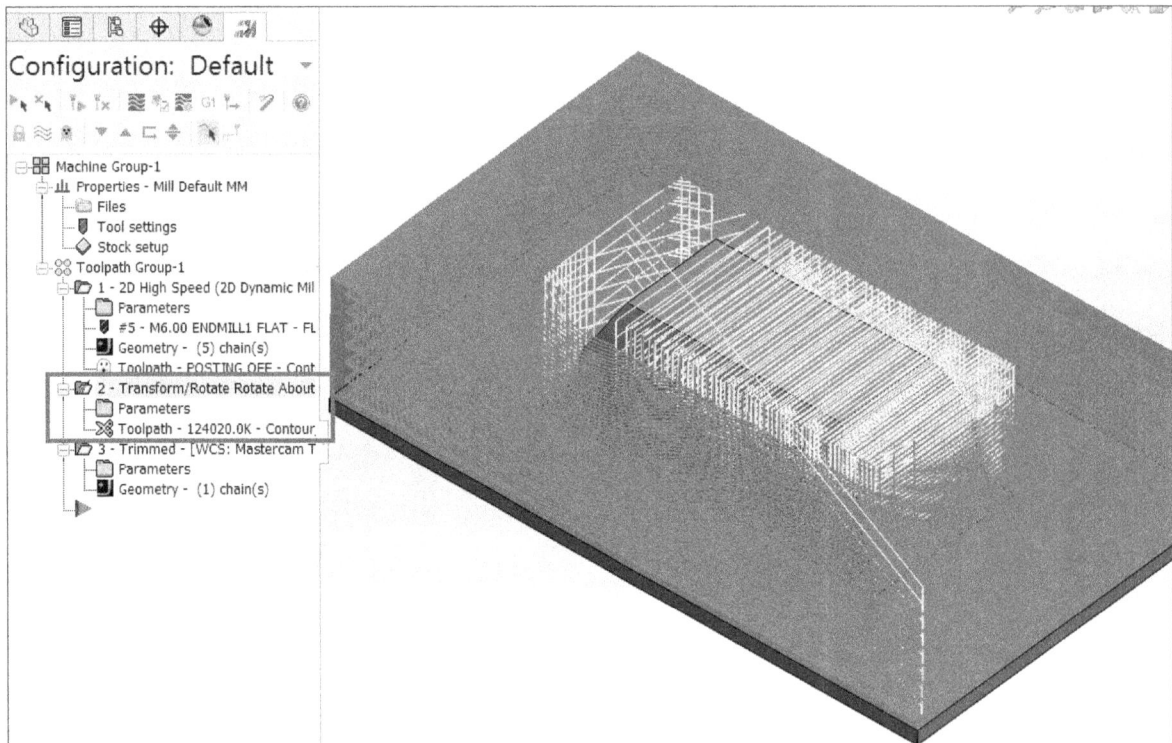

Figure-52. Trimmed toolpath

NESTING TOOLPATHS

Nesting is done to create multiple parts from the same workpiece. In this way, we can reduce material wastage and overall cycle time for multiple parts. The procedure to use **Nesting** tool is given next.

- Click on the **Nesting** tool from the **Mastercam2022 CommandManager**. A warning message will be displayed tell you that toolpaths created using Top view can only be used for nesting. Click on the **OK** button from the dialog box. The **Nesting** dialog box will be displayed; refer to Figure-53. By default, the **Sheets** tab is selected in the dialog box. The options of this tab are discussed next.

Sheets tab

- Specify desired length and width of sheet in the **Size** edit boxes.
- Select the **Create necessary quantity** check box to automatically specify the number of toolpaths to be fitted in the sheet.
- Click in the **X** and **Y** edit boxes for **Origin** option to define location for start point of toolpaths.
- Click on the **Material** button to define the material of workpiece. Based on selected material, the feed rate and other cutting parameters will be decided.

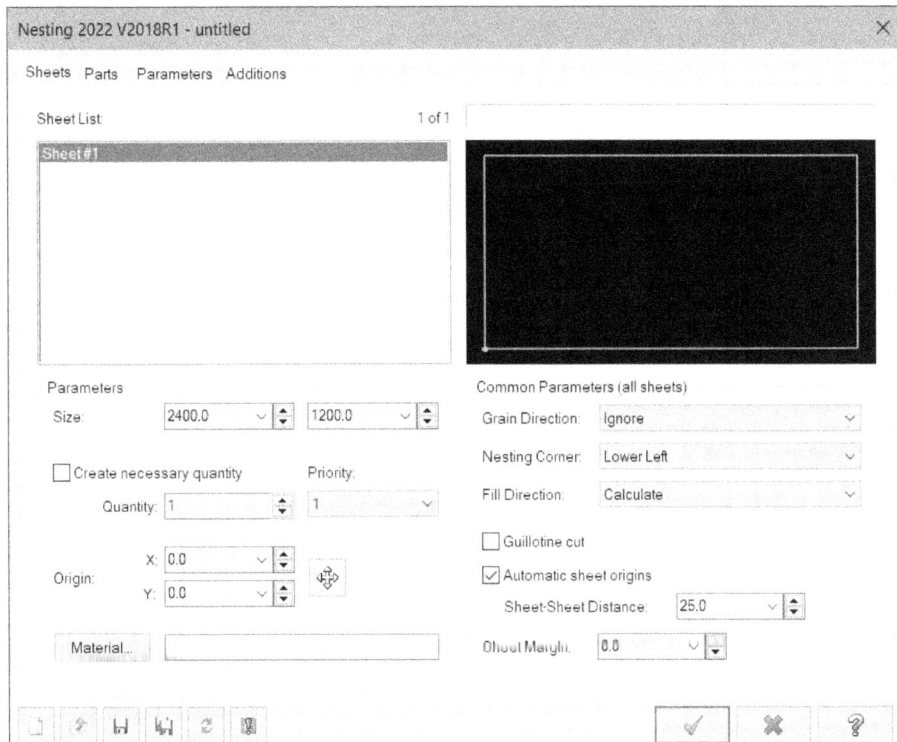

Figure-53. Nesting dialog box

- Select desired option from the **Grain Direction** drop-down to define grain direction for easy cutting of workpiece, if there is a noticeable difference in grains on surface of workpiece. Select the **Horizontal** or **Vertical** option from the drop-down to use respective direction as grain direction. Select the **Ignore** option, if grain direction does not affect cutting.
- Select desired option from the **Nesting Corner** drop-down to define start point for nesting of toolpaths.
- Select desired option from the **Fill Direction** drop-down to define the direction to be used for placing instances of the toolpath. You can select **Horizontal** or **Vertical** option from the drop-down to use respective direction or you can use the **Calculate** option and let software to decide automatically the direction for nesting.
- Select the **Guillotine cut** check box to arrange the parts in such a way that straight cuts can be made on the workpieces.
- Select the **Automatic sheet origins** check box to allow Mastercam to place origins of multiple sheets automatically.
- Specify desired value in **Sheet-Sheet Distance** edit box to define gap between two consecutive sheets.
- Specify desired value in the **Sheet Margin** edit box to define margin gap around the sheets.

Parts tab

The options in the **Parts** tab are used to add, remove, and arrange parts; refer to Figure-54. The options of this tab are discussed next.

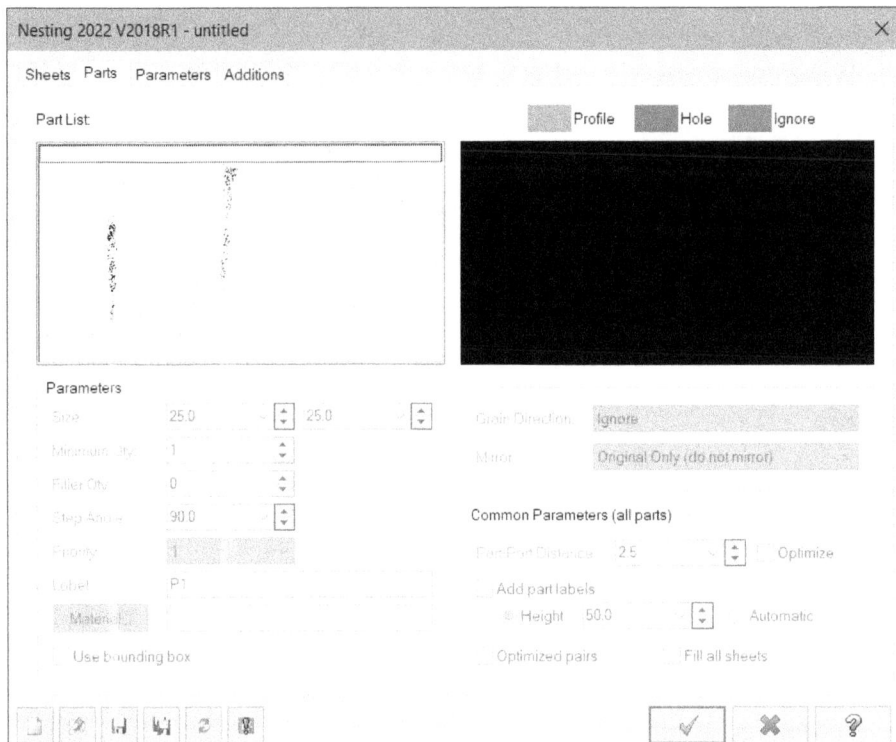

Figure-54. Parts tab

- Right-click in the **Part List** area to select parts/operations to be nested. A shortcut menu will be displayed; refer to Figure-55. Select the **Add operations** option from the menu to add new operations in the list. The **Select Operations** dialog box will be displayed with the list of operations applied in current file; refer to Figure-56.

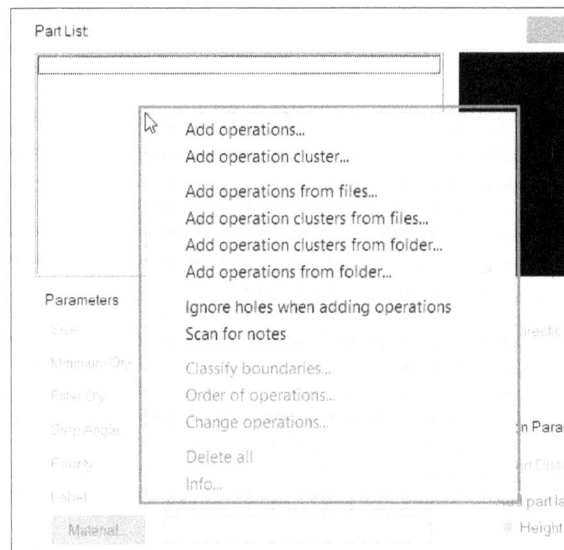

Figure-55. Shortcut menu displayed

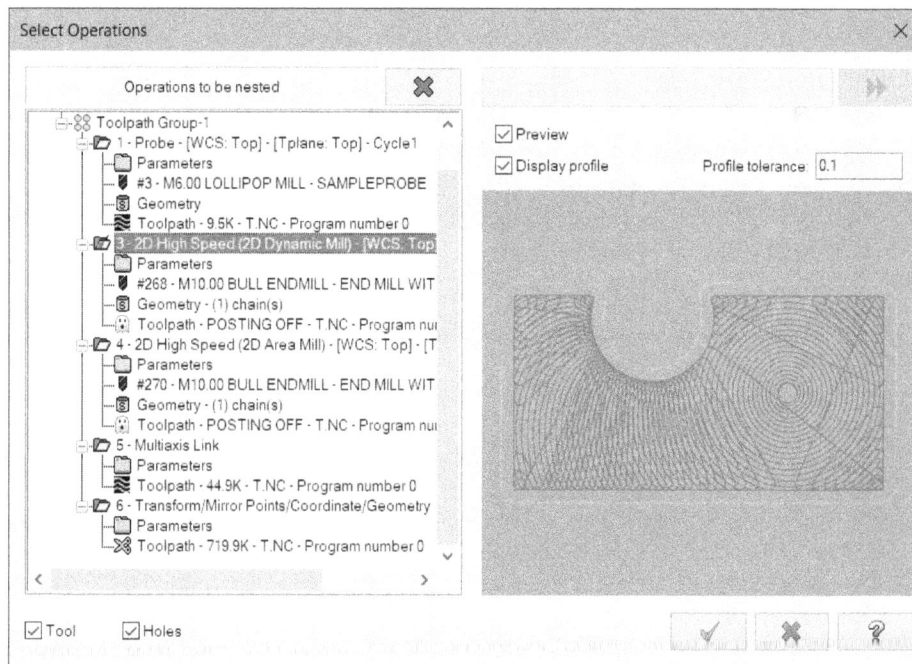

Figure-56. Select Operations dialog box

- Select desired operation(s) to be added in the part list and click on the **OK** button from the dialog box. The selected operations will be added in the list. Similarly, you can use the other options of the shortcut menu to add operations.
- Specify desired value in the **Minimum Qty** edit box to define minimum number of copies of selected operation to be added in the sheet.
- Specify desired value in the **Filler Qty** edit box to define number of filler part copies to be added in the sheet after main parts have been added.
- Specify desired value in the **Step Angle** edit box to define angle steps to be used for rotating parts when nesting.
- Select desired option from the **Priority** drop-down to define which parts will be placed on sheet for nesting. By default, larger parts are given preference in nesting.
- Select the **Use bounding box** check box to use the bounding boxes around parts to define their size on sheet.
- Select the **Use part labels** check box to place the label of part along with part on the sheet. After selecting this check box, specify the label value in **Label** edit box and height of label in the **Height** edit box below the check box. You can select the **Automatic** radio button below the check box to let Mastercam automatically decide the height of label.
- Select the **Optimize pairs** check box to create close fit pairs of parts in nesting for efficient use of space.
- Select the **Fill all sheets** check box to automatically fill the spaces in all sheets with main parts and filler parts as possible.
- Set the other parameters of this tab as discussed earlier.

Parameters tab

The options in the **Parameters** tab are used to define parameters like sorting order, conditions for stopping nesting toolpaths, and work offsets; refer to Figure-57. The options in this tab are discussed next.

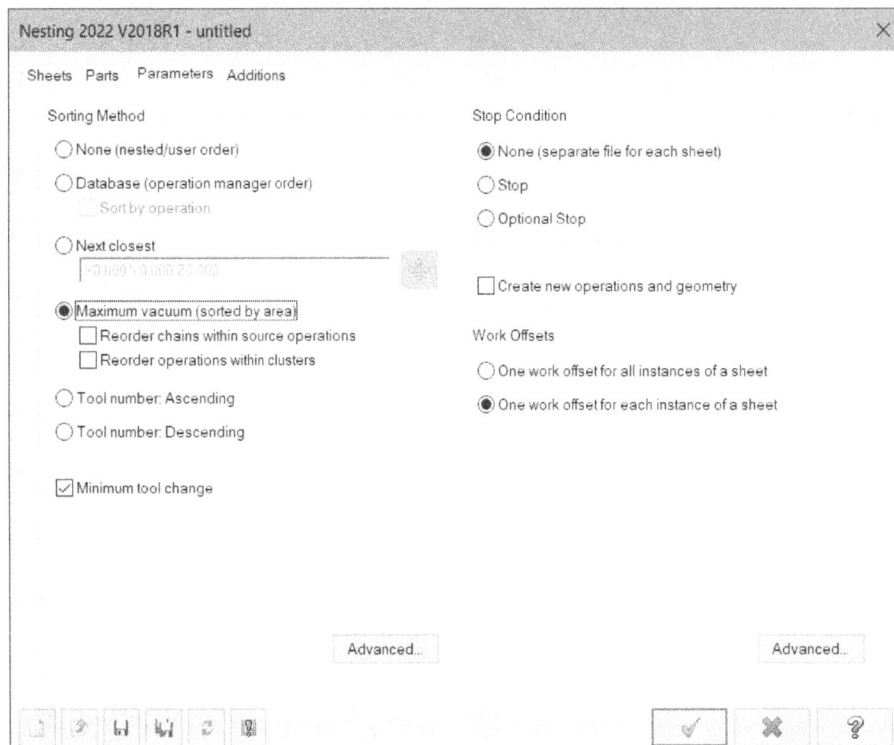

Figure-57. Parameters tab

- Select desired radio button from the **Sorting Method** area to define how operations in various sheets will be arranged for machining. Select the **None (nested/user order)** radio button to use the same order for machining as specified for nesting. Select the **Database (operation manager order)** radio button to use same order in which operation are present in the **Toolpaths Manager (Operations Manager)** for machining. Select the **Next closest** radio button from the **Sorting Method** area to arrange the operations as per their distance from specified point. Select the **Maximum vacuum (sorted by area)** radio button to machine those operations first which have more free area around them for operation. Select the **Tool number: Ascending** radio button from the dialog box if you want to perform operations as per the ascending number of tools used in those operations. Similarly, select the **Tool number: Descending** radio button to sort operations as per descending number of used tools.

- Select the **Minimum tool change** check box to sort the operations by tool, so that all parts using one tool are cut before changing to the next tool.

- Click on the **Advanced** button from the **Sorting Method** area to divide the sheet into regions for user defined sorting. The **Advanced Sorting** dialog box will be displayed; refer to Figure-58. Select the **Sort by region** check box and specify related parameters in the dialog box to define sorting order for regions. After setting parameters, click on the **OK** button from the dialog box.

- Select desired radio button from **Stop Condition** area of the dialog box to specify when will the machine stop for changing material. Select the **None (separate file for each sheet)** radio button to output toolpaths as per the sheets without program stops. Select the **Stop** radio button to insert program stop code (M00) at the end of each sheet. Select the **Optional Stop** radio button to insert optional stop code (M01) at the end of each sheet.

- Select the **Create new operations and geometry** check box to copy geometry and operations from original operations for each location in the nesting sheets.

- Select the **One work offset for all instances of a sheet** radio button to use a single work offset number for same tool operations in all the sheets. Select the **One work offset for each instance of a sheet** radio button to use different work offset numbers for same operations in different sheets.

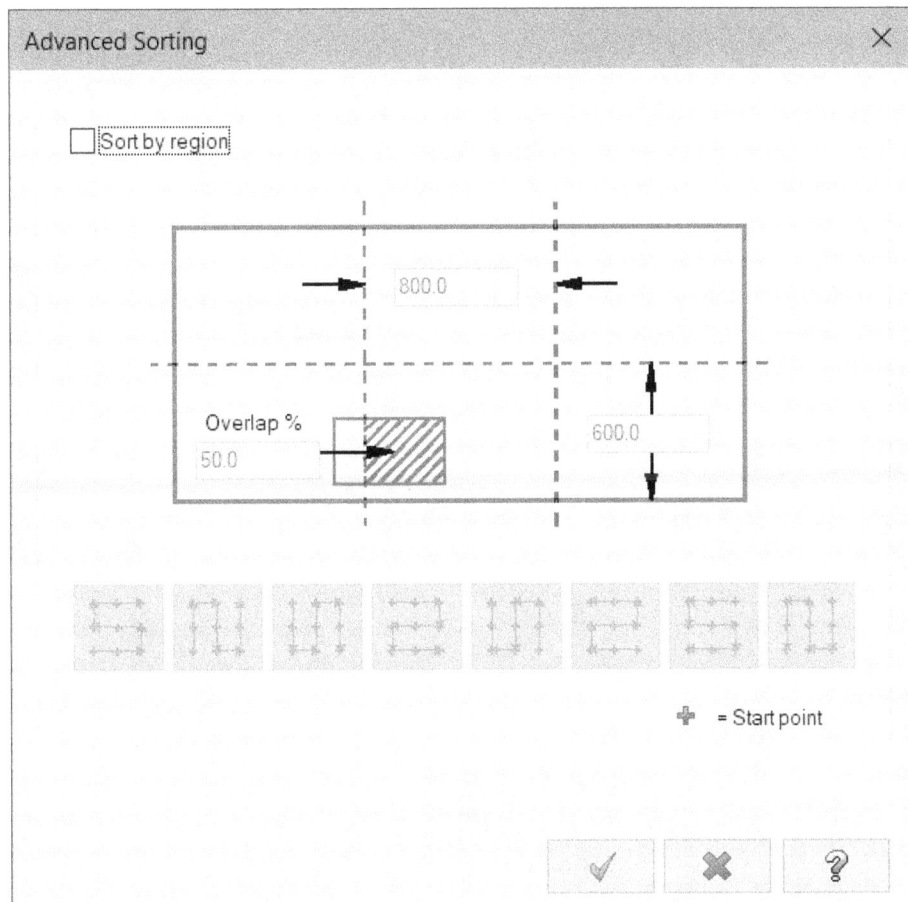

Figure-58. Advanced Sorting dialog box

- Click on the **Advanced** button from the **Work Offsets** area of the dialog box to specify advanced parameters related to work offset. The **Advanced Offsets** dialog box will be displayed; refer to Figure-59. Select the **Automatic** radio button from the dialog box to use next available work offset automatically. Select the **Maintain source operation's** radio button to use offset numbers as specified in operation data. Select the **Assign new** radio button to create a new set of offset numbers with specified increment parameters. After setting desired parameters, click on the **OK** button.

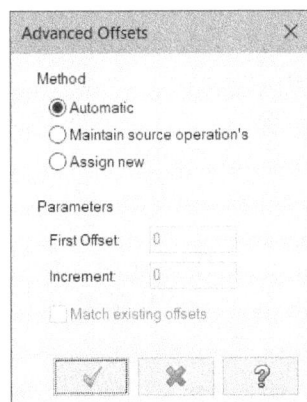

Figure-59. Advanced Offsets dialog box

Additions tab

The options in the **Additions** tab are used to specify parameters related to stock and toolpath links; refer to Figure-60. The parameters in this tab are discussed next.

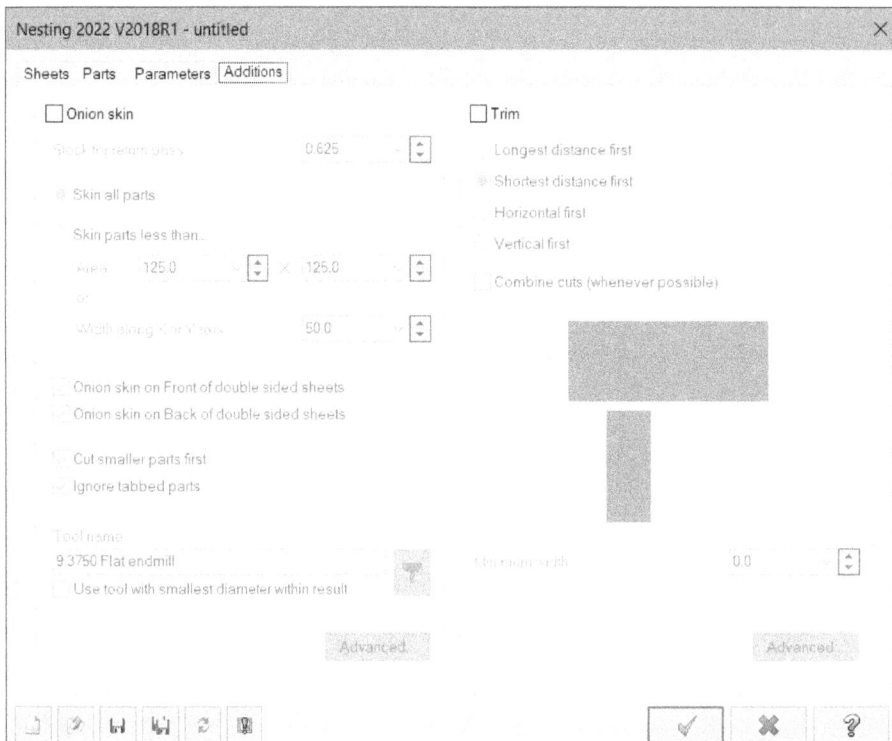

Figure-60. Additions tab

- Select the **Onion skin** check box from the dialog box to create a final return cutting pass after performing operations on all the parts of sheets. On selecting the check box, options in the **Onion skin** area will become active.
- Specify the value of stock to be removed by return pass in the **Stock for return pass** edit box.
- Select the **Skin all parts** radio button to include all the parts for performing skin contour machining operation.
- Select the **Skin parts less than** radio button to include only those parts for skin contour machining operation which have area less than values specified in the edit boxes below the radio button. You can either specify the area parameters or width parameters in these edit boxes.
- Select the **Onion skin on Front of double sided sheets** check box to create operation for front side of double sided sheets. Double sided sheets are those which have parts on their both sides.
- Select the **Onion skin on Front of double sided sheets** check box to create operation for back side of double sided sheets.
- Select the **Cut smaller parts first** check box to start cutting small size parts first and then machine larger parts. By default, the larger parts are machined first.
- Click on the **Select library tool** button next to **Tool name** edit box for selecting cutting tool to perform return cut.
- If you want to use smallest cutting tool earlier used in performing various operations on the sheet then select the **Use tool with smallest diameter within result** check box.

- Click on the **Advanced** button from the **Onion skin** area of dialog box to define advanced toolpath parameters for onion skin machining. The **2D Toolpaths - Onion skin** dialog box will be displayed; refer to Figure-61.

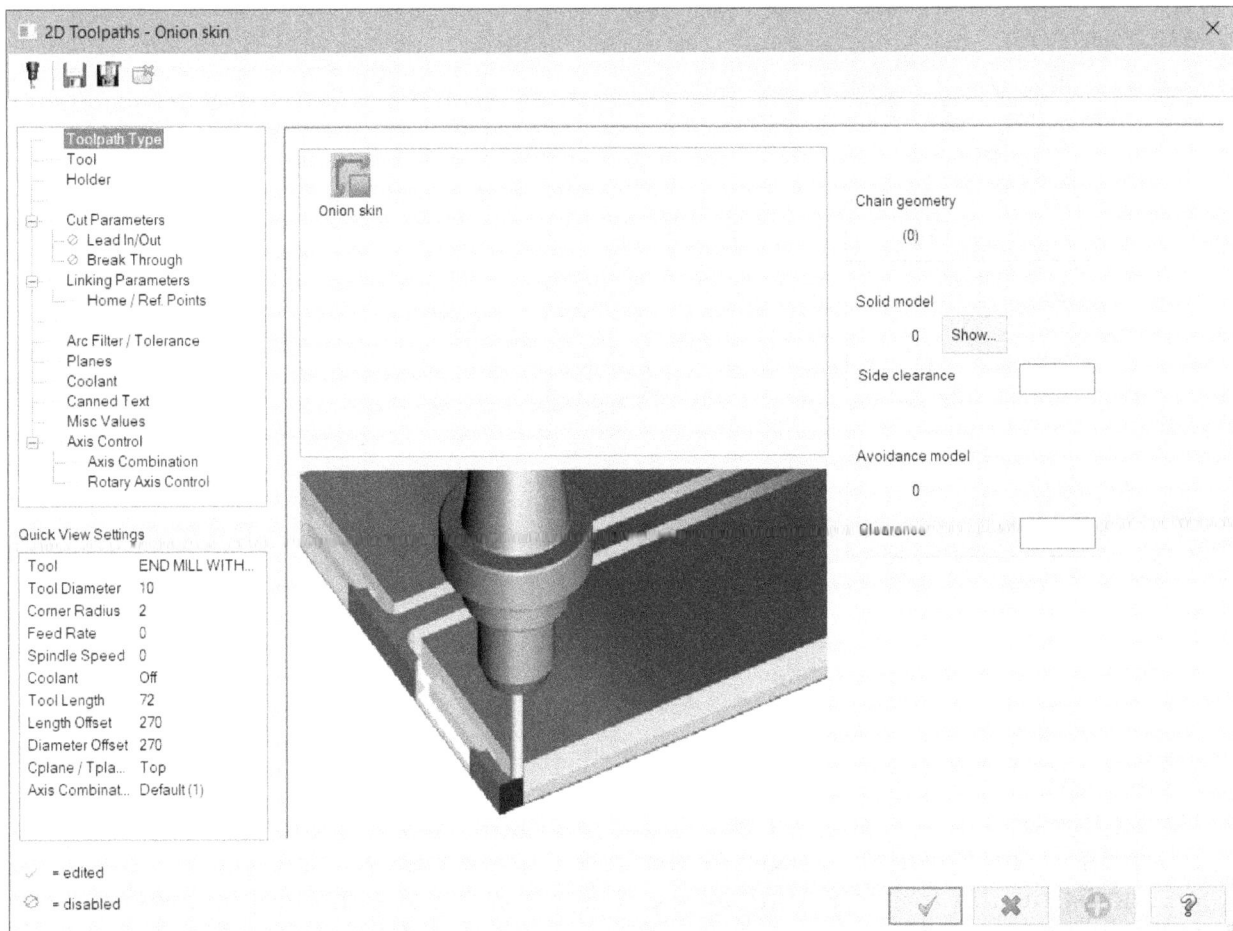

Figure-61. 2D Toolpaths-Onion skin dialog box

- The options in this dialog box are same as discussed for 2D toolpaths in previous chapters. Set the parameters as desired and click on the **OK** button.
- Select the **Trim** check box to create toolpath for machine remnants from the sheet. Note that you will need saw operation to perform trimming. On selecting this check box, the options in the **Trim** area will become active.
- Select the **Longest distance first** radio button to cut the sheet from longest side first. Select the **Shortest distance first** radio button to cut the sheet from shortest side first. Select the **Horizontal first** radio button to cut from horizontal side first. Select the **Vertical first** radio button to cut from vertical side first.
- Select the **Combine cuts (whenever possible)** check box if you want to combine different cut sides if necessary/possible for cutting sheet.
- Specify desired value in the **Minimum width** edit box to define minimum width to be considered for saw cutting.
- After setting desired parameters, click on the **OK** button from the **Nesting** dialog box. The **Nesting Results** dialog box will be displayed showing arrangement of parts in the sheet; refer to Figure-62.

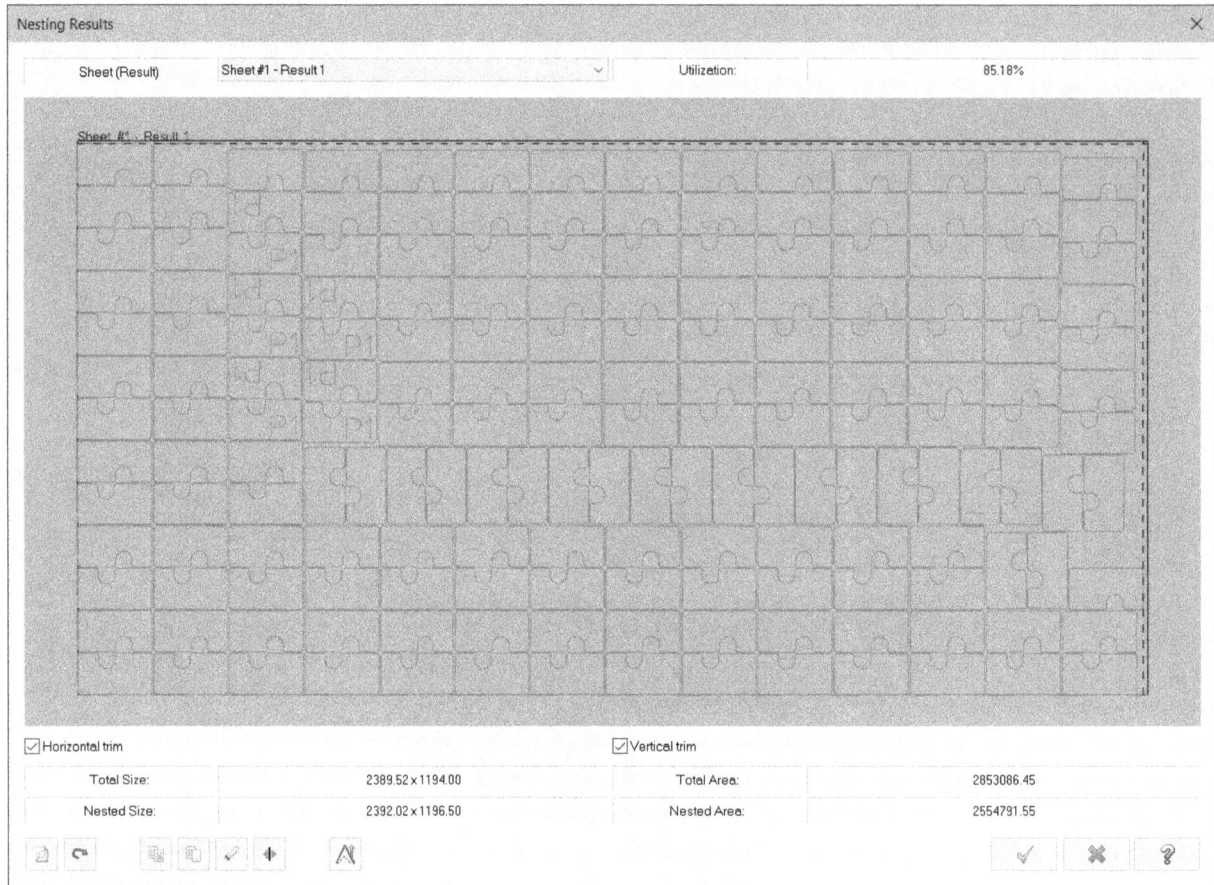

Figure-62. Nesting Results dialog box

- Using the buttons at the bottom in the dialog box, you can save, export, copy, delete, sort, and drag the report as desired. After setting desired parameters, click on the **OK** button from the dialog box. The nested toolpaths will be created; refer to Figure-63.

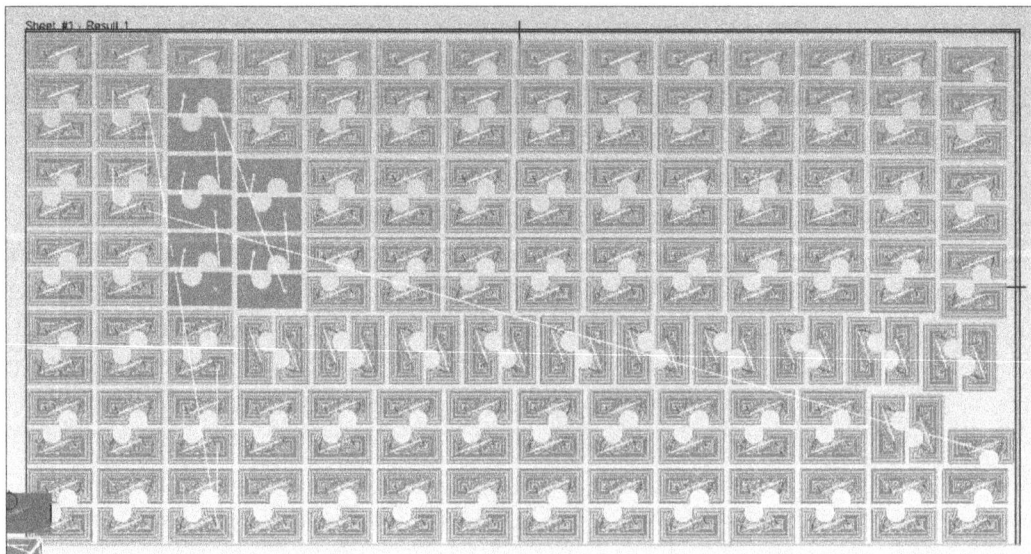

Figure-63. Nested toolpaths created

CREATING PROBE TOOLPATH

The **Probe** tool is used to create toolpath for measuring various coordinates to define fixture offsets, orientation, and critical dimensions. The procedure to use this tool is given next.

• Click on the **Probe** tool from the **Utilities** drop-down in the **Ribbon**. The **Select probe** dialog box will be displayed; refer to Figure-64.

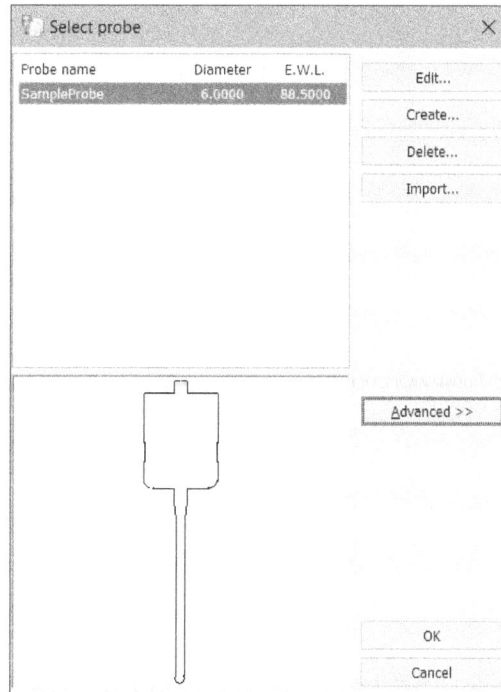

Figure-64. Select probe dialog box

• Select desired probe tool from the list to be used for identifying coordinates.
• Click on the **Advanced >>** button to check advanced parameters of probe tool. The options will be displayed as shown in Figure-65.
• If you want to edit the parameters of probe tool then click on the **Edit** button from the dialog box. The **Edit Probe** dialog box will be displayed; refer to Figure-66.

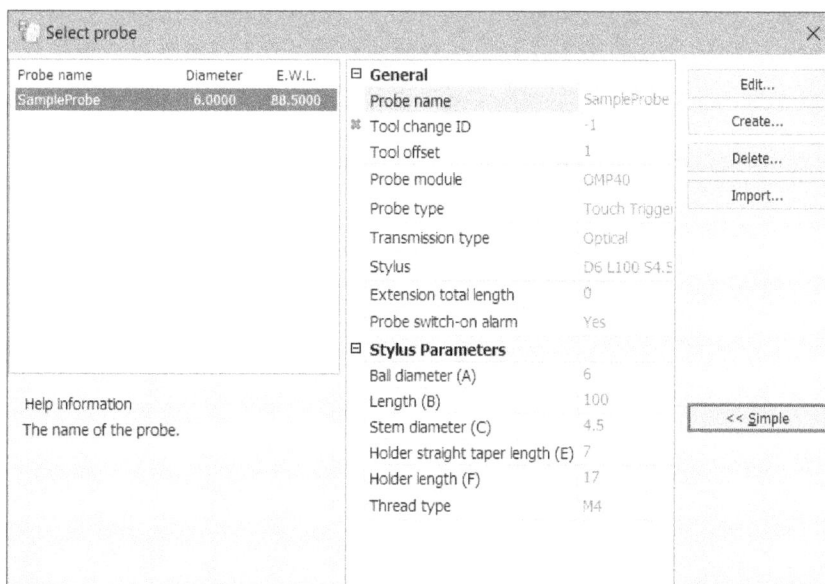

Figure-65. Advanced options for selected probe tool

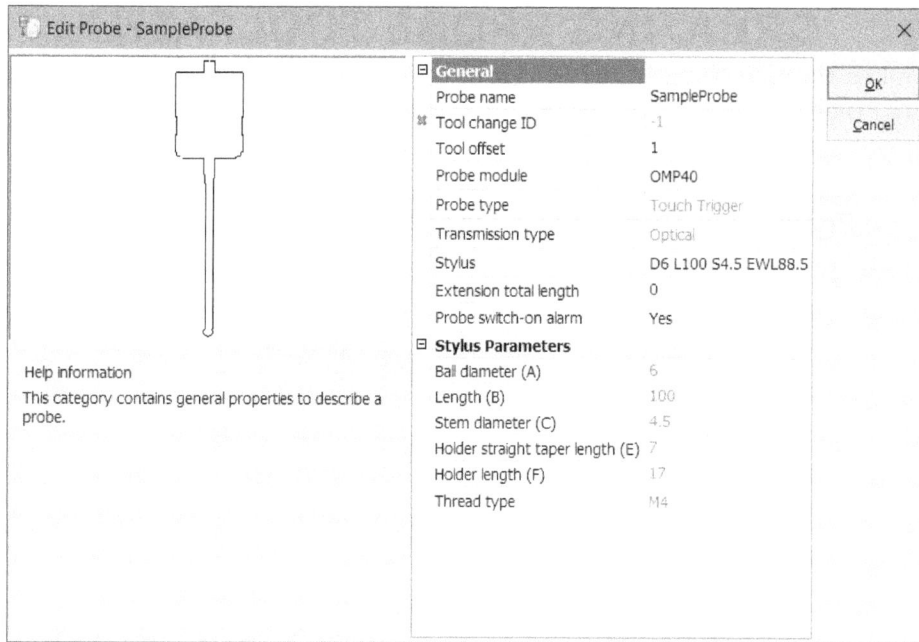

Figure-66. Edit Probe dialog box

- Set desired values in the **Probe name**, **Tool change ID**, and **Tool offset** edit boxes.
- Click in the field for **Probe module** option from the dialog box and select desired assembly of probe module from the drop-down. Some of the common probe assemblies are available in this drop-down.
- Click in the field for **Stylus** and select desired probe tool. Note that generally, name of stylus defines the size of stylus. For example, D6 L100 S4.5 EWL8 represents stylus of diameter 6, total length 100, stem diameter 4.5, and holder's taper length as 8.
- Set the other parameters as desired and click on the **OK** button from the **Edit Probe** dialog box. The **Select probe** dialog box will be displayed again.
- Click on the **OK** button from the dialog box to select desired probe tool. If the probe is modified then the **Database modified** dialog box will be displayed.
- Click on the **Save** button from the dialog box to apply changes. The **Probing Dialog** will be displayed; refer to Figure-67.

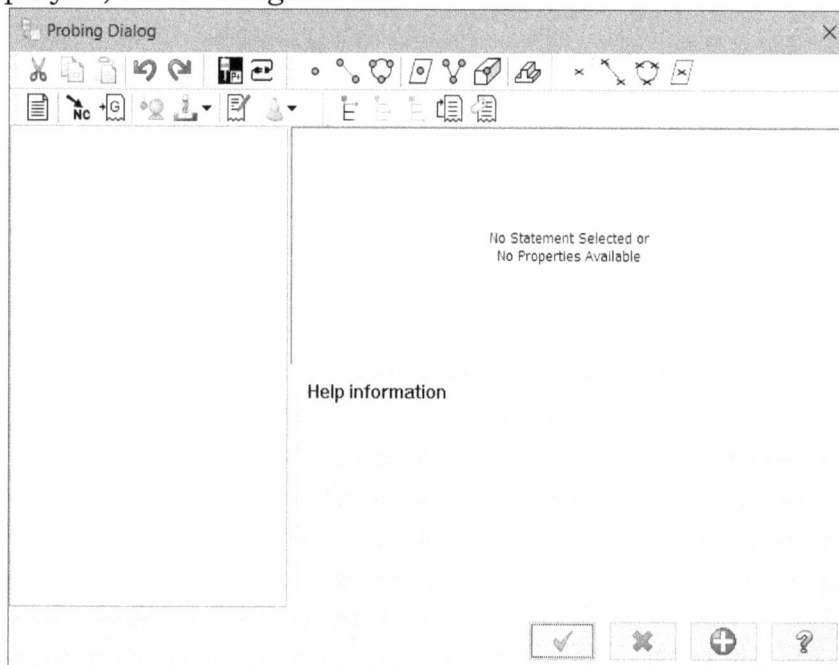

Figure-67. Probing Dialog

Configuring Probe

- Click on the **Configure** button from the **Probing Dialog**. The **Probe Configuration** dialog box will be displayed; refer to Figure-68.

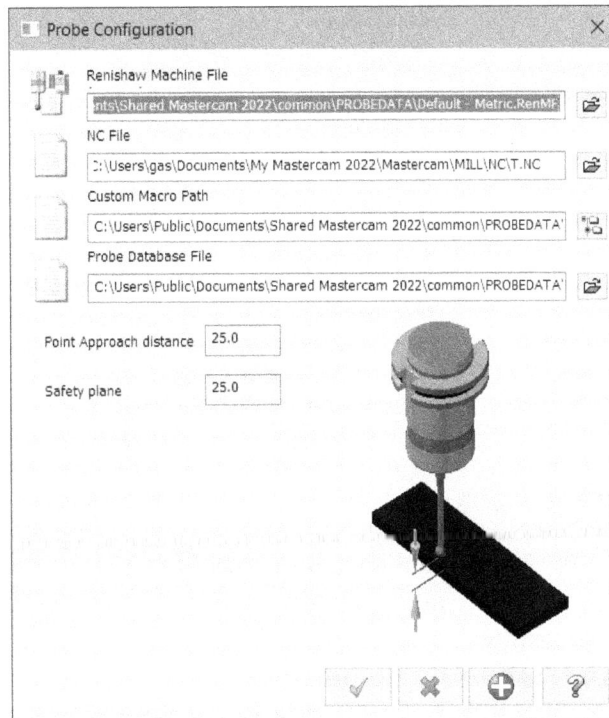

Figure-68. Probe Configuration dialog box

- Set desired parameters in the edit boxes of dialog box to define locations for various files.
- Click in the **Point Approach** distance edit box and define distance from which probe will start to approaching the model to be measured.
- Click in the **Safety plane** edit box and specify distance of safety plane from top face of model.
- Click on the **OK** button from the **Probe Configuration** dialog box to apply desired configuration.

Measuring Objects

There are various buttons available in the toolbar of **Probing Dialog**. These buttons are discussed next.

Measuring Points of Model

- The **Measure Point** button is used to measure coordinates of selected point. Click on the **Measure Point** button from the dialog box and select the points to be measured.
- After selecting points to be measured, press **ENTER**. The points for probing will be displayed in the dialog box; refer to Figure-69.

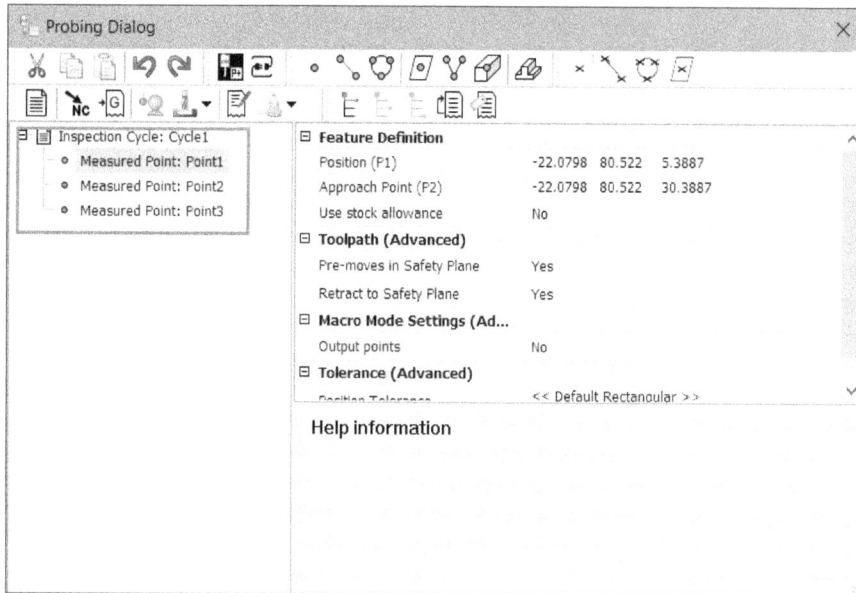

Figure-69. Points for measurement

Measuring Lines of Model

- The **Measure Line** button is used to measure parameters of selected line on the model. Click on the **Measure Line** button from the top in the dialog box. You will be asked to select line/arc to be measured.
- Select desired lines or arcs to be measured and press **ENTER**. A cycle of measurement points will be displayed in the dialog box.

Measuring Circles of Model

- The **Measure Circle** button is used to measure points on a circle using the probe. Click on the **Measure Circle** button from the top in the dialog box. You will be asked to select circular edges of the model.
- Select desired circular edges and press **ENTER**. The Measured Circle nodes will be added in the inspection cycle of dialog box.

Measuring Plane

- The **Measure Plane** button is used to measure various points of a plane. Click on the **Measure Plane** button from the top in the dialog box. A new plane will be added in the inspection cycle of dialog box; refer to Figure-70.

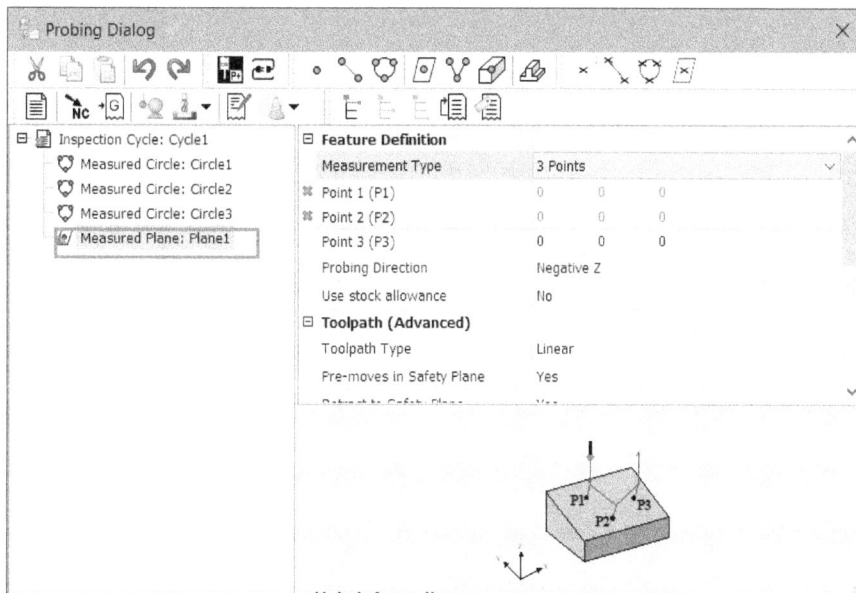
Figure-70. Measure plane added in inspection cycle

- Click in the **Point 1** field and then click on the ⬚ button to select location of first point of plane. Select desired point from the model to define the point.
- Similarly, set the locations for **Point 2** and **Point 3** of plane to define parameters of plane to be measured.
- Set the other parameters as desired for plane measurement in the right area of the dialog box.

Measuring 2D Corner

- The **Measure 2D Corner** button is used to measure a diagonal line joining two points of a face. Click on the **Measure 2D Corner** button from the top in the dialog box. A new 2D corner feature will be added in the inspection cycle for measurement in the dialog box; refer to Figure-71.

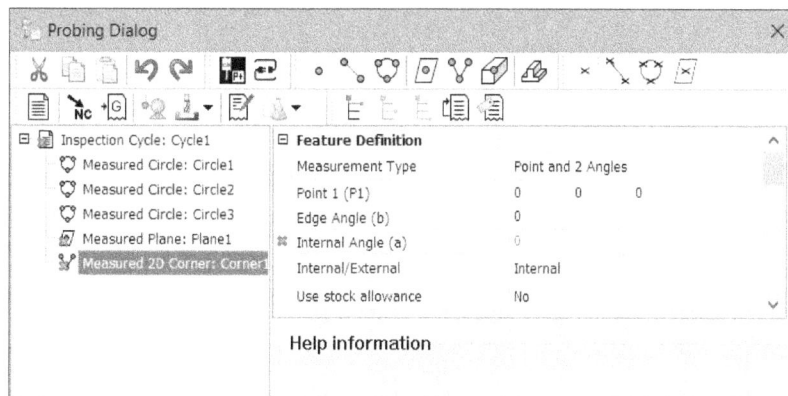
Figure-71. Measure 2D corner feature added

- You can modify the parameters like location of start point, edge angle, internal/external angle, toolpath depth, and so on in the right area of the dialog box in respective fields.

Measuring 3D Corner

- The **Measure 3D Corner** button is used to measure corner formed by three intersecting faces/surfaces at a single point. Click on the **Measure 3D Corner** button from the top in the dialog box. A new measure corner point will be added in the inspection cycle.

- Set desired parameters in the dialog box as discussed earlier.

Measuring Web Pocket

- The **Measure Web Pocket** button is used to measure key coordinates of a pocket in the model. Click on the **Measure Web Pocket** button from the top in the dialog box. A web pocket will be added in the inspection cycle; refer to Figure-72.

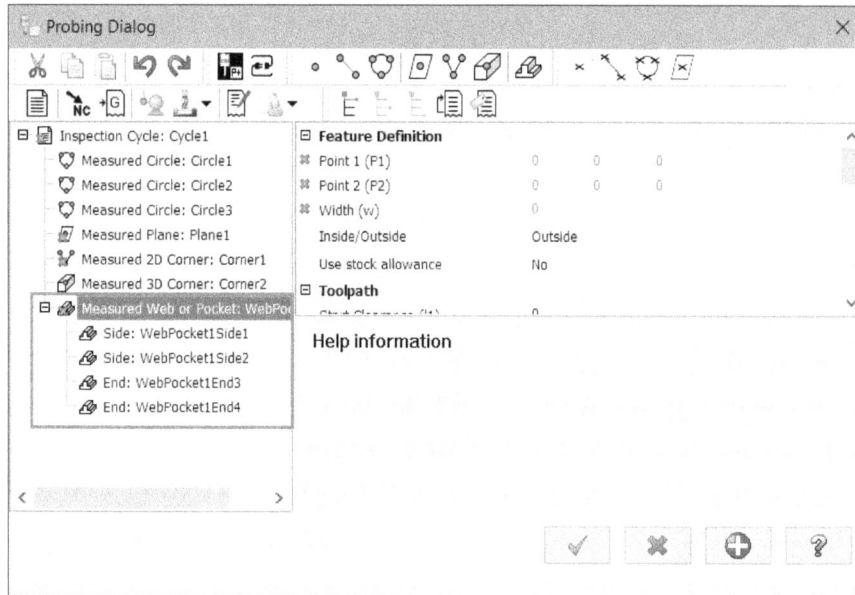

Figure-72. Web Pocket added for measurement

- Specify the measurement parameters for web pocket as discussed earlier.

Constructing Point for Measurement

- The **Constructed Point** button in the dialog box is used to measure a point created by specified parameters. Note that generally, we select a point from body to measure but in this case, we will create a point on model to be measured. To do so, click on the **Constructed Point** button from the top in the dialog box. A constructed point will be added in the list of inspection cycle; refer to Figure-73.

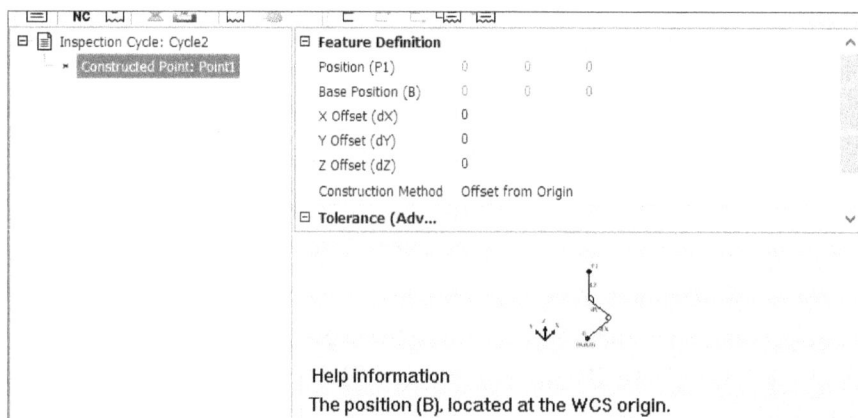

Figure-73. Constructed Point added in inspection cycle

- Specify desired parameters in the **X Offset (dX)**, **Y Offset (dY)**, and **Z Offset (dZ)** edit boxes to define location of measurement point with respect to origin.
- Click on the ... button for **Position Tolerance** from the right area in the dialog box to define tolerance allowed for position measurement.

You can use the **Constructed Line**, **Constructed Circle**, and **Constructed Plane** buttons in the same way.

Creating Inspection Cycle

- Click on the **Inspection Cycle** button from the top in the dialog box. A new inspection cycle will be added in the list. You can add points, lines, curves, and so on in the inspection cycle for measurement.

Performing Machine Update

- Click on the **Machine Update** button from the top in the dialog box. The machine update feature will be added in the list; refer to Figure-74. Using this feature, you can perform changes in probe machine like WCS update, tool length change, tool diameter change, and so on.

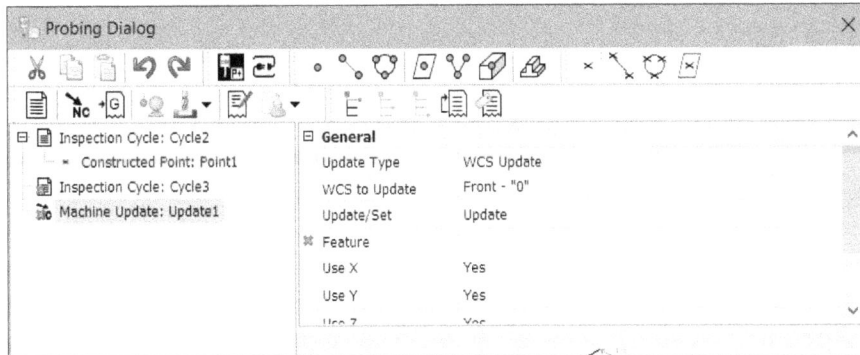

Figure-74. Machine update feature

- Click in the **Update Type** field and select desired type of update from the drop-down. You can select **WCS Update**, **Tool Length**, **Tool Diameter**, **Machine Variable**, or **Rotation Update** option from the drop-down.
- Set desired parameters for selected update from the right area in the dialog box.

Inserting G-Code Block

- Click on the **G-Code Block** button from the toolbar in the dialog box. A new block for manually inserting G-Codes of probing will be added in the inspection cycle; refer to Figure-75.

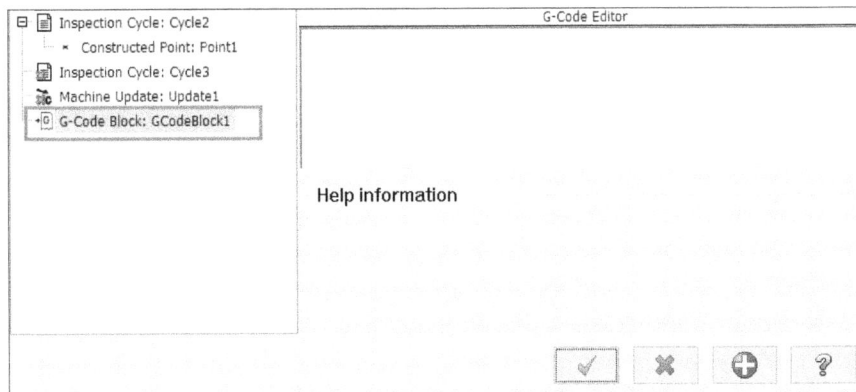

Figure-75. G-code block added

- Specify desired G-codes in the **G-Code Editor** area of the dialog box to manually move the probe tool to desired locations in model for measurement.

Performing Probe Calibration

- Click on the **Probe Calibration** button from the toolbar in the dialog box to center the probe properly for accurate measurements. The **Probe Calibration** feature will be added in the machining sequence; refer to Figure-76.

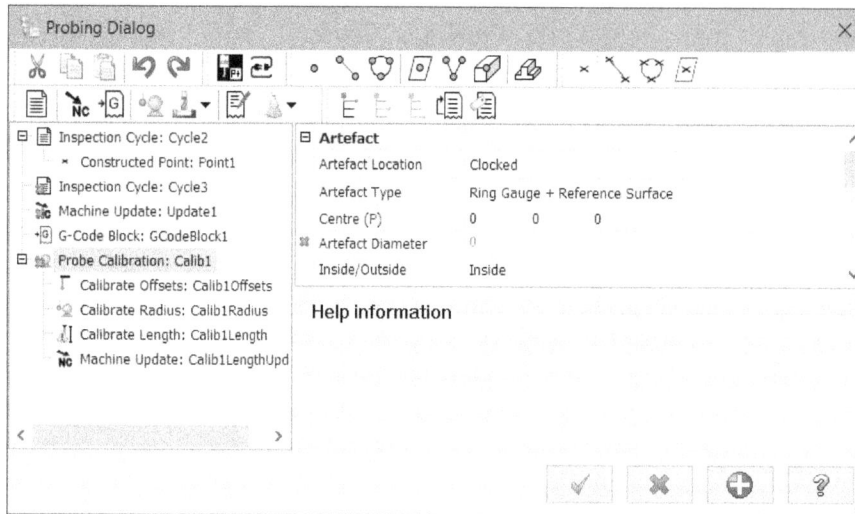

Figure-76. Probe Calibration feature

- Specify the parameters like center location, artefact diameter, safety plane height, and so on in the right area of the dialog box. Note that you need to check the deviation of probe physically so that you can specify correct parameters.

Similarly, you can set the other parameters for probe measurement. After setting desired parameters, click on the **OK** button from the dialog box. The probe toolpath will be generated; refer to Figure-77.

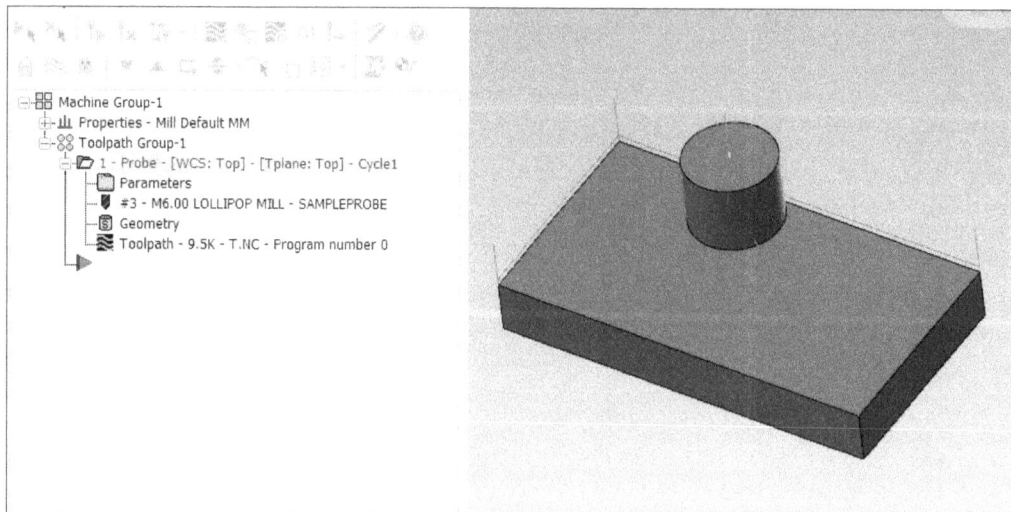

Figure-77. Probe toolpath generated

MULTIAXIS LINKING

The **Multiaxis Linking** tool is used to create a toolpath at safe distance from model which links two or more toolpaths if you are using same cutting tool for two consecutive cutting strategies. The procedure to use this tool is given next.

- Click on the **Multiaxis Linking** tool from the **Utilities** drop-down in the **Ribbon**. The **Multiaxis Link** dialog box will be displayed; refer to Figure-78.

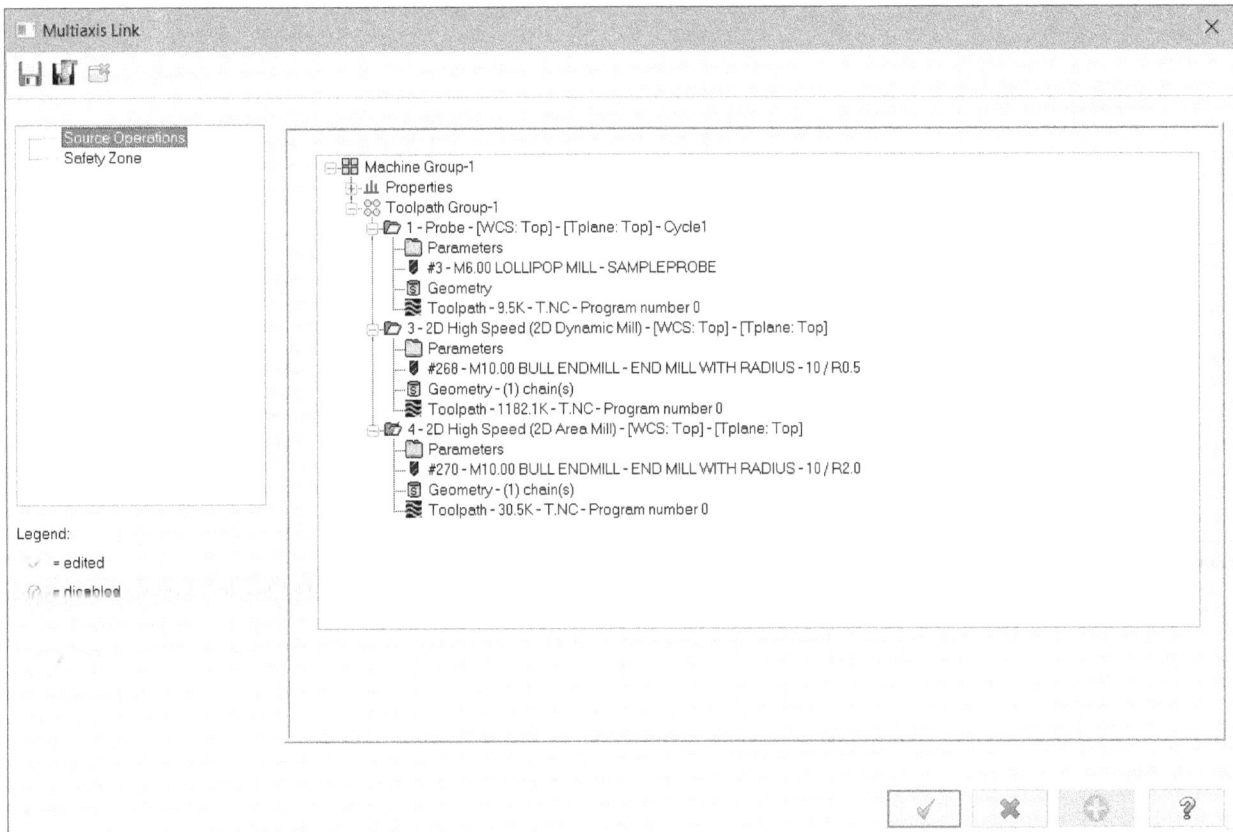

Figure-78. Multiaxis Link dialog box

- Select the toolpaths you want to link from the right area in the dialog box while holding the **CTRL** key.
- Select the **Safety Zone** option from the left area in the dialog box to define parameters related to safe distance where cutting tool will move while linking the toolpaths. The **Safety Zone** page will be displayed in the dialog box; refer to Figure-79.
- Select desired option from the **Axis of Rotation** drop-down to define axis along which cutting tool can rotate/tilt and specify related parameters in **Tool Motion** area.
- Click on the **Define Shape** button from the **Shape** area of the dialog box to define safety zone. The **Safety Zone Manager** will be displayed; refer to Figure-80 and you will be asked to select the model.

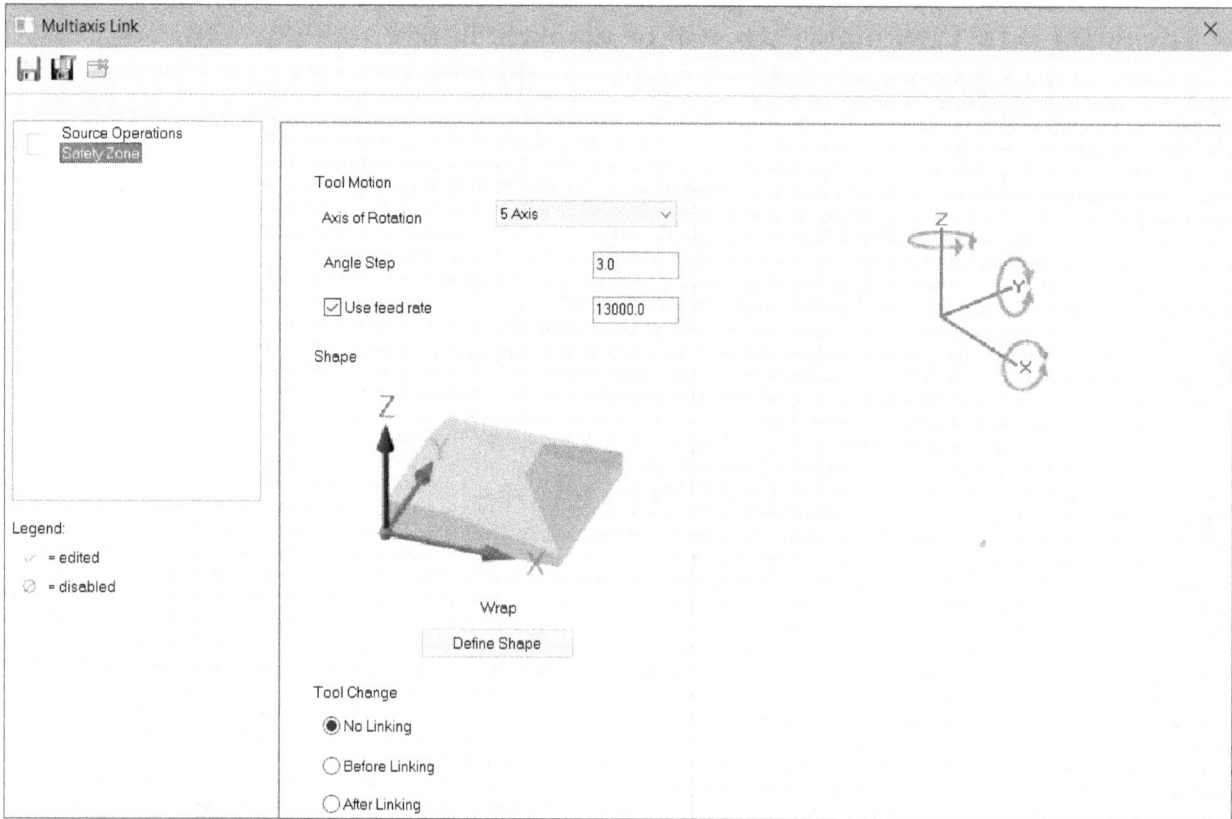

Figure-79. Safety Zone page

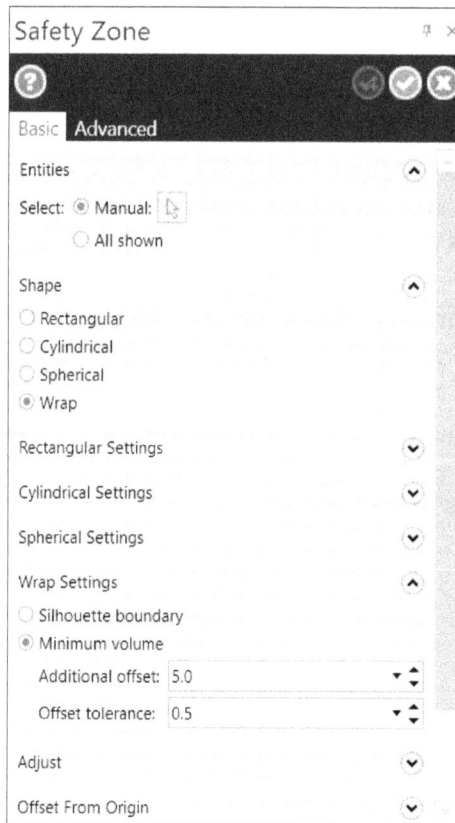

Figure-80. Safety Zone Manager

- Select the model to be used as reference for creating wrap shaped safety zone. Preview of the shape will be displayed; refer to Figure-81.

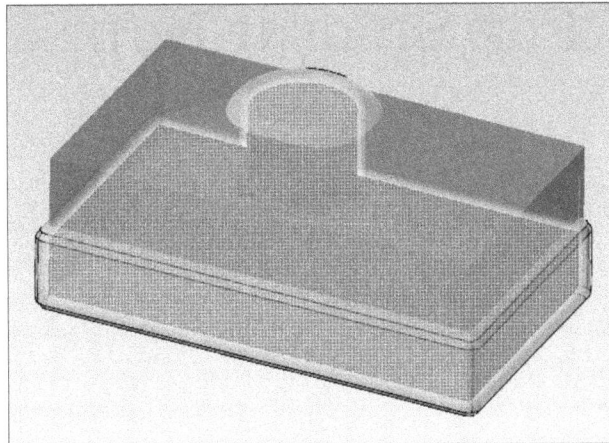

Figure-81. Preview of wrap safety zone

- You can select desired radio button from the **Shape** rollout of the **Manager** to modify shape of safety zone.
- Specify the other parameters as discussed earlier and click on the **OK** button to confirm the shape of safety zone. The **Multiaxis Link** dialog box will be displayed again.
- Select desired radio button from the **Tool Change** area of the dialog box to define what will happen to linking when cutting tool is changed from one toolpath to another. Select the **No Linking** radio button if you want to break the link between toolpaths when tool is changed. Select the **Before Linking** radio button to change tool before creating link. Select the **After Linking** radio button to perform tool change after creating link.
- After setting desired parameters, click on the **OK** button from the dialog box. The linking toolpath will be generated connecting end point of first toolpath with start point of next toolpath; refer to Figure-82.

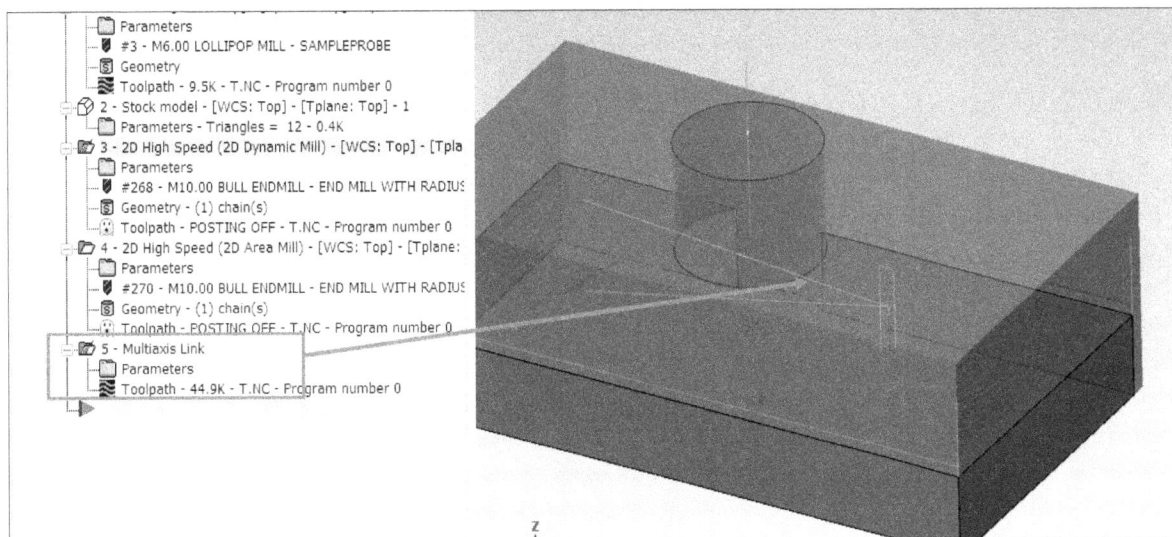

Figure-82. Multiaxis link toolpath

FOR STUDENT NOTES

Chapter 9

Multiaxis Milling Toolpaths

Topics Covered

The major topics covered in this chapter are:

- *Introduction.*
- *Curve Toolpath*
- *Swarf Milling Toolpath*
- *Parallel Multiaxis Toolpath*
- *Multiaxis Along Curve Toolpath, Multiaxis Morph Toolpath, Multiaxis Flow Toolpath*
- *Multiaxis Multisurface Toolpath, Multiaxis Port Toolpath, Multiaxis Triangular Mesh Toolpath*
- *Deburr Toolpath*
- *Multiaxis Pocketing Toolpath*
- *3+2 Automatic Roughing Toolpaths, Multiaxis Project Curve Toolpath*
- *Multiaxis Rotary Toolpath and Multiaxis Rotary Advanced Toolpath*
- *Multiaxis Swarf Toolpath*
- *Verifying Toolpath with Fixture*

INTRODUCTION

In previous chapters, you have learned to create 2D and 3D toolpaths. In this chapter, you will learn to create multiaxis toolpaths for machining irregular 3D faces of the model. The tools to create multiaxis toolpaths are available in the **Multiaxis** drop-down in the **Ribbon**; refer to Figure-1. Various tools for multiaxis toolpath creation are discussed next.

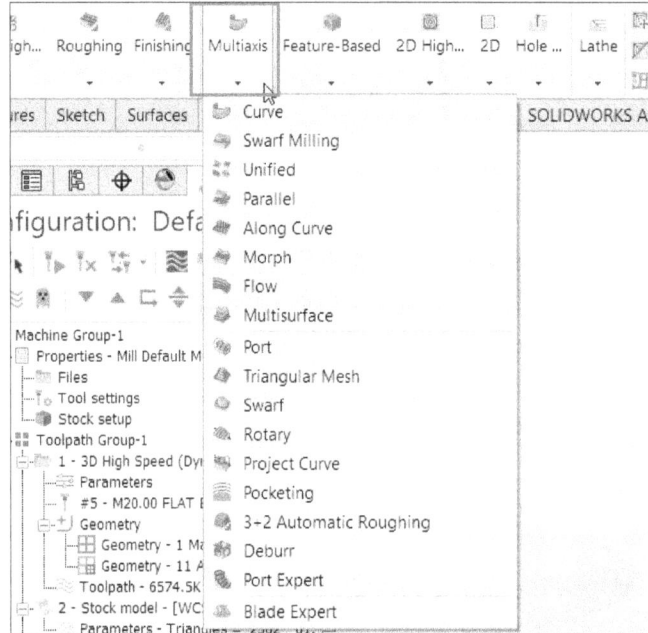

Figure-1. Multiaxis tool

CREATING CURVE TOOLPATH

The **Curve** tool is used to create toolpath for machining 3D curves. The procedure to use this tool is given next.

- Click on the **Curve** tool from the **Multiaxis** drop-down in the **Ribbon**. The **Multiaxis Toolpath - Curve** dialog box will be displayed; refer to Figure-2.
- Select desired cutting tool and tool holder using the options in the dialog box. Generally, a ball end mill cutting tool is used to perform multiaxis milling operations.
- If you want to use stock left from any previous operation then select the **Stock** option from the left area of the dialog box and specify related parameters.
- Click on the **Cut Pattern** option from the left area of the dialog box to define how cutting tool will move on the 3D curve path; refer to Figure-3.

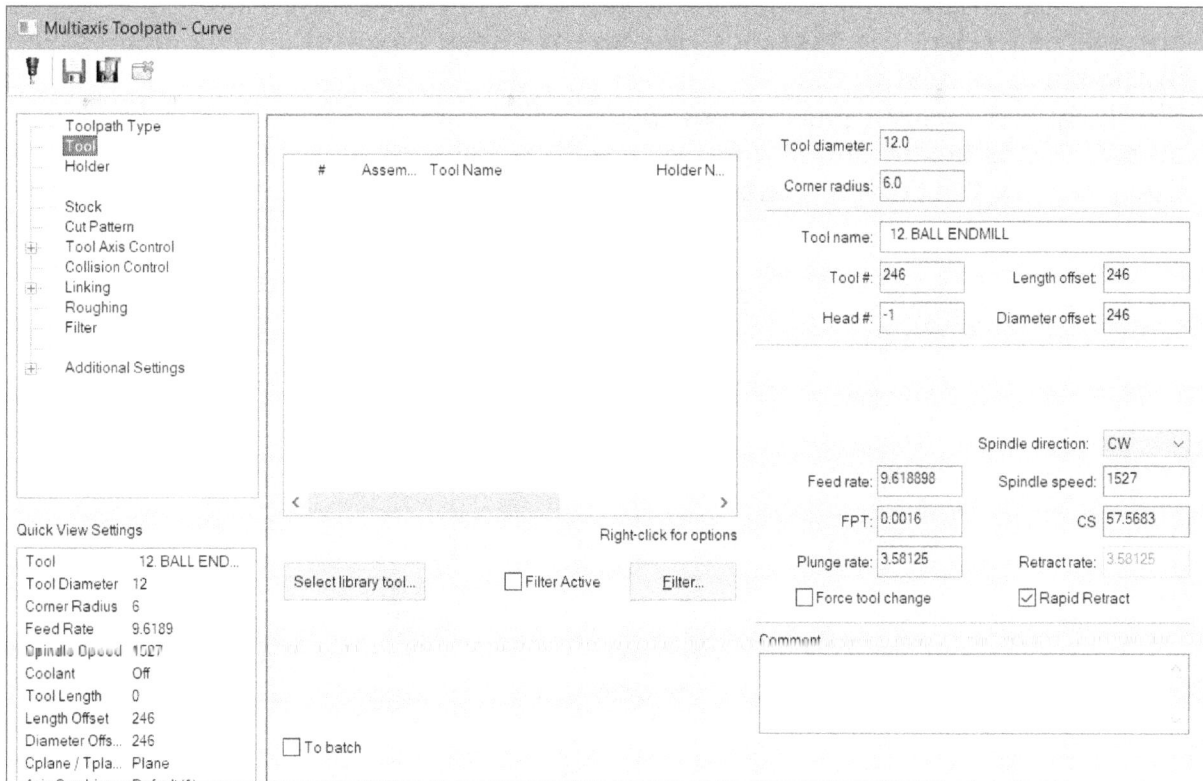

Figure-2. Multiaxis Toolpath-Curve dialog box

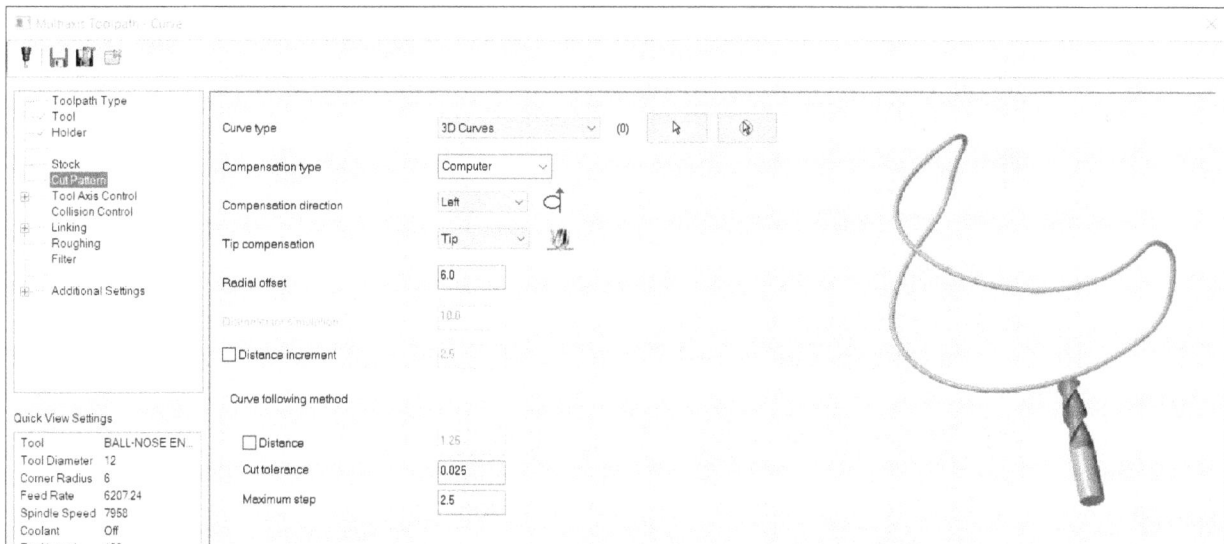

Figure-3. Cut Pattern options for 3D Curve toolpath

- Select the **3D Curves** option from the **Curve type** drop-down if you want to select 3D curves created by using selected wireframe curves or edges of the model. Select the **Surface edge - all** option from the **Curve type** drop-down if you want to select all the edges of selected surface for creating toolpath. Select the **Surface edge - single** option from the drop-down if you want to select edges one by one for defining path of curves. After selecting desired option from drop-down, click on the **Select** button next to the drop-down to select the entities.

- Set desired parameters in **Compensation type**, **Compensation direction**, and **Tip compensation** drop-downs as discussed earlier.

- Set desired value in the **Radial offset** edit box to offset cutting tool by specified value away from the selected curves.

- If you want to specify an increment value for cutting passes along the selected curves then select the **Distance increment** check box and specify desired value in the edit box next to it.
- Similarly, set desired parameters in the edit boxes of **Curve following method** area of dialog box to define how curve will be followed while cutting.
- Click on the **Collision Control** option from the left area of the dialog box to define parameters related to collision avoidance by cutting tool; refer to Figure-4.

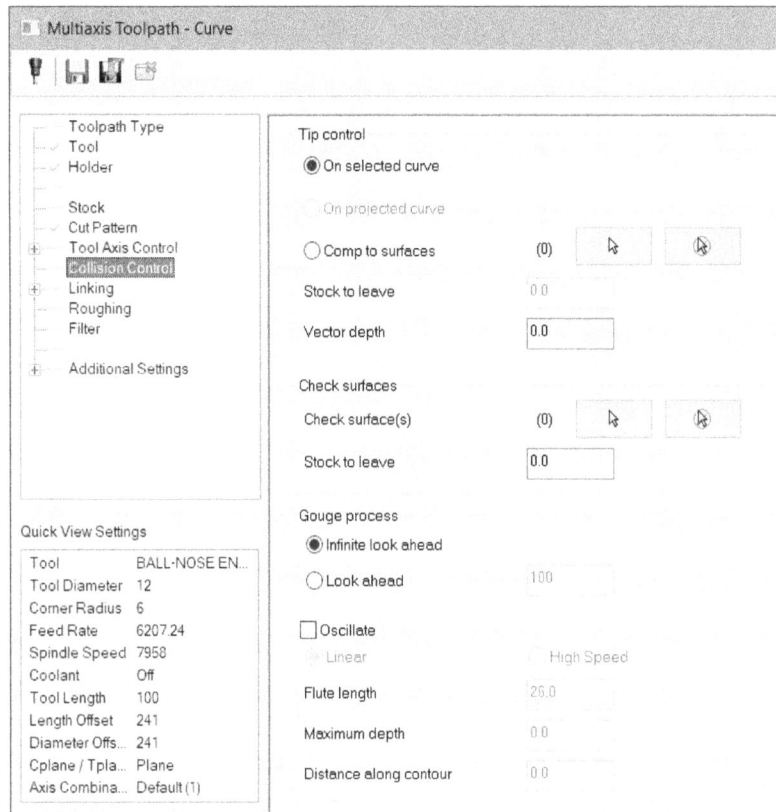

Figure-4. Collision Control options

- Select the **On selected curve** radio button to move tool tip over selected curves.
- Select the **On projected curve** radio button to move tool tip on curves projected on surfaces.
- Select the **Comp to surfaces** radio button to put tool tip on selected compensation surfaces.
- Specify desired value in **Stock to leave** edit box to define how much stock will be left after performing operation.
- Specify desired value in the **Vector depth** edit box to offset tool tip by specified value in the vector direction. A positive value in the edit box will raise the tool tip and negative value will lower the tool tip.
- Click on the **Select check surfaces** button from the **Check surfaces** area of the dialog box to specify surfaces to be avoided while machining.
- Set the other parameters as desired and click on the **OK** button to generate the toolpath. The curve toolpaths will be created; refer to Figure-5.

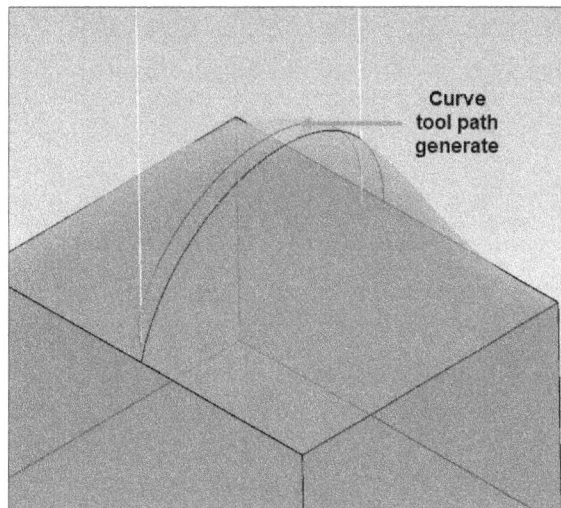
Figure-5. Curve toolpath generated

CREATING SWARF MILLING TOOLPATH

The **Swarf Milling** tool is used to machine the geometry using side edge of cutting tool. You can use this milling path to rough machine blades of turbines and other objects. The procedure to use this tool is given next.

- Click on the **Swarf Milling** tool from the **Multiaxis** drop-down in the **Ribbon**. The **Multiaxis Toolpath - Swarf Milling** dialog box will be displayed; refer to Figure-6.

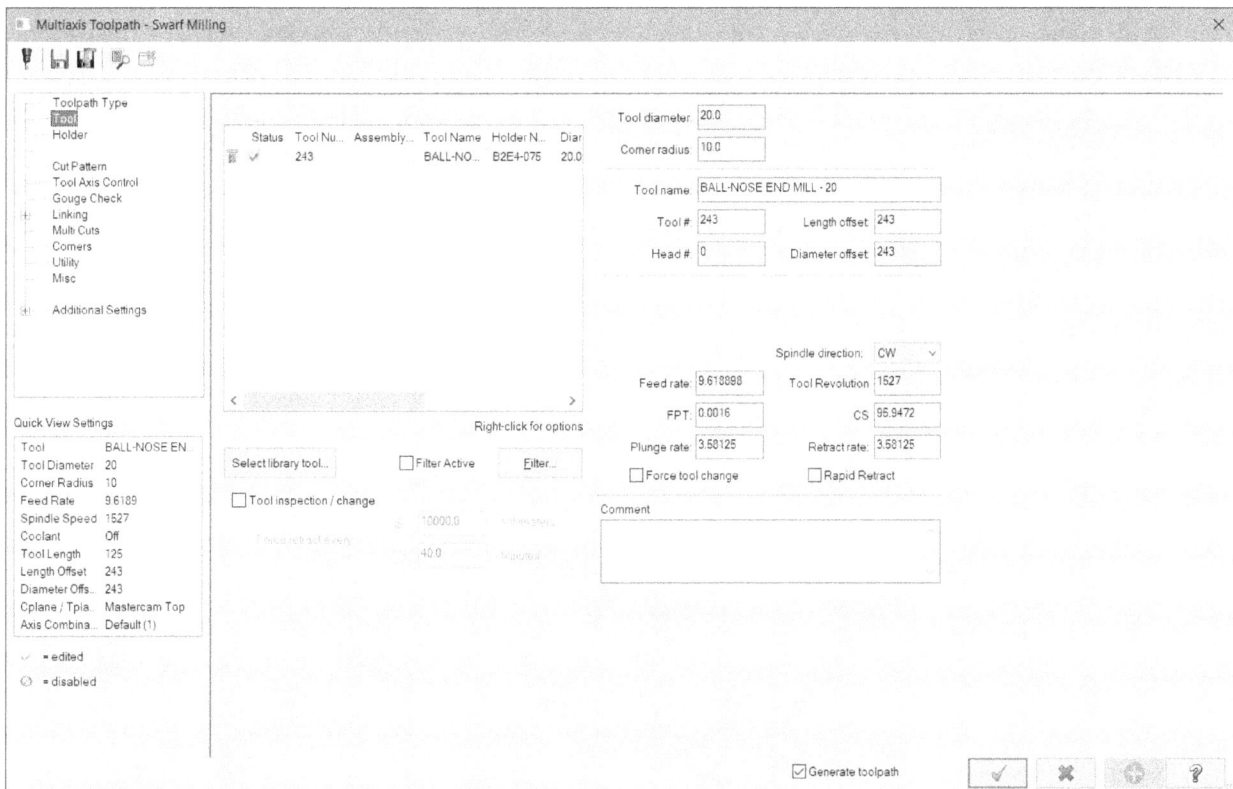
Figure-6. Multiaxis Toolpath-Swarf Milling dialog box

- Select the cutting tool and tool holder as discussed earlier.
- Click on the **Cut Pattern** option from the left area to set parameters related to cutting pattern for swarf milling; refer to Figure-7.

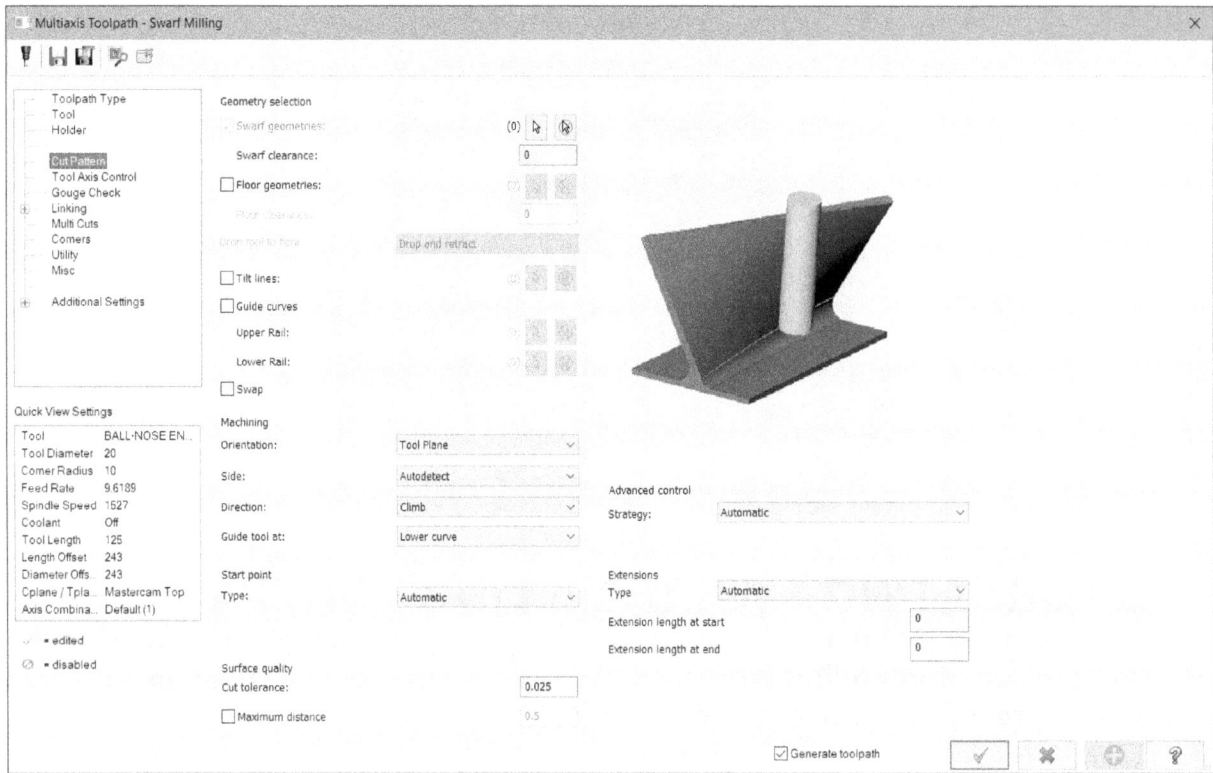

Figure-7. Cut pattern for Multiaxis Toolpath-Swarf Milling dialog box

- Click on the **Select** button for the **Swarf surfaces** section and select surfaces to be machined by swarf milling.
- If you want to machine floor of surface as well then select the **Floor surfaces** check box and select desired faces/surfaces for floor. Select the **Drop and retract** option from **Drop tool to floor** drop-down if you want the cutting tool to reach to floor while cutting and then move back after it touches the floor (This may cause machining marks on the floor). Select the **Retract only** option from the drop-down if you want the cutting tool to shy away from even touching the floor!!
- Select the **Tilt lines** check box to define axis direction of cutting tool and select desired lines.
- Select the **Guide curves** check box to select upper rail curve and lower rail curve; refer to Figure-8.

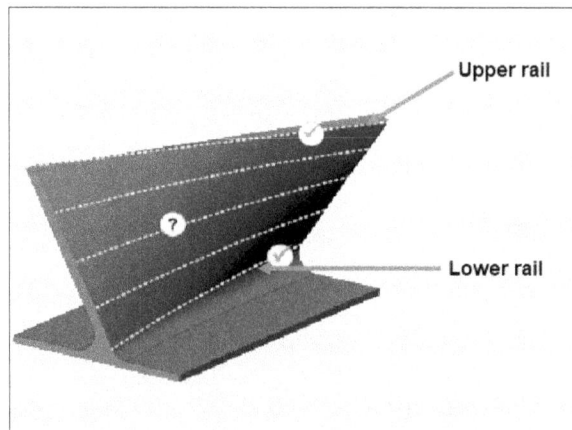

Figure-8. Rail curves for swarf toolpath

- Select the **Swap** check box to switch upper rail with lower rail.

- Select desired option from the **Orientation** drop-down to define orientation of tools for operations.
- Select desired option from the **Side** drop-down to define which side of cutting tool will be used for performing milling operation. Specifying a correct option in this drop-down is important when cutting walls.
- Set the other parameters as desired in the **Cut Pattern** page.
- Click on the **Tool Axis Control** option from the left area to specify output format for milling operation. By default, **5 axis** option is selected in the **Output format** drop-down, so cutting tool can move freely along all five axes while cutting. Select the **4 axis** option from the **Output format** drop-down to generate 4 axis toolpath output using selected tilt axis. Select the **3 axis** option from the drop-down to generate 3 axis toolpath.
- Click on the **Gouge Check** option from the left area to define parameters for avoiding collision and perform gouge check; refer to Figure-9.

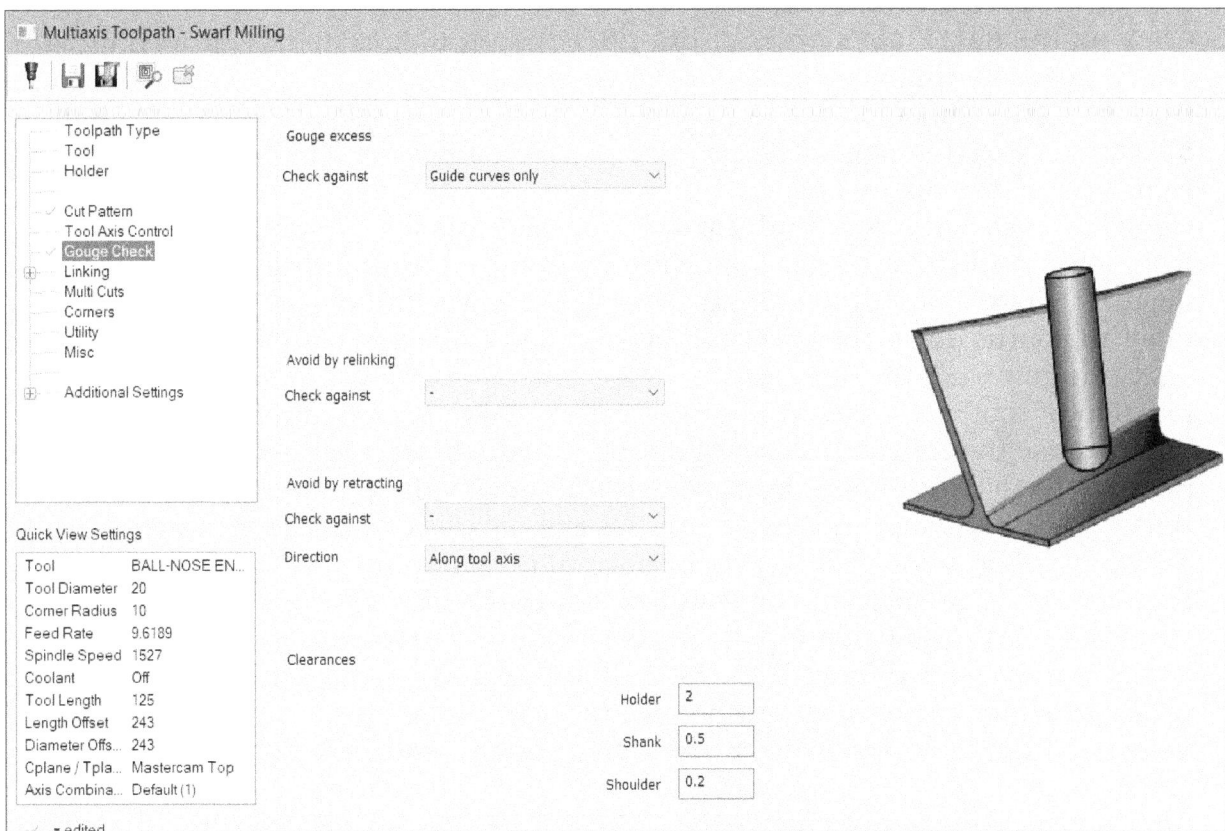

Figure-9. Gouge Check options

- Select desired option from the **Check against** drop-down in **Gouge excess** area to define reference to be used for checking collision of tool and holder with workpiece (This collision of tool & holder with workpiece is called gouging). Select the **Guide curves only** option if you want to perform gouge check using guide curves as reference. Select the **Swarf geometries** option to perform gouge check using selected swarf geometries as reference. Select the **Additional geometries** option if you want to use additional selected geometries of the model for gouge check. Select the **Swarf and additional geometries** option if you want to use both swarf geometries and additional selected geometries for gouge check.

- If you have selected **Swarf geometries** option or **Swarf and additional geometries** option in the **Check against** drop-down then Error handling drop-down will be displayed in **Gouge excess** area. Select desired option from the **Error handling** drop-down to define how the tool will behave when a possible collision can occur. Select the **Degouge** option to move tool away from swarf surfaces by a small distance. Select the **Balance** option to mark gouge amount of material as rest material for next toolpath. Select the **Balance within allowance** to define the limit of material that can be marked as rest material on gouging.

- In Gouge allowance edit box, you can specify the amount of material that can gouge while machining. Specify the amount that you can trust, your cutting tool side edge will remove.

- Set desired options in the **Avoid by relinking** and **Avoid by retracting** drop-downs to define how the geometry of part to be avoided by respective method.

- Click on the **Linking** option from the left area in the dialog box and specify the entry/exit path for machining.

- Click on the **Multi Cuts** option from the left area to specify the pattern in which cutting passes will be generated; refer to Figure-10.

- Specify the number of cutting passes and the cutting pattern in the **Pattern slices** area of this page. The preview of cutting passes will be displayed in the right area of the dialog box.

- Click in the **To** edit box of **Tool guidance** area in this page and specify the distance from lower rail to be used as starting location for generating cutting passes. A negative value will make toolpaths below the lower rail and a positive value will move toolpaths above the lower rail by specified value. Select the **Gradual for each slice** option from the **Tool shift** drop-down if you want to set default starting location of toolpath away from upper rail by value specified in **From** edit box.

- You can use the **Toolpath damping** option to reduce irregular peaks in the toolpath for smooth tool movement by specified tolerance value.

- Specify desired value in **Number of layers** edit box of **Pattern layers** area to define number of toolpath layers in normal direction to swarf surfaces.

- You can anytime click on the **Preview toolpath** button at the top in the dialog box to check preview of toolpath while working on parameters.

- Select the **Corners** option from the dialog box and specify whether you want to generate sharp corners or round corners.

- Select the **Utility** option from the dialog box to set feed rate and direction of cutting.

- After setting all desired parameters, click on the **OK** button from the dialog box. The toolpath will be generated; refer to Figure-11.

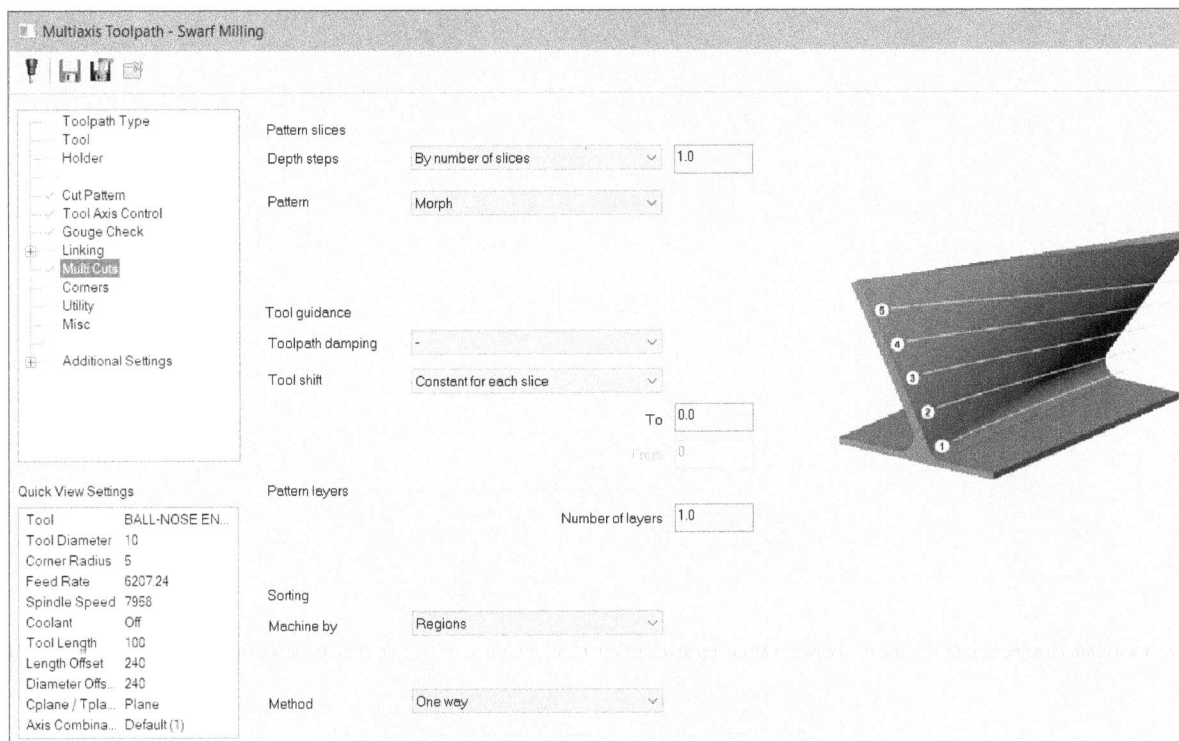

Figure-10. Multi Cuts page

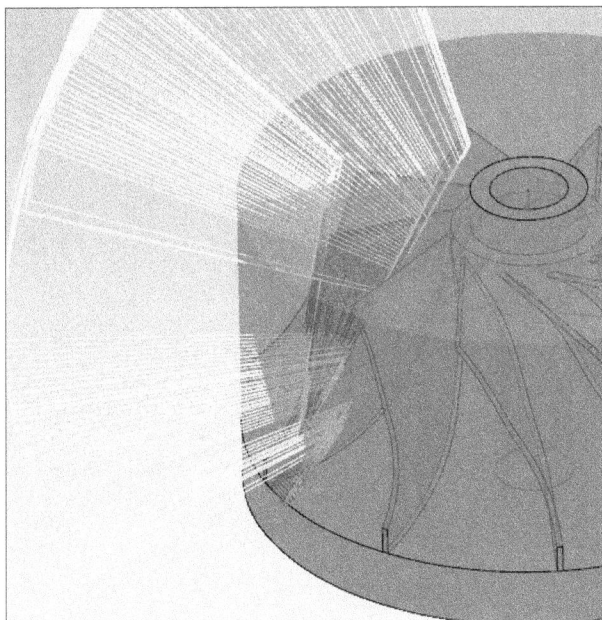

Figure-11. Swarf toolpath generated

CREATING UNIFIED MULTIAXIS TOOLPATH

The **Unified** tool is used to create multiaxis toolpath whose pattern is controlled by selected guide curves/faces/surface. This toolpath combines the benefits of Parallel and Morph toolpaths. The procedure to use this tool is given next.

- Click on the **Unified** tool from the **Multiaxis** drop-down in the **Ribbon**. The **Multiaxis Toolpath - Unified** dialog box will be displayed; refer to Figure-12.

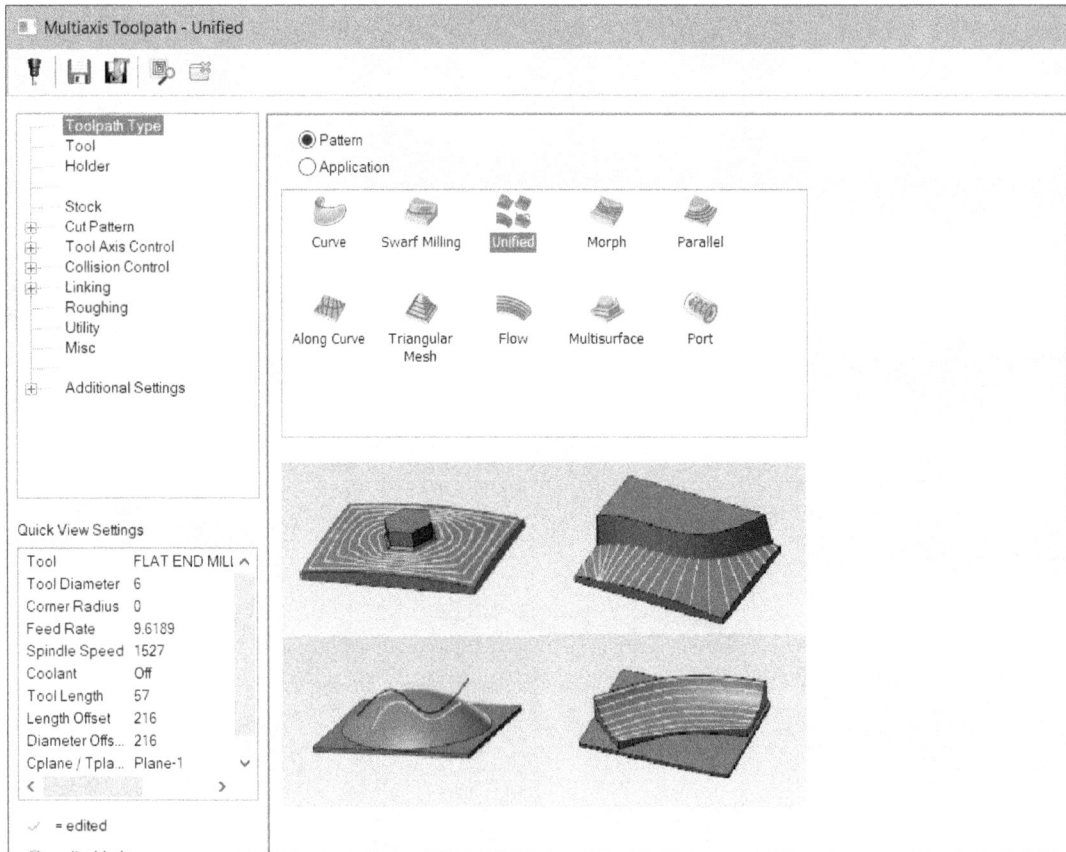

Figure-12. Multiaxis Toolpath-Unified dialog box

- Set the cutting tool, holder, and stock parameters as discussed earlier.
- Click on the **Cut Pattern** option from the left in the dialog box. The options in the dialog box will be displayed as shown in Figure-13.

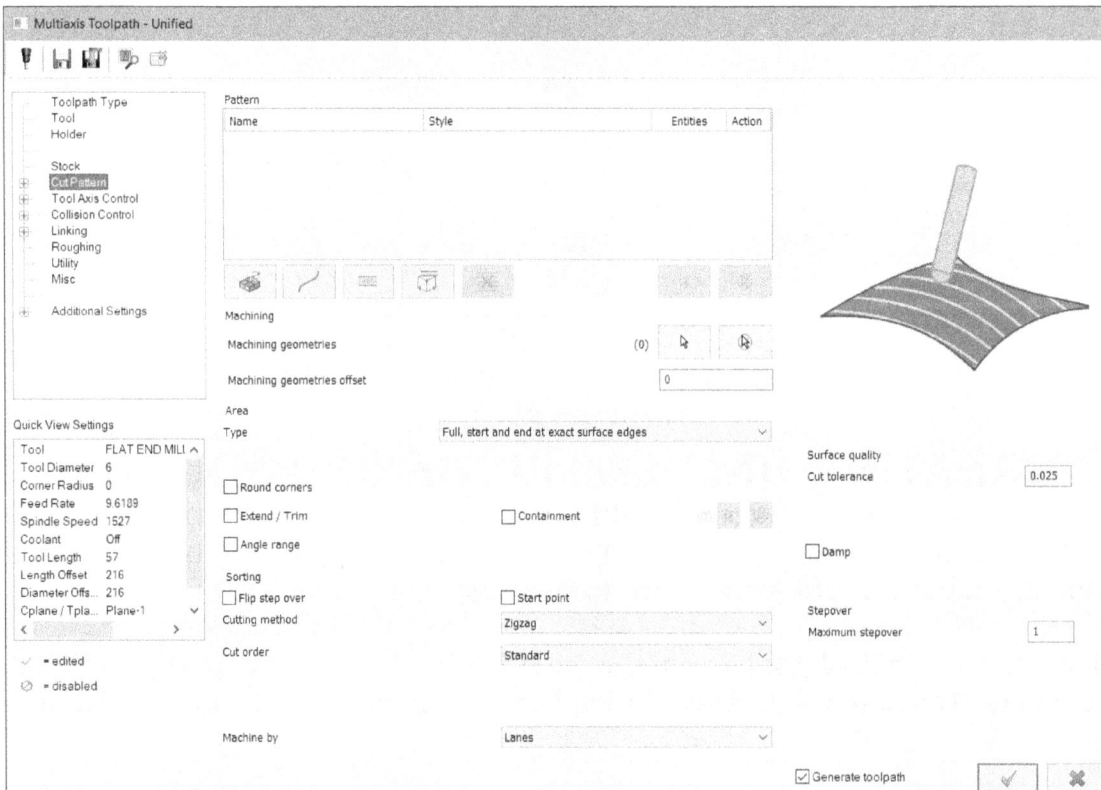

Figure-13. Cut Pattern options for Unified toolpath

- Click on the **Select** button for `Machining geometries` option in the `Machining` area of dialog box and select the entity to be machined from the model.
- Select desired button from the **Pattern** area in the dialog box to define the pattern in which toolpaths will be created. These options are discussed next.

Patterns for Unified Toolpath

- Click on the **Add Automatic row** button from the **Pattern** area to automatically generate pattern of toolpath based on selected cutting strategy. On selecting this option, an entry of Automatic pattern will be added in the table.
- Click in the **Style** drop-down to select cutting strategy; refer to Figure-14. Previews of some toolpaths for different automatic styles as shown in Figure-15.

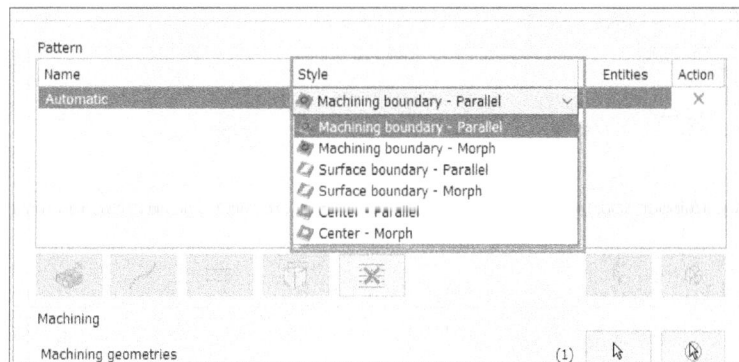

Figure-14. Style drop-down

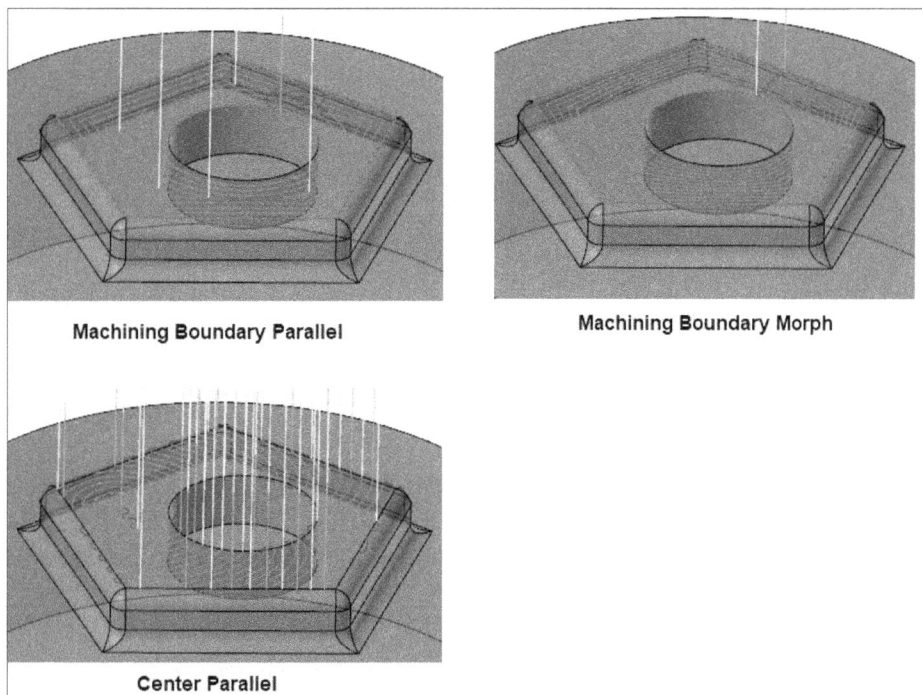

Figure-15. Preview of Automatic pattern toolpaths

- Click on the **Add Curve row** button from the **Pattern** area to use curves as references for defining cutting pattern. Using this option, you can create cutting passes parallel to curve, perpendicular to curve, use selected curve as guide, or project the curve on machining surface to cutting pass; refer to Figure-16.
- After setting desired curve options, click on the **Select** button at the bottom in **Pattern** area and select desired curve; refer to Figure-17.

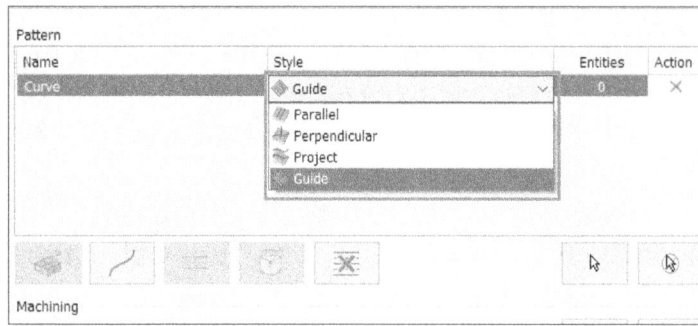

Figure-16. Style drop-down for curve

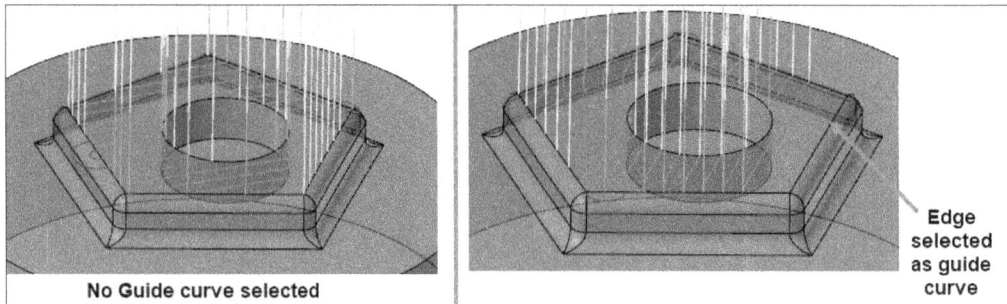

Figure-17. Using curve as guide for toolpath

- Select the **Parallel** option from the **Style** drop-down for selected curve to make cutting passes parallel to selected curve; refer to Figure-18. Similarly, you can use the **Perpendicular** option from the drop-down to create cutting passes perpendicular to selected curve. Select the **Project** option to machine only projection of selected curve of the model surface.

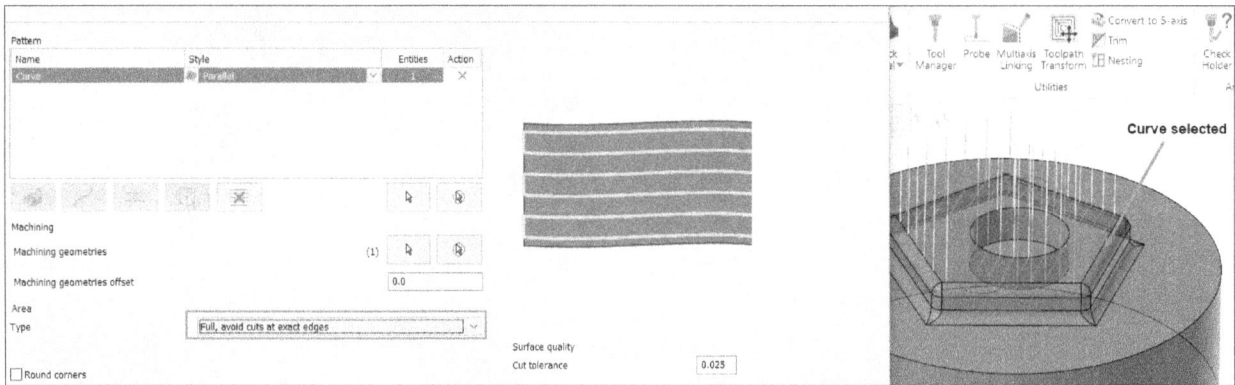

Figure-18. Cutting passes parallel to selected curve

- Select the **Add Surface row** button from the **Pattern** area to use selected surfaces for defining cutting pass pattern. You can use selected surface as guide for cutting passes, create cutting passes parallel to selected surface, and create flowline U or V cutting passes; refer to Figure-19.

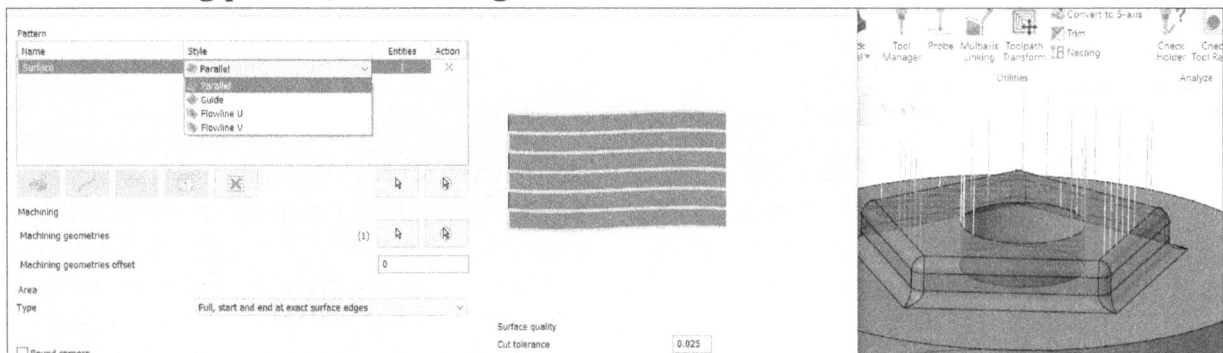

Figure-19. Using surface for cutting pattern

- Select the **Add Plane row** button to use selected plane for defining direction of cutting passes. Select desired option from the **Style** drop-down to define reference direction for cutting passes; refer to Figure-20.

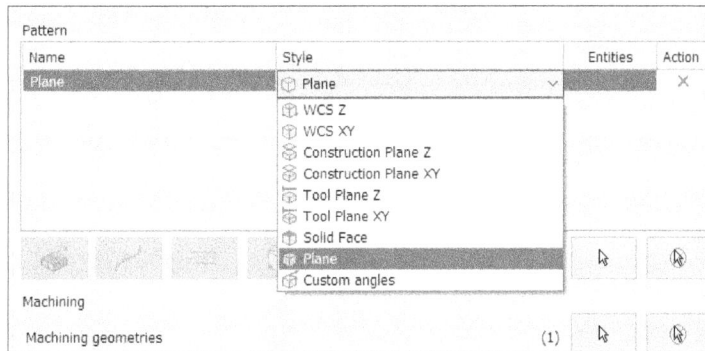

Figure-20. Style drop-down for plane

- Specify other parameters as discussed earlier and click on the **OK** button to create the toolpath.

CREATING PARALLEL MULTIAXIS TOOLPATH

The Parallel multiaxis toolpath is used to generate multiaxis toolpaths parallel to each other following curvature of surface. The procedure to generate this toolpath is given next.

- Click on the **Parallel** tool from the **Multiaxis** drop-down in the **Ribbon**. The **Multiaxis Toolpath - Parallel** dialog box will be displayed; refer to Figure-21.

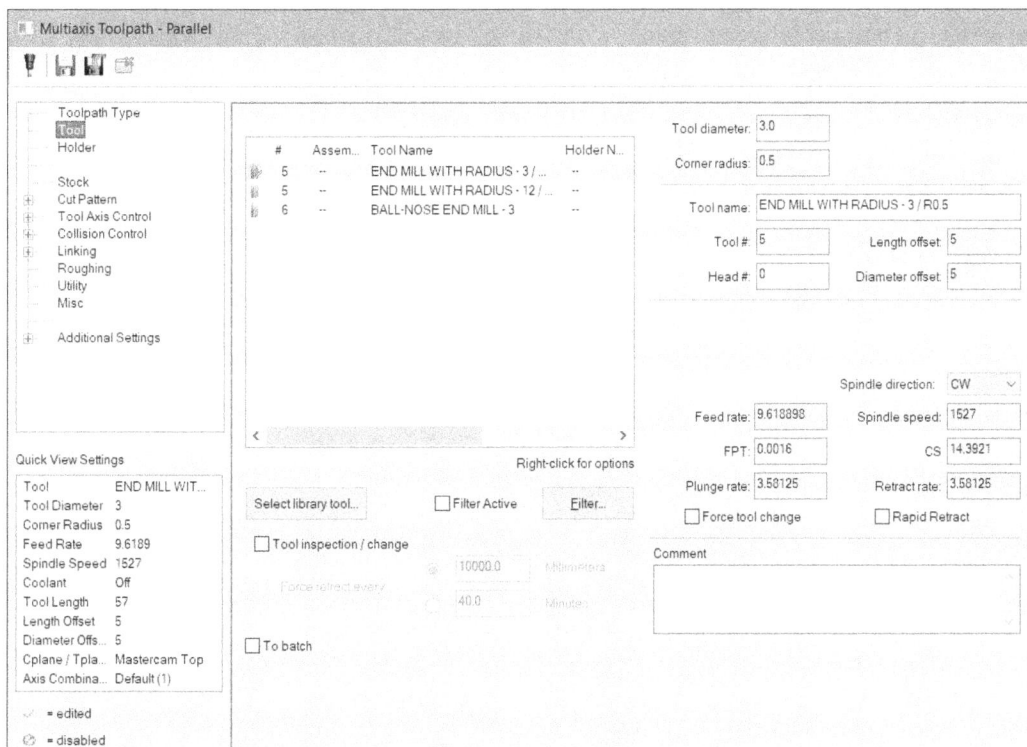

Figure-21. Multiaxis Toolpath-Parallel dialog box

Most of the options in this dialog box are same as discussed earlier. The options different from others are discussed next.

- Set the parameters related to tool, holder, and stock as discussed earlier.

Cut Pattern Options

- Click on the **Cut Pattern** option from the left area of the dialog box. The options will be displayed as shown in Figure-22.

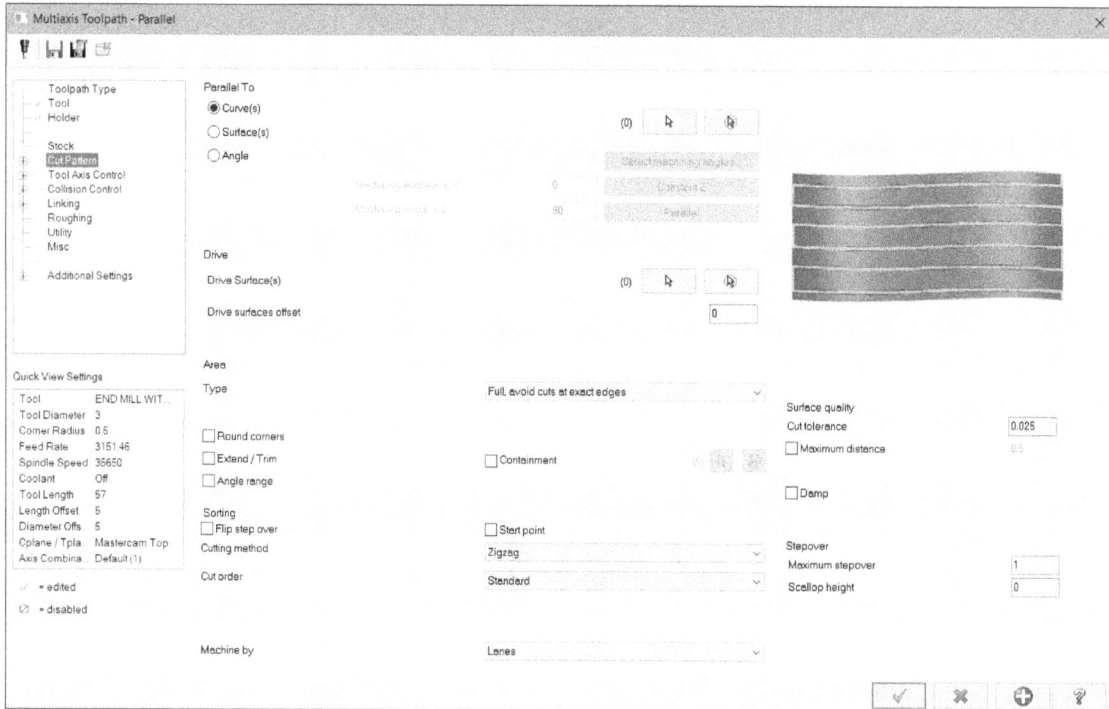

Figure-22. Cut Pattern options for Parallel toolpath

- Select desired radio button from the **Parallel To** area of the dialog box to define whether toolpath will be parallel to curves, surfaces, or angle value. The most common option used for defining parallel toolpath is using curves. So, we have selected **Curve(s)** radio button. After selecting the radio button, click on the **Select** button next to **Curve(s)** radio button and select desired curve; refer to Figure-23. After selecting curve(s), click on the **OK** button from the **Wireframe Chaining** dialog box.

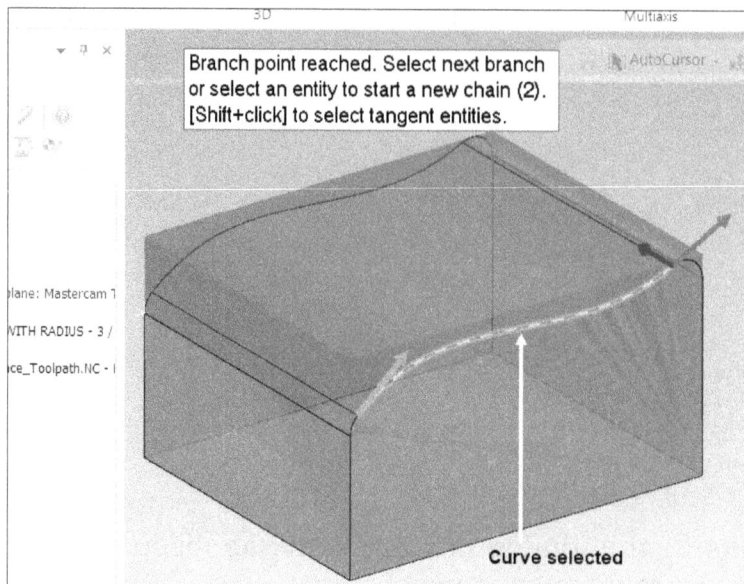

Figure-23. Curve selected for defining parallel direction

- Click on the **Select** button for **Drive Surface(s)** option from the **Drive** area of the dialog box and select the face(s) on which machining will be performed; refer to Figure-24.
- After selecting faces, click on the **End Selection** button.

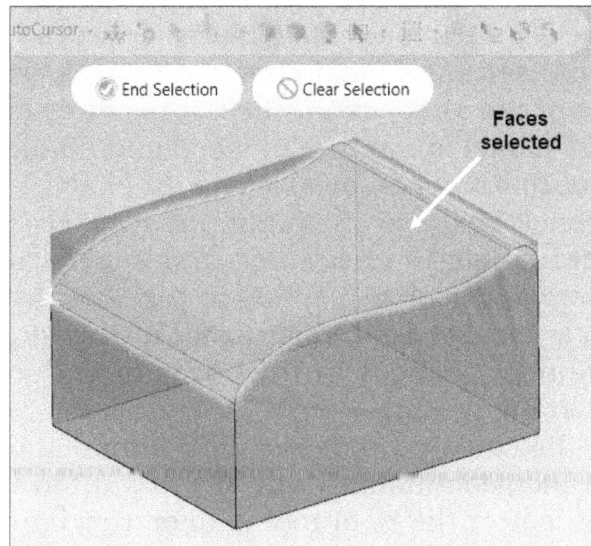

Figure-24. Drive faces selected

- Set the other parameters like cutting method, cut order, machining order, and so on as desired in this page.
- Click on the **Advanced Options For Surface Quality** option from the **Cut Pattern** node to define parameters for smoothening the toolpath. The options will be displayed as shown in Figure-25.

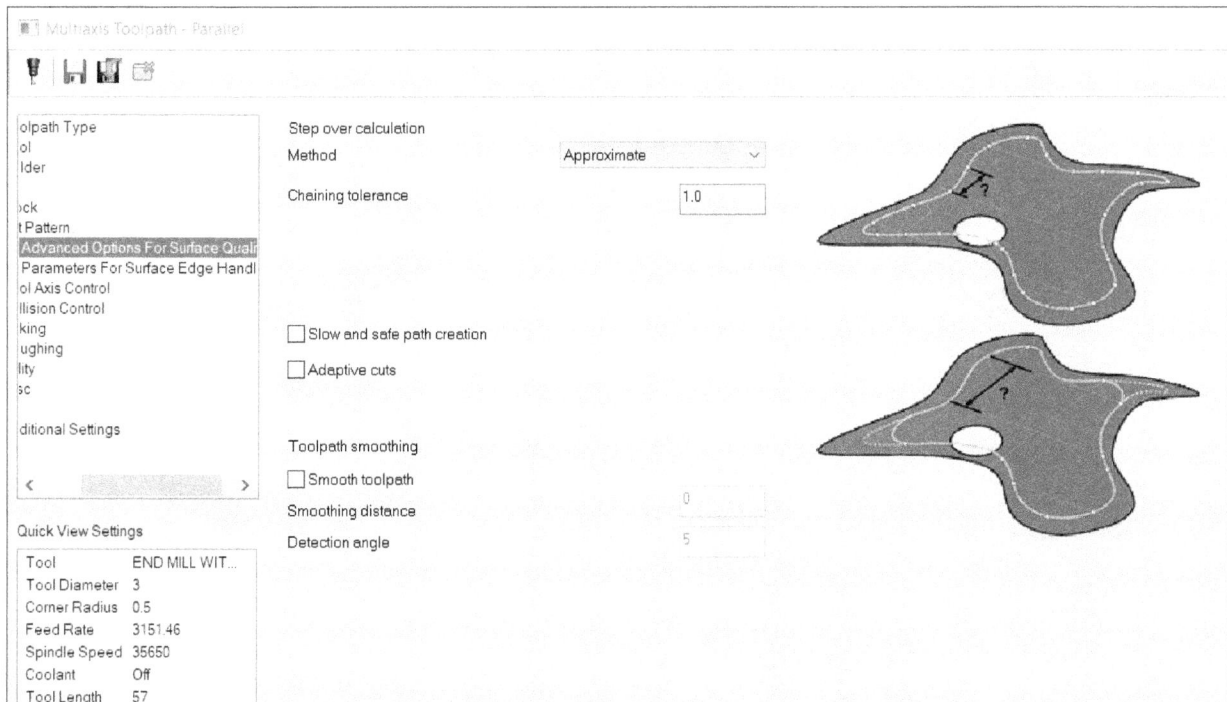

Figure-25. Advanced Options For Surface Quality page

- Select the **Approximate** option from the **Method** drop-down if you want to calculate stepover points using the entire surface as reference. Select the **Exact** option from the **Method** drop-down if you want to calculate stepover points individually for each cutting pass which results in a smoother cutting pass. If you have selected **Approximate** option then you can specify chaining tolerance in the respective edit box. You can also select **Slow and safe path creation** check box to create a smoother toolpath and select the **Adaptive cuts** check box to create a constant stepover irrespective of the shape of surface. It is better to select **Adaptive cuts** check box when your model is mixture of steep and shallow surfaces.

- Select the **Smooth toolpath** check box to smooth sharp corners in the toolpath. After selecting the check box, specify minimum angle to be considered as sharp corner in the **Detection angle** check box and set the distance value allowed between sharp corners and toolpath spline in the **Smoothing distance** edit box.

- Select the **Parameters For Surface Edge Handling** option from the **Cut Pattern** node to smoothen abrupt changes in the edges of surface. The options will be displayed as shown in Figure-26.

- Set desired value in the edit box to define minimum distance value of gap below which the gap will be merged automatically. If you want to specify gap with respect to tool diameter then select the % of tool diameter radio button and specify the value in respective edit box.

- Select the **Maintain outside sharp edges** check box to move tool away from sharp corner while cutting so that the sharp corners are not damaged by machining. On selecting this check box, you will be asked to specify radius of outer loop and angle value above which all corners will be considered as sharp edges.

- Select the **Extend edge curve** check box if you are not getting proper machining of metal at the end locations of cutting passes; refer to Figure-27.

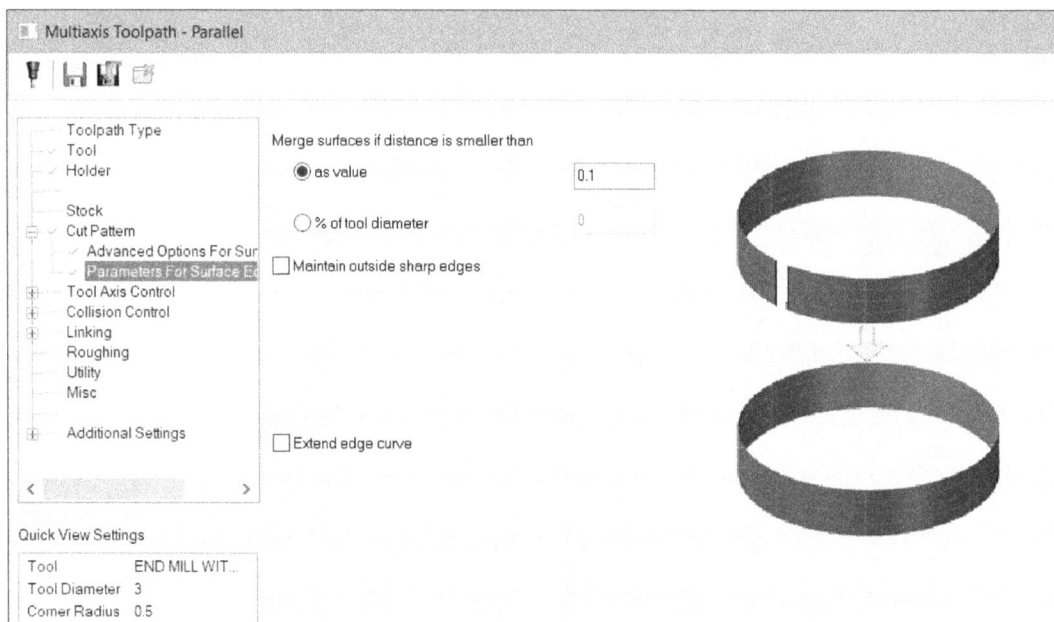

Figure-26. Parameters For Surface Edge Handling options

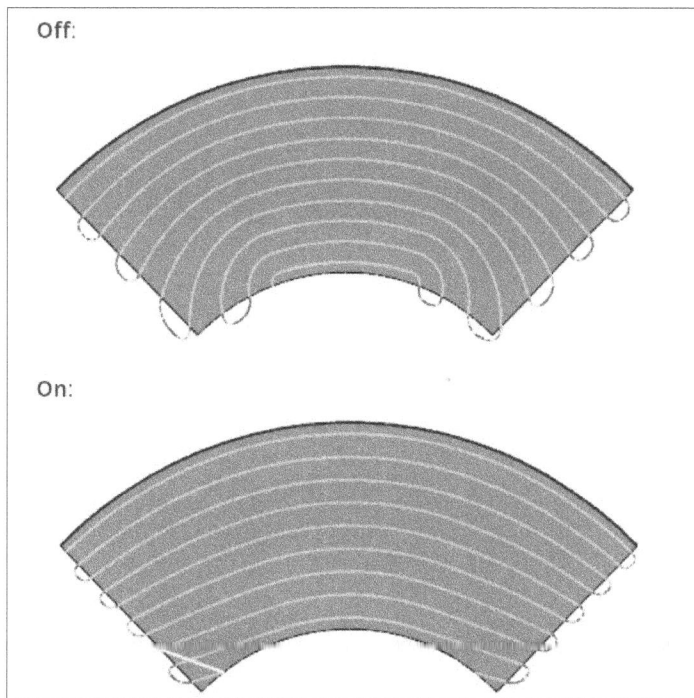

Figure-27. Extended edge curves on and off

Collision Control Options

- Select the **Collision Control** option from the left area to define parameters for avoiding collision while performing machining; refer to Figure-28.
- Select desired check boxes from the **Check** area to define which components of cutting tool will be considered while creating strategy for avoiding collision. You can select **Flute**, **Shoulder**, **Shank**, and **Holder** check boxes to use them as reference for avoiding collision.
- Select desired options from the drop-downs in the **Strategy and parameters** area of the dialog box to define how the collision will be avoided. You can set the tool to retract, tilt, trim & relink toolpath, stop cutting, or just report collision.
- Select desired check boxes from the **Geometry** area to define while surfaces will be used from the model to avoid collision. By default, the **Drive surfaces** check box is selected so the faces/surfaces selected for machining are automatically avoided from collision. If you select the **Check surfaces** check box then you can select desired surfaces to be avoided from collision.
- Select **Cylindrical** radio button from the **Clearance type** area to define clearance distance from cylindrical components by specified values. Select the **Conical** radio button if the components are conical and you want to specify lower as well as upper offset values for clearance. After selecting the radio button, specify related parameters in the **Clearances** area of the dialog box.
- Set the other parameters as desired. Note that by default, the **1** check box is selected so one strategy is created for avoiding collision. You can create up to 4 different strategies to avoid collision by using the options in **Additional Collision Control Strategies** page of the dialog box.
- Set the other parameters as discussed earlier and click on the **OK** button to create the toolpath. The toolpath will be generated; refer to Figure-29.

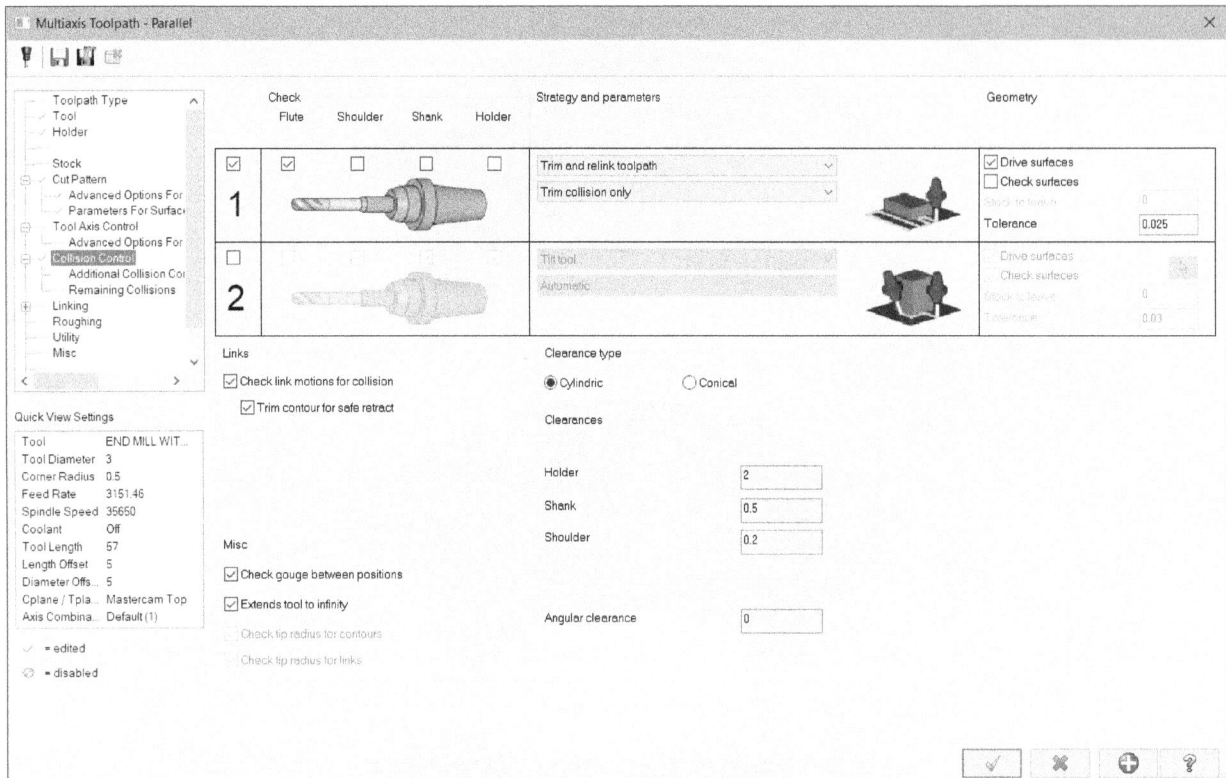

Figure-28. Collision Control options for Parallel toolpath

Figure-29. Parallel toolpath generated

CREATING MULTIAXIS ALONG CURVE TOOLPATH

The **Along Curve** tool is used to perform machining perpendicular to selected curve. The procedure to use this tool is given next.

- Click on the **Along Curve** tool from the **Multiaxis** drop-down in the **Ribbon**. The **Multiaxis Toolpath - Along Curve** dialog box will be displayed; refer to Figure-30.

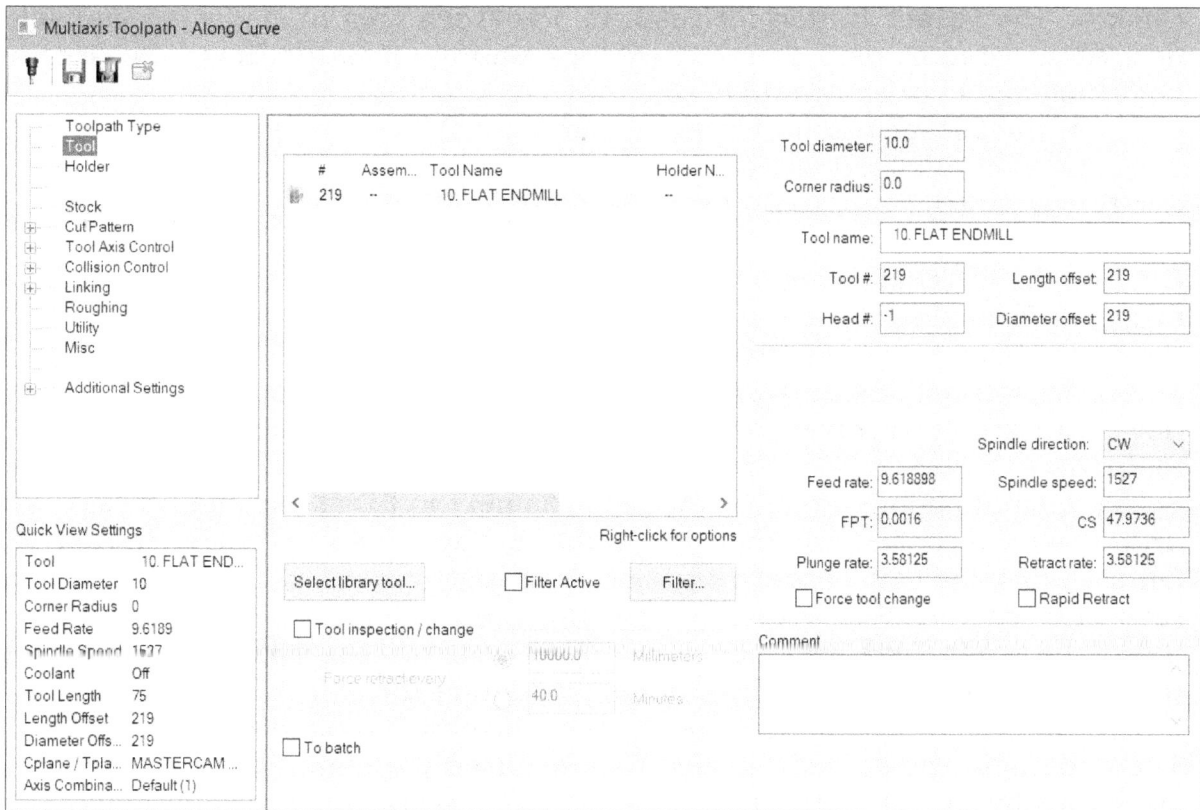

Figure-30. Multiaxis Toolpath-Along Curve dialog box

- Select desired cutting tool, holder, and stock material from the dialog box as discussed earlier.
- Click on the **Cut Pattern** option from left in the dialog box to define pattern of cutting passes. The options in the dialog box will be displayed as shown in Figure-31.

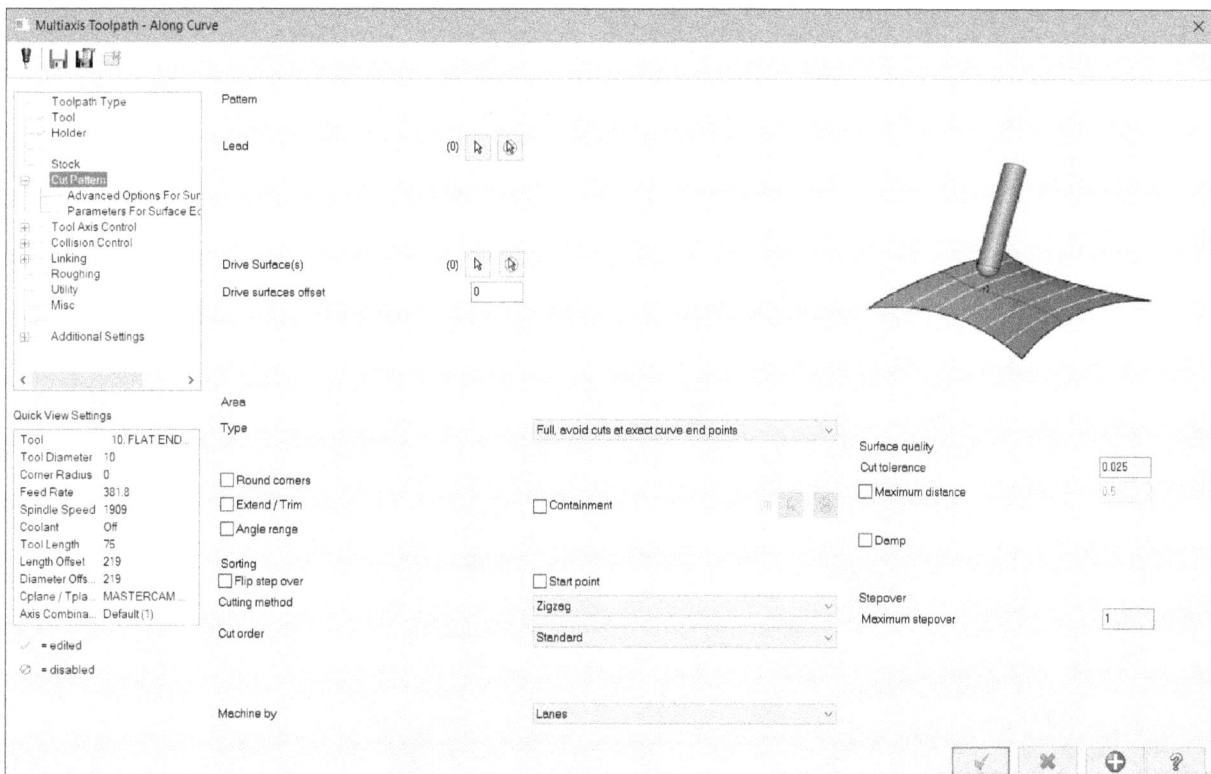

Figure-31. Cut Pattern options for Along Curve toolpath

- Click on the **Select** button for **Lead** from **Pattern** area in this page and select desired curve to be used as reference for creating toolpath. The tool passes will be orthogonal to selected curve.
- Click on the **Select** button for **Drive Surface(s)** from the **Pattern** area in this page and select desired faces to be machined.
- Set the parameters in **Tool Axis Control** and **Collision Control** pages as discussed earlier.
- Click on the **Linking** option from the left area of the dialog box to define entry and exit path for toolpath and cutting passes. The **Linking** page will be displayed in the dialog box; refer to Figure-32.
- Select desired options in the **First entry** and **Last exit** drop-downs to define how cutting tools will enter the workpiece and exit the workpiece while generating toolpath.
- Select desired options from the drop-downs next to **First entry** and **Last exit** drop-downs to define whether you want to apply lead in/out curves or not. On selecting the **Use Lead-In** and **Use Lead-Out** options from the drop-downs, respective pages will be added in the dialog box.
- Select the **Start from home position** and **Return to home position** check boxes to start moving cutting tool from home position at the start of toolpath and then return the cutting tool to home position at the end of toolpath, respectively.
- The options in the **Default links** area are used to define how small and large gaps in toolpaths will be filled by linking toolpath segments. Select desired options from the **Small gaps** and **Large gaps** drop-downs to define linking segments.
- Specify desired value in the **Small gap size** edit box to define size of small gap in percentage of cutting tool diameter.
- Set the other parameters as discussed earlier and click on the **OK** button from the dialog box. The toolpath will be generated; refer to Figure-33.

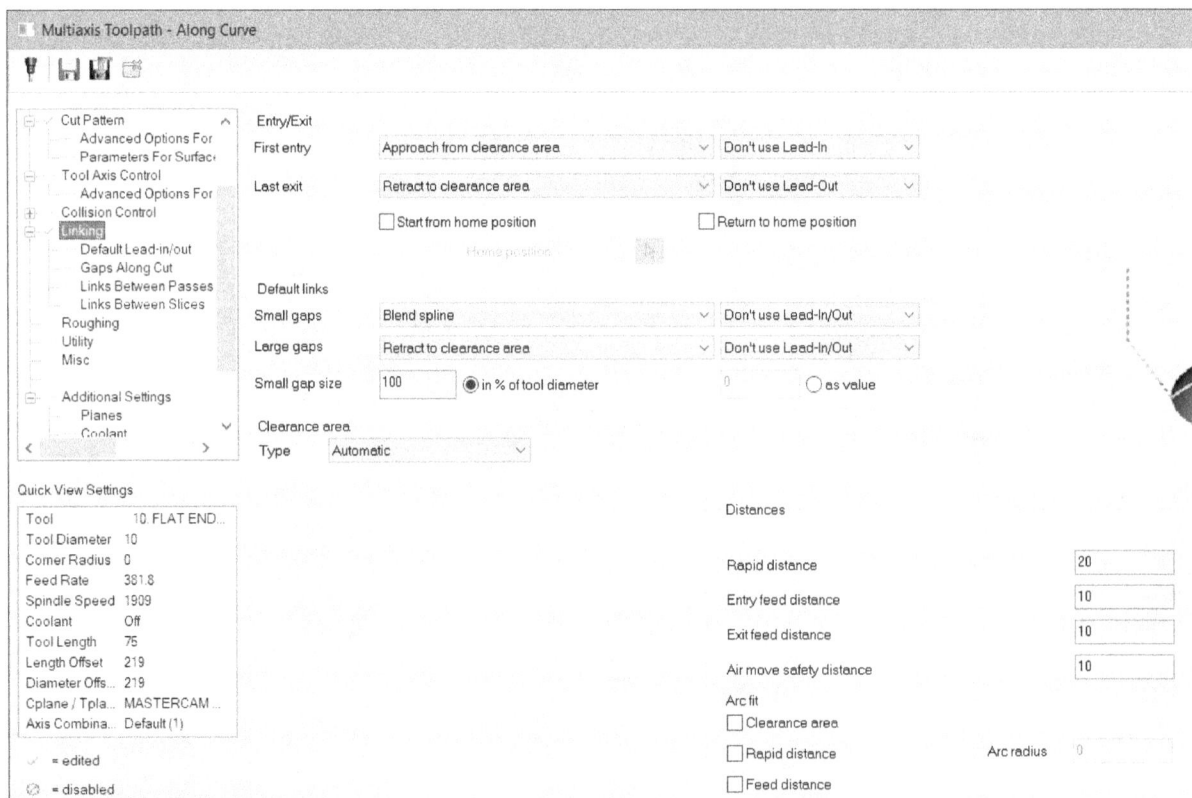

Figure-32. Linking page

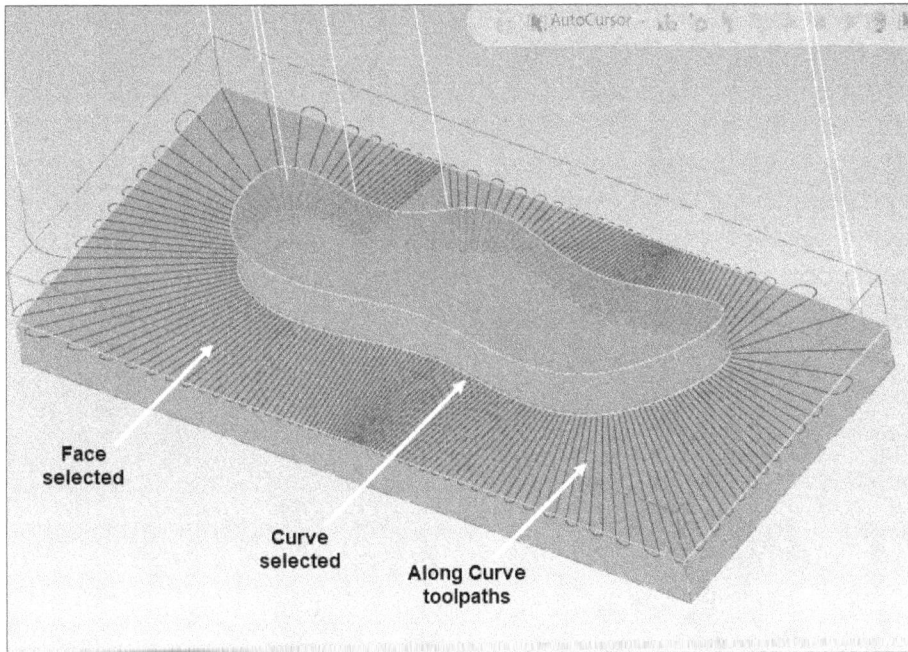

Figure-33. Along Curve toolpaths created

CREATING MULTIAXIS MORPH TOOLPATH

The **Morph** tool is used to perform machining on surface trapped between two leading curves. The procedure to use this tool is given next.

- Click on the **Morph** tool from the **Multiaxis** drop-down in the **Ribbon**. The **Multiaxis Toolpath - Morph** dialog box will be displayed; refer to Figure-34.

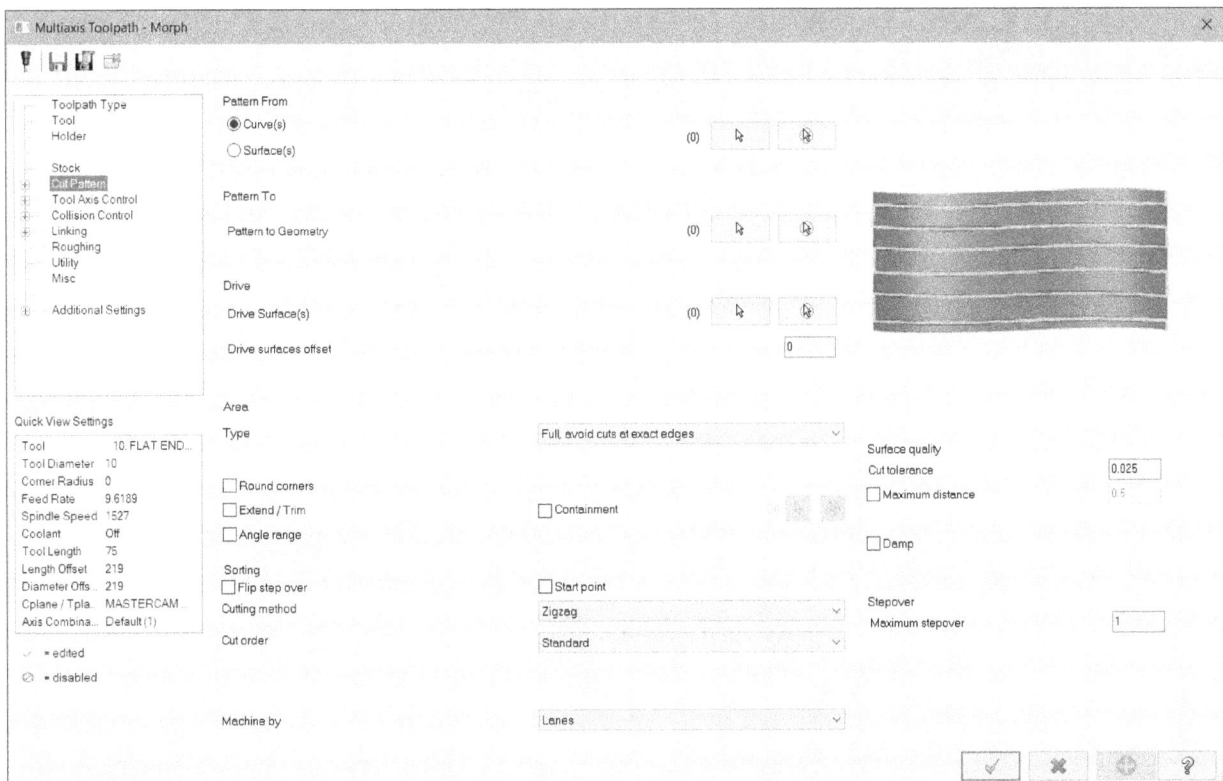

Figure-34. Multiaxis Toolpath-Morph dialog box

- Specify desired parameters in **Tool**, **Holder**, and **Stock** pages as discussed earlier.

- Select the **Curve(s)** or **Surface(s)** radio button and select a set of two chains to be used as boundary for the toolpath.
- Click on the **Select** button from the **Pattern To** area and select curve to be used for defining pattern for cutting toolpath.
- Click on the **Select** button for **Drive Surface(s)** from the **Drive** area and select desired surfaces to be used as reference for machining.
- Set the other parameters in the dialog box as discussed earlier and click on the **OK** button to create the toolpath. The morph toolpath will be created; refer to Figure-35.

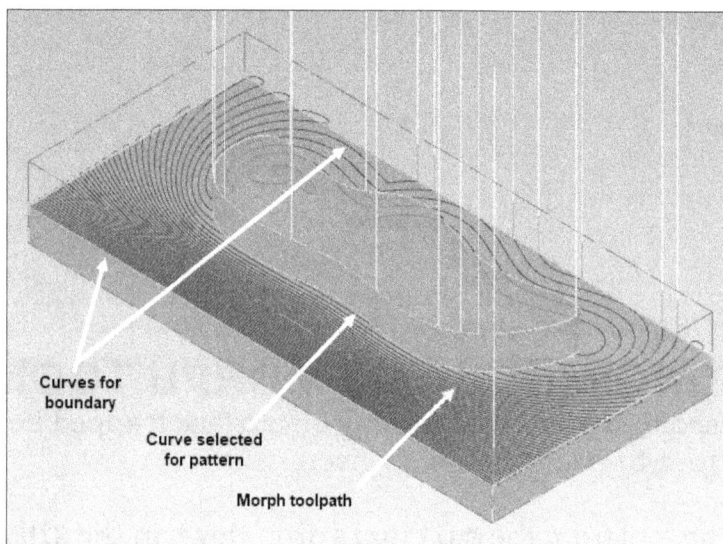

Figure-35. Morph toolpath created

CREATING MULTIAXIS FLOW TOOLPATH

The **Flow** tool is used to create toolpath on selected surfaces using U (horizontal) and V (vertical) lines as cutting passes. The procedure to generate this toolpath is given next.

- Click on the **Flow** tool from the **Multiaxis** drop-down in the **Ribbon**. The **Multiaxis Toolpath - Flow** dialog box will be displayed; refer to Figure-36.

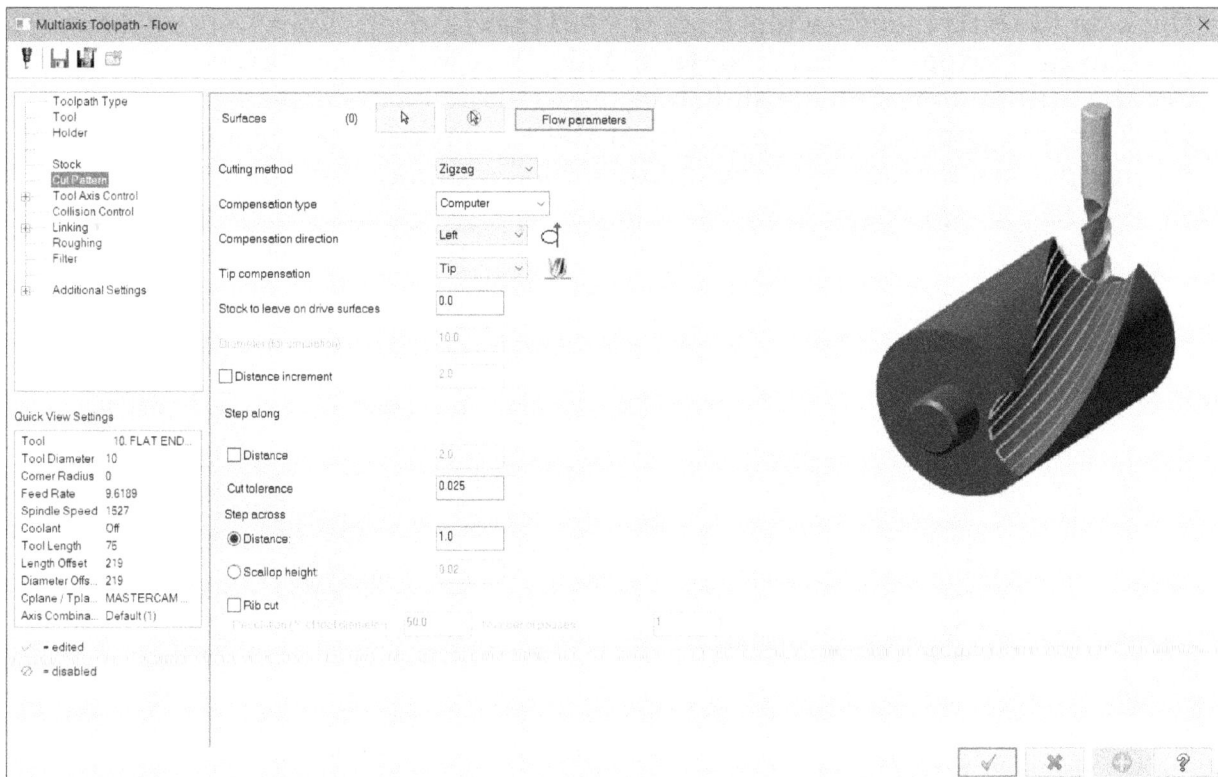

Figure-36. Multiaxis Toolpath-Flow dialog box

- Click on the **Cut Pattern** option from the left area to define cutting pattern for toolpath. Click on the **Select** button from the **Surfaces** area of the dialog box and select desired surfaces to be used for machining.
- After selecting surfaces, click on the **End Selection** button. The **Flowline options PropertyManager** will be displayed; refer to Figure-37.

Figure-37. Flowline options PropertyManager

- Specify the parameters as desired in the dialog box. The options of this dialog box have been discussed earlier.
- Set the other parameters as discussed earlier and click on the **OK** button from the dialog box. The toolpaths will be generated.

CREATING MULTIAXIS MULTISURFACE TOOLPATH

The **Multisurface** tool is used to create toolpaths for machining 3D surfaces with uneven shapes. The procedure to generate multisurface toolpath is given next.

- Click on the **Multisurface** tool from the **Multiaxis** drop-down in the **Ribbon**. The **Multiaxis Toolpath - Multisurface** dialog box will be displayed.
- Set the cutting tool, holder, and stock parameters as discussed earlier.
- Click on the **Cut Pattern** option from the left area in the dialog box to define cutting pattern for toolpath. The options will be displayed as shown in Figure-38.

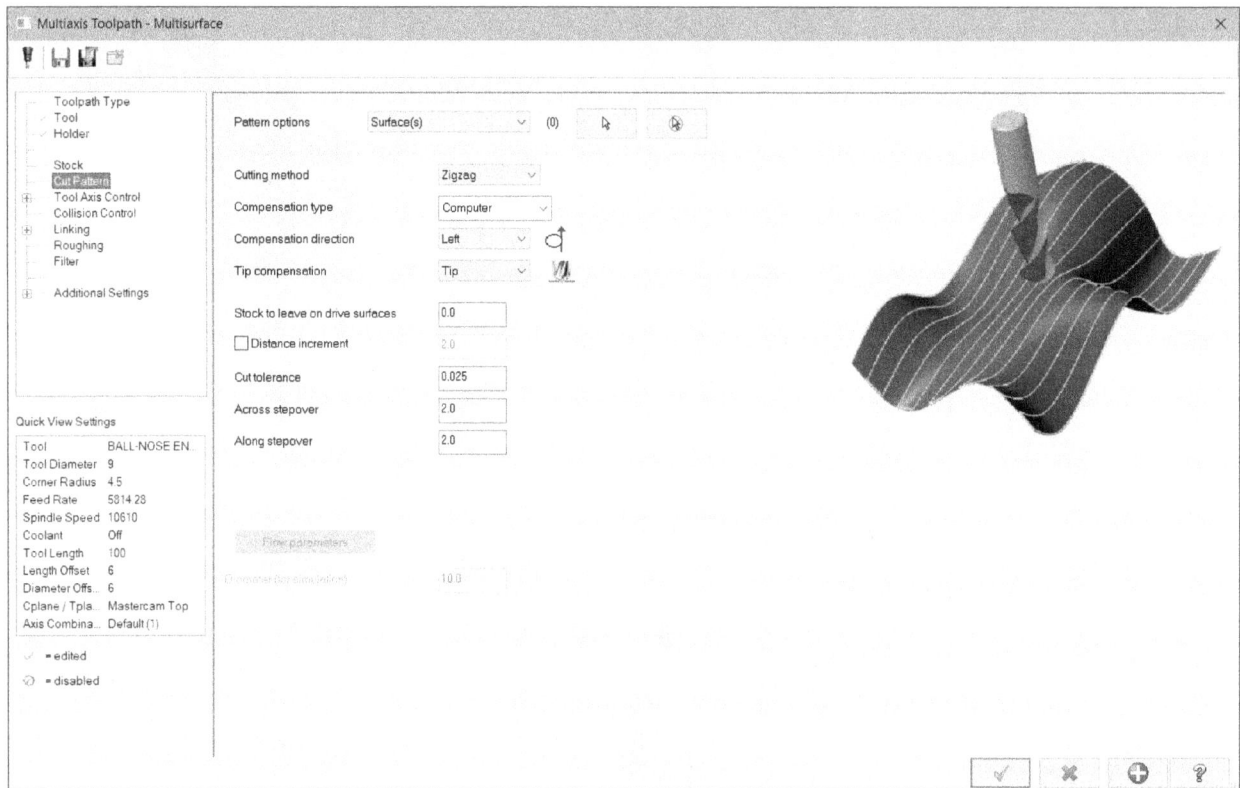

Figure-38. Multiaxis Toolpath–Multisurface dialog box

- Select desired option from the **Pattern Options** drop-down to define shape of toolpath. By default, **Surface(s)** option is selected in this drop-down, so you can select desired surfaces to be machined. Select the **Cylinder/Sphere/Box** option from the drop-down to use respective shape for generating cutting toolpath. If the **Cylinder**, **Sphere**, or **Box** option is selected in the drop-down then click on the **Select** button next to drop-down for defining parameters. The option dialog box for respective shape will be displayed; refer to Figure-39.

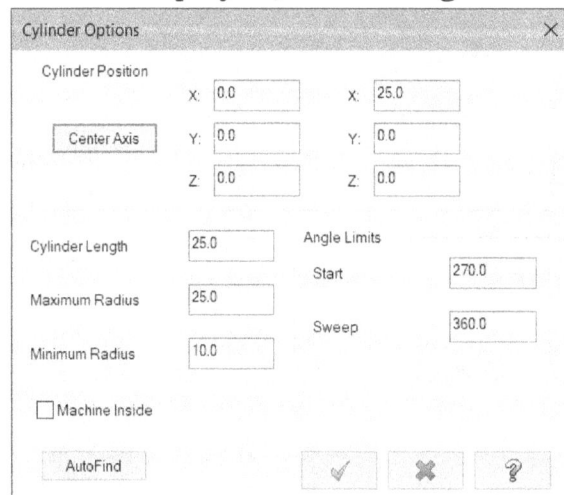

Figure-39. Cylinder Options dialog box

- Set desired parameters in the dialog box and click on the **OK** button if you want to use predefined shape for toolpath generated.
- Select the **Filter** option from the left area to define points to be filtered from the toolpath if they are within tolerance range. The **Filter** page will be displayed; refer to Figure-40.

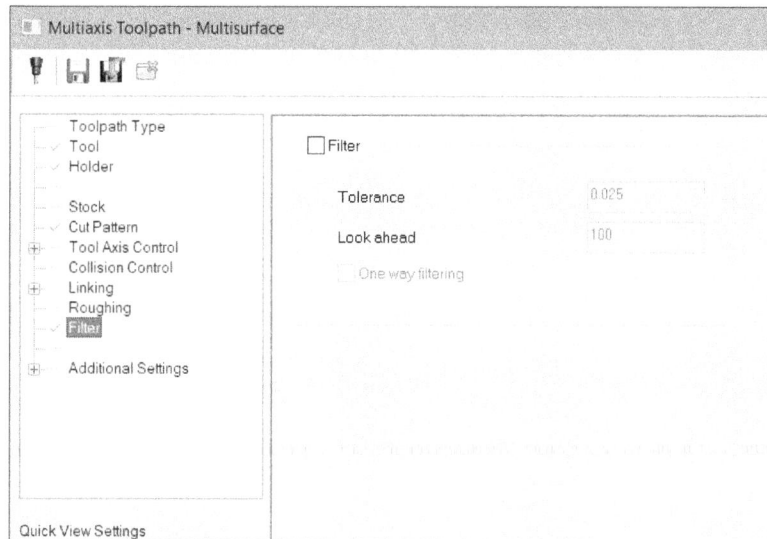

Figure-40. Filter page

- Select the **Filter** check box from the dialog box to define tolerance parameters for filtering points on toolpaths.
- Specify the distance value in **Tolerance** edit box below which the points will be filtered if distance between two points is less than tolerance value.
- Set the other parameters as desired in the **Filter** area.
- Specify other parameters as discussed earlier and click on the **OK** button to create the toolpath.

CREATING MULTIAXIS PORT TOOLPATH

The **Port** tool is used to machine inside of ports (tube like complex structures) in the model. The procedure to use this tool is given next.

- Click on the **Port** tool from the **Multiaxis** drop-down in the **Ribbon**. The **Multiaxis Toolpath - Port** dialog box will be displayed; refer to Figure-41.
- Click on the **Select** button from the **Surfaces** area of **Cut Pattern** page in the dialog box to select internal surfaces of the port (tube).
- After selecting faces/surfaces, click on the **End Selection** button. The **Flowline data** dialog box will be displayed. The options of this dialog box have already been discussed.
- Click on the **OK** button from the **Flowline data** dialog box after applying changes. The **Multiaxis Toolpath** dialog box will be displayed again. Set the parameters as discussed earlier.
- These is a very high chance of tool colliding with workpiece when using this toolpath if you do not provide proper tool axis control. Click on the **Tool Axis Control** option from the left in the dialog box. The **Tool Axis Control** page will be displayed; refer to Figure-42.

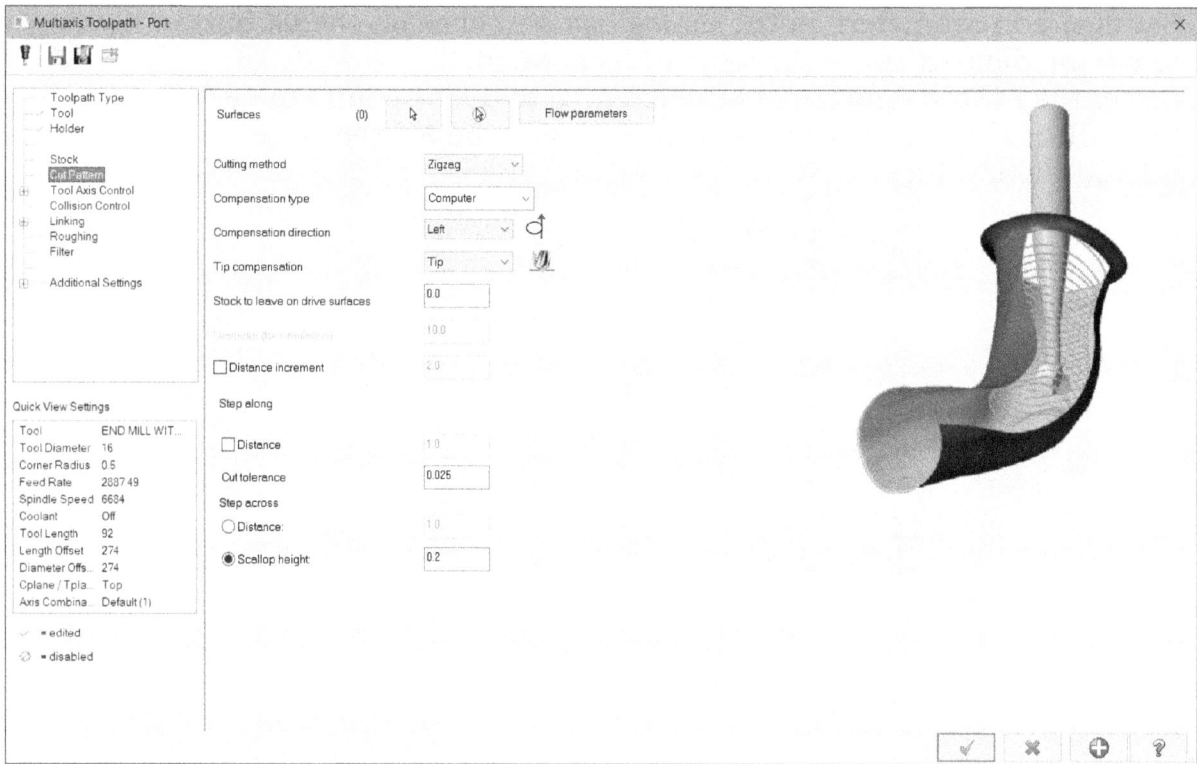

Figure-41. Multiaxis Toolpath-Port dialog box

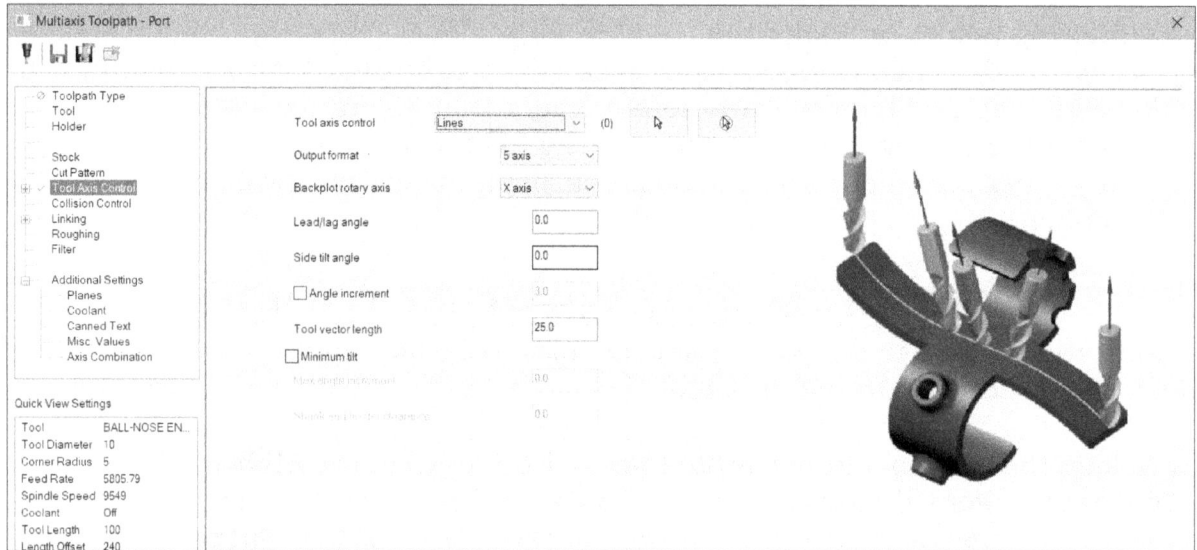

Figure-42. Tool Axis Control page

- Select desired option from the **Toolaxis control** drop-down to define how cutting tool will orient when performing operation. Select the **Lines** option if you want to use selected line(s) for aligning cutting tool and use the selection button next to drop-down to select the lines. Select the **Pattern surface** option to keep tool normal to selected pattern surface. Select the **Plane** option to keep cutting tool normal to selected plane. Select the **From point** option to keep tool aligned to selected point such that selected point is collinear with tool axis while cutting. Note that selected point should be at the starting of toolpath. This option is useful when machining inside of a pipe; refer to Figure-43. Select the **To point** option to keep tool aligned with selected point and the point is at end of toolpath. Select the **Chain** option to keep cutting tool axis aligned to normal of selected curve chain. This option is useful for irregular surfaces where no tool obstruction is present above the selected chain. Select the **Boundary** option to keep tool axis aligned normal to boundary chain (you can select face also for this option). Out of all the tool axis control options, the **From point** and **To point** give maximum freedom of tool axis rotation when performing port machining as cutting tool can form conical shape about selected point while machining.

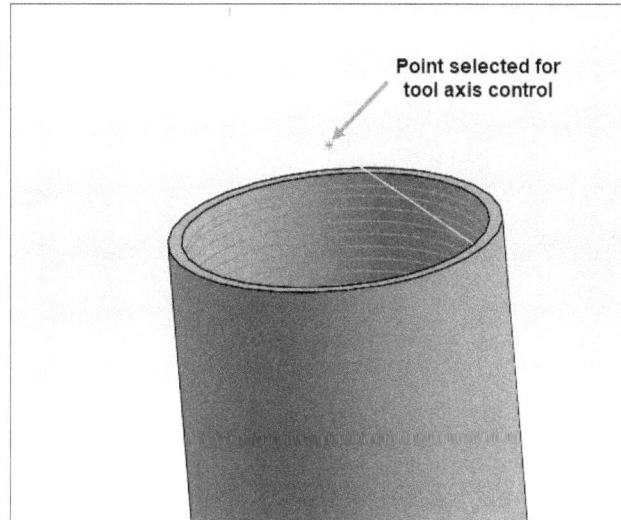
Figure-43. Using point for tool axis control

• Specify the other parameters as discussed earlier and click on the **OK** button to generate the toolpath; refer to Figure-44.

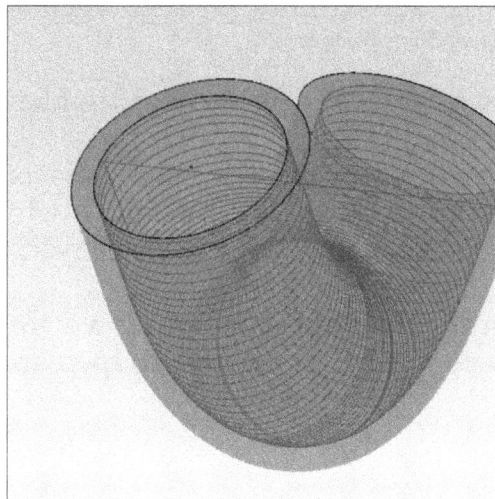
Figure-44. Port toolpath generated

CREATING MULTIAXIS TRIANGULAR MESH TOOLPATH

The **Triangular Mesh** tool is used to machine objects in pyramid shapes. In this toolpath, the cutting will be in continuous contact with the body. The procedure to use this tool is given next.

- Click on the **Triangular Mesh** tool from the **Multiaxis** drop-down in the **Ribbon**. The **Multiaxis Toolpath - Triangular Mesh** dialog box will be displayed; refer to Figure-45.

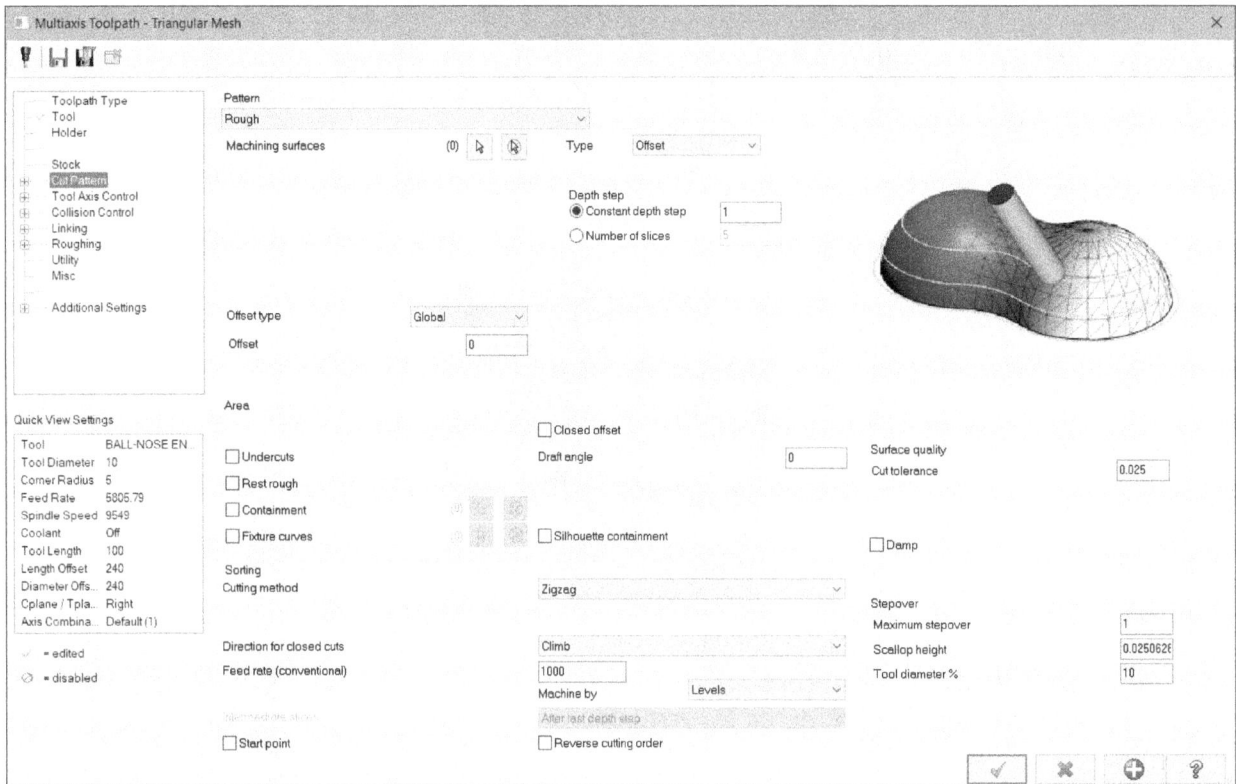

Figure-45. Multiaxis Toolpath–Triangular Mesh dialog box

- Specify the parameters related to cutting tool, holder, and stock as discussed earlier.
- Click on the **Cut Pattern** option from the left area of the dialog box to define parameters related to how cutting tool will move while performing machining on the model. The **Cut Pattern** page will be displayed in the dialog box; refer to Figure-45.
- Select desired option from the **Pattern** drop-down to define cutting strategies. Various cutting strategies with their related shapes are shown in Figure-46.

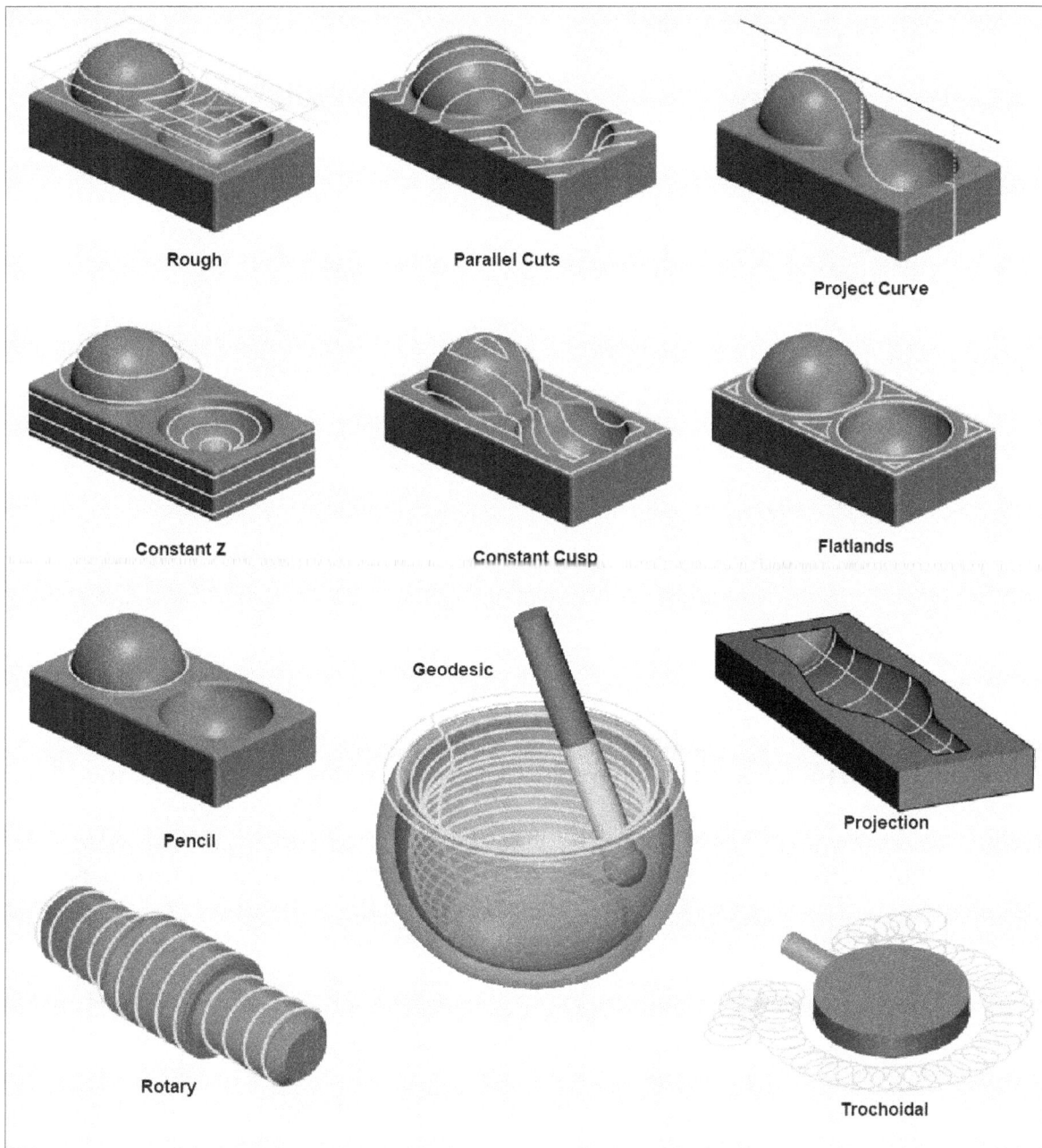

Figure-46. Triangular cut patterns

- Click on the **Select** button for **Machining surfaces** and select desired surfaces to be machined. Based on selected pattern type, you will be asked to select different type of curves like projection curves, drive curves, lines, and so on.
- Set the other parameters as discussed earlier and click on the **OK** button from the dialog box to create the toolpath.

CREATING DEBURR TOOLPATH

The Deburr toolpath is created to machine broken edges and burrs in the model. The procedure to create this toolpath is given next.

- Click on the **Deburr** tool from the **Multiaxis** drop-down in the **Ribbon**. The **Multiaxis Toolpath - Deburr** dialog box will be displayed; refer to Figure-47.

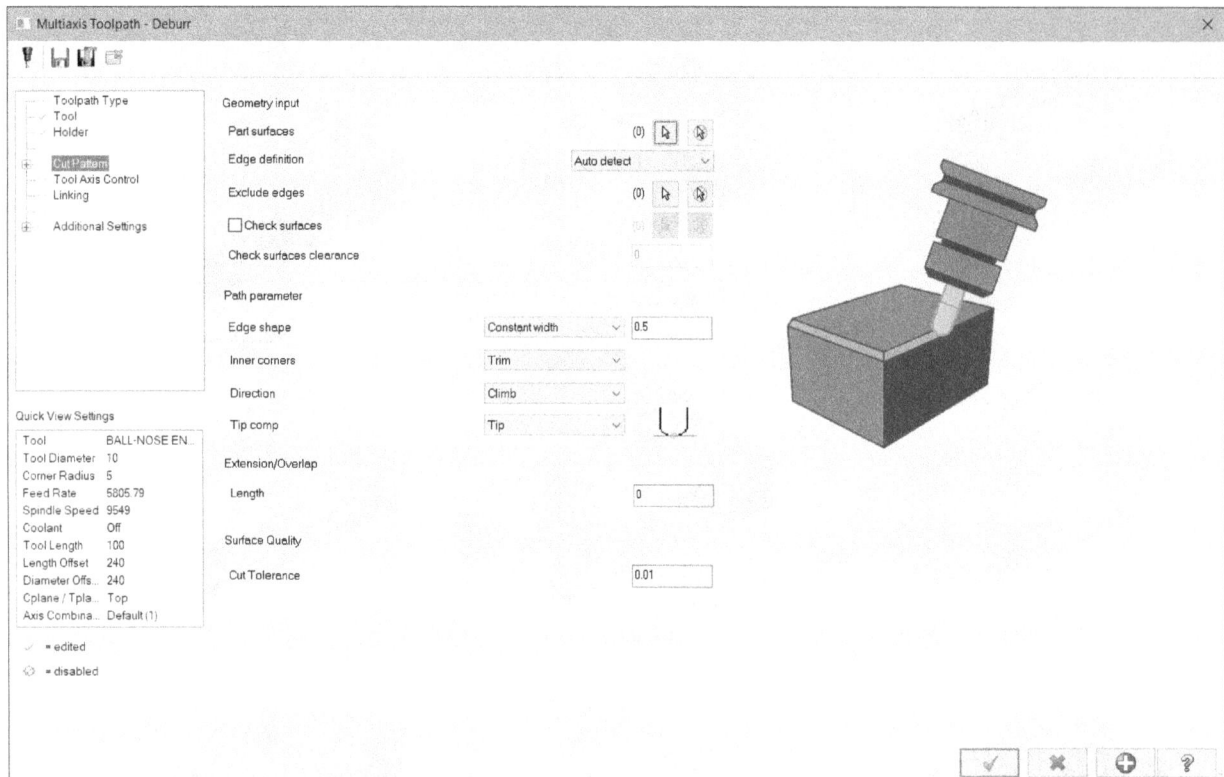

Figure-47. Multiaxis Toolpath-Deburr dialog box

- Set desired cutting tool and tool holder as discussed earlier.
- Click on the **Select** button for **Part surfaces** from **Cut Pattern** page and select the faces of model whose edges will be machined for deburring.
- Select desired option from the **Edge definition** drop-down to define which edges will be deburred or excluded. If the **Auto detect** option is selected in the drop-down then all the edges will be selected automatically. Click on the **Select** button for **Exclude edges** option and select desired edges to be excluded from machining. If the **User defined** option is selected in the drop-down then click on the **Select** button for **User defined edges** option and select desired edges to be machined.
- Select the **Check surfaces** check box if you want to avoid faces/surfaces during machining.
- Select desired option from the **Edge shape** drop-down to define size of cut on edges. Select the **Constant width** option from the drop-down if you want to specify width of cut and select the **Constant depth** option from the drop-down if you want to specify depth of cut at the edges.
- Set the other parameters as discussed earlier and click on the **OK** button to create the toolpath.

CREATING MULTIAXIS POCKETING TOOLPATH

The **Pocketing** tool is used to machine pockets with undercuts and irregular shapes. The procedure to use this tool is given next.

- Click on the **Pocketing** tool from **Multiaxis** drop-down in the **Ribbon**. The **Multiaxis Toolpath - Pocketing** dialog box will be displayed; refer to Figure-48.

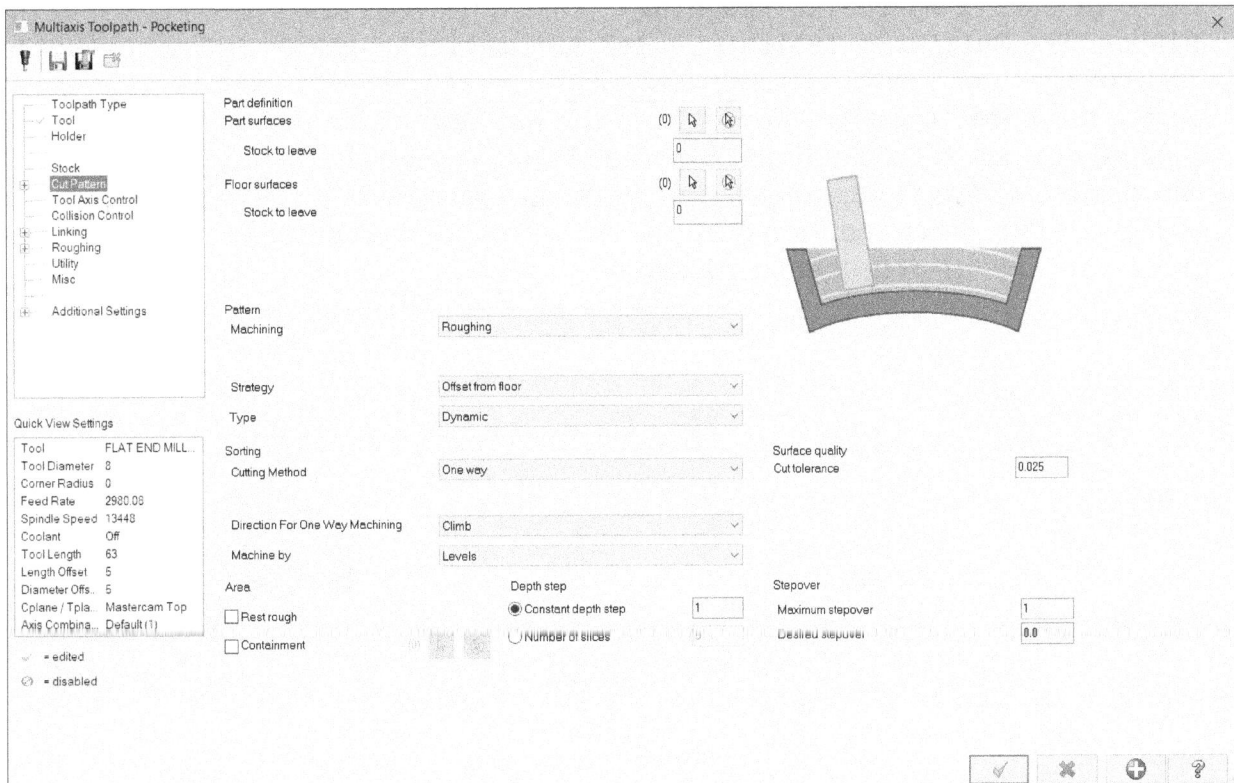

Figure-48. Multiaxis Toolpath-Pocketing dialog box

- Set desired cutting tool, holder, and stock parameters.
- Click on the **Cut Pattern** option from the left area to modify cutting pattern parameters. The **Cut Pattern** page will be displayed in the dialog box.
- Click on the **Select** button for **Part surfaces** option from the **Part definition** area of the dialog box and select the faces to be machined.
- Click on the **Select** button for **Floor surfaces** option from the dialog box and select the faces to define floor faces up to which the machining will be performed.
- Specify desired values in the **Stock to leave** edit boxes to define amount of stock left after performing the pocketing toolpath. If **0** value is specified in the edit boxes then finishing toolpath will be performed.
- Select desired option from the **Machining** drop-down in the **Pattern** area of the dialog box to define which faces are to be machined. Select the **Roughing** option to machine both floors as well as walls in the pocket. Select the **Floor finishing** option from the drop-down if you want to machine floors only. Select the **Wall finishing** option from the drop-down to machine walls of pocket only.
- Specify desired parameters in the **Strategy** drop-down based on options selected in the **Machining** drop-down. Note that preview of toolpath will be displayed in the dialog box based on selected options in the drop-down.
- Set the other parameters as discussed and click on the **OK** button to generate toolpath.

CREATING 3+2 AUTOMATIC ROUGHING TOOLPATH

The 3+2 Automatic Roughing tool is used to create roughing toolpaths for 3 translation axes (X, Y, and Z) and 2 tilt axes (A and B). In simple terms, this tool creates roughing toolpaths using tilt axes along with general X, Y, and Z axes. The procedure to create this toolpath is given next.

- Click on the **3+2 Automatic Roughing** tool from the **Multiaxis** drop-down in the **Ribbon**. The **Multiaxis Toolpath - 3+2 Automatic Roughing** dialog box will be displayed.
- Select the **Model Geometry** option from the left area of the dialog box and select desired entities to be machined & avoided as discussed in previous chapters.
- Set desired cutting tool, holder, and stock parameters as discussed earlier.
- Select the **Cut Pattern** option from the left area to define pattern in which cutting passes will be created; refer to Figure-49.

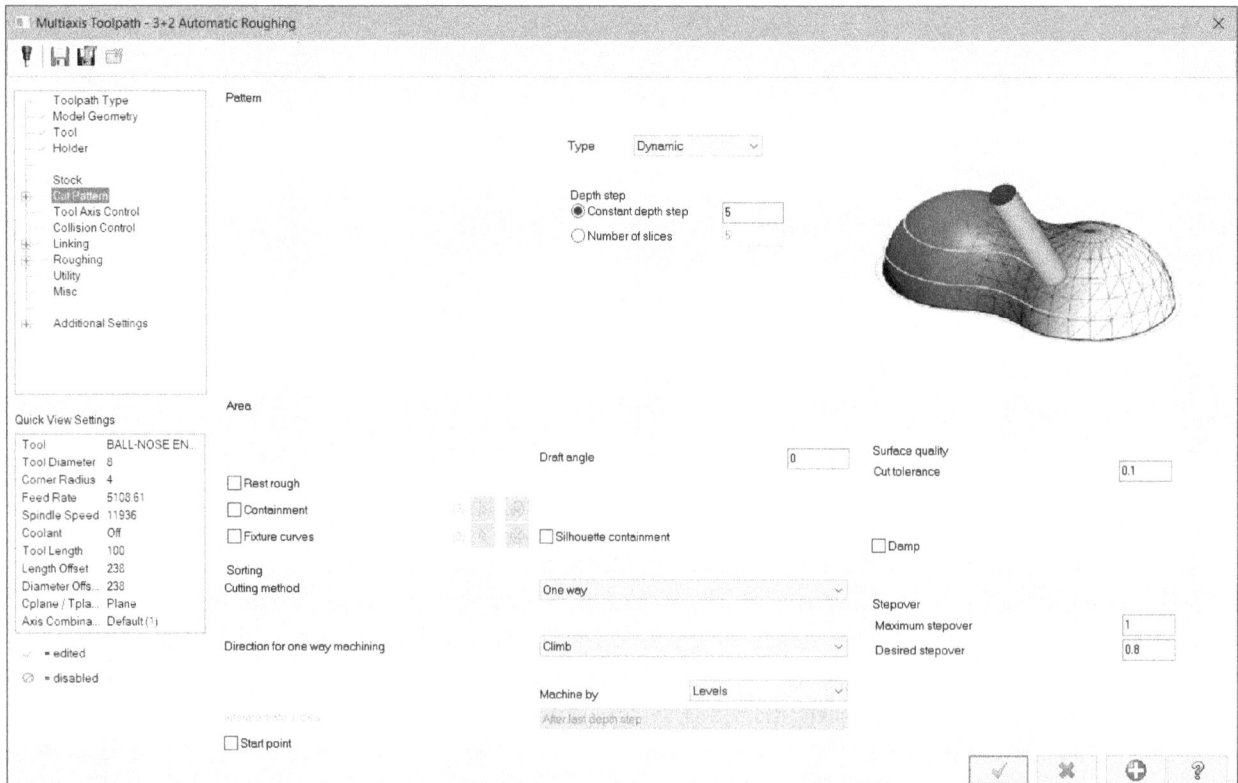

Figure-49. Cut Pattern page

- The options in this dialog box have been discussed earlier in previous chapters. Set the parameters and click on the **OK** button to create the toolpaths.

CREATING MULTIAXIS PROJECT CURVE TOOLPATH

The **Project Curve** tool is used to machine curves projected on the 3D surface. The procedure to use this tool is given next.

- Click on the **Project Curve** tool from the **Multiaxis** drop-down in the **Ribbon**. The **Multiaxis Toolpath - Project Curve** dialog box will be displayed.
- Select desired cutting tool, holder, and stock as discussed earlier.
- Select the **Cut Pattern** option from the left area. The **Cut Pattern** page will be displayed in the dialog box for projected curves; refer to Figure-50.
- Click on the **Select** button for **Projection** from the **Pattern** area of the dialog box and select desired curves to be projected.

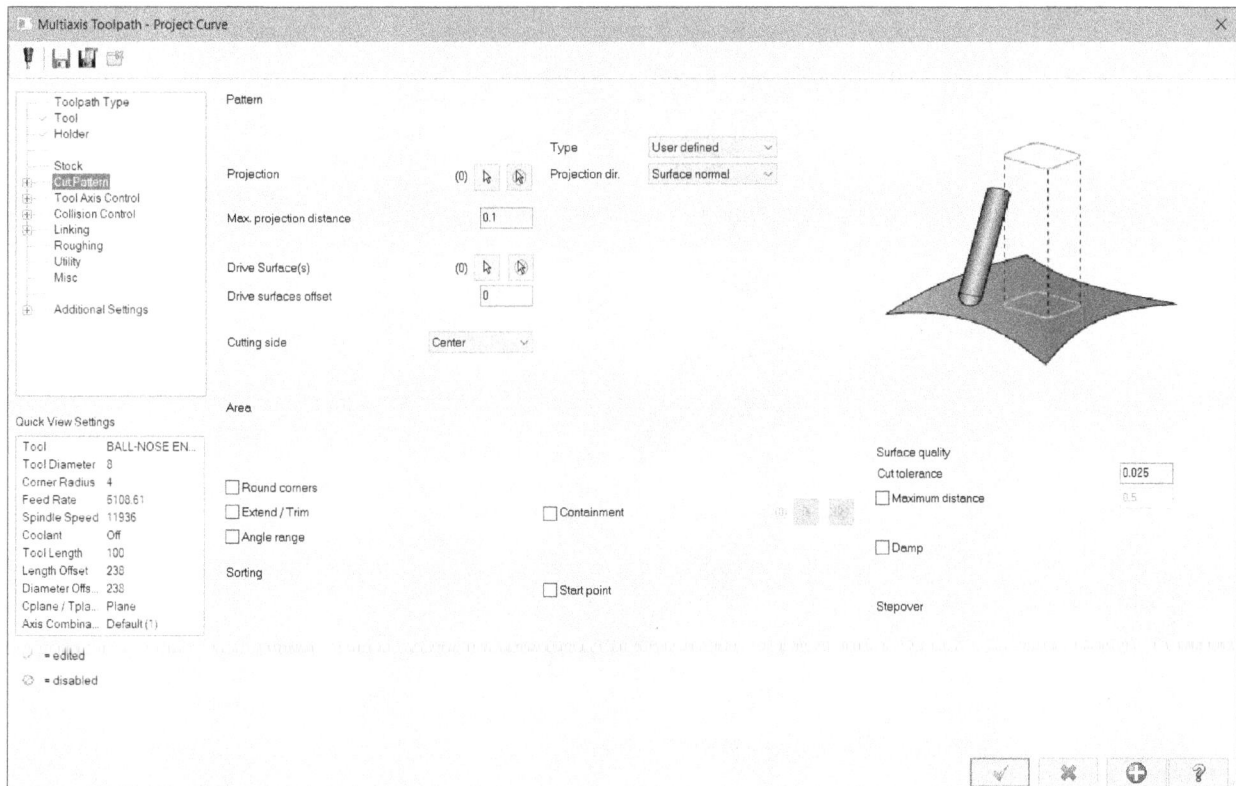

Figure-50. Multiaxis Toolpath-Project Curve dialog box

- Click on the **Select** button for **Drive Surface(s)** option and select the face(s) on which curve will be projected.
- Select desired option from the **Type** drop-down to define the pattern in which curves will be projected on selected surfaces. Select the **User defined** option from the drop-down to project curves as they are created on the surface. Select the **Radial** option from the drop-down to project curves in radial pattern and set desired parameters in **Center point**, **Radius**, and **Angle** areas. Select the **Spiral** option from the drop-down to project curves in spiral motion and specify related parameters. Select the **Offset** option from the drop-down and specify offset direction with number of cuts to create offset copies of projection curves on selected drive surface.
- Specify the other parameters as discussed earlier and click on the **OK** button to create the toolpath; refer to Figure-51.

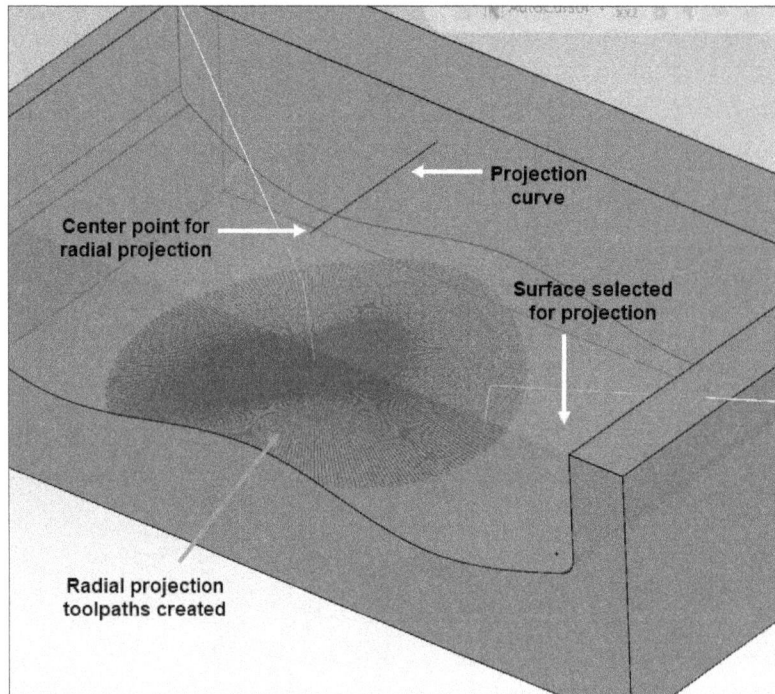

Figure-51. Projected curve toolpath created

CREATING MULTIAXIS ROTARY TOOLPATH

The **Rotary** tool is used to machine 3D models with irregular surfaces. General shapes of these models is cylindrical with irregular surfacing. The procedure to use this tool is given next.

- Click on the **Rotary** tool from the **Multiaxis** drop-down in the **Ribbon**. The **Multiaxis Toolpath - Rotary** dialog box will be displayed.
- Set desired cutting tool and tool holder parameters as discussed earlier.
- Click on the **Cut Pattern** option from the left area of the dialog box. The options to define pattern of cutting passes will be displayed in the **Cut Pattern** page of the dialog box; refer to Figure-52.

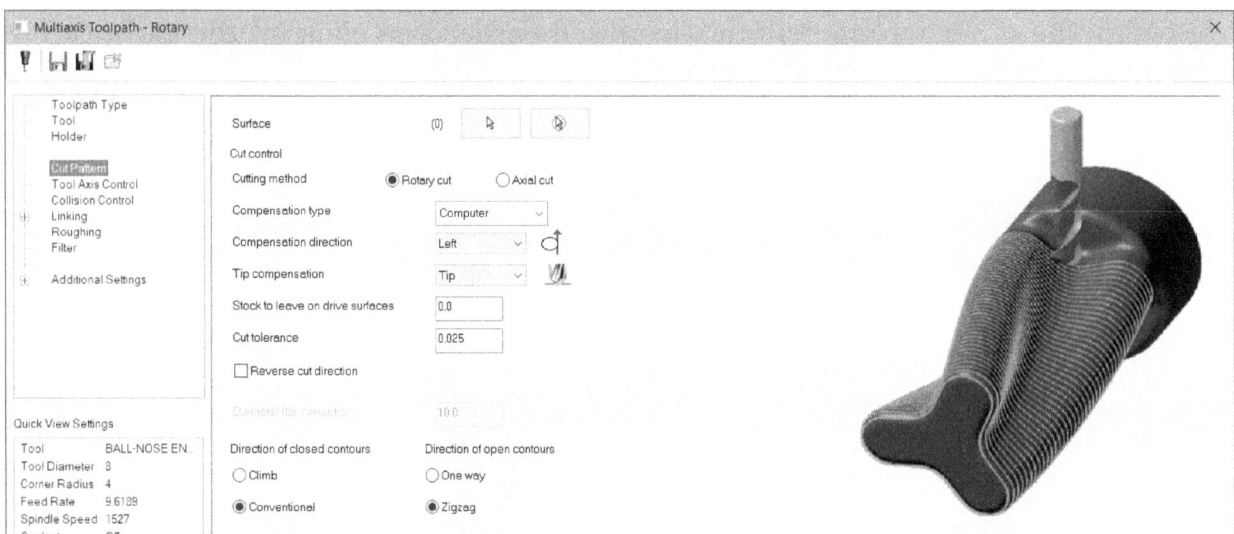

Figure-52. Multiaxis Toolpath-Rotary dialog box

- Click on the **Select** button for **Surface** option from the dialog box and select desired faces (generally, cylindrical surfaces) of the model to be machined.
- Select desired radio button for **Cutting method** to define direction of cut. Select **Rotary cut** radio button to remove material while moving cutting tool in radial direction. Select the **Axial cut** radio button to remove material while moving cutting tool parallel to axis of cylindrical model.
- Set the other parameters as desired and click on the **OK** button. The toolpath will be created; refer to Figure-53.

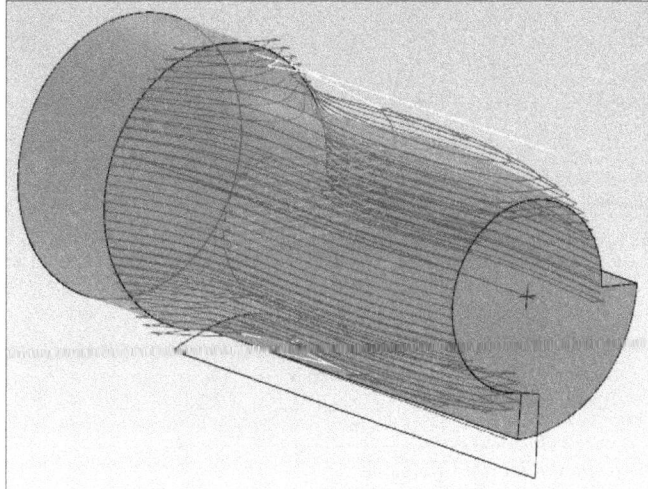

Figure-53. Rotary toolpath created

CREATING MULTIAXIS ROTARY ADVANCED TOOLPATH

The **Rotary Advanced** tool is used to create toolpaths for machining faces of helical gear like structures. The procedure to use this tool is given next.

- Click on the **Rotary Advanced** tool from the **Multiaxis** drop-down in the **Ribbon**. The **Multiaxis Toolpath - Rotary Advanced** dialog box will be displayed.
- Select desired cutting tool, holder, and stock parameters in the dialog box as discussed earlier.
- Click on the **Cut Pattern** option from the left area of the dialog box to define pattern in which cutting passes will be created. The **Cut Pattern** page will be displayed in the dialog box; refer to Figure-54.
- Select desired option from the **Machining** drop-down to define whether you want to create roughing toolpath or finishing toolpath.
- Set desired option in the **Slice pattern** drop-down to define how cutting passes will be placed on the model. The **Radius constant** option will apply same radius in cutting passes.
- Set the parameters as discussed earlier in this page.

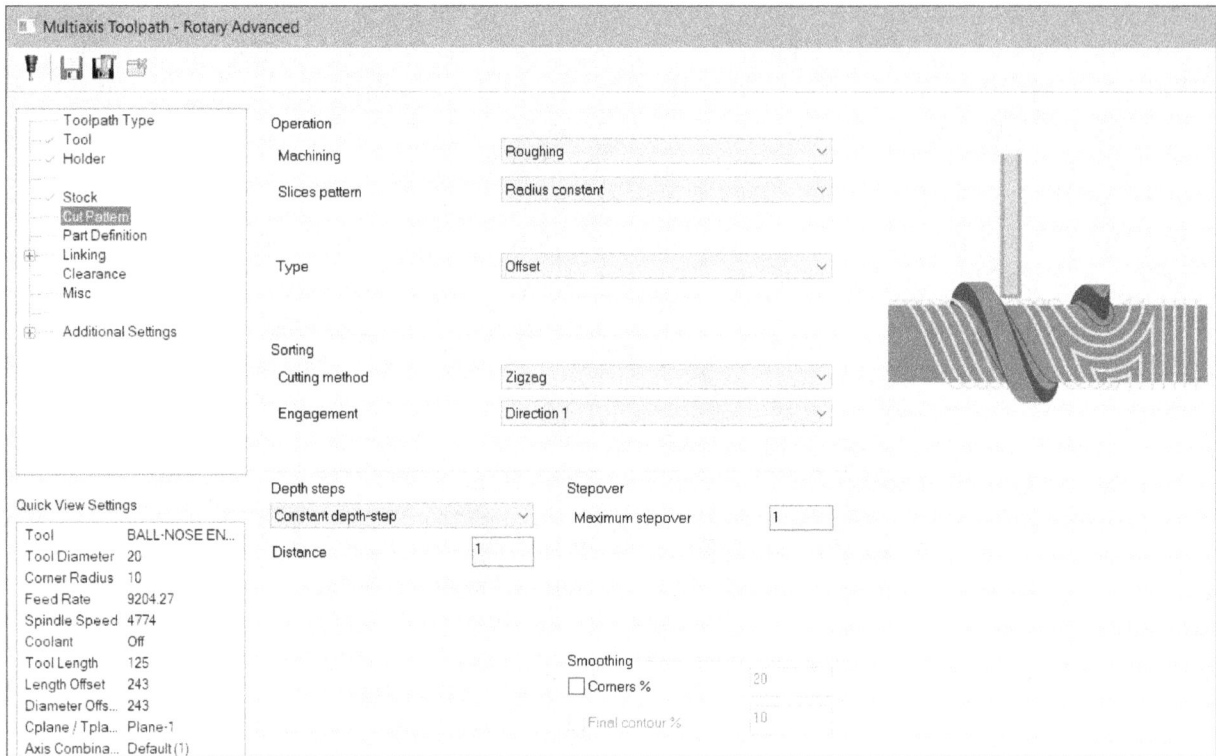

Figure-54. Multiaxis Toolpath-Rotary Advanced dialog box

- Click on the **Part Definition** option from the left area of the dialog box to select surfaces/faces to be machined. The **Part Definition** page will be displayed in dialog box; refer to Figure-55.

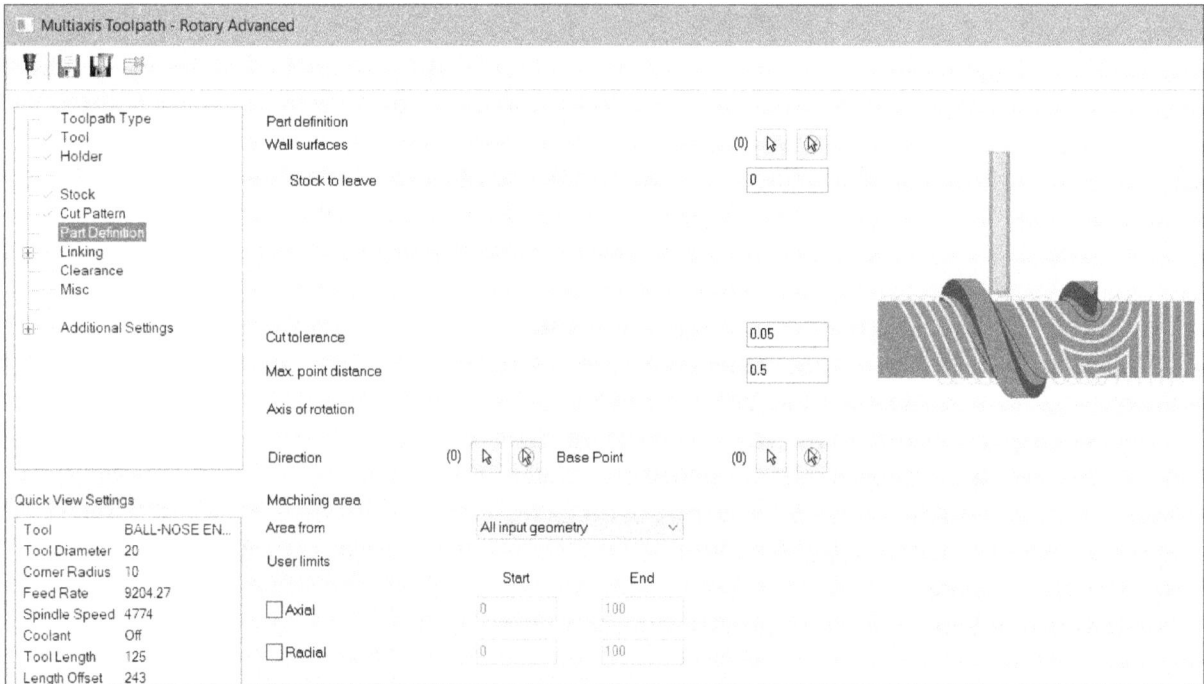

Figure-55. Part Definition page

- Click on the **Select** button for **Wall surfaces** option from the **Part definition** area of the dialog box and select desired faces to define walls of part to be machined; refer to Figure-56.

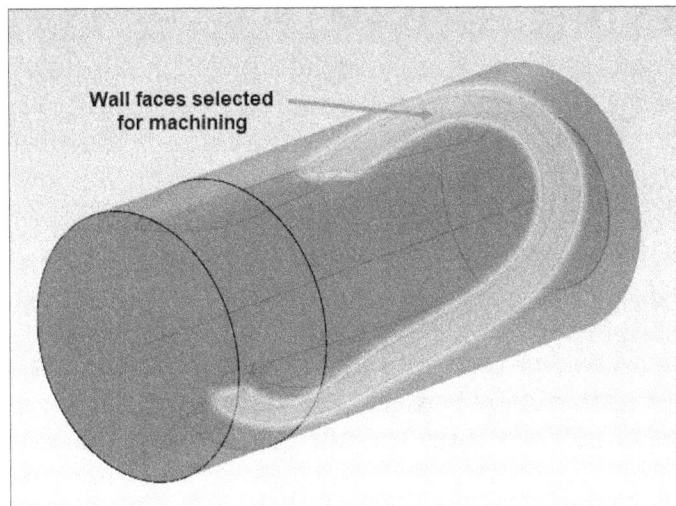

Figure-56. Wall faces selected for machining

- Click on the **Select** button for **Direction** option from the **Axis of rotation** area of the dialog box to define axis to be used for defining axis of model.
- Click on the **Select** button for **Base Point** option from the **Axis of rotation** area to define base point of axis.
- Set the other parameters as discussed earlier and click on the **OK** button to create the toolpath; refer to Figure-57.

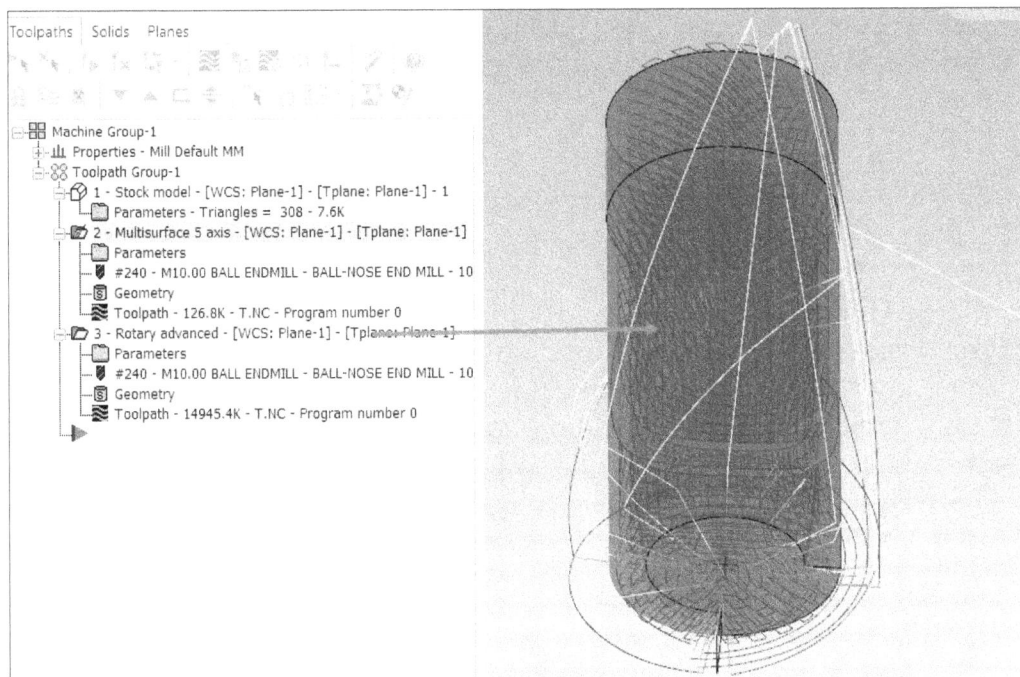

Figure-57. Advanced Rotary toolpath created

CREATING MULTIAXIS SWARF TOOLPATH

The **Swarf** tool is used to create toolpaths for machining side walls of model while keeping the tool in contact with model. The procedure to use this tool is similar to **Swarf Milling** tool discussed earlier.

Similarly, you can use **Port Expert** and **Blade Expert** tools to create toolpaths.

VERIFYING TOOLPATH WITH FIXTURE

Once you have generated toolpaths and verified it by simulation, there can be a need of fixture on the machine to hold your part during machining. There may be chances that cutting tool gets collided with the fixture. You can simulate the toolpath with fixture attached to stock during simulation by using the a separate body created in model or by using a CAD model file. The procedure to use separate body for fixture during simulation is discussed next. You can apply same procedure for using CAD model file.

- Click on the **Simulation Options** tool in toolbar of **Mastercam Toolpaths Manager**. The **Simulator Options** dialog box will be displayed.
- Select the **Fixtures** check box from the bottom in the dialog box; refer to Figure-58. The options to define fixture during simulation will become active.

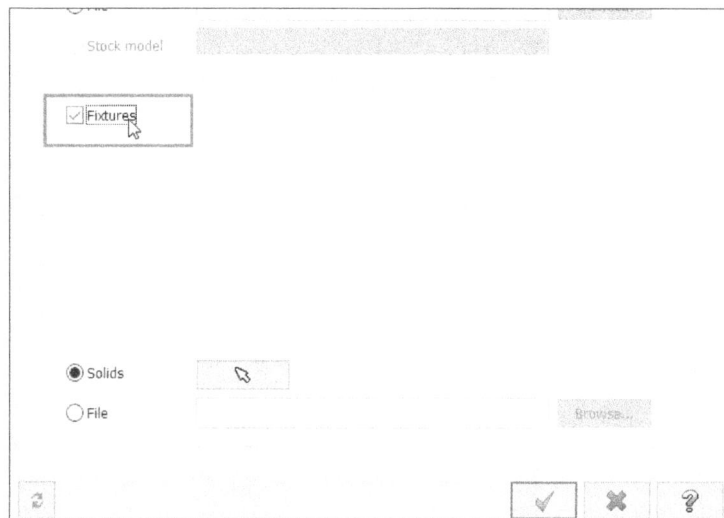

Figure-58. Fixtures check box

- Select the **Solids** radio button and click on the **Select** button next to it. The **Selection PropertyManager** will be displayed.
- Select the body that you want to use a fixture; refer to Figure-59 and click on the **OK** button. The **Select fixtures** dialog box will be displayed.

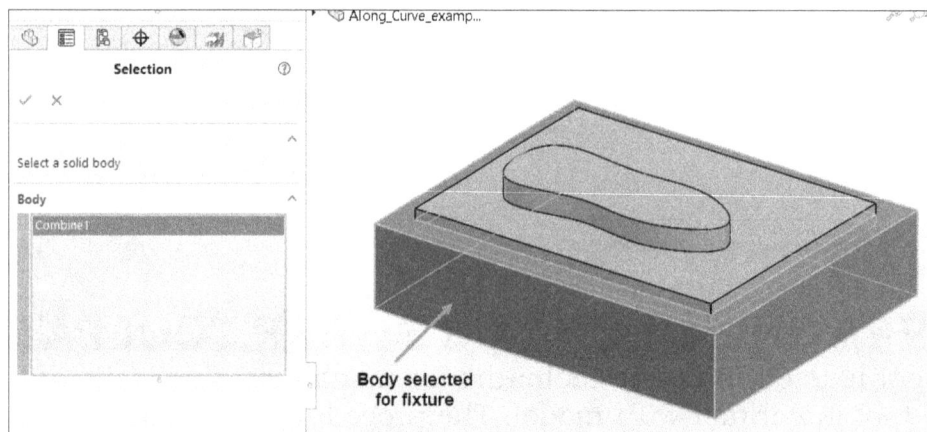

Figure-59. Selecting body for fixture

- Click on the **OK** button from the dialog box. The **Simulator Options** dialog box will be displayed again. You can change the machine to be used for simulation using the options in **Simulation** tab of the dialog box; refer to Figure-60.

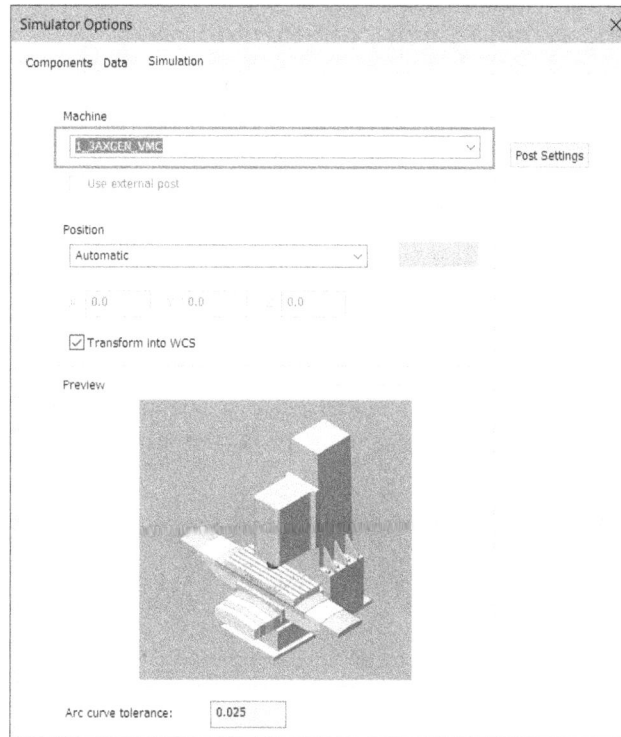

Figure-60. Machine for simulation

- Click on the **OK** button from the dialog box to apply the changes. If you now verify the toolpath then fixture will be displayed in the verification window; refer to Figure-61.

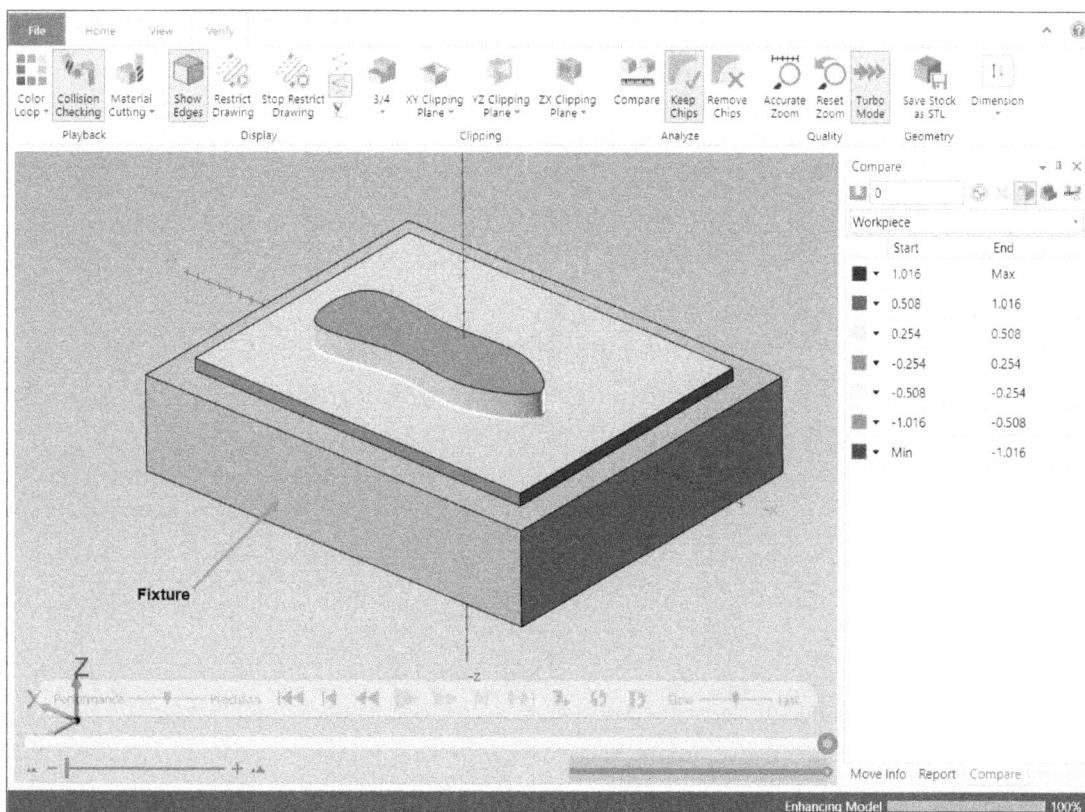

Figure-61. Toolpath simulation with fixture

Note that you can compare the machined part with final model using **Compare** tool in **Analyze** panel of **Verify** tab in the **Ribbon** in Mastercam Simulation application window; refer to Figure-62. The values of stock range and colors can be changed as per your tolerance requirements in the table of **Compare** panel. Click on the **Disable Compare** button from the panel to exit comparison mode.

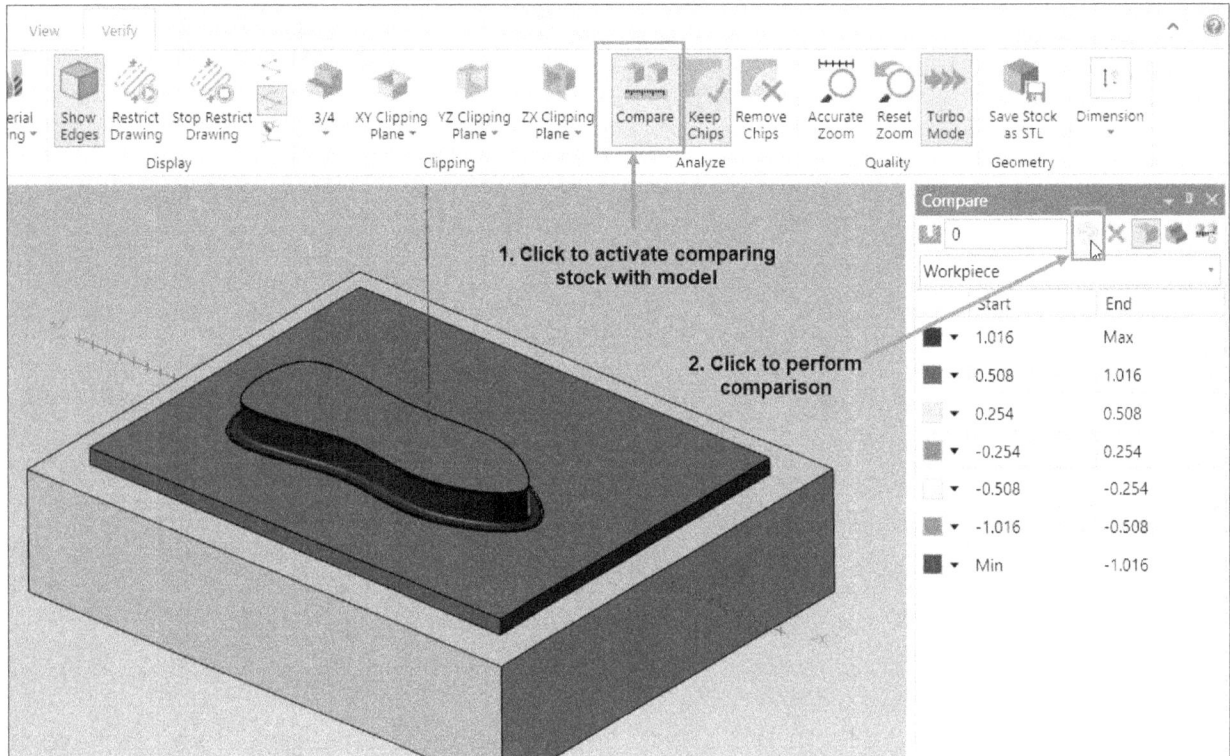

Figure-62. Comparing machined stock with model

Practice

Topics Covered

The major topics covered in this chapter are:

- *Practice 1-3 for Lathe*
- *Practice 1-12 for Milling*

PRACTICE 1 - 3 FOR TURNING

Create machining setup for models shown in Figure-1 to Figure-3. The part files are available in resource kit. The stock material is Aluminum alloy 6061 and use cylindrical bounding box stock.

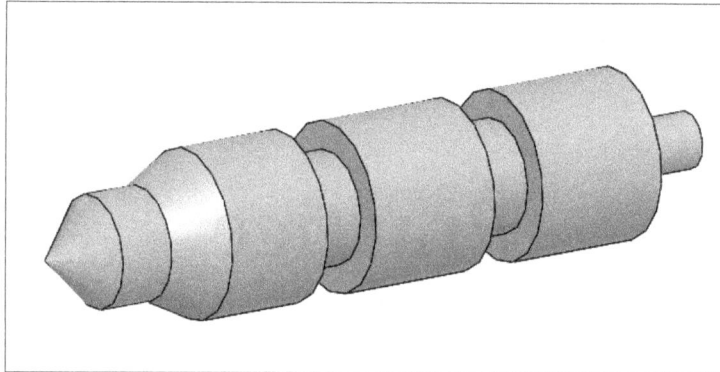

Figure-1. Practice 1

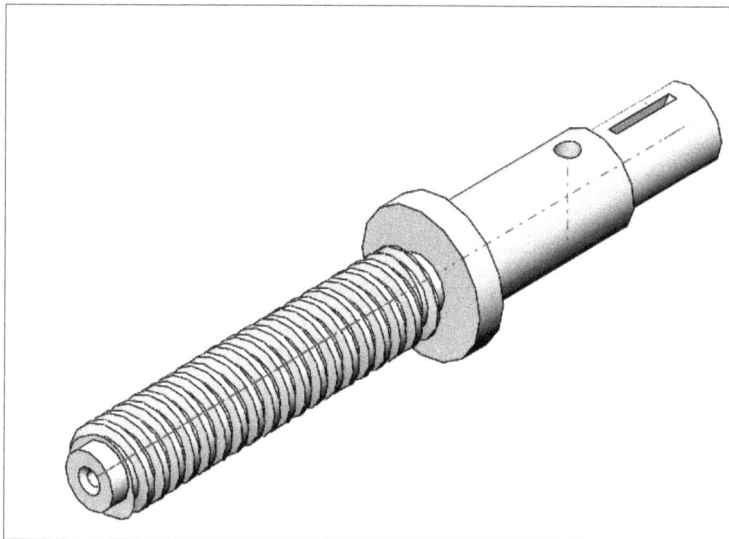

Figure-2. Practice 2

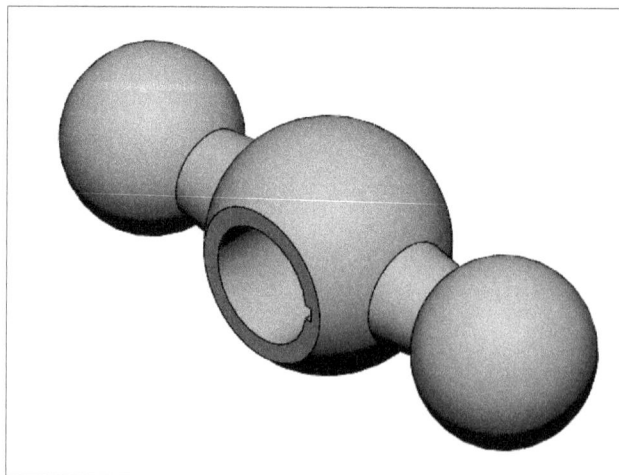

Figure-3. Practice 3

PRACTICE 1 - 12 FOR MILLING

Create machining setup for models shown in Figure-4 to Figure-15. The part files are available in resource kit. The stock material is Aluminum alloy 6061 and use rectangular bounding box stock.

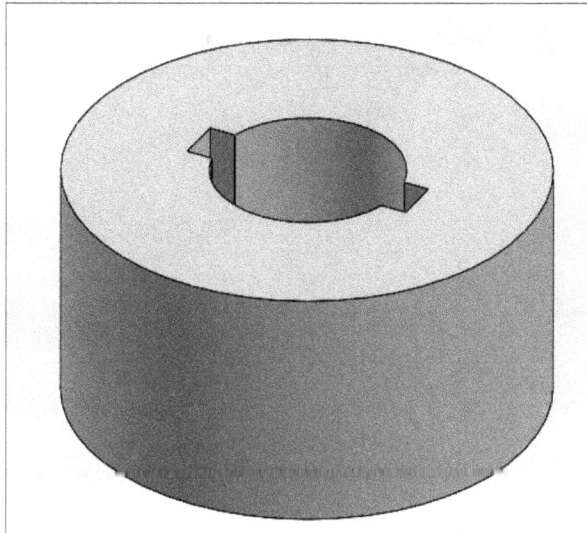

Figure-4. Practice 1 milling

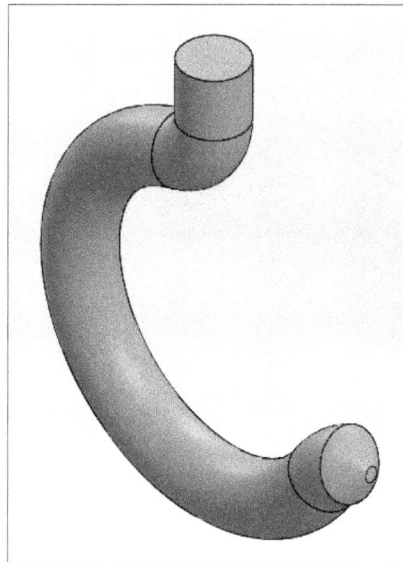

Figure-5. Practice 2 milling

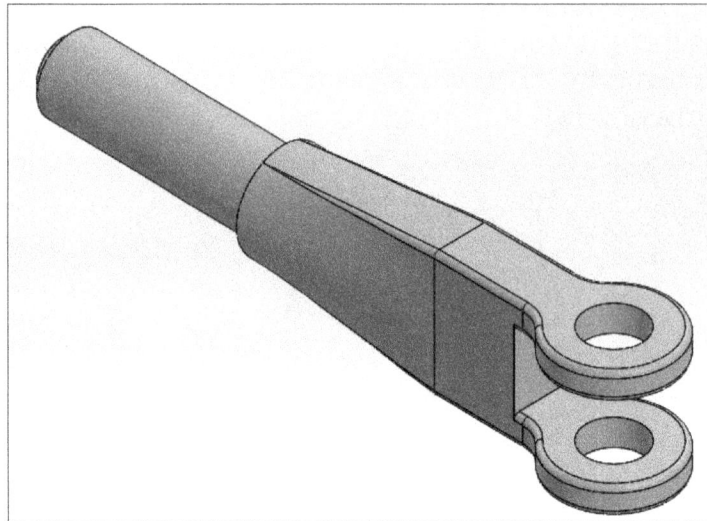

Figure-6. Practice 3 milling

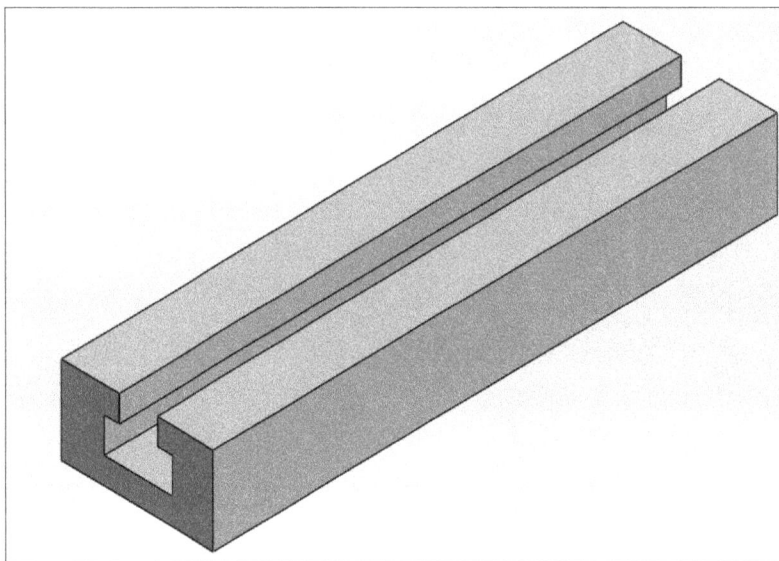

Figure-7. Practice 4

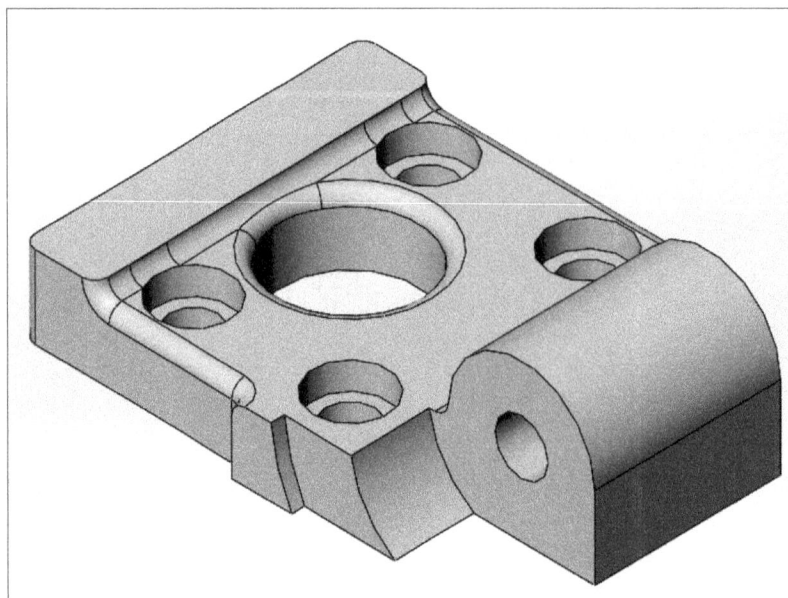

Figure-8. Practice 5

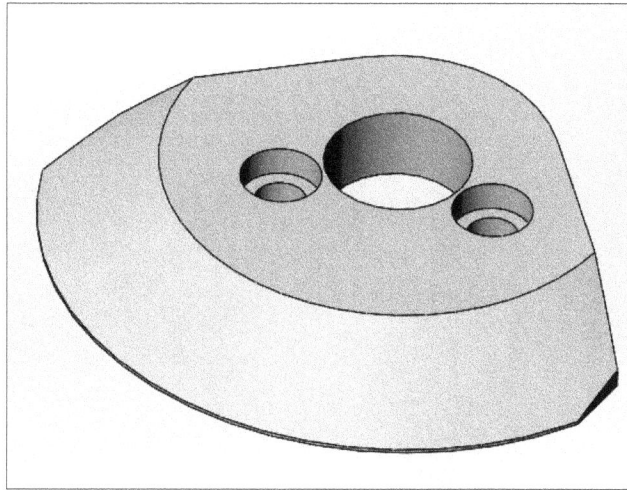

Figure-9. Practice 6

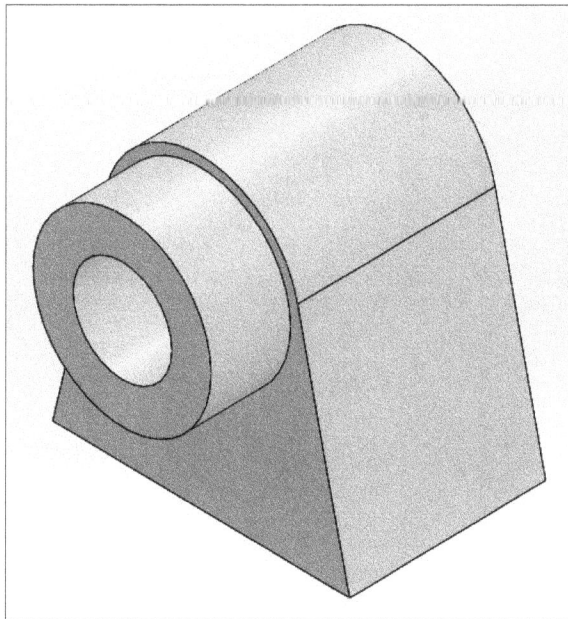

Figure-10. Practice 7

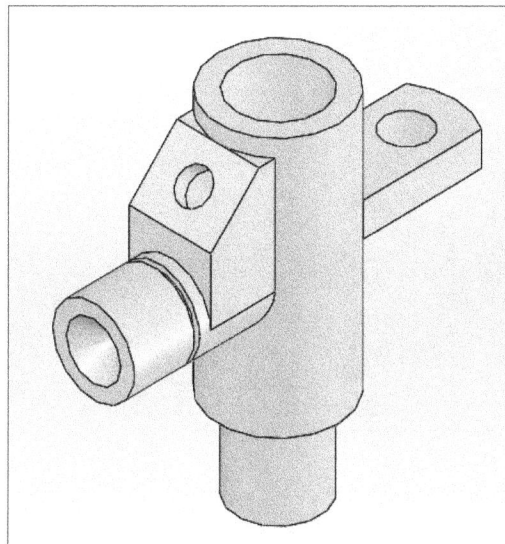

Figure-11. Practice 8

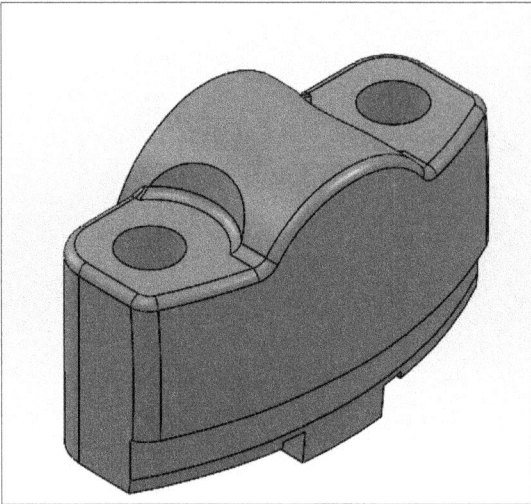

Figure-12. Practice 9

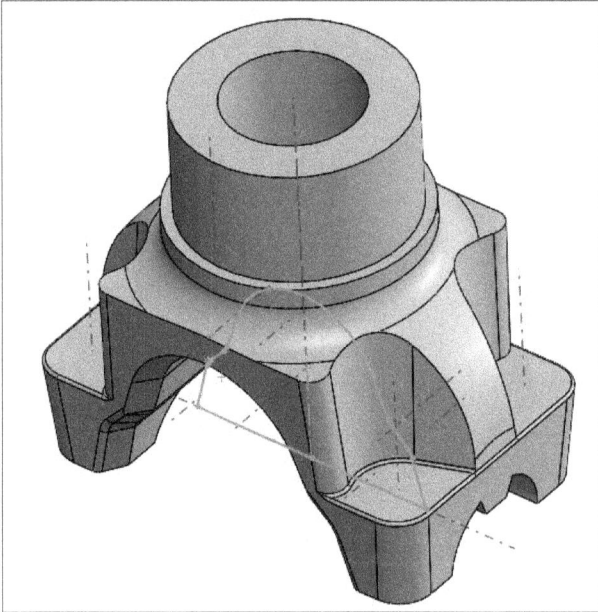

Figure-13. Practice 10

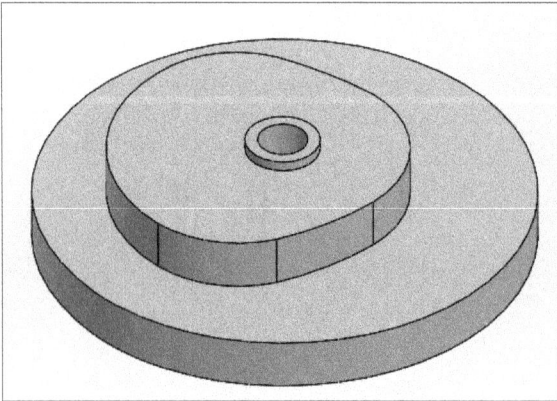

Figure-14. Practice 11

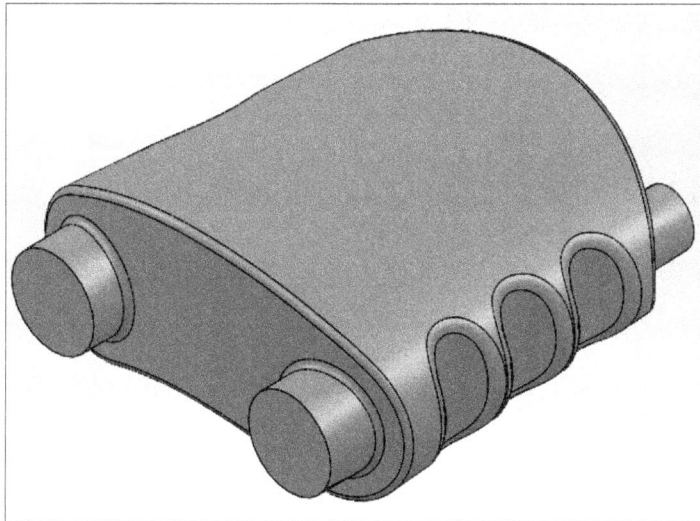

Figure-15. Practice 12

Once you have generated machining setup and verified toolpaths, try to reduce your total machining time while keeping the machining quality same.

FOR STUDENT NOTES

FOR STUDENT NOTES

Index

Ethics of an Engineer

• Engineers shall hold paramount the safety, health and welfare of the public and shall strive to comply with the principles of sustainable development in the performance of their professional duties.

• Engineers shall perform services only in areas of their competence.

• Engineers shall issue public statements only in an objective and truthful manner.

• Engineers shall act in professional manners for each employer or client as faithful agents or trustees, and shall avoid conflicts of interest.

• Engineers shall build their professional reputation on the merit of their services and shall not compete unfairly with others.

• Engineers shall act in such a manner as to uphold and enhance the honor, integrity, and dignity of the engineering profession and shall act with zero-tolerance for bribery, fraud, and corruption.

• Engineers shall continue their professional development throughout their careers, and shall provide opportunities for the professional development of those engineers under their supervision.

www.ingramcontent.com/pod-product-compliance
Lightning Source LLC
Chambersburg PA
CBHW081800200326
41597CB00023B/4099